Library of Congress Cataloging-in-Publication Data

Murty, Katta G.
  Operations research: deterministic optimization models
  p. cm.
  Includes bibliological references and index.
  ISBN: 0-13-056517-2
  1. Linear programming. 2. Mathematical optimization. I. Title.
T57.74.M88   1995
519.7--dc20
                        94-19150
                        CIP

Acquisitions editor: MARCIA HORTON
Editorial/production supervision: BARBARA KRAEMER
Copy editor: MICHAEL SCHIAPARELLI
Buyer: PHIL ZOLIT
Cover designer: DELUCA DESIGN
Editorial assistant: DOLORES MARS

 ©1995 by Prentice-Hall, Inc.
A Simon & Schuster Company
Englewood Cliffs, New Jersey 07632

The author and publisher of this book have used their best efforts in preparing this book. These efforts
include the development, research, and testing of the theories and programs to determine their effectiveness.
The author and publisher make no warranty of any kind, expressed or implied, with regard to these programs
or the documentation contained in this book. The author and publisher shall not be liable in any event for
incidental or consequential damages in connection with, or arising out of, the furnishing, performance, or use
of these programs.

Printed in the United States of America

10 9 8 7 6 5 4 3 2 1

ISBN   0-13-056517-2

Prentice-Hall International (UK) Limited, London
Prentice-Hall of Australia Pty. Limited, Sydney
Prentice-Hall Canada Inc., Toronto
Prentice-Hall Hispanoamericana, S.A., Mexico
Prentice-Hall of India Private Limited, New Delhi
Prentice-Hall of Japan, Inc., Tokyo
Simon & Schuster Asia Pte. Ltd., Singapore
Editora Prentice-Hall do Brasil, Ltda., Rio de Janeiro

Operations Research
Deterministic Optimization Models

Katta G. Murty

Department of Industrial and Operations Engineering
University of Michigan, Ann Arbor

PRENTICE HALL, Englewood Cliffs, New Jersey 07632

Contents

Preface

Optimization is a branch of science in which humans had an abiding interest since prehistoric times. Now-a-days all the decisions that we make at work, and those affecting our personal lives, usually have the goal of optimizing some desirable characteristics.

In large organizations, decisions are usually complex, with possibly many hundreds or thousands of individual components. For example, consider a company's production planner who has to determine the production levels for the various products under different brand names that the company produces. The production level of each of these is a component in the planner's overall decision.

The decision making process that is most commonly used can be characterized by one word: "myopic". In this process each separate component of the decision is arrived at in isolation, through a limited local search for the best decision that can be taken at the time that component is considered. The overall decision reached through such a patchwork scheme may be very far from optimal for the problem as a whole. An example of this is given in Section 1.4. The problem considered there is one of maximizing profit, and this myopic approach actually leads to a solution that yields the minimum possible profit! This example shows that decision making is a tricky business, and highlights the importance of using a reliable optimization procedure on a valid mathematical model for the problem to reach a good solution.

Most undergraduate students in the sciences take calculus. Calculus is one of the tools they need to handle decision making when they join the workforce; it is a means to an "end" − "optimization". By teaching them calculus and not making them take a course in optimization techniques, we are giving them the means but forgetting the important "end". Also, calculus alone does not provide the necessary tools to solve the many discrete decision making problems they may encounter. To survive in the fiercely competitive business climate of today, these students have to go beyond calculus and learn about optimization. I believe that knowledge of optimization methods − what they are, how they work, where and how to use them, and how to draw useful conclusions by applying them − is as essential as calculus is, particularly for students in engineering, mathematics, and business administration.

The purpose of this book is to serve as a basic text on optimization methods and their various applications, for the upper class undergraduate and first year graduate students, following the linear algebra and calculus sequence. The objectives of this book are:

- To present the techniques for modeling real world decision making problems using appropriate mathematical models of the linear, integer, combinatorial, or nonlinear programming types, and to present many modeling examples that not only illustrate the use of modeling techniques, but also present a variety of problem classes to which these techniques are applicable.

- To present a comprehensive treatment of the various types of algorithms to solve the wide variety of mathematical programming models, not with detailed proofs, but with explanations on how each algorithm works, and what useful conclusions can be drawn from its output. Limitations of each algorithm, together with what can and cannot be solved efficiently by existing methods, will also be discussed.

Preview

Chapter 1 begins with a brief history of optimization and its importance to mankind. It explains the classification of optimization problems into three categories, and the typical multi-characteristic approach for handling Category 1 problems in which the best among a small number of alternatives is to be selected.

Chapter 2 deals with the modeling of problems in which linearity assumptions hold to a reasonable degree of approximation, as linear programs (LPs). A variety of applications of the LP model are presented. To make sure that the reader does not get carried away and try to model every problem as an LP, an example in which the linearity assumptions are quite unreasonable is presented, for which the LP model leads to grossly unrealistic conclusions. The concept of marginal values associated with resources is introduced, and their importance in providing valuable planning information is discussed. The chapter concludes with an accounting of all the useful information that can be derived from the solution of an LP model, illustrated with many examples.

Chapter 3 reviews the necessary parts of linear and matrix algebra that every student of optimization must know. They include the concept of linear independence; the definitions of geometric objects like convex sets, etc.; and pivotal algorithms for checking linear independence, for computing the inverse of a nonsingular square matrix, and for solving systems of linear equations. It summarizes all this information in a simple manner to serve as a convenient reference when needed, or to be used for a review in a course if necessary.

Chapter 4 presents the derivation of the dual of an LP through economic arguments. It gives the interpretation of the optimal dual solution as the vector of marginal values, when the primal problem has a nondegenerate optimal basic feasible solution. The chapter concludes with a discussion of the optimality conditions for LP.

The aim of Chapter 5 is to show how the optimality conditions discussed in Chapter 4 can be used directly to develop an algorithm for solving a very special LP problem, namely the assignment problem. This results in the Hungarian method which is discussed completely. This chapter also emphasizes the importance of simple and efficient techniques that can compute a lower bound for the minimum objective value in a minimization problem. These bounding techniques will become very important later in the branch and bound approach for general combinatorial optimization problems. The Hungarian method illustrates the use and the power of optimality conditions. It is a special algorithm for a highly structured LP that is very easy to understand.

Chapter 6 treats a slightly more general problem, the transportation problem, and presents the simplified version of the primal simplex algorithm for it. The techniques of marginal and sensitivity analyses are illustrated and their various applications presented with many examples.

Chapter 7 discusses the revised simplex method for general LP and illustrates the applications of the infeasibility, marginal, and sensitivity analyses in this model with many examples. With the organization of the material in Chapters 5 to 7 in this order, the book moves from simple, elegant algorithms for problems with very special structure, to somewhat complicated algorithms for general problems.

Chapter 8 focusses on multiobjective models and discusses the two most popular techniques: combining the various objective functions linearly with positive weights reflecting their importance, and goal programming.

Chapter 9 presents techniques for modeling integer and combinatorial optimization problems, and illustrates their application to many areas with lots of examples. To heighten the importance of integer programming, it draws examples from classics in literature.

The branch and bound approach for combinatorial optimization is discussed in Chapter 10. The major issues in developing a successful branch and bound algorithm for a problem are discussed in detail, and illustrated with applications to several types of problems. The advantages and limitations of this approach are also discussed.

The mathematically rich and elegant NP-Completeness theory, in development since the late 1960s, has shown that many of the combinatorial optimization problems encountered in applications belong to a class of hard problems called "NP-Hard". Nobody has so far been able to develop a provably efficient algorithm for these problems. Practical experience indicates that large scale versions of these problems take an enormous amount of time to solve exactly by any of the available algorithms. Efforts to transform an NP-Hard problem into a simpler problem for which there may be an efficient algorithm have so far been fruitless. All transformations of an NP-Hard problem seem to lead only to other NP-Hard problems. In this respect, the reader may be interested in the following Telugu poem by Enugu Lakshmanakavi, dating back to 1790 from his translation of the Sanskrit book *Subhashitalu* by the Indian poet Bhartruhari, written around the year 450 AD. The word "Subhashitalu" is a little hard to translate exactly into English, but "Words of Wisdom" is a close equivalent. This book received world-wide acclamation; several translations into English verse and prose are available (see *Bhartruhari* by

Harold G. Coward, Twayne Publishers, 1976; my thanks to P. Ramakrishna, and
S. Narayanaswamy for this historical information).

తివిరి ఇసుమున తైలంబు దీయవచ్చు

తవిలి మృగతృష్ణలో నీరు (దావవచ్చు

తిరిగి కుందేటి కొమ్ము సాధించవచ్చు

చేరి క్లిష్ట సమస్యను ఛేధించ లేము.

The meaning of this poem is:

> It may be possible to get oil by squeezing dry beach sand hard,
> It may be possible to drink water by searching far in a mirage,
> It may be possible to find a rabbit's horn by looking all over,
> But an intractable problem can never be transformed into a tractable one.

The realization that there may be no efficient exact algorithms to handle these
problems has forced researchers to seek heuristic approaches with the aim of obtaining the best possible approximate solution within a reasonable time. Practitioners
have noticed that even though there are no guarantees, well designed heuristic
methods seem to produce satisfactory solutions for many hard problems. Heuristics are the subject of Chapter 11, with a complete discussion of many common
approaches that have worked well in practice, and a discussion on how to design
good heuristics.

Chapter 12 deals with recursive dynamic programming methods for solving deterministic multistage decision problems. Chapter 13 discusses the very important
critical path methods for project management. Finally, in Chapter 14 we consider
nonlinear programming problems − how to construct them, what are the optimality
conditions for them, and what are some of the most useful algorithms for solving
them.

The Current Status of Optimization

In terms of its effectiveness in giving satisfaction to the clients, the current
status of optimization as a science can be compared to that of many other branches
of science. In optimization there are many hard problems for which efficient algorithms are unknown, just as there are unsolved problems in other branches of
science. But by spending time to design a good exact or heuristic method, we can
expect to generate reasonable solutions to many optimization problems encountered
in applications.

How to Use the Book in a Course

There is more material in the book than can fit comfortably into a one semester course. But usually a course in linear algebra is a prerequisite for an optimization course. In this case, Chapter 3 needs to be reviewed only briefly; and Chapters 1, 2, 4 to 10, 12, 13 can form the basic core of a course on "Introduction to optimization methods", with the students expected to read Chapters 11 and 14 on their own later on. The simplex method for general LP in Chapter 7 is widely taught in many courses, and usually many students end up taking one of these courses. If this is the case, Chapters 1, 2, 4, 5, 6, 8 to 14, can form the basis for a course on optimization methods. In business schools and branches of engineering other than IE, the instructor may want to omit any one or two chapters of his/her personal choice to fit the course into a semester; and expect the students to read the omitted material on their own afterwards. The easy-to-follow style of the book makes it possible for students to explore some topics on their own once they are initiated into the subject.

Comments for the Reader

Now some comments directed towards the reader. If you are already familiar with the basic concepts of linear independence, rank, bases, pivot steps, and pivotal methods for solving systems of linear equations, you can skip Chapter 3, or refer to it whenever you need to refresh your memory.

The basic concepts of mathematical modeling are provided in Chapters 1, 2 (in the context of problems that can be analyzed by linear models), Chapter 9 (in the context of problems that can be analyzed through discrete or combinatorial models), and Chapter 14 (in the context of problems that can be analyzed through nonlinear models).

Chapter 4 provides the optimality conditions for linear programming models, which form the foundation for the development of algorithms discussed in Chapters 5, 6, and 7. To aid the comprehension of these algorithms, we provide detailed flow charts for them in each of those chapters.

Chapter 8 discusses two popular and important methods for handling multiobjective models, and provides many applications of the multiobjective linear programming problem.

After studying Chapters 5, 7, and 9, you are ready for the branch and bound and heuristic methods for discrete and combinatorial optimization models, discussed in detail in Chapters 10 and 11.

Chapter 12 on dynamic programming is an independent chapter by itself. The dynamic programming method discussed in Chapter 12 is applied to project management problems in Chapter 13.

Chapter 14 is again an independent chapter by itself. Here we provide the available optimality conditions for nonlinear programming, and a discussion of the limitations of existing nonlinear programming algorithms. This chapter concludes with a brief discussion of a couple of the most important nonlinear programming algorithms to get the reader interested in exploring the vast literature in this area.

Numerical examples illustrating the various possibilities are provided under each method (except for some of the methods not amenable for hand computation in Chapters 11 and 14). And a variety of exercises are provided in each chapter. You can sharpen your skills by doing these exercises.

References

In keeping with the level of this book, references at the end of each chapter are mostly restricted to popular books on the subject, and a few publications from which examples and results quoted in the chapter are drawn.

Acknowledgments

My thanks to several friends who helped me with this book in several ways. In particular, my thanks go to Faiz Al-Khayyal, David Jacobsen, and Jeffrey K. Cochran for reviewing the manuscript; and to Hossam Al-Mohammad, John Birge, Soo Chang, Jong Chow, Salih Duffuaa, Jolene Glaspie, Santosh Kabadi, Steve Pollock, K. N. Rao, Cosimo Spera, Samer Takriti, and Miguel Taube for pointing out corrections, contributing exercises, and being helpful in many other ways. I thank my sister Geeta Bandla for help in typesetting the index.

Finally, I thank my mother Adilakshmi Katta, and wife Vijaya Katta, for their affection and encouragement. This book is dedicated to both of them.

Katta Gopalakrishna Murty
Ann Arbor, Michigan.

Other books by Katta G. Murty:

→ *Linear and Combinatorial Programming*, first published in 1976, available from R. E. Krieger Publishing Co., Inc., P. O. Box 9542, Melbourne, FL 32901.

→ *Linear Programming*, published in 1983, available from John Wiley & Sons, Inc. Publishers, 605 Third Avenue, New York, NY 10158.

→ *Linear Complementarity, Linear and Nonlinear Programming*, published in 1988, available from Heldermann Verlag, Nassauische Str. 26, D-1000 Berlin 31, Germany.

→ *Network Programming*, published in 1992, available from Prentice Hall, Englewood Cliffs, NJ 07632.

Glossary

Figures, equations, exercises, tableaus, arrays, comments, etc. are numbered serially within each chapter. So, for each of these, the entity $i.j$ refers to the jth entity in Chapter i.

The *size* of an optimization problem is a parameter that measures how large the problem is. Usually it is the number of digits in the data in the problem when it is encoded in binary form. When n is some measure of how large a problem is (either the size, or some quantity which determines the number of data elements in the problem), a finitely terminating algorithm for solving it is said to be of order n^r or $O(n^r)$, if the worst case computational effort required by the algorithm grows as αn^r, where α is a number that is independent of the size and the data in the problem. An algorithm is said to be *polynomially bounded* if the computational effort required by it to solve an instance of the problem is bounded above by a fixed polynomial in the size of the problem.

AOA	Activity-on-arc project network, or also called arrow diagram.
AON	Activity-on-node project network.
BFS	Basic feasible solution for a linear program.
BV	Branching variable used in the branching operation in a branch and bound algorithm.
B&B	Branch and bound approach or algorithm .
CP	Candidate problem in a branch and bound algorithm.
CPM	Critical path method for project scheduling.
DP	Dynamic programming.
$ES(i,j)$, $EF(i,j)$ $LS(i,j)$, $LF(i,j)$	Early (late) start and finish times associated with the job corresponding to arc (i,j) in project scheduling.

Euclidean TSP	A traveling salesman problem in which the distance matrix comes from the Euclidean distances between cities on a two dimensional plane.
Exact algorithm	An algorithm that is guaranteed to find an optimum solution, if one exists.
FIFO	First in first out strategy for selecting objects from a queue.
GA	Genetic algorithm.
Head(e), tail(e)	The head and tail nodes on an arc e.
I/O coefficient	Input-output coefficient in the constraint coefficient matrix of a linear programming model.
iff	If and only if.
IP	Integer program.
KKT Conditions	The Karush-Kuhn-Tucker necessary optimality conditions for a point to be a local minimum to a nonlinear program.
KOP	Kilo operations per second, in Chapter 11.
LB	Lower bound for the minimum objective value in a candidate problem in the branch and bound algorithm.
LIFO	Last in first out strategy for selecting objects from a queue.
LP	linear program.
MDR	Minimum daily requirement for a nutrient in a diet model, see Section 2.3.
MIP	Mixed integer program.
NLP	Nonlinear program.
Oc.R	Octane rating of gasoline in Section 2.2.
OR	Operations research.
OVF	Optimum value function in dynamic programming.
PD	Positive definite - a property of a square matrix, see Chapter 14.
PMX	Partially matched crossover operation in genetic algorithms for permutation or tour problems.
Predecessor index or predecessor label	A node label in a labeling technique for storing chains from an origin in a directed network, or for storing a tree.
PSD	Positive semidefinite - a property of a square matrix, see Chapter 14.

RHS constant	Right hand side constants vector in a linear programming model.
RM	Raw material.
RO	Relaxed optimum in the lower bounding strategy in a branch and bound algorithm.
R.O.A.	Relaxed optimum assignment for a candidate problem in the branch and bound algorithm for the traveling salesman problem in Chapter 10.
SA	Simulated annealing algorithm.
TPB	Three point bracket for a local minimum in line search algorithms, see Chapter 14.
TSP	Traveling salesman problem.
wrt	With respect to.
$0-1$ variable	A variable that is constrained to take only values of 0 or 1.
$1, 2, \ldots, n; 1$	A tour for a TSP, indicating the order in which the cities are visited.
$\mathcal{N}$	The finite set of nodes in a network.
$\mathcal{A}$	The set of arcs in a network.
$G = (\mathcal{N}, \mathcal{A})$	A network with node set $\mathcal{N}$ and arc set $\mathcal{A}$.
$\mathcal{C}$	A chain in a directed network from an origin to a destination node.
n, m	The number of variables and constraints respectively in an LP, or equality constrained NLP. In a transportation problem, these denote the number of markets (columns in the array), and the number of sources (rows in the array), respectively. The symbol n is also used to denote the order of an assignment problem, the number of cities in a TSP, the number of objects in a knapsack problem, etc.
(i, j)	An arc joining node i to node j in a network, or the cell in row i and column j of a transportation or assignment array.
x_j, x_{ij}, x	Decision variables. x_j denotes the jth decision variable in an LP, IP, or NLP. x_{ij} is the decision variable associated with cell (i, j) in an assignment or transportation problem, or a TSP. x is the vector or matrix of these decision variables.
a_i, b_j	In a transportation problem, these are the amounts of material to be shipped out of source i (or into sink j).

c_{ij}, c_j, c	The original unit cost coefficient in cell (i, j) in an assignment, transportation, or traveling salesman problem. c_j is usually the original cost coefficient of a the variable x_j in an LP. c denotes the matrix of c_{ij}s, or vector of c_js.
$z(x), z_c(x)$	Usually denotes the objective function. $z_c(x)$ is the objective value of assignment x wrt cost matrix c.
$\mathcal{B}$	A basic set of cells in a transportation array.
u_i, u	The dual variables associated with row i of a transportation or assignment array in Chapters 4, 5, and 6; u is the vector of u_i.
v_j, v	The dual variable associated with column j of a transportation or assignment array in Chapters 4, 5, and 6; v is the vector of v_js.
$\bar{c}_{ij}, \bar{c}_j, \bar{c}$	The reduced or relative cost coefficient in cell (i, j),wrt a dual solution $(u = (u_i), v = (v_j))$ in an assignment or transportation problem, it is $c_{ij} - (u_i + v_j)$. $\bar{c}_j$ is the relative cost coefficient of the variable x_j in an LP wrt the present basic vector. $\bar{c}$ is the matrix of $\bar{c}_{ij}$s, or vector of $\bar{c}_j$s.
π_i, π	The dual variable associated with the ith constraint in a linear program; or the Lagrange multiplier associated with an inequality constraint in a nonlinear program. π is the vector of π_i.
μ_i, μ	The Lagrange multiplier associated with the ith equality constraint in a nonlinear program. μ is the vector of μ_i.
$\blacksquare$	Symbol indicating end of a proof.
s. t.	Such that.
c.s. conditions	The complementary slackness optimality conditions.
x^T, A^T	Transposes of vector x, matrix A.
K, E, F, B, Γ	These bold face symbols usually denote sets.
$\mathbb{R}^n$	Real Euclidean n-dimensional vector space.
e	The column vector of all 1s in $\mathbb{R}^n$.
$|\alpha|$	Absolute value of real number α.
$|\mathbf{F}|$	Cardinality of the set **F**, it is the number of elements in **F**.
$\lceil \alpha \rceil$	Ceiling of real number α, smallest integer $\geqq \alpha$. e.g., $\lceil -4.3 \rceil = -4, \lceil 4.3 \rceil = 5$.
$\lfloor \alpha \rfloor$	Floor of real number α, largest integer $\leqq \alpha$. e.g., $\lfloor -4.3 \rfloor = -5, \lfloor 4.3 \rfloor = 4$.

$n^{\mathbf{r}}$ n to the power of $\mathbf{r}$. Exponents are set in this type style to distinguish them from ordinary superscripts.

$O(n^{\mathbf{r}})$ A positive valued function $g(n)$ of the nonnegative variable n is said to be $O(n^{\mathbf{r}})$ if there exists a constant α s. t. $g(n) \leqq \alpha n^{\mathbf{r}}$ for all $n \geqq 0$. For meaning in the context of computational complexity, see the beginning of this glossary.

$n!$ n factorial.

$||x||$ Euclidean norm of the vector x. For $x = (x_1, \ldots, x_n)$ it is $+\sqrt{x_1^2 + \ldots + x_n^2}$.

∞ Infinity.

$\in$ Set inclusion symbol. $a \in \mathbf{D}$ means that a is an element of $\mathbf{D}$. $b \notin \mathbf{D}$ means that b is not an element of $\mathbf{D}$.

$\subset$ Subset symbol. $\mathbf{E} \subset \mathbf{F}$ means that set $\mathbf{E}$ is a subset of $\mathbf{F}$, i.e., every element of $\mathbf{E}$ is an element of $\mathbf{F}$.

$\cup$ Set union symbol.

$\cap$ Set intersection symbol.

$\emptyset$ The empty set.

$\setminus$ Set difference symbol. $\mathbf{D} \setminus \mathbf{H}$ is the set of all elements of $\mathbf{D}$ that are not in $\mathbf{H}$.

$\geqq$ Greater than or equal to.

$\leqq$ Less than or equal to.

$\sum$ Summation symbol.

$\sum(x_j : \text{over } j \in \mathbf{J})$ Sum of x_j over j from the set $\mathbf{J}$.

I unit matrix.

A, B, D, M Usually matrices. In Chapters 4 and 7, B usually denotes a basis.

B^{-1} The inverse of the matrix B.

$A_{i.}$ The ith row vector of the matrix A.

$A_{.j}$ The jth column vector of the matrix A.

x_B, x_D The vectors of basic (dependent), nonbasic (independent) variables.

θ Usually the minimum ratio in the simplex algorithm for solving a transportation problem, or a general LP.

$\theta(x), f(x), h(x), g(x)$ Real valued functions defined over $\mathbb{R}^n$. Sometimes these symbols are also used to denote a vector of such functions.

$\nabla\theta(x)$ The row vector of partial derivatives of the function $\theta(x)$ at x.

$H(\theta(x))$ The hessian matrix, or the matrix of second partial derivatives of the function $\theta(x)$ at x.

$J(h(x))$ The Jacobian matrix of the vector of functions $h(x)$ at x

$L_2(a)$ Least squares measure of deviation as a function of the parameter vector a in curve fitting, see Chapter 14.

$\exp(.)$ The exponential function. $\text{Exp}(x) = e^x$ where e is the base of the natural logarithm.

Chapter 1

Introduction

1.1 What Is Operations Research?

Operations Research (to be abbreviated OR) is the branch of science dealing with techniques for optimizing the performance of systems. It is a scientific method that provides executives a quantitative and rational basis for taking decisions, especially those dealing with the allocation of resources. The subject is also called by names such as *management science, scientific decision making, decision analysis,* etc. OR originated during World War II when the British government hired scientists from various disciplines to assist with the operational problems of the war, hence the name Operations Research. The focus of the subject is on scientific methods of decision making that seek to understand the complex operations of any system to predict its behavior and improve its performance. OR applications deal with making decisions or taking actions that are **optimal** in the sense of helping the system performance to reach its ideal level. The word "system" here refers to any department, organization, or entity such as a manufacturing company making and selling a variety of products, a hospital delivering health services, a university educating students, the directorate of a dam on a river responsible for managing the water releases from the reservoir, etc. The OR approach for improving any system typically takes the following steps.

(i) **Identifying the decision variables.** Observe the operation of the system carefully and identify the parameters whose values can be controlled, and which affect its performance. For example, if the system is a chemical reactor for manufacturing a chemical, these parameters may be the temperature and pressure in the reactor, the concentrations of the various inputs and catalysts etc., the flow rates or the amount of time for which the reaction is allowed to continue, etc. These parameters are called the **decision variables**. The important properties of a decision variable are that its value has an affect on system performance, and there are mechanisms available by which we can

1

change its value to any desired value within its specified operating range.

(ii) Construct a mathematical model of system operation and objectives.
Identify measures of effectiveness of system performance, and express each of them as a mathematical function of the decision variables.

If higher values of a measure of performance are more desirable (such a measure could be considered as a **profit measure**) we seek to attain the maximum or highest possible value for it. If lower values of a measure of performance are more desirable (such a measure could be interpreted as a **cost measure**) we seek to attain the minimum or the lowest possible value for it. The various measures of performance are usually called **the objective function(s)** in the mathematical model for the system. To **optimize** an objective function means to either maximize or minimize it as desired.

If there is only one measure of performance (such as yearly total profit, or production cost per unit, etc.) the model will be a **single objective model**. When there are several measures of performance, we get a **multiobjective model** in which two or more objective functions are required to be optimized simultaneously.

Then identify all the relationships that must hold among the decision variables and the various static or dynamic structural elements, by the nature of system operation, using mathematical equations or inequalities. And identify all the bounds on the decision variables or functions of decision variables, in order to account for the physical limitations under which the system must operate. Each of these leads to a constraint in the model.

When the objective function(s) and all the constraints are put together, we get the **mathematical model** describing the system, the conditions under which it must operate, and our aspiration(s) for it. This type of model is usually known as a **mathematical program**, or an **optimization problem**.

(iii) Solve the model for an optimum solution. A solution refers to a numerical vector giving values to each decision variable. A solution is said to be a **feasible solution** if it satisfies all the constraints in the model. An **optimum solution** is a feasible solution that has the most desirable value(s) for the objective function(s) among all feasible solutions. Solving the model means finding an optimum solution for it. Except for the most trivial types of models, solving requires the use of an efficient algorithm. An **algorithm** is a step by step procedure where each step is described so precisely that it can be set up for execution by a computer, that leads unambiguously from one step to the next, until in a finite number of steps it terminates with an optimum solution of the model if one exists.

This text essentially discusses various algorithms for arriving at optimum (or at least satisfactory) solutions to a variety of optimization problems.

(iv) Perform sensitivity analyses. These analyses determine the sensitivity of the optimum solution to model specifications. They determine how robust the optimum solution is under inaccuracies in input data and structural assumptions. This type of analysis is essential to the validation process.

(v) Implementing the findings and updating the model. In this final phase, the optimum solution is implemented. This requires checking it for practical feasibility, making necessary modifications in the model and solving it again if it is found to be impractical for some reason; and repeating the whole process as needed.

Often the output from the model is not implemented as is. It provides insight to the decision maker who combines it with his/her practical knowledge and transforms it into an implementable solution.

We thus see that optimization of system performance is the basic aim of OR. Almost all scientific, management, or planning activity deals with some type of optimization. Optimization is the main pillar of OR, and the performance of most systems can be improved through intelligent use of optimization algorithms.

In the past, tremendous emphasis has been placed on keeping models small, simple, and compact since only such models could be solved by the techniques and computing equipment available then. With the development of many techniques and algorithms, and the amazing development of highly sophisticated computing equipment in recent years, large scale models are now being built and solved routinely.

1.2 Optimization Models

Optimization is one of the earliest branches of science investigated by human beings. More than 100,000 years ago, ancient humans at that time worried about problems like, "What is the shortest route to the river from my cave?" And they worked out correct solutions by trial and error. See Figure 1.1. Since there are no constraints on the route, this is an **unconstrained optimization problem**.

The unconstrained shortest route between the cave and the river may pass through risky areas, for example those infested with tigers and other dangerous animals. See Figure 1.2. The cave dweller then naturally considers the modified problem, "What is the shortest route from my cave to the river that avoids areas infested with tigers?" This is a **constrained optimization problem**.

Thus there are unconstrained and constrained optimization problems. Since there are usually limits on the quantities of resources that can be drawn, and/or specifications on the solution obtained, most real world applications lead to constrained optimization models. However, unconstrained optimization continues to be extremely important in optimization theory, because it is often advantageous to solve constrained optimization models using unconstrained optimization techniques, either by ignoring the constraints in the model, or by transforming the constrained problem into an unconstrained one using penalty function methods or Lagrangian methods. Such transformations are most commonly used for solving nonlinear models in continuous variables.

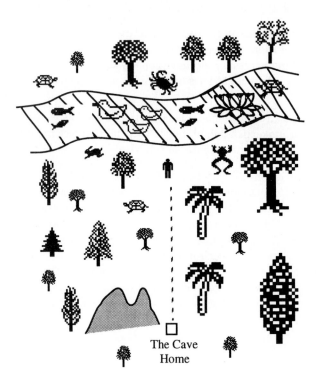

The Cave
Home

Figure 1.1 An unconstrained shortest route problem considered by our ancestors thousands of years ago. They figured out that the shortest route from home (a cave in those days) to the river is the straight line joining the cave to the nearest point to it on the river, marked with a dashed line.

With the passage of time, humanity explored the world around them and ac-cumulated more and more knowledge. Farming was invented and it soon replaced hunting as the primary means of supplying the human population with the neces-sary food. Then, humans had to deal with the following types of problems.

"How much of my 100 acres of farmland should I allocate to wheat, corn, and oats to maximize my income?" Since the amount of land allotted to a crop can be a continuous variable (i.e., one which can assume all possible real values within the bounds on it imposed by the constraints in the model), problems of this type are called **continuous variable optimization problems**. In this class are the **linear programs** (these are models in which all the functions used are linear functions,

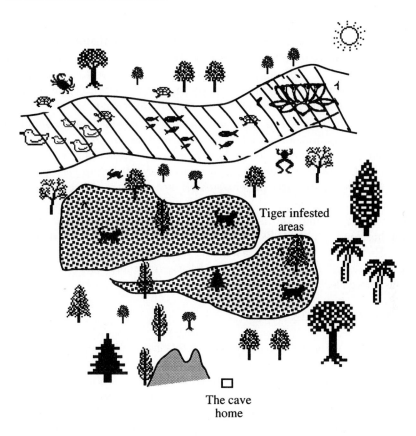

Tiger infested
areas

The cave
home

Figure 1.2 A constrained optimization problem, to find a shortest
route from the cave to the river that avoids areas infested by tigers,
considered by our ancestors thousands of years ago.

see Chapter 2 for examples), and the **nonlinear programs** (in these models at
least one function, either the objective function, or one of the constraint functions
is not a linear function, see Chapter 14 for examples). Among nonlinear programs,
there are the **smooth nonlinear programs** (here all the functions in the model
have derivatives everywhere), and the **nonsmooth nonlinear programs** (in these
models, even though all the functions used are continuous functions, one or more
of them do not have derivatives at some points in the space whose measure is zero).
In this book we only discuss smooth nonlinear programs in Chapter 14.

"How many cows and lambs should I grow to maximize my income?" This
involves the number of cows and lambs grown, which can only assume integer values.
Integer variables are a special case of **discrete valued variables**, those which are
restricted to assume a value from a specified discrete set. For an example of a general
discrete variable, consider a paint which is sold only in cans of capacity either 1/2
gallon, or 1 gallon. Then the variable, the amount of this paint purchased, is a

discrete variable that should have a value from the discrete set $\{0, 1/2, 1, 3/2, 2, \ldots\}$.
Problems involving such variables are called **discrete optimization problems**,
or **integer programs** if all the discrete variables are integer variables. Also, if a
problem involves some continuous variables and some discrete variables, it is known
as a **mixed discrete optimization problem**, or a **mixed integer program**, as
appropriate. See Chapters 9 and 10 for examples of integer programming models.

"I need to visit markets 1, 2, 3, and 4 today. In which order should I visit
them in order to minimize the distance I need to walk for these visits?" Or, "We
have decided to set up 3 new rice mills. There are 15 sites where rice mills can be
located. Which subset of sites should we select for setting up the mills?" These
problems involve **combinatorial choices**. Hence they are called **combinatorial
optimization problems**.

"What is the optimum farming policy for my farm for each year over the next
five year horizon?" These involve decisions for each period over a multi-period
horizon. Hence they are known as **multi-period or dynamic programming
problems**. In contrast, all the problems discussed earlier, which model one period
situations, are called **static problems**.

The Objective Function

In optimization problems, there is a criterion function called the **objective
function**, which is required to be optimized (i.e., minimized or maximized as desired). For example in the shortest route problems in Figures 1.1 and 1.2, the
objective function is the length of the route. If the objective function is a cost
function (i.e., one for which higher values are undesirable), then we would like to
minimize it. If it is a profit function (i.e., one for which higher values are desirable),
then we would like to maximize it. Fortunately, it is not necessary to consider minimization and maximization problems separately, since any minimization problem
can be transformed directly into a maximization problem and vice versa. For example, to maximize a function $f(x)$ of decision variables x, is equivalent to minimizing
$-f(x)$ subject to the same system of constraints, and both these problems have the
same set of optimum solutions. Also, we can use

$$\begin{array}{l} \text{Maximum value of } f(x) \text{ subject} \\ \text{to some constraints} \end{array} = -\left(\begin{array}{l} \text{Minimum value of } -f(x) \text{ sub-} \\ \text{ject to the same constraints} \end{array} \right)$$

For this reason, we will discuss algorithms for minimization only in this book.

Single Versus Multi-Objective Models

In the discussion so far we assumed that there is a uniquely identified objective
function to be optimized. Problems with this feature are called **single objective
optimization problems**.

In real world applications one often encounters problems in which there are two or more objective functions which need to be optimized simultaneously. For example in determining its corporate policy, a company may want to minimize its production + marketing costs, but at the same time strive to maximize its market share. There are two objective functions here, problems like this are called **multi-objective optimization problems**. Often, the various objective functions conflict with each other (i.e., optimizing one of them usually tends to move another towards undesirable values), for solving such models one needs to know how many units of one function can be sacrificed to gain one unit of another. In other words, one is forced to determine the best compromise that can be achieved. That's why multi-objective optimization problems are not precisely stated mathematical problems. Techniques for handling them usually involve trial and error using several degrees of compromises among the various objective functions until a consensus is reached that the present solution looks reasonable from the point of view of all the objective functions.

Most real world applications actually involve several objective functions. In some applications, there is one among the various objective functions which is considered to be of the highest priority by all the decision makers. One then treats it as a single objective optimization problem with the highest priority function as the one to optimize. In Chapters 2 to 7, and 9 to 14 we restrict our discussion to single objective problems.

In multi-objective models, it is convenient to pose each objective function in its minimization form as discussed above. Also assume that all the objective functions are in common units such as dollars, or scores on comparable scales, etc., so that adding two objective functions makes sense. One popular technique develops a weight for each objective function that represents its relative importance. A weighted linear combination of all the objective functions using these weights gives a **composite function**. Optimizing this composite function using single objective optimization techniques yields a compromise solution for the original multi-objective problem (the solution obtained may depend critically on the weights selected for the various objective functions, so, some trial and error with different sets of weights may be necessary before reaching a satisfactory solution). We provide a brief discussion of this and other methods for handling multi-objective problems in Chapter 8.

Static Versus Dynamic Models

Models that deal with a one-shot situation are known as **static models**. These include models which involve determining an optimum solution for a one period problem. For example, consider the production planning problem in a company making a variety of products. To determine the optimum quantities of each product that this company should produce in a single year, leads to a static model.

However, planning does involve the time element, and if an application is concerned with a situation that lasts over several years, the same types of decisions may have to be made in each year. In the production planning problem discussed above, if a planning horizon of 5 years is being considered, it is necessary to determine the optimum quantities of each product to produce, in each year of the planning horizon. Models that involve a sequence of such decisions over multiple periods are called **dynamic models**.

When planning for a multi-period horizon, if there is no change in the data at all from one period to the next, then the optimum solution for the first period determined from a static model for that period, will continue to be optimal for every period of the planning horizon. Thus multi-period problems in which the changes in the data over the various periods are small, can be handled through a static one period model, by repeating the same optimum solution in every period. Even when changes in the data from one period to the next are significant, many companies find it convenient to construct a static single period model for their production planning decisions, which they solve at the beginning of every period with the most current estimates of data, for the optimum plan for that period. This points out the importance of static models, even though most real world problems are dynamic.

In most multi-period problems, data changes from one period to the next are not insignificant. In this case the optimum decisions for the various periods may be different, and the sequence of decisions will be interrelated, i.e., a decision taken during a period may influence the state of the system for several periods in the future. Optimizing such a system through a sequence of single period static models solved one at a time, may not produce a policy that is optimal over the planning horizon. However, constructing a dynamic model with the aim of finding a sequence of decisions (one for each period) that is optimum for the planning horizon as a whole, requires reasonably accurate estimates of data for each period of the planning horizon. When such data is available, a dynamic model tries to find the entire sequence of interrelated decisions that is optimal for the system over the planning horizon.

In this book we will discuss both static and dynamic models. We begin with techniques for finding optimum solutions to static models, and then discuss how to extend these to handle dynamic models. Chapter 12 deals with dynamic programming, which is a basic approach for solving problems that are posed as multi-period or multistage decision problems.

Stochastic Versus Deterministic Models

In a single objective static optimization model, the objective function can be interpreted as the yield or profit that is required to be maximized. The objective function expresses the yield as a function of the various decision variables. In real world applications, the yield is almost never known with certainty, typically it is a random variable subject to many random fluctuations that are not under our control. For example the yield may depend on the unit profit coefficients of the various goods manufactured by the company (these are the **data elements** in the model) and these things fluctuate randomly. To analyze the problem treating the yield as a random variable requires the use of complicated **stochastic programming models**. Instead, one normally analyses the problem using a deterministic model in which the random variables in the yield function are replaced by either their most likely values, or expected values. The solution of the deterministic approximation often gives the decision maker an excellent insight for making the best choice. We can also perform sensitivity analysis on the deterministic model. This involves a study of how the optimum solution varies as the data elements in the model vary within a small neighborhood of their current values. Decision makers combine all this information with their human judgment to come up with the best decision to implement.

Some people may feel that even though it is more complicated, a stochastic programming model treating the data elements as random variables (which they are), leads to more accurate solutions than a deterministic approximation obtained by substituting expected values and the like for the data elements. In most cases this is not true. To analyze the stochastic model one needs the probability distributions of the random data elements. Usually, this information is not available. One constructs stochastic models by making assumptions about the nature of probability distributions of random data elements, or estimating these distributions from past data. The closeness of the optimum solution obtained from the model may depend on how close the selected probability distributions are to the true ones. In the world of today, economic conditions and technology are changing constantly, and probability distributions estimated in a month may no longer be valid in the next. Because of this constant change, many companies find it necessary almost in every period to find a new optimum solution by solving the model with new estimates for the data elements. In this mode, an optimum solution is in use only for a short time (one period), and the solution obtained by solving a reasonable deterministic approximation of the real problem is quite suitable for this purpose. For all these reasons most real world optimization applications are based on deterministic models. In this book we discuss only methods for solving deterministic optimization models.

1.3 Optimization in Practice

Optimization is concerned with decision making. Optimization techniques provide tools for making optimal or best decisions. To maintain their market position, or even to continue to exist in business these days, businesses everywhere have to organize their operations to deliver products on time and at the least possible cost; offer services that consistently satisfy customers at the smallest possible price; and introduce new and efficient products and services that are cheaper and faster than competitors. These developments indicate the profound importance of optimization techniques. The organizations that master these techniques are emerging as the new leaders. All the countries in the world today that have a thriving export trade in manufactured goods have achieved it by applying optimization techniques in their manufacturing industries much more vigorously than the other countries.

Optimization problems arising in practice can be classified into three broad categories as in Figure 1.3.

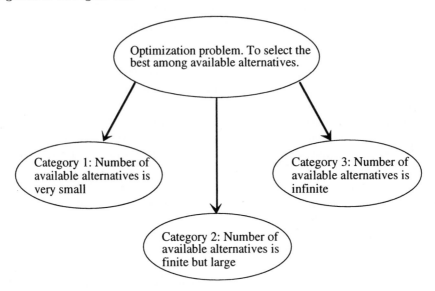

Figure 1.3 Three categories of optimization problems.

Category 1: Number of Alternatives to Compare is Small

Life is full of these problems. We encounter them almost daily. Examples can be found in daily living, in school, at work, almost everywhere. Also, in these problems there are usually several objective functions or characteristics to consider.

For example, consider a boy who has to decide which girl he has to go steady with, among three girls that he knows well. This is an optimization problem in which the best among three alternatives is to be selected. How does the boy solve it? He makes a list of the various attributes of each girl (beauty, friendliness, physical attributes such as height, weight, etc., intelligence, ambition, drive, etc., this is a typical multi-objective situation) and decides how important each attribute is to his personal liking. He combines all this information and develops a collective score for each girl. This score is a measure of how much he likes that girl over all her attributes. He then selects the girl with the highest score.

For another example consider the problem of an automobile manufacturer who needs to decide whether to use a cast iron engine block or an aluminum engine block in their new car models. Here there are two alternatives to evaluate. The aluminum block increases the production costs, but it is lighter and helps the car achieve better gas mileage. To make an objective assessment of the two alternatives one must evaluate the effect of each on the manufacturability of the block, the availability of needed raw material supplies, the repair and maintenance costs of the car, engine life, customer appeal, market share, the profit per car, and on the reputation of the company as a technology leader. After making a careful evaluation on each aspect, a combined overall score for each type of block is developed, and the company then selects the block with the highest score.

We will discuss one more example of a location problem in this category. Consider a city which has decided to build a new airport. Land is available at four different sites, one of which has to be selected for building the airport. Evaluating each site is a fairly complicated problem. The location of an airport at a site inevitably accelerates development in its neighborhood, and environmental effects have to be considered very carefully. The location determines flight paths, noise pollution and the size of the population likely to be affected by it has to be estimated. The existence of transit facilities for people and goods between the city and the airport, and the feasibility of expanding them to meet the projected demands need to be considered. Implications to public safety if a crash occurs at the airport have to be evaluated. Even though there are only four sites to compare, evaluating each of them on all the relevant factors is a complicated and time consuming job. Again, after all the data is in, a combined score is developed for each site, and the one with the best score is selected.

There is not much theory needed to solve problems in this category, and most optimization books do not even mention this category of problems. One has to make a list of all the relevant factors, and develop a combined score for each alternative taking its likely performance on each factor into consideration.

Optimizing with simulation

Simulation is a very handy tool for solving problems in this category in which there are many characteristics to consider, and the yield on them under each of the available alternatives to evaluate is highly stochastic and difficult to assess analytically. Evaluating each alternative analytically may require too much simplification which may not be reasonable. In this situation simulation modeling enables the decision makers to get a meaningful evaluation of the performance of each alternative much more conveniently.

Category 2: Number of Alternatives to Compare is Finite, But Large

An example in this category is the following problem known as the **assignment problem**. A company has divided their marketing area into n zones based on the characteristics of the shoppers, their economic status, etc. They want to appoint a director for each zone to run the marketing effort there. They have already selected n candidates to fill the positions. The total annual sales in a zone would depend on which candidate is appointed as director there. Based on the candidates skills, demeanor, and background, it is estimated that $\$c_{ij}$ million in annual sales will be generated in zone j if candidate i is appointed as director there, and this (c_{ij}) data is given. The problem is to decide which zone each candidate should be assigned to, to maximize the total annual sales in all the zones (each zone gets one candidate and each candidate goes to one zone). We provide the data for a sample problem with $n = 6$ in Table 1.1 given below.

Table 1.1

	c_{ij} = annual sales volume in \$million if candidate i is assigned to zone j					
Zone j =	1	2	3	4	5	6
Candidate i = 1	1	2	6	10	17	29
2	3	4	8	11	20	30
3	5	7	9	12	22	33
4	13	14	15	16	23	34
5	18	19	21	24	25	35
6	26	27	28	31	32	36

In this problem candidate 1 can go to any one of the n zones (so, n possibilities for candidate 1). Then candidate 2 can go to any one of the other zones (so, $n - 1$ possibilities for candidate 2 after candidate 1's zone is fixed). And so on. So, the total number of possible ways of assigning the candidates to the zones is $n \times (n-1) \times (n-2) \times \ldots \times 2 \times 1 = n!$. For the example problem with $n = 6$ there are $6! = 720$ ways of assigning candidates to jobs. As n grows, $n!$ grows very rapidly. Real world applications of the assignment problem typically lead to problems with $n = 100$ to 10,000, and the number of possible assignments in these models is finite but very very large. So, it is not practical to evaluate each alternative separately to select the best as we did in Category 1. So, problems in this category can only be handled by efficient algorithms which are guaranteed to obtain a solution within a reasonable time.

Integer programs and combinatorial optimization problems typically belong in this category.

Category 3: Number of Alternatives is Infinite

An example in this category is the problem of allocating farm land to the various crops discussed earlier. A farmer has 100 acres of farmland on which he can grow wheat, corn, or oats. From past experience on the likely demand for the various grains, the farmer has decided to allot between 20 to 60 acres to wheat, between 30 to 50 acres to corn, and between 10 to 30 acres to oats. Each acre allotted to wheat, corn, and oats yields a profit of \$100, \$120, and \$60 per season, respectively. What is the optimum allocation of the available land to the crops that maximizes the total profit per season subject to the bounds on the amount of land allotted to each crop mentioned above? The decision variables in this problem are the acres of land allotted to wheat, corn, and oats. All these are continuous variables, and this problem can be modeled as a linear program (see Chapter 2). There are an infinite number of feasible solutions to it, and therefore it belongs to this category.

Models involving continuous variables, such as linear programs and nonlinear programs, typically belong to this category.

To solve a problem in Category 2 or 3, we need to construct a mathematical model for it, and solve the model using the appropriate algorithm discussed later on.

1.4 Optimization Problems Are Tricky

People in responsible positions face decision making problems at work almost daily. When you talk to such people, they tell you stories about how much money they are saving for their company. They feel proud about their work and how it is contributing to the company's survival. However, many of them may not be using methods that are guaranteed to give optimum decisions, and many decisions made are far from optimal.

As an example consider the problem from [Jenkyns 1986] of allotting candidates to marketing director positions in zones for which data is given in Table 1.1. Many people would consider this a very easy problem to solve, and would use a method which can aptly be called the **greedy method**.

The method proceeds this way. Since we want to maximize total sales, look for the cell in the table fetching the largest annual sales, this is the cell corresponding to the assignment of Candidate 6 to Zone 6, yielding an annual sales rate of $36 million. So, assign Candidate 6 to Zone 6. This selection achieves the maximum possible sales at this stage with this one allocation, hence it is known as a **greedy choice**. Since Candidate 6 is already assigned to Zone 6, strike off the row of Candidate 6 and the column of Zone 6 from the table; and handle the remaining problem the same way, one allocation at a time, using the greedy choice at each stage. Verify that the solution obtained by this method is $\{(C_1, Z_1), (C_2, Z_2), (C_3, Z_3), (C_4, Z_4), (C_5, Z_5), (C_6, Z_6)\}$, where for $i = 1$ to 6, C_i denotes Candidate i, Z_i denotes Zone i, and (C_i, Z_i) indicates that Candidate i is assigned to Zone i. This fetches a total annual sales rate of $36 + 25 + 16 + 9 + 4 + 1 = \$91$ million.

Let us examine another possible solution to this problem, say the assignment $\{(C_1, Z_6), (C_2, Z_5), (C_3, Z_4), (C_4, Z_3), (C_5, Z_2), (C_6, Z_1)\}$, this leads to an annual sales volume of $29 + 20 + 12 + 15 + 19 + 26 = \$131$ million, far higher than that under the solution obtained by the greedy method. In fact it turns out that in this problem, any arbitrary assignment yields a total sales volume higher than that under the solution produced by the greedy method. Using the theory discussed in Chapter 4, it can be verified that the assignment produced by the greedy method actually minimizes the total sales volume (see Exercise 4.9), i.e., it is the worst possible solution for this problem since our aim is to maximize the total sales volume.

This shows that optimization problems can be very tricky. Heuristic methods such as the greedy method based on common sense may lead to very poor solutions. And yet, many decision makers use such common sense techniques every day, and do not even bother to check whether the solution obtained is reasonable before implementing it. Using a method such as the greedy method to handle every optimization problem, is analogous to a physician prescribing aspirin for every disease.

For some problems such as the assignment problem, linear programs, reasonable size combinatorial optimization problems, and some types of nonlinear programs, we now have efficient algorithms that are guaranteed to yield truly optimum solutions when they exist. The example discussed above points out the need to learn these algorithms in earnest, and to use them whenever they can solve the problem on hand. Unfortunately, there are some hard problems for which efficient algorithms guaranteed to solve them optimally are not known. We need to resort to heuristic methods to handle such hard problems. Even the greedy method has its place as it solves some problems optimally, and produces reasonably good solutions on many others. But when using a heuristic method to solve a problem, a study should be carried out to check the quality of the solution obtained.

In general, optimization problems tend to be complex. A solution that may seem to be optimal based on a cursory analysis, may in fact be very poor. It usually requires a deeper analysis to check how good an alleged optimum solution is. In problems involving many objective functions in particular, a solution may be fantastic for an objective function of minor importance, but very poor overall when all the factors are taken into account.

1.5   Are Optimization Techniques Used Heavily?

This may not be the case. Many people are not aware of the techniques and of the benefits they can bring. In some workplaces, decisions are sometimes taken to suit the vested interests and personal benefits of the decision makers. The extent to which personal benefit sometimes figures in decision making is illustrated by the following popular story. The story reports a conversation between two high level decision makers.

First decision maker: "Did you hire a candidate for the management position you talked about last month?"

Second decision maker: "Yes, three candidates applied. I interviewed all three personally. I asked each of them the same simple question: what is 5 + 5? The first candidate said 10. The second candidate looked surprised, thought for a while and said 20. The third candidate said 12. It is clear that the first candidate goes by the rules, he showed no imagination, so, I rejected him. By his answer the second candidate showed imagination, but I rejected him because his imaginative ability is too wild. I selected the third candidate for the position because she displayed just the right amount of imagination by her answer, and besides, she is my daughter-in-law".

And in some places there is apathy or lack of motivation. But companies that use optimization techniques are thriving. It is hoped that more companies will take advantage of these techniques soon. And the widespread availability of computers, and easy to use software, will definitely help.

The aim of this book is to provide the basic fundamentals of some optimization techniques, and to inspire all the readers to apply optimization techniques vigorously when they join the workforce.

1.6 Optimization Languages and Software Packages

In the following chapters we discuss linear programming, integer programming, and nonlinear programming models; we also discuss algorithms for solving them, and their applications. These three types of models are the major classes of optimization models that show up most frequently in applications. Several high quality, user friendly software packages and languages for solving these models are available commercially on most common platforms. These are designed and implemented to help people use computers to develop and apply optimization models. Most universities and many companies make these packages available to their employees, researchers, and students, through licensing agreements. We provide references to the most popular packages at the end of this chapter, and summarize their main features here briefly.

AMPL [R. Fourer, D. M. Gay, and B. W. Kernighan, 1993] is a modeling language for mathematical programming which promotes a natural form of input for linear, mixed integer, and nonlinear mathematical models. AMPL is a relatively recent entry into the field of algebraic modeling languages for mathematical programming. It emphasizes the general model that can be used to describe large scale optimization problems. The book is accompanied by a PC student version of AMPL and representative solvers, enough to easily handle problems of a few hundred variables and constraints. Versions that support much larger problems, additional solvers, and other operating systems are also available from the publisher. It can be used to solve general linear programming models, integer programs, piecewise-linear programs, and nonlinear programs. AMPL uses the MINOS solver for linear and nonlinear models, and the OSL and CPLEX for linear and mixed integer models.

GAMS [A. Brooke, D. Kendric, and A. Meeraus, 1988] is a high level language that is designed to make the construction and solution of large and complex mathematical programming models straightforward for programmers, and more comprehensible to users of models. It can solve linear, integer, and nonlinear programming problems. A student (restricted) version and a professional version are available.

LINDO [L. Schrage, 1987] is an interactive system that can be used to solve linear, quadratic and integer programs. Versions of LINDO running on PCs, workstations, and mainframes, under several operating systems, are available.

GINO [J. Liebman, L. Lasdon, L. Schrage, and A. Waren, 1986] is a modeling program which can be used to solve nonlinear programs, and sets of simultaneous linear and nonlinear equations and inequalities.

LINGO [K. Cunningham, and L. Schrage, 1990] is a mathematical modeling language that provides an environment with which mathematical models can be developed, modified, and run. The LINGO language is essentially a superset of the GINO language, and the linear solver used by LINGO is LINDO.

OSL [IBM, 1990] The IBM Optimization Subroutine Library is a collection of high performance mathematical subroutines for solving linear, mixed-integer, and quadratic programming models.

1.7   Exercises

1.1 For the fall campaign the democratic presidential candidate has to decide how to allocate his advertising budget among the four media: TV, radio, newspapers and magazines, and billboards. The expenses have already been worked out by his campaign manager, and the choice for him narrowed to two levels, low (L), or high (H), in each media. If he chooses the high level for TV advertising, his budget would only permit low level advertising in each of the other three media. On the other hand if he chooses the low level for TV advertising, his budget would permit him to advertise at the high level in two of the other three media and at the low level in the remaining.

His statistical advisers came up with estimates in the table given below for the reach of the various media.

Each person who is positively (negatively) influenced by the advertisements is expected to discuss and positively (negatively) influence an additional 0.5 (0.3) persons in the same age group through personal conversations. Only 25% of the people in the age group 20-30 years, 50% of the people in the age group 30-60 years, and 70% of the people in the age group 60 years and up, are expected to vote; the corresponding fractions are the weights for the three age groups in developing a combined score for each alternative. The overall score for any alternative is the weighted average over the different age groups of   (the number of positively influenced people − the number of negatively influenced people) summed over all the media. Determine the best advertising strategy for the candidate.

Medium	Advertising level	Estimated number of people (millions) in age group who are influenced					
		20-30 years		30-60 years		60 and up	
		P	N	P	N	P	N
TV	L	5	1	12	3	5	2
	H	9	1.5	20	4	8	3
Radio	L	2.5	0.9	6	1	1.8	0.6
	H	5	0.4	12	1.8	4	0.8
Newspapers and magazines	L	1.6	0.2	4	0.4	1.5	0.3
	H	3	0.1	8	0.6	3	0.2
Billboards	L	0.7	0.2	2	0.2	0.5	0.1
	H	1.2	0.3	4	0.3	0.7	0.2

P = positively influenced, N = negatively influenced

1.2 A person has decided to buy a new car. Seven cars are investigated. The criteria are price, comfort, gas mileage, and looks. Data is given below. Determine which car the person should buy.

Characteristic	car							Weight
	1	2	3	4	5	6	7	
Price ($ 1000 units)	15	13.5	12.5	13	12	12	11	5
Comfort	E	E	A	A	A	W	W	4
MPG	20	17	22	24	18	25	28	3
Looks	D	D	D	O	D	D	O	3

E = Excellent, A = Average, W = Weak, D = Distinctive, O = Ordinary

Weight is the importance person attaches to characteristic, the higher, the more important

1.3 A girl named Anita is in college pursuing a masters degree in engineering. She has been dating 4 boys off and on over the last 3 years, and has come to know each of them very well. The table given below contains her ratings (on a scale in which 1 = least desirable, and 10 = highly desirable) of each boy on characteristics that she considers very important. Also given is a weight between 1 to 10 for each characteristic, which measures how important she considers it to be (higher weight means more important). She needs to decide which boy she should go steady with. Who among the four boys would be her best choice?

Characteristic	Rating of				Weight for this characteristic
	Bill	Raj	Tom	Dick	
Ability to support a family	8	6	4	5	6
Friendliness	7	4	6	8	7
Honesty and trustworthiness	6	5	7	4	10
Respect for women's rights	5	4	6	7	9
Handsomeness	4	7	6	8	5
Interest in decent appearance	7	3	4	9	6
Degree of reciprocity of love	6	8	5	6	8

1.4 A grocery chain is considering opening two new stores in the Ann Arbor area

Site →	1	2	3	4	5
1. Cost of land at site (in $1000 units)	300	500	700	250	600
2. Expected no. of customers coming into new store per hour in first year	140	180	200	120	150
3. % of customers in 2. who are switching from existing store of chain	20	10	5	5	10
4. Average expected profit ($) from each customer in 2.	2	3	4	1.5	3

during the next two years. They have identified 5 different sites where they can buy land to build these stores. Market analysis has provided the data given above.

It is expected that at sites 1, 2, 3 the number of customers coming per hour will grow about 3% per year in the first 5 years, then at 2% per year in the next 5 years, and then level off. Corresponding figures for sites 4, 5 are 2% for first 5 years, and 1% in next 5 years.

Identify the various objective functions that are important for this decision. Determine the best pair of sites to locate the stores. If you need any further information, assume an appropriate value for it and carry out your analysis.

1.5 A city development commissioner has to determine the airconditioning system to be installed in the city library proposed to be built. The contractor offered five feasible airconditioning plans called A_1 to A_5 which might be adapted to the proposed structure of the library. These alternatives are to be evaluated under three major impacts: economic, functional, and operational. Two monetary attributes and six non-monetary attributes emerged from these impacts. The ratings of each alternative wrt each attribute are given below (the higher the rating, the more the preference for that alternative by that attribute; in fact a rating of 1 denotes the worst alternative, and a rating of 10 denotes the best). The weights for the various attributes in combining the ratings into the composite index evaluating the alternative are also given in the following table. Determine which airconditioning plan should be selected for the proposed library.

Attribute	Weight	Rating by attribute for alternative				
		A_1	A_2	A_3	A_4	A_5
1. Economic						
1.1 Installation cost	0.0455	4.2	2	3.5	4	3
1.2 Monthly operational cost	0.0911	6.4	5.2	4.7	5	5.5
2. Functional						
2.1 Performance	0.1297	9	5	5	8	7
2.2 Comfort (noise level etc.)	0.1297	3.5	7	6.5	4	5.5
3. Operational						
3.1 Maintainability	0.1749	4	4	8	9	6
3.2 Reliability	0.2216	9.5	7.5	8.5	9	9
3.3 Flexibility	0.0426	4	8	8	7	5
3.4 Safety	0.1647	7	5	6	8	9

([K. Yoon, July 1989]).

1.6 The setting of this exercise is *Mahabharata*, the great Indian epic that is dated earlier than 5000 BC. It is about a beautiful princess Satyabhama who is trying to select a prince to marry.

Satya was very progressive for her time. While most of her girl friends looked forward to getting married and having lots of children and a large family, she considered that not suitable as a goal for women. Of course she was not opposed to having one or a maximum of two children, but she felt very strongly that women should develop a passionate interest in something more worthwhile than bringing up a lot of children. Even in those days she was quite concerned that the human population growth was contributing to the destruction of nature. She used to go hiking in the forest on the outskirts of her father's capital city, and she particularly admired a rare flowering bush called *Parijata* in that forest. Every morning it used to blossom forth with what appeared to be a million flowers with very bright orange stems and a heavenly fragrance. To her great grief that Parijata bush was devastated in a recent spate of house building as the city expanded, and she was very concerned that it may have become extinct.

Trait	Score on trait of				Weight
	Krishna	Sisupala	Jarasandha	Rukmi	
1. Easygoing nature, friendly disposition	80	70	60	50	10
2. Being a lively and animated companion	90	95	70	65	8
3. Sharing her concern about destruction of nature	40	30	20	45	10
4. Willingness to limit family size to two children	50	30	60	25	10
5. Archery skill	60	70	80	70	7
6. Skill in negotiating deals with opposing parties	80	75	70	60	6
7. Concern for people, particularly those of other tribes etc.	60	45	45	45	5
8. Willingness to let females to join in wars	60	40	40	40	8

She learned horse riding, and driving chariots, and became an expert at these skills; quite unusual for a woman in those days. She learned to launch arrows using a bow with deadly accuracy, and could compete with the best archers in her kingdom. Fighting little wars was almost daily work for kings in those days, and she made up her mind that after marriage she would join her husband in any wars that he may have to fight.

In those days in India, marriages for princesses used to be organized through a function called *swayamvara* (which literally means "self-chosen"). All the eligible princes would be invited to a gala party. There would be sumptuous meals followed by dancing where the princess dances and chats with each visiting prince. There would be contests in archery etc. where the various suitors display their skills.

During this entire process the princess is gathering information about each suitor and weighing her choices. When her decision is finalized, she would come out with a garland of flowers with which she would adorn the prince of her choice, and then the wedding would be celebrated.

At Satya's swayamvaram there were four suitors, Krishna, Sisupala, Jarasandha, and Rukmi. The personality traits that she considered important in her future husband are listed in the left hand column in the above table. She scored each suitor on each trait on a scale of 0 to 100 (the higher the score, the more desirable the suitor is on that trait). In the rightmost column of the table we provide the weight for each trait which measures the relative importance she attached to that trait (again, the higher the weight, the more important she considered that trait to be).

Help Satya choose her fiance from among the four suitors.

1.8 References

A. BROOKE, D. KENDRIC, and A. MEERAUS, 1988, *GAMS: A User's Guide*, The Scientific Press, San Francisco.

K. CUNNINGHAM, and L. SCHRAGE, 1989, *The LINGO Modeling Language*, Lindo Systems, Inc., Chicago, IL.

R. FOURER, D. M. GAY, and B. W. KERNIGHAN, 1993, *AMPL A Modeling Language for Mathematical Programming*, The Scientific Press, San Francisco.

IBM, 1990, *Optimization Subroutine Library Guide and Reference*, IBM Corp., NY.

T. A. JENKYNS, 1986, "The Greedy Algorithm is a Shady Lady", *Congressus Numerantium*, 51(209-215).

J. LIEBMAN, L. LASDON, L. SCHRAGE, and A. WAREN, 1986, *Modeling and Optimization with GINO*, The Scientific Press, San Francisco.

L. SCHRAGE, 1987, *User's manual for Linear, Integer, and Quadratic Programming with LINDO*, 3rd edition, The Scientific Press, San Francisco.

K. YOON, July 1989, "The Propagation of Errors in Multiple Attribute Decision Analysis: A Practical Approach", *Journal of the Operational Research Society*, 40, no. 7 (681-686).

Chapter 2

Modeling Linear Programs

In all the problems discussed in this chapter (and also the rest of this book) there are variables called the **decision variables**, $x_1, \ldots, x_n$, say. We denote by x the column vector of these decision variables, written as the transpose of a row vector $(x_1, \ldots, x_n)^T$ to conserve space.

In most applications the values of the decision variables are required to satisfy certain constraints. Each constraint is either of the form $g_i(x) = b_i$ (called an **equality constraint**) or of the form $g_i(x) \geqq b_i$ or $g_i(x) \leqq b_i$ (called an **inequality constraint**), where $g_i(x), b_i$ are a given function of the decision variables, and a given constant, respectively. The function $g_i(x)$ is called the **constraint function** and b_i the **right hand side constant** or the **RHS constant**) in this constraint. There may be constraints of the form $x_j \geqq b_j$ or $x_j \leqq b_j$ in which the constraint function involves only one variable, these are **bound constraints** or **bounds, lower or upper, on individual variables**. In particular, lower bound constraints of the form $x_j \geqq 0$ present in most models, are the **nonnegativity restrictions** on individual variables.

A vector $x = (x_1, \ldots, x_n)^T$ which satisfies all the constraints is called a **feasible solution** for the problem. We are also given either a **cost function** to be minimized, or a **profit function** to be maximized, this is the **objective function** in the problem. The aim is to find an **optimum solution** which is a feasible solution that has the best value for the objective function among all the feasible solutions.

A **linear function** is a function of the form $c_1 x_1 + \ldots + c_n x_n$ where $c_1, \ldots, c_n$ are given constants known as the **coefficients of the variables in the function** (some of these could be zero). For example, $x_2 - 7x_4$ is a linear function of the variables x_1 to x_4 with the coefficient vector $(0, 1, 0, -7)$.

A **linear program** (LP in short) is an optimization problem in which all the constraint functions and the objective function are linear functions. Linear programming is an extremely important branch of optimization that finds many applications in industry, business, and government. Its importance is even recognized by the Nobel prize committee when it awarded the 1975 Nobel prize in economics to L. Kantorovitch and T. C. Koopmans for developing the application of linear programming to the economic problem of allocating resources. In this chapter we present elementary methods for formulating problems as linear programs with many illustrative examples.

There are many different classes of applications of linear programming models. We present some of them in the following sections.

2.1 A Product Mix Problem

Product mix problems are an extremely important class of problems that manufacturing companies face. Normally the company can make a variety of products using the raw materials, machinery, labor force, and other resources available to them. The problem is to decide how much of each product to manufacture in a period, to maximize the total profit subject to the availability of needed resources. To model this, we need data on the units of each resource necessary to manufacture one unit of each product, any bounds (lower, upper, or both) on the amount of each product manufactured per period, any bounds on the amount of each resource available per period, and the cost or net profit per unit of each product manufactured. Assembling this type of reliable data is one of the most difficult jobs in constructing a product mix model for a company, but as we will see later on, it is very worthwhile. A product mix model can be used to derive extremely useful planning information for the company. The process of assembling all the needed data is sometimes called **input-output analysis** of the company. The coefficients, which are the resources necessary to make a unit of each product, are called **input-output coefficients**, or **technology coefficients**.

Table 2.1

Item	Tons required to make one ton of		Maximum amount of item available daily (tons)
	Hi-ph	Lo-ph	
RM 1	2	1	1500
RM 2	1	1	1200
RM 3	1	0	500
Net profit ($) per ton made	15	10	

As an example, consider a fertilizer company that makes two kinds of fertilizers called Hi-phosphate (Hi-ph) and Lo-phosphate (Lo-ph). The manufacture of these fertilizers requires three raw materials called RM 1, 2, 3. These raw materials come from the company's own quarry, and hence the maximum amount of them available daily is limited by how much the quarry can supply, which is 1500 tons, 1200 tons, 500 tons for RM 1, 2, and 3, respectively. There are no bounds on the amount of Hi-ph, Lo-ph manufactured daily since the market seems to be able to absorb any amounts that the company makes. The remaining data is summarized in Table 2.1.

So, in this example, the Hi-ph manufacturing process can be imagined as a black box which takes as input a packet consisting of 2 tons RM 1, 1 ton RM 2, and 1 ton RM 3; and outputs 1 ton of Hi-ph. See Figure 2.1. A similar interpretation can be given for the Lo-ph making process.

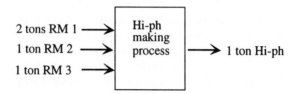

Figure 2.1 Input-Output representation of the Hi-ph making process.

Constructing a mathematical model for the problem involves the following steps.

Step 1: Make a list of all the decision variables

The list must be complete in the sense that if an optimum solution providing the values of each of the variables is obtained, the decision maker should be able to translate it into an optimum policy that can be implemented. In product mix models, there is one decision variable for each possible product the company can produce, it measures the amount of that product made per period.

In our example, there are clearly two decision variables; these are:

$$x_1 = \text{the tons of Hi-ph made per day}$$
$$x_2 = \text{the tons of Lo-ph made per day}$$

Associated with each variable in the problem is an **activity** that the decision maker can perform. The activities in this example are:

$$\text{Activity 1} : \text{to make 1 ton of Hi-ph}$$
$$\text{Activity 2} : \text{to make 1 ton of Lo-ph}$$

The variables in the problem just define the **levels** at which these activities are carried out. So, one way of carrying out this step is to make a list of all the possible activities that the company can perform, and associate a variable that measures the level at which it is carried out, for each activity.

When the constraints binding the variables are written, it will be clear that there are additional variables (called **slack variables** in what follows) imposed on the problem by the inequality constraints. Slack variables are discussed later.

Even though it is mathematically convenient to denote the decision variables by symbols x_1, x_2, etc., practitioners find it very cumbersome to look up what each of these variables represents in the practical problem. For this reason they give the decision variables suggestive names, for example x_1, x_2 here would be called Hi-ph, Lo-ph instead.

Step 2: Verify that the linearity assumptions hold

Proportionality Assumption

It requires 2 tons of RM 1 to make 1 ton of Hi-ph. From this, the **proportionality assumption** implies that $2x_1$ tons of RM 1 are required to make x_1 tons of Hi-ph for any $x_1 \geqq 0$. In general, the proportionality assumption guarantees that if a_{ij} units of the ith item are consumed (or produced) in carrying out activity j at unit level, then $a_{ij}x_j$ units of this item are consumed (or produced) in carrying out activity j at level x_j for any $x_j \geqq 0$. Similarly if $\$\ c_j$ of profit is earned in carrying out activity j at unit level, then $\$\ c_jx_j$ must be the contribution of activity j to the profit function when it is carried out at level x_j, if the proportionality assumption holds for the profit function.

Additivity Assumption

It requires 2 tons of RM 1 to make 1 ton of Hi-ph, and 1 ton of the same to make 1 ton of Lo-ph. From this, the **additivity assumption** implies that $2x_1 + x_2$ tons of RM 1 are required to make x_1 tons of Hi-ph and x_2 tons of Lo-ph for any $x_1, x_2 \geqq 0$. The additivity assumption generally implies that the total consumption (or production) of an item is equal to the sum of the various quantities of the item consumed (or produced) in carrying out each individual activity at its specified level. When the additivity assumption holds for the objective function $z(x_1, \ldots, x_n) = z(x)$, it must be **separable in the variables**, i.e., it can be written as a sum of n functions each of which involves only one variable (as $z_1(x_1) + \ldots + z_n(x_n)$, where $z_j(x_j)$ is the contribution of the variable x_j to the objective function).

The proportionality and the additivity assumptions together are known as the **linearity assumptions**. Formulation of the problem as an LP requires that these assumptions must hold. These assumptions imply that the objective function is linear, and that all the constraints are either linear equations or inequalities.

In most real world applications the linearity assumptions may not hold exactly. For example, the selling price per unit normally decreases, with increasing discounts as the number of units purchased increases. However, it has been observed that in many of these applications, the linearity assumptions provide reasonably good approximations. This, and the relative ease with which LPs can be solved, have made it possible for linear programming to find a vast number of applications.

Step 3: Verify that decision variables are all continuous variables

Verify that each decision variable in the model can take all the real values in its range of variation.

In some applications, variables may be restricted to take only integer values (e.g., if the variable represents the number of empty buses transported from one location to another). Such restrictions make the problem an **integer program**. However, sometimes people ignore the integer restrictions on integer variables and treat them as continuous variables. If the linearity assumptions hold, this leads to the **LP relaxation** of the integer program.

Step 4: Construct the objective function

Since the profit (or cost) coefficients associated with each product are given, the objective here is either to maximize the total profit (or to minimize the total cost). By the linearity assumptions the objective function is a linear function, it is obtained by multiplying each decision variable by its profit (or cost) coefficient and summing up.

In our example problem, the objective function is the total net daily profit, $z(x) = 15x_1 + 10x_2$, and it has to be maximized.

Step 5: Identify the Constraints on the Decision Variables

Nonnegativity constraints

In product mix models the decision variables are the amounts of various products made per period; these have to be nonnegative to make any practical sense. In linear programming models in general, the nonnegativity restriction on the variables is a natural restriction that occurs because certain activities (manufacturing a product, etc.) can only be carried out at nonnegative levels.

The nonnegativity constraint is a lower bound constraint. Sometimes it may be necessary to impose a positive lower bound on a variable. This occurs if we have a commitment to make a minimum quantity, ℓ_j units say, where $\ell_j > 0$, of product j. Then the lower bound constraint on the decision variable $x_j =$ amount of product j manufactured, is $x_j \geqq \ell_j$.

There may be an upper bound constraint on a variable too. This occurs if we know that only a limited quantity, say u_j units, of product j can be either sold in a period or stored for use later on, then $x_j \leqq u_j$ is the upper bound constraint on $x_j =$ the amount of product j made in that period.

On some decision variables there may be both a lower and an upper bound constraint.

In our example problem the bound constraints are: $x_1, x_2 \geq 0$.

Items and the associated constraints

There may be other constraints on the variables, imposed by lower or upper bounds on certain goods that are either inputs to the production process or outputs from it. Such goods that lead to constraints in the model are called **items**. Each item leads to a constraint on the decision variables, and conversely every constraint in the model is associated with an item. Make a list of all the items that lead to constraints.

In the fertilizer problem each raw material leads to a constraint. The amount of RM 1 used is $2x_1 + x_2$ tons, and it cannot exceed 1500, leading to the constraint $2x_1 + x_2 \leq 1500$. Since this inequality compares the amount of RM 1 used to the amount available, it is called a **material balance inequality**. The material balance equations or inequalities corresponding to the various items are the constraints in the problem.

When all the constraints are obtained, the formulation of the problem as an LP is complete. The LP formulation of the fertilizer product mix problem is given below.

$$
\begin{array}{llcccccl}
\text{Maximize} & z(x) = 15x_1 & + & 10x_2 & & & & \text{Item} \\
\text{Subject to} & 2x_1 & + & x_2 & \leq & 1500 & & \text{RM 1} \\
& x_1 & + & x_2 & \leq & 1200 & & \text{RM 2} \qquad (2.1) \\
& x_1 & & & \leq & 500 & & \text{RM 3} \\
& x_1 & \geq 0, & x_2 & \geq & 0 & &
\end{array}
$$

Practitioners usually name each decision variable by an acronym of the corresponding activity, and each constraint by an acronym of the corresponding item.

Slack Variables

Each material balance inequality in the model contains in itself the definition of another nonnegative variable known as a **slack variable**. For example, the RM 1 constraint in the fertilizer product mix problem, $2x_1 + x_2 \leq 1500$, can be written as $1500 - 2x_1 - x_2 \geq 0$. If we define $x_3 = 1500 - 2x_1 - x_2$, then x_3 is the slack variable corresponding to this constraint, it represents the amount of RM 1 remaining unutilized in the daily supply; and the constraint can be written in the equivalent form $2x_1 + x_2 + x_3 = 1500, x_3 \geq 0$.

In the same manner, if there is a constraint of the form $3x_1 + 2x_2 \geq 100$, the slack variable corresponding to it will be $s_1 = 3x_1 + 2x_2 - 100$, and after introducing it the constraint can be written in the equivalent form $3x_1 + 2x_2 - s_1 = 100, s_1 \geq 0$.

Remember that each inequality constraint in the problem leads to a different slack variable, and that all slack variables are always nonnegative variables. By introducing all the necessary slack variables, the LP model can be written as one in which all the variables are either unrestricted or nonnegative variables, and all the other constraints are equality constraints.

After introducing x_3, x_4, x_5, the slack variables for RM 1, 2, 3 constraints, the fertilizer product mix problem (2.1) is the following in detached coefficient form.

Tableau 2.2

x_1	x_2	x_3	x_4	x_5	=	
2	1	1	0	0	1500	
1	1	0	1	0	1200	
1	0	0	0	1	500	
15	10	0	0	0	$z(x)$	maximize

$$x_1 \text{ to } x_5 \overset{\geq}{=} 0$$

An LP in which all the variables are nonnegative variables, and all the other constraints are equality constraints, is said to be in **standard form**. Tableau 2.2 gives the fertilizer product mix problem in standard form. Every LP can be transformed into standard form, this is discussed in Chapter 7.

In real world applications, typically after each period there may be changes in the profit or cost coefficients, the RHS constants (availabilities of items), and technology coefficients. Also, new products may come on stream and some old products may fade out. So, most companies find it necessary to revise their product mix model and solve it afresh at the beginning of each period.

To model any problem as an LP we need to go through the same Steps 1 to 5 given above. We will now discuss examples from different classes of applications.

2.2   A Blending Problem

This is another large class of problems in which linear programming is applied heavily. Blending is concerned with mixing different materials called the **constituents** of the mixture (these may be chemicals, gasolines, fuels, solids, colors, foods, etc.) so that the mixture conforms to specifications on several properties or characteristics. To model a blending problem as an LP, the linearity assumptions must hold. This implies that the value for a characteristic of a mixture is the weighted average of the values of that characteristic for the constituents in the mixture; the weights being the proportions of the constituents. As an example, consider a mixture consisting of 4 barrels of fuel 1 and 6 barrels of fuel 2, and suppose the characteristic of interest is the octane rating (Oc.R). If linearity assumptions hold, the Oc.R of the mixture must be equal to (4 times the Oc.R of fuel 1 + 6 times the Oc.R of fuel 2)/(4 + 6). These linearity assumptions hold to a reasonable degree of precision for many important characteristics of blends of gasolines, of crude oils, of paints, of foods, etc. That's why linear programming is used extensively in optimizing gasoline blending, in the manufacture of paints, cattle feed, beverages, etc.

The decision variables in a blending problem are usually either the quantities or the proportions of the constituents.

If a specified quantity of the blend needs to be made, then it is convenient to take the decision variables to be the quantities of the various constituents blended; in this case one must include the constraint that the sum of the quantities of the constituents = the quantity of the blend desired.

If there is no restriction on the amount of blend made, but the aim is to find an optimum composition for the mixture, it is convenient to take the decision variables to be the proportions of the various constituents in the blend; in this case one must include the constraint that the sum of all these proportions is 1.

As an example we consider a gasoline blending problem. A refinery takes four raw gasolines, blends them to produce three types of fuel. Here are the data.

Raw gas type	Octane rating (Oc.R)	Available daily (barrels)	Price per barrel
1	68	4000	$31.02
2	86	5050	33.15
3	91	7100	36.35
4	99	4300	38.75

Fuel type	Minimum Oc.R	Selling price (barrel)	Demand
1	95	$45.15	At most 10,000 barrels/day
2	90	42.95	No limit
3	85	40.99	At least 15,000 barrels/day

The company sells raw gasoline not used in making fuels at $38.95/barrel if its Oc.R is > 90, and at $36.85/barrel if its Oc.R is ≤ 90. The problem is to determine the compositions of the three fuels to maximize total daily profit.

To model this problem, we assume that the linearity assumptions stated above hold. Since three different fuels are under consideration, it is convenient to use a double subscript notation to denote the blending decision variables as given below.

$$x_{ij} = \begin{cases} \text{barrels of raw gasoline type } i \text{ used in making fuel type } j \\ \text{per day, } i = 1 \text{ to } 4, \ j = 1,2,3 \end{cases}$$

$$y_i = \text{barrels of raw gasoline type } i \text{ sold as is}$$

So, the total amount of fuel type 1 made daily is $x_{11} + x_{21} + x_{31} + x_{41}$. If this is > 0, by the linearity assumptions its Oc.R is $(68x_{11} + 86x_{21} + 91x_{31} + 99x_{41})/(x_{11} +$

$x_{21} + x_{31} + x_{41}$). This is required to be $\geqq 95$. Thus, if $x_{11} + x_{21} + x_{31} + x_{41} > 0$, we must have

$$\frac{68x_{11} + 86x_{21} + 91x_{31} + 99x_{41}}{x_{11} + x_{21} + x_{31} + x_{41}} \geqq 95$$

In this form the constraint is not a linear constraint since the constraint function is not a linear function. So, if we write the constraint in this form the model will not be an LP. However we see that this constraint is equivalent to

$$68x_{11} + 86x_{21} + 91x_{31} + 99x_{41} - 95(x_{11} + x_{21} + x_{31} + x_{41}) \geqq 0 \qquad (2.2)$$

and this is a linear constraint. Also, if the amount of fuel type 1 made is zero, all of $x_{11}, x_{21}, x_{31}, x_{41}$ are zero, and (2.2) holds automatically. Thus, the Oc.R constraint on fuel type 1 can be represented by the linear constraint (2.2). Proceeding in a similar manner, we obtain the following LP formulation for this problem.

Maximize $\quad 45.15(x_{11} + x_{21} + x_{31} + x_{41}) + 42.95(x_{12} + x_{22} + x_{32} + x_{42})$
$\qquad\qquad + 40.99(x_{13} + x_{23} + x_{33} + x_{43}) + y_1(36.85 - 31.02) + y_2(36.85$
$\qquad\qquad - 33.15) + y_3(38.95 - 36.35) + y_4(38.95 - 38.75) - 31.02(x_{11}$
$\qquad\qquad + x_{12} + x_{13}) - 33.15(x_{21} + x_{22} + x_{23}) - 36.35(x_{31} + x_{32} + x_{33})$
$\qquad\qquad - 38.75(x_{41} + x_{42} + x_{43})$

Subject to $\quad 68x_{11} + 86x_{21} + 91x_{31} + 99x_{41} - 95(x_{11} + x_{21} + x_{31} + x_{41}) \geqq 0$
$\qquad\qquad 68x_{12} + 86x_{22} + 91x_{32} + 99x_{42} - 90(x_{12} + x_{22} + x_{32} + x_{42}) \geqq 0$
$\qquad\qquad 68x_{13} + 86x_{23} + 91x_{33} + 99x_{43} - 85(x_{13} + x_{23} + x_{33} + x_{43}) \geqq 0$

$$
\begin{aligned}
x_{11} + x_{12} + x_{13} + y_1 &= 4000 \\
x_{21} + x_{22} + x_{23} + y_2 &= 5050 \\
x_{31} + x_{32} + x_{33} + y_3 &= 7100 \\
x_{41} + x_{42} + x_{43} + y_4 &= 4300 \\
x_{11} + x_{21} + x_{31} + x_{41} &\leqq 10,000 \\
x_{13} + x_{23} + x_{33} + x_{43} &\geqq 15,000 \\
x_{ij}, y_i &\geqq 0, \text{ for all } i, j
\end{aligned}
$$

Blending models are economically significant in the petroleum industry. The blending of gasoline is a very popular application. A single grade of gasoline is normally blended from about 3 to 10 individual components, no one of which meets the quality specifications by itself. A typical refinery might have 20 different components to be blended into 4 or more grades of gasoline, and other petroleum products such as aviation gasoline, jet fuel, and middle distillates; differing in Oc.R and properties such as pour point, freezing point, cloud point, viscosity, boiling characteristics, vapor pressure, etc., by marketing area.

2.3 A Diet Problem

A **diet** is a combination of available foods to be consumed daily. The constraints on a diet are that it must meet the minimum daily requirements (MDR) of each nutrient identified as being important for the individual's well-being. The diet problem is concerned with finding a minimum cost diet meeting all the nutrient requirements.

It is a classic problem, one among the earliest problems formulated as an LP. The first paper on it was published by G. J. Stigler under the title "The Cost of Subsistence" in the *Journal of Farm Economics*, vol. 27, 1945. Those were the war years, food was expensive, and the problem of finding a minimum cost diet was of more than academic interest. Nutrition science was in its infancy in those days, and after extensive discussions with nutrition scientists Stigler identified nine essential nutrient groups for his model. His search of the grocery shelves yielded a list of 77 different available foods. With these, he formulated a diet problem which was an LP involving 77 nonnegative decision variables subject to nine inequality constraints.

Stigler did not know of any method for solving his LP model at that time, but he obtained an approximate solution using a trial and error search procedure that led to a diet meeting the MDR of the nine nutrients considered in the model at an annual cost of $39.93 in 1939 prices! After Dantzig developed the simplex algorithm for solving LPs in 1947, Stigler's diet problem was one of the first nontrivial LPs to be solved by the simplex method on a computer, and it gave the true optimum diet with an annual cost of $39.67 in 1939 prices. So, the trial and error solution of Stigler was very close to the optimum.

The Nobel prize committee awarded the 1982 Nobel prize in economics to Stigler for his work on the diet problem and later work on the functioning of markets and the causes and effects of public regulation.

Nutrient	Nutrient units/kg. of grain type		MDR of nutrient in units
	1	2	
Starch	5	7	8
Protein	4	2	15
Vitamins	2	1	3
Cost ($/kg.) of food	0.60	0.35	

The units for measuring the various nutrients and foods may be very different, for example carrots may be measured in pounds, chestnuts in kilograms, milk in gallons, orange juice in liters, vitamins in IU, minerals in mg., etc. The data in the diet problem consists of a list of nutrients with the MDR for each, a list of available foods with the price and composition (i.e., information on the number of units of each nutrient in each unit of food) of every one of them; and the data defining any other constraints the user wants to place on the diet.

As an example we consider a very simple diet problem in which the nutrients are starch, protein, and vitamins as a group; and the foods are two types of grains with data given above.

The activities and their levels in this model are: for $j = 1, 2$

Activity j: to include one kg. of grain type j in the diet, associated level $= x_j$

So, x_j is the amount in kg. of grain j included in the daily diet, $j = 1, 2,$ and the vector $x = (x_1, x_2)^T$ is the diet. The items in this model are the various nutrients, each of which leads to a constraint. For example, the amount of starch contained in the diet x is $5x_1 + 7x_2$, which must be $\geqq 8$ for feasibility. This leads to the formulation given below.

$$
\begin{array}{llllllll}
\text{Minimize} & z(x) = 0.60x_1 & + & 0.35x_2 & & & \text{Item} \\
\text{Subject to} & 5x_1 & + & 7x_2 & \geqq & 8 & \text{Starch} \\
& 4x_1 & + & 2x_2 & \geqq & 15 & \text{Protein} \\
& 2x_1 & + & x_2 & \geqq & 3 & \text{Vitamins} \\
& x_1 & \geqq 0, & x_2 & \geqq & 0 &
\end{array}
$$

This simple model contains no constraints to guarantee that the diet is palatable, and does not allow any room for day-to-day variations that contributes to eating pleasure, and hence the solution obtained from it may be very hard to implement for human diet. The basic model can be modified by including additional constraints to make sure that the solution obtained leads to a tasteful diet with ample scope for variety. This sort of modification of the model after looking at the optimum solution to determine its reasonableness and implementability, solving the modified model, and even repeating this whole process several times, is typical in practical applications of optimization.

We human beings insist on living to eat rather than eating to live. And if we can afford it, we do not bother about the cost of food. It is also impossible to make a human being eat a diet that has been determined as being optimal. For all these reasons, it is not practical to determine human diet using an optimization model.

However, it is much easier to make cattle and fowl consume the diet that is determined as being optimal for them. Almost all the companies in the business of making feed for cattle, other animals, birds, etc. use linear programming extensively to minimize their production costs. The prices and supplies of various grains, hay, etc. are constantly changing, and feed makers solve the diet model frequently with new data values, to make their buy-decisions and to formulate the optimum mix for manufacturing the feed.

2.4   A Transportation Problem

An essential component of our modern life is the shipping of goods from where they are produced to markets worldwide. Nationally, companies spend billions of dollars annually in transporting goods. The aim of this problem is to find a way of carrying out this transfer of goods at minimum cost. Historically it is among the first linear programming problems to be modeled and studied. The Russian economist L. V. Kantorovitch studied this problem in the 1930's and published a book on it *Mathematical Methods in the Organization and Planning of Production* in Russian in 1939. In the USA, F. L. Hitchcock published a paper "The Distribution of a Product From Several Sources to Numerous Localities" in the *Journal of Mathematics and Physics*, vol. 20, 1941, where he developed an algorithm similar to the primal simplex algorithm for finding an optimum solution to the problem. And T. C. Koopmans published a paper "Optimum Utilization of the Transportation System" in *Econometrica*, vol. 17, 1949, in which he developed an optimality criterion for a basic solution to the transportation problem in terms of the dual basic solution (see Chapter 4 for a discussion of this criterion). The early work of L. V. Kantorovitch and T. C. Koopmans in these publications is part of their effort for which they received the 1975 Nobel prize in economics.

The classical transportation problem is concerned with a set of nodes or places called **sources** which have a commodity available for shipment, and another set of places called **sinks** or **demand centers** or **markets** which require this commodity. The data consists of the **availability** at each source (the amount available there to be shipped out), the **requirement** at each market, and the cost of transporting the commodity per unit from each source to each market. The problem is to determine the quantity to be transported from each source to each market so as to meet the requirements at minimum total shipping cost.

	c_{ij} (cents/ton)			Availability at
	$j = 1$	2	3	mine (tons) daily
Mine $i = 1$	11	8	2	800
2	7	5	4	300
Requirement at plant (tons) daily	400	500	200	

As an example, we consider a small problem where the commodity is iron ore, the sources are mines 1 and 2 that produce the ore, and the markets are three steel plants that require the ore. Let c_{ij} = cost (cents per ton) to ship ore from mine i to steel plant j, $i = 1, 2$, $j = 1, 2, 3$. The data is given above. To distinguish between different data elements, we show the cost data in normal size letters, and the supply and requirement data in larger size letters.

The activities in this problem are: to ship one ton of the commodity from source i to market j. It is convenient to represent the level at which this activity is carried out by the double subscripted symbol x_{ij}. In this example x_{ij} represents the amount of ore (in tons) shipped from mine i to plant j.

The items in this model are the ore at various locations. Consider ore at mine 1. There are 800 tons of it available, and the amount of ore shipped out of this mine, $x_{11} + x_{12} + x_{13}$, cannot exceed the amount available, leading to the constraint $x_{11} + x_{12} + x_{13} \leqq 800$. Likewise, considering ore at steel plant 1, at least 400 tons of it is required there, so the total amount of ore shipped to this plant has to be $\geqq 400$, leading to the constraint $x_{11} + x_{21} \geqq 400$. The total amount of ore available at both mines 1, 2 together is $800 + 300 = 1100$ tons daily; and the total requirement at plants 1, 2, 3 is also $400 + 500 + 200 = 1100$ tons daily. Clearly, this implies that all the ore at each mine will be shipped out, and the requirement at each plant will be met exactly; i.e., all the constraints will hold as equations. Therefore we have the following LP formulation for this problem.

Min. $z(x) =$	$11x_{11}$	$+ 8x_{12}$	$+ 2x_{13}$	$+ 7x_{21}$	$+ 5x_{22}$	$+ 4x_{23}$		Item
S. to	x_{11}	$+ x_{12}$	$+ x_{13}$				$= 800$	Ore/mine 1
				x_{21}	$+ x_{22}$	$+ x_{23}$	$= 300$	Ore/mine 2
	x_{11}			$+ x_{21}$			$= 400$	Ore/plant 1
		x_{12}			$+ x_{22}$		$= 500$	Ore/plant 2
			x_{13}			$+ x_{23}$	$= 200$	Ore/plant 3

$$x_{ij} \geqq 0 \quad \text{for all } i = 1, 2, j = 1, 2, 3$$

The Special Structure of the Transportation Problem

As an LP, the transportation problem has a very special structure. It can be represented very compactly in a two dimensional array in which row i corresponds to source i; column j corresponds to demand center j; and (i, j), the cell in row i and column j, corresponds to the shipping route from source i to demand center j. Inside the cell (i, j), record the decision variable x_{ij} which represents the amount of commodity shipped along the corresponding route, and enter the unit shipping cost on this route in the lower right-hand corner of the cell. The objective function in this model is the sum of the variables in the array multiplied by the cost coefficient in the corresponding cell. Record the availabilities at the sources in a column on the right-hand side of the array; and similarly the requirements at the demand centers in a row at the bottom of the array. Then each constraint other than any bound constraints on individual variables is a constraint on the sum of all the variables either in a row or a column of the array, and it can be read off from the array as shown below for the iron ore example.

Array Representation of the Transportation Problem

	Steel Plant			
	1	2	3	
Mine 1	x_{11} \qquad 11	x_{12} \qquad 8	x_{13} \qquad 2	$= 800$
Mine 2	x_{21} \qquad 7	x_{22} \qquad 5	x_{23} \qquad 4	$= 300$
	$= 400$	$= 500$	$= 200$	

$x_{ij} \geqq 0$ for all i, j. Minimize cost.

Supplies, requirements in large size numbers

Any LP, whether it comes from a transportation or a different context, that can be represented in this special form of a two dimensional array is called a **transportation problem**. The constraints in the example problem are equations, but in general they may be equations or inequalities.

Integer Property in the Transportation Model

In a general LP, even when all the data are integer valued, there is no guarantee that there will be an optimum integer solution. However, the special structure of the transportation problem makes the following theorem possible.

Theorem 2.1 *In a transportation model, if all the availabilities and requirements are positive integers, and if the problem has a feasible solution, then it has an optimum solution in which all the decision variables x_{ij} assume only integer values.*

This theorem follows from the results discussed in Chapter 6. In fact in that chapter we discuss the primal simplex algorithm for the transportation problem, which terminates with an integer optimum solution for it when the conditions mentioned in the theorem hold.

A word of caution. The statement in Theorem 2.1 does not claim that an optimum solution to the problem must be an integer vector when the conditions stated there hold. There may be many alternate optimum solutions to the problem and the theorem only guarantees that at least one of these optimum solutions will be an integer vector.

The practical importance of the integer property will become clear from the next example.

The Balanced Transportation Problem

As mentioned above, the constraints in a transportation problem may be equations or inequalities. However, when the following condition holds

$$\left.\begin{array}{l}\text{total material available}\\ \text{= sum of availabilities at}\\ \text{all sources}\end{array}\right\} = \left\{\begin{array}{l}\text{total material required = sum}\\ \text{of the requirements at all the}\\ \text{markets}\end{array}\right. \qquad (2.3)$$

to meet the requirements at the markets, all the material available at every source will be shipped out and every market will get exactly as much as it requires, i.e., all constraints hold as equations. That's why (2.3) is a **balance condition**, and when it holds, and all the constraints are equations, the problem is called a **balanced transportation problem**. As formulated above, the iron ore problem is a balanced transportation problem.

2.5   A Marriage Problem

This problem was proposed as an application of linear programming to sociology in a paper "The Marriage Problem" in *American Journal of Mathematics*, vol. 72, 1950, by P. R. Halmos and H. E. Vaughan. It is concerned with a club consisting

	c_{ij} for woman $j =$				
	1	2	3	4	5
man $i = 1$	78	−16	19	25	83
2	99	98	87	16	92
3	86	19	39	88	17
4	−20	99	88	79	65
5	67	98	90	48	60

of an equal number of men and women (say n each), who know each other well. The data consists of a rating (or happiness coefficient) c_{ij} which represents the amount of happiness that man i and woman j acquire when they spend a unit of time together, $i, j = 1$ to n. The coefficients c_{ij} could be positive, 0, or negative. If $c_{ij} > 0$, man i and woman j are happy when together. If $c_{ij} < 0$, they are unhappy when together, and so acquire unhappiness only. Hence, in this setup unhappiness is a negative value for happiness and vice versa. To keep the model simple, it is assumed that the remaining life of all club members is equal, and time is measured in units of this lifetime. The problem is to determine the fraction of this lifetime that man i, woman j should spend together for $i, j = 1$ to n, to maximize the overall club's happiness.

As an example, we consider a club consisting of 5 men and 5 women and the happiness ratings (c_{ij}) given above. These happiness ratings are on a scale of -100 to $+100$ where -100 represents "very unhappy" and $+100$ represents "very happy".

There are 25 activities in this model. These are, for $i, j = 1$ to 5

Activity: Man i and woman j to spend one unit of time together. Associated level $= x_{ij}$.

	Woman					
	1	2	3	4	5	
Man 1	x_{11} 78	x_{12} -16	x_{13} 19	x_{14} 25	x_{15} 83	$= 1$
2	x_{21} 99	x_{22} 98	x_{23} 87	x_{24} 16	x_{25} 92	$= 1$
3	x_{31} 86	x_{32} 19	x_{33} 39	x_{34} 88	x_{35} 17	$= 1$
4	x_{41} -20	x_{42} 99	x_{43} 88	x_{44} 79	x_{45} 65	$= 1$
5	x_{51} 67	x_{52} 98	x_{53} 90	x_{54} 48	x_{55} 60	$= 1$
	$= 1$	$= 1$	$= 1$	$= 1$	$= 1$	

$x_{ij} \geqq 0$ for all i, j. Maximize objective.

Thus x_{ij} is the fraction of their lifetime that man i and woman j spend together. The items in this model are the lifetimes of the various members of the club. Halmos and Vaughan made the "monogamous assumption", i.e., that at any instant of time a man can be with only one woman and vice versa. Under this assumption, man 1's lifetime leads to the constraint that the sum of the fractions of his lifetime that he spends with each woman should be equal to 1, i.e., $x_{11} + x_{12} + x_{13} + x_{14} + x_{15} = 1$. Similar constraints result from other members of the club.

Under the linearity assumptions, the club's happiness is $(\sum c_{ij}x_{ij} : \text{over } i, j = 1$ to 5). These things lead to the conclusion that the marriage problem in this example is the transportation problem given above. It is a special transportation problem in which the number of sources is equal to the number of demand centers, all the availabilities and requirements are 1, and the constraints are equality constraints. Since all variables are $\geqq 0$, and the sum of all the variables in each row and column of the array is required to be 1, all variables in the problem have to lie between 0 and 1. By Theorem 2.1 this problem has an optimum solution in which all x_{ij} take integer values, and in such a solution all x_{ij} should be 0 or 1. One such solution, for example, is given in (2.4).

$$x = (x_{ij}) = \begin{pmatrix} 0 & 0 & 0 & 0 & 1 \\ 1 & 0 & 0 & 0 & 0 \\ 0 & 0 & 0 & 1 & 0 \\ 0 & 1 & 0 & 0 & 0 \\ 0 & 0 & 1 & 0 & 0 \end{pmatrix} \qquad (2.4)$$

In this solution man 1 (corresponding to row 1) spends all his lifetime with woman 5 (corresponding to column 5). So in this solution we can think of man 1 **being assigned** to woman 5, etc. Hence an integer solution to this problem is known as an **assignment**, and the problem of finding an optimum solution to this problem that is integral is called the **assignment problem**. In the optimum assignment, each man lives ever after with the woman he is assigned to and vice versa, and there is never any divorce!

An assignment problem of order n is a transportation problem in which there are exactly n sources with one unit of material available at each, n sinks with one unit of material required at each, and an optimum solution that is integral (i.e., a 0-1 solution) is required for the problem. The problem of allocating candidates to zones as marketing directors discussed in Chapter 1 is an assignment problem of order 6. In this problem and in the other applications of the assignment problem of which there are many, a solution in which variables have fractional values does not make any practical sense, hence the integer requirement on the variables is a critical one. Theorem 2.1 guarantees that this problem always has an integer optimum solution; that's what is needed. We discuss an efficient method for solving the assignment problem in Chapter 5.

For the marriage problem the conclusion that there exists an optimum marriage policy that maximizes the overall club's happiness without any divorce is extremely interesting. Extending this logic to the whole society itself, one can argue that there exists a solution pairing each man with a woman in society that maximizes the society's overall happiness without any divorce. Natural systems have a tendency to move towards an optimum solution, and if such a divorceless optimum solution exists, one would expect it to manifest itself in nature. Why, then, is there so much divorce going on, and why is the frequency of divorce increasing rather than declining? This seems to imply that the conclusion obtained from the model - that there exists an optimum marriage policy that maximizes society's happiness without any divorce - is false. If it is false, some of the assumptions on which the model is based must be invalid. The major assumptions made in constructing the model are the linearity assumptions needed to express the club's overall happiness as the linear function $\sum(c_{ij}x_{ij} : \text{over } i, j = 1 \text{ to } n)$. Let us examine the proportionality and additivity assumptions that lead to the choice of this objective function carefully.

The proportionality assumption states that the happiness acquired by a couple is proportional to the time they spend together, i.e., it behaves as the function represented by the thin line in Figure 2.2. In practice though, a couple may begin their life together in utter bliss, but develop a mutual dislike for each other as they get to know each other over time. After all, the proverb says: "Familiarity breeds contempt". The actual happiness acquired by the couple as a function of the time spent together often behaves as the nonlinear curve in Figure 2.2. Thus the proportionality assumption is not reasonable for the marriage problem.

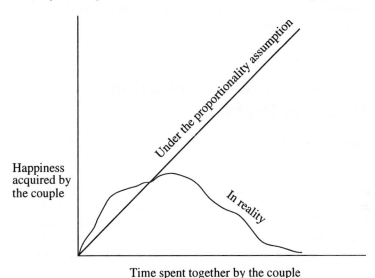

Figure 2.2 The inappropriateness of the proportionality assumption in the marriage problem.

The additivity assumption states that the society's happiness is the sum of the happiness acquired by the various members in it. In particular, this states that a person's unhappiness cancels with another person's happiness. In reality these things are quite invalid. History has many instances of major social upheavals just because there was one single unhappy person. The additivity assumption is quite inappropriate for determining the society's happiness as a function of the happiness of its members.

Finally the choice of the objective of maximizing society's happiness is itself quite inappropriate. In determining their marriage partners, most people are guided by the happiness they expect to acquire, and do not care what impact it will have on society. It is extremely hard to force people to do something just because it is good for the society as a whole. Political ideologies based on societal objective functions that ignore individual objectives have turned out to be miserable failures. This is clear from the recent collapse of communist governments in many countries even after intense efforts to make them successful over a period of 75 years.

In summary, for studying the marriage problem, the linearity assumptions and the choice of the objective of maximizing the club's happiness seem very inappropriate, and hence mathematical models for this problem based on these assumptions can lead to very unnatural or even wrong conclusions. We discussed this problem here mainly to provide an example where the linearity assumptions are totally inappropriate.

This points out the importance of checking the validity of the mathematical model very carefully after it is constructed. Things to review are: Is the objective function appropriate? Are all the constraints relevant or essential, or can some of them be eliminated? Are any decision variables missing? Is the data fairly reliable? Etc.

2.6 A Multi-Period Production Planning Problem

So far we have discussed a variety of static one period problems. Now we will discuss a **multi-period problem**. As an example we will consider the problem of planning the production, storage, and marketing of a product whose demand and selling price vary seasonally. An important feature in this situation is the profit that can be realized by manufacturing the product in seasons during which the production costs are low, storing it, and putting it in the market when the selling price is high. Many products exhibit such seasonal behavior, and companies and businesses take advantage of this feature to augment their profits. A linear programming formulation of this problem has the aim of finding the best production-storage-marketing plan over the planning horizon, to maximize the overall profit. For constructing a model for this problem we need reasonably good estimates of the demand and the expected selling price of the product in each period of the planning horizon. We also need data on the availability and cost of raw materials, labor, machine times etc. necessary to manufacture the product in each period; and the availability and cost of storage space.

Period	Total Pro-duction cost ($/ton)	Production capacity (tons)	Demand (tons)	Selling price ($/ton)
1	20	1500	1100	180
2	25	2000	1500	180
3	30	2200	1800	250
4	40	3000	1600	270
5	50	2700	2300	300
6	60	2500	2500	320

As an example, we consider the simple problem of a company making a product subject to such seasonal behavior. The company needs to make a production plan for the coming year divided into 6 periods of 2 months each, to maximize net profit (= sales revenue − production and storage costs). Relevant data is in the table given above. The production cost there includes the cost of raw material, labor, machine time etc., all of which fluctuate from period to period. And the production capacity arises due to limits on the availability of raw material and hourly labor.

Product manufactured during a period can be sold in the same period, or stored and sold later on. Storage costs are \$2/ton of product from one period to the next. Operations begin in period 1 with an initial stock of 500 tons of the product in storage, and the company would like to end up with the same amount of the product in storage at the end of period 6.

The decision variables in this period are, for period $j = 1$ to 6

$$
\begin{aligned}
x_j &= \text{product made (tons) during period } j \\
y_j &= \text{product left in storage (tons) at the end of period } j \\
z_j &= \text{product sold (tons) during period } j
\end{aligned}
$$

In modeling this problem the important thing to remember is that inventory equations (or material balance equations) must hold for the product for each period. For period j this equation expresses the following fact.

Amount of product in storage at the beginning of period j + the amount manufactured during period j $\Big\} = \Big\{$ Amount of product sold during period j + the amount left in storage at the end of period j

The LP model for this problem is given below.

Maximize $\qquad 180(z_1 + z_2) + 250z_3 + 270z_4 + 300z_5 + 320z_6$
$\qquad\qquad -20x_1 - 25x_2 - 30x_3 - 40x_4 - 50x_5 - 60x_6$
$\qquad\qquad -2(y_1 + y_2 + y_3 + y_4 + y_5 + y_6)$

Subject to $\qquad x_j, y_j, z_j \geqq 0$ for all $j = 1$ to 6
$\qquad\qquad x_1 \leqq 1500, x_2 \leqq 2000, x_3 \leqq 2200, x_4 \leqq 3000, x_5 \leqq 2700, x_6 \leqq 2500$
$\qquad\qquad z_1 \leqq 1100, z_2 \leqq 1500, z_3 \leqq 1800, z_4 \leqq 1600, z_5 \leqq 2300, z_6 \leqq 2500$

$$
\begin{aligned}
500 + x_1 - (y_1 + z_1) &= 0 \\
y_1 + x_2 - (y_2 + z_2) &= 0 \\
y_2 + x_3 - (y_3 + z_3) &= 0 \\
y_3 + x_4 - (y_4 + z_4) &= 0 \\
y_4 + x_5 - (y_5 + z_5) &= 0 \\
y_5 + x_6 - (y_6 + z_6) &= 0 \\
y_6 &= 500
\end{aligned}
$$

2.7   Minimizing the Sum of Absolute Deviations

In some applications, we are given several linear objective functions of the decision variables, $z_t(x) = c_1^t x_1 + \ldots + c_n^t x_n$, $t = 1$ to k say; and desired values d_t, $t = 1$ to k for them. In addition the decision variables may be required to satisfy a specified system of linear constraints, which we denote by (P). The problem is then to find a solution x satisfying the constraints in (P), and

$$z_t(x) = c_1^t x_1 + \ldots + c_n^t x_n = d_t, \quad t = 1 \text{ to } k \tag{2.5}$$

If a solution satisfying (P) and (2.5) exists, it is the best solution for this problem. However, it is possible that (P) and (2.5) together have no feasible solution, in this case it is impossible to find a feasible solution of the constraints (P) that exactly attains the desired value for each of the k objective functions. In this case, a feasible solution of (P) which minimizes

$$a(x) = \sum_{t=1}^{k} |c_1^t x_1 + \ldots + c_n^t x_n - d_t|$$

known as a **minimum absolute deviation solution**, could be considered as the best solution for this problem.

For $t = 1$ to k, $u_t = c_1^t x_1 + \ldots + c_n^t x_n - d_t$ is the **deviation** in the objective function $z_t(x)$ from its desired value of d_t at the point x, and $a(x) = \sum_{t=1}^{k} |u_t|$ is the sum of absolute deviations. The problem of finding the minimum absolute deviation solution can be posed as an LP by the following technique. The deviation u_t may be positive, zero, or negative, and it can always be expressed as the difference of two nonnegative variables which we denote u_t^+, u_t^-. So, we have for $t = 1$ to k

$$u_t = c_1^t x_1 + \ldots + c_n^t x_n - d_t \;\; = \;\; u_t^+ - u_t^-$$

$$\tag{2.6}$$

$$u_t^+, u_t^- \;\; \geqq \;\; 0$$

By defining

$$u_t^+ \;\; = \;\; \begin{cases} 0 & \text{if } u_t \leqq 0 \\ u_t & \text{if } u_t > 0 \end{cases}$$

$$\tag{2.7}$$

$$u_t^- \;\; = \;\; \begin{cases} 0 & \text{if } u_t \geqq 0 \\ -u_t & \text{if } u_t < 0 \end{cases}$$

we can guarantee that

$$u_t^+ u_t^- = 0 \tag{2.8}$$

When defined as in (2.7) to satisfy (2.8), u_t^+ is known as the **positive part** of the deviation u_t, and u_t^- is known as the **negative part** of u_t. As an example

if $u_t = 4.7$, its positive part $u_t^+ = 4.7$, negative part $u_t^- = 0$
if $u_t = -5.8$, its positive part $u_t^+ = 0$, negative part $u_t^- = 5.8$
if $u_t = 0$, both its positive and negative parts $u_t^+ = u_t^- = 0$

Clearly, $u_t = u_t^+ - u_t^-$, and when u_t^+, u_t^- also satisfy (2.8) we have $|u_t| = u_t^+ + u_t^-$. Consider the following LP.

$$\text{Minimize} \quad \sum_{t=1}^{k} (u_t^+ + u_t^-)$$

$$\text{subject to} \quad c_1^t x_1 + \ldots + c_n^t x_n - u_t^+ + u_t^- = d_t, \quad t = 1 \text{ to } k \tag{2.9}$$

$$u_t^+, u_t^- \geqq 0 \quad \text{for all } t$$

and the constraints in (P)

It can be proved (see [K. G. Murty, 1983]) that if $(\hat{x}, (\hat{u}_t^+), (\hat{u}_t^-))$ is an optimum solution of the LP (2.9), then $(\hat{u}_t^+), (\hat{u}_t^-)$ satisfy (2.8), and $\hat{x}$ is a minimum absolute deviation solution to the problem, i.e., it minimizes the sum of absolute deviations $a(x)$.

In constructing the cost function minimized in (2.9), we assumed that all the k objective functions, $z_1(x), \ldots, z_k(x)$ are equally important. In some applications, some of these objectives may have higher priority than others. To reflect these priorities, suppose we are given positive weights w_t corresponding to $z_t(x)$ for $t = 1$ to k. The higher the weight of an objective function, the higher its priority, i.e., the more important it is that its value be closer to its desired value. In this case, the best solution to the problem is the one that minimizes the weighted cost function $\sum_{t=1}^{k} w_t(u_t^+ + u_t^-)$ subject to the constraints in (2.9).

As an example, consider the product mix problem of the fertilizer manufacturer discussed in Section 2.1. In Section 2.1, we modeled the problem of finding the product mix that maximizes the daily profit. Instead, suppose the fertilizer manufacturer would like to find a feasible product mix that achieves the following goals as closely as possible.

- The daily profit $15x_1 + 10x_2$ should be \$13,000.

- The market share of the company, measured by the total daily sales revenue, $300x_1 + 175x_2$ should be \$220,000.

- The hi-tech market share of the company, measured by the daily sales revenue from Hi-ph fertilizer sales, $300x_1$, should be \$80,000.

Suppose the weights reflecting the priorities for the daily profit, market share, hi-tech market share, are given to be 10, 6, and 8, respectively. Then the best product mix that achieves these goals as closely as possible, is an optimum solution of the following LP.

$$\text{Minimize} \quad 10(u_1^+ + u_1^-) + 6(u_2^+ + u_2^-) + 8(u_3^+ + u_3^-)$$

$$
\begin{array}{rlrlrcl}
\text{subject to} \quad 15x_1 & +10x_2 & -u_1^+ & +u_1^- & = & 13{,}000 \\
300x_1 & +175x_2 & -u_2^+ & +u_2^- & = & 220{,}000 \\
300x_1 & & -u_3^+ & +u_3^- & = & 80{,}000 \\
2x_1 & +x_2 & & & \leq & 1500 \\
x_1 & +x_2 & & & \leq & 1200 \\
x_1 & & & & \leq & 500 \\
x_1, & x_2 & & & \geq & 0 \\
& & u_t^+, & u_t^- & \geq & 0,\ t=1,2,3
\end{array}
$$

As another example, suppose we are trying to solve a system of linear equations

$$
\begin{array}{rrrcr}
2x_1 & & +x_3 & = & 11 \\
& 2x_2 & -x_3 & = & 19 \\
-x_1 & & +2x_3 & = & 13 \\
x_1 & -x_2 & +x_3 & = & -7 \\
-x_1 & +2x_2 & -3x_3 & = & -9
\end{array}
$$

(which may contain more equations than variables) that has no solution. Then, a minimum absolute deviation solution for it minimizes the sum of absolute deviations between the left hand side and the RHS constant in all the equations, and it can be taken as a reasonable approximation for this problem. For a numerical example, consider the system of 5 equations in 3 variables given above. The minimum absolute deviation solution for this system is the solution of the LP given at the top of the next page. An optimum solution of this LP can be taken as the best approximate solution for the above system of linear equations.

The transformation into an LP of the problem of finding a minimum absolute deviation solution, is the basis for an approach called **goal programming approach** for handling multiobjective linear programs. This approach is discussed in full detail in Chapter 8. It is one of the most popular approaches for multiobjective linear programming.

$$\text{Minimize} \qquad \sum_{t=1}^{5}(u_t^+ + u_t^-)$$

$$
\begin{array}{rrrrrrr}
\text{subject to} \quad 2x_1 & & +x_3 & -u_1^+ & +u_1^- & = & 11 \\
& 2x_2 & -x_3 & -u_2^+ & +u_2^- & = & 19 \\
-x_1 & & +2x_3 & -u_3^+ & +u_3^- & = & 13 \\
x_1 & -x_2 & +x_3 & -u_4^+ & +u_4^- & = & -7 \\
-x_1 & +2x_2 & -3x_3 & -u_5^+ & +u_5^- & = & -9 \\
\end{array}
$$

$$u_t^+, u_t^- \geqq 0 \text{ for all } t$$

2.8   Solving LPs in Two Variables

LPs involving only two variables can be solved geometrically by drawing a diagram of the feasible region (i.e., the set of feasible solutions) in $\mathbb{R}^2$ = the two dimensional Cartesian plane. The optimum solution is identified by tracing the line corresponding to the set of feasible solutions that give a specific value to the objective function and then moving this line parallel to itself in the optimal direction as far as possible.

In $\mathbb{R}^2$ a linear equation in the variables represents a straight line, hence the set of all points giving a specific value to the objective function is a straight line. Each straight line divides $\mathbb{R}^2$ into two half-spaces, and every linear inequality represents a half-space.

As an example, consider the fertilizer product mix problem (2.1). The constraint $2x_1 + x_2 \leqq 1500$ requires that any feasible solution (x_1, x_2) to the problem should be on one side of the line represented by $2x_1 + x_2 = 1500$, the side that contains the origin (because the origin makes $2x_1 + x_2 = 0 < 1500$). This side is indicated by an arrow on the line in Figure 2.3. Likewise, all the constraints can be represented by the corresponding half-spaces in the diagram. The feasible region is the set of points in the plane satisfying all the constraints; i.e., the shaded region in Figure 2.3.

Let $z(x)$ be the linear objective function that we are trying to optimize. Select any point, $\bar{x} = (\bar{x}_1, \bar{x}_2)$ say, in the feasible region, and compute the objective value at it, $z(\bar{x}) = \bar{z}$, and draw the straight line represented by $z(x) = \bar{z}$. This straight line has a nonempty intersection with the feasible region since the feasible point $\bar{x}$ is contained on it. For any value $z_0 \neq \bar{z}$, $z(x) = z_0$ represents a straight line which is parallel to the line represented by $z(x) = \bar{z}$.

If $z(x)$ is to be maximized, move the line $z(x) = z_0$ in a parallel fashion by increasing the value of z_0 beginning with $\bar{z}$, as far as possible while still maintaining a nonempty intersection with the feasible region. If $\hat{z}$ is the maximum value of z_0 obtained in this process, it is the maximum value of $z(x)$ in the problem, and the set of optimum solutions is the set of feasible solutions lying on the line $z(x) = \hat{z}$.

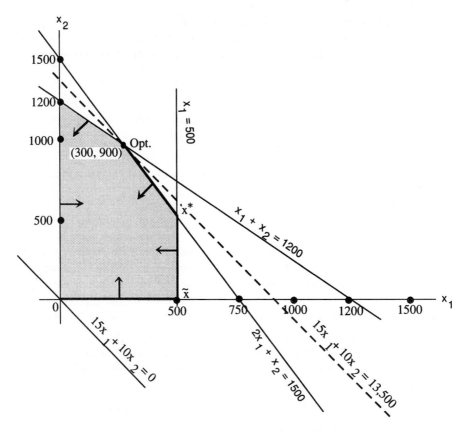

Figure 2.3 Fertilizer product mix problem.

On the other hand, if the line $z(x) = z_0$ has a nonempty intersection with the feasible region for every $z_0 \geq \bar{z}$, then $z(x)$ is **unbounded above** on this set. In this case $z(x)$ can be made to diverge to $+\infty$ on the feasible region, and the problem has no finite optimum solution.

If the aim is to minimize $z(x)$, then decrease the value of z_0 beginning with $\bar{z}$ and apply the same kind of arguments.

In the fertilizer product mix problem we start with the feasible point $\bar{x} = (0, 0)$ with an objective value of 0. As z_0 is increased from 0, the line $15x_1 + 10x_2 = z_0$ moves up keeping a nonempty intersection with the feasible region, until the line coincides with the dashed line $15x_1 + 10x_2 = 13,500$ in Figure 2.3 passing through the point of intersection of the two lines

$$
\begin{aligned}
2x_1 + x_2 &= 1500 \\
x_1 + x_2 &= 1200
\end{aligned}
$$

which is $\hat{x} = (300, 900)$. For any value of $z_0 > 13,500$ the line $15x_1 + 10x_2 = z_0$ does not intersect with the feasible region. Hence, the optimum objective value in this problem is \$13,500, and the optimum solution of the problem is $\hat{x} = (300, 900)$. Hence the fertilizer maker achieves his maximum daily profit of \$13,500 by manufacturing 300 tons of Hi-ph, and 900 tons of Lo-ph daily.

The feasible region of LPs involving n variables is a subset of $\mathbb{R}^n$. So, if $n \geqq 3$, it is hard to visualize geometrically. Hence this simple geometric method cannot be used to solve LPs involving 3 or more variables. Fortunately there are now efficient computational algorithms to solve LPs involving any number of variables. We discuss these in later chapters, and LPs in higher dimensional spaces can be solved efficiently using them.

When the objective function is $z(x)$, the main idea in the geometric method described above is to identify the straight line $z(x) = z_0$ for some z_0, and to move this line parallel to itself in the desired direction, keeping a nonempty intersection with the feasible region. In LPs with $n \geqq 3$ variables, $z(x) = z_0$ defines a hyperplane in $\mathbb{R}^n$ and not a straight line (a hyperplane in $\mathbb{R}^2$ is a straight line). An approach for solving LPs in these higher dimensional spaces, based on the above idea of parallel sliding of the objective plane, would be very efficient. However, checking whether the hyperplane still intersects the feasible region after a small parallel slide requires full-dimensional visual information which is not available currently for $n \geqq 3$. So it has not been possible to adopt this parallel sliding of the objective hyperplane to solve LPs in spaces of dimension $\geqq 3$. The simplex algorithm for solving LPs discussed in the sequel uses an entirely different approach. It takes a path along line segments called **edges** on the boundary of the feasible region, moving from one corner point to an adjacent one along an edge in each move, using local one-dimensional information collected in each step. As an example, to solve the fertilizer product mix problem starting with the feasible point 0, the simplex algorithm takes the edge path marked by thick lines (0 to $\tilde{x}$, then from $\tilde{x}$ to $x*$, and finally from $x*$ to the optimum solution) in Figure 2.3.

2.9   What Planning Information Can Be Derived from an LP Model?

Finding the Optimum Solutions

We can find an optimum solution for the problem, if one exists, by solving the model using the algorithms discussed later on. These algorithms can actually identify the set of all the optimum solutions if there are alternate optimum solutions. This may be helpful in selecting a suitable optimum solution to implement (one that satisfies some conditions that may not have been included in the model, but which may be important).

For the fertilizer product mix problem, we found out that the unique optimum solution is to manufacture 300 tons Hi-ph, and 900 tons Lo-ph, leading to a maximum daily profit of $13,500.

Infeasibility Analysis

We may discover that the model is **infeasible** (i.e., it has no feasible solution). If this happens, there must be a subset of constraints that are mutually contradictory in the model (maybe we promised to deliver goods without realizing that our resources are inadequate to manufacture them on time). In this case the algorithms will indicate how to modify the constraints in order to make the model feasible. After making the necessary modifications, the new model can be solved.

Values of Slack Variables at an Optimum Solution

The values of the slack variables at an optimum solution provide useful information on which supplies and resources will be left unused and in what quantities, if that solution is implemented.

For example, in the fertilizer product mix problem, the optimum solution is $\hat{x} = (300, 900)$. At this solution, RM 1 slack is $\hat{x}_3 = 1500 - 2\hat{x}_1 - \hat{x}_2 = 0$, RM 2 slack is $\hat{x}_4 = 1200 - \hat{x}_1 - \hat{x}_2 = 0$, and RM 3 slack is $\hat{x}_5 = 500 - \hat{x}_1 = 200$ tons. Thus, if this optimum solution is implemented, the daily supply of RM 1 and RM 2 will be completely used up, but 200 tons of RM 3 will be left unused. This shows that the supplies of RM 1, RM 2 are very critical to the company, and that there is currently an oversupply of 200 tons of RM 3 that cannot be used in the optimum operation of the Hi-ph and Lo-ph fertilizer processes.

Marginal Values and Their Uses

Each constraint in an LP model is the material balance constraint of some item, the RHS constant in that constraint being the availability or the requirement of that item. The **marginal value** of that item (also called the marginal value corresponding to that constraint) is defined to be the rate of change in the optimum objective value of the LP, per unit change in the RHS constant in the associated constraint.

For example, in the fertilizer product mix problem, the marginal value of RM 1 (and of the corresponding first constraint in (2.1)) is the rate of change in the maximum daily profit per unit change in the supply of RM 1 from its present value of 1500. These rates are also called **dual variables**, or the **shadow prices of the items**. These are the variables in another linear programming problem that is in **duality relationship** with the original problem. In this context the original problem is called the **primal problem**, and the other problem is called the **dual problem**. The derivation of the dual problem is discussed in Chapter 4.

If $b = (b_1, \ldots, b_m)^T$ is the vector of RHS constants in an LP model, and $f(b)$ denotes the optimum objective value in the LP as a function of the RHS constants vector, then the marginal value corresponding to constraint 1 is therefore the limit of $[f((b_1 + \epsilon, b_2, \ldots, b_m)^T) - f(b)]/\epsilon$ as $\epsilon \to 0$. So, one crude way of getting this marginal value is to select a small nonzero quantity ϵ, and then take $[f((b_1 + \epsilon, b_2, \ldots, b_m)^T) - f(b)]/\epsilon$ as an approximation to this marginal value.

As an example, let us consider the fertilizer product mix problem (2.1) again. The present RHS constants vector is $(1500, 1200, 500)^T$, and we computed the optimum objective value to be $f((1500, 1200, 500)^T) = \$13,500$. To get the marginal value of RM 1 (item corresponding to the first constraint) we can change the first RHS constant to 1501 and solve the new problem by the same geometric method discussed above. The optimum solution of the new problem is $x^1 = (301, 899)$, with an optimum objective value of $f((1501, 1200, 500^T) = \$13,505$. So, the crude approach discussed above suggests that the marginal value of RM 1 in this problem is $\pi_1 = f((1501, 1200, 500)^T) - f((1500, 1200, 500)^T) = \$13,505 - 13,500 = \$5$.

In the same way, if we change the RHS constants vector in the original (2.1) to $(1500, 1201, 500)^T$ and solve the new problem, we get the optimum solution $x^2 = (299, 902)$, with an optimum objective value of $f((1500, 1201, 500)^T) = \$13,505$. So, by the same argument, we conclude that the marginal value of RM 2 in this problem is $\pi_2 = (f((1500, 1201, 500)^T) - f((1500, 1200, 500)^T)) = \$13,505 - 13,500 = \$5$.

If we change the RHS constants vector in the original (2.1) to $(1500, 1200, 501)^T$ and solve the new problem, we find that there is no change in the optimum solution, $\hat{x} = (300, 900)$ continues to be the optimum with the optimum objective value of $f((1500, 1200, 501)^T) = \$13,500$. This is to be expected since in the original (2.1), 200 of the 500 tons of RM 3 available is left unused by the optimum solution, so increasing the supply of RM 3 alone would not help increase the daily profit. So, the marginal value of RM 3 is $\pi_3 = f((1500, 1200, 501)^T) - f((1500, 1200, 500)^T) = \$13,500 - 13,500 = 0$.

So, using the crude approach outlined above, we found that the marginal values of RM 1, RM 2, RM 3 in this problem are $\pi_1 = \$5, \pi_2 = \$5, \pi_3 = 0$ respectively.

This indicates that each additional ton of RM 1 supply is worth \$5 in additional profits to the company at this stage. So, if an outside supplier offers to sell RM 1 at a price of $\$(\alpha+$ the production cost per ton of RM 1 at company's quarry) per ton, then the company will benefit by buying RM 1 from that supplier as long as $\alpha < 5$. A similar statement holds for RM 2. RM 3 is already in excess supply, so it does not pay to get additional supplies of RM 3.

This type of analysis is called **marginal analysis**. It helps companies to determine what their most critical resources are, and how the requirements or resource availabilities can be modified to arrive at much better objective values than those possible under the existing requirements and resource availabilities.

Evaluating the Profitability of New Products

One major use of marginal values is in evaluating the profitability of new products. It helps to determine whether they are worth manufacturing, and if so at what level they should be priced so that they are profitable in comparison with existing product lines.

We will illustrate this again using the fertilizer product mix problem. Suppose the company's research chemist has come up with a new fertilizer that he calls *lushlawn*. Its manufacture requires as inputs 3 tons of RM 1, 2 tons of RM 2, and 2 tons of RM 3 per ton. How much profit (in $per ton made) should lushlawn fetch before the company considers manufacturing it? Using the marginal values, we find that a packet consisting of 3 tons RM 1, 2 tons RM 2, and 2 tons RM 3, is equivalent to $3\pi_1 + 2\pi_2 + 2\pi_3 = \$25$ in profits, this packet contains the inputs necessary to make one ton of lushlawn. So, the **breakeven profit** for lushlawn is $25/ton. That is, lushlawn is worth manufacturing if it can be sold at a price that leads to a profit $\geqq \$25$/ton made. The company can conduct a market survey and determine whether the market will accept lushlawn at a price $\geqq$ this breakeven level. Once this is known, the decision whether to produce lushlawn is obvious.

The crude method outlined above for computing the marginal values is not reliable in general. It is given here only to illustrate the application of marginal values in decision making. Later on we will see that if an optimum solution is obtained when an LP is solved by the simplex algorithm, the algorithm also gives all the marginal values in the form of the optimum dual solution automatically.

By providing this kind of valuable planning information, the linear programming model becomes a highly useful decision making tool.

2.10 Exercises

2.1 A nonferrous metals company makes four different alloys from two metals. The requirements are given below. Formulate the problem of finding the optimal product mix that maximizes gross revenue as an LP.

	proportion of metal in alloy				Availability
Metal	1	2	3	4	per day
1	0.5	0.6	0.3	0.1	25 tons
2	0.5	0.4	0.7	0.9	40 tons
Alloy price ($/ton)	750	650	1200	2200	

2.2 Oil Refinery Optimization A refinery has a distillation capacity of 100,000 barrels of crude/day in its fractionater. Here crude oil is basically heated and as the temperature increases different products called DN (distillation naptha), DHO (distillation heating oil), DGO (distillation gas oil), and P (pitch), are given off in vapor form, and are collected at various levels. The refinery gets crude oil from three different countries, these are called crudes 1, 2, 3. All the crudes and the various products are measured by volume in barrels. The output statistics from the distillation of each of the available crudes are tabulated below.

Distillation output	Yield (barrels/barrel) from distillation of		
	Crude 1	Crude 2	Crude 3
DN	0.19	0.16	0.02
DHO	0.27	0.32	0.24
DGO	0.38	0.27	0.26
P	0.05	0.13	0.39
Price ($/barrel)	23.25	22.00	20.50
Available (barrels/day)	60,000	90,000	80,000

It can be verified that the total volume of outputs from the distillation of one barrel of crude is < 1. The loss is due to evaporation and unusable heavy residuals.

DHO can be sold directly as heating oil. DGO can be sold directly as diesel fuel. Sale prices of these products are given below.

DHO and DGO can also be processed further in a catalytic cracker. The catalytic cracker can either process a maximum of 100,000 barrels/day of DHO, or a maximum of 50,000 barrels/day of DGO, or a combination of these in proportion of these levels adding up to 1. Also, when processing DHO the catalytic cracker can be run either at a normal level or at a high severity level. The high severity level helps to convert more of the DHO into naptha as seen from the table below. In processing DGO the catalytic cracker is run at normal level only and never on high severity level.

Catalytic cracker outputs	Output (barrels/barrel of feed)		
	DHO normal level	DHO high severity level	DGO normal level
Catalytic naptha (CN)	0.18	0.32	0.48
Catalytic heating oil (CHO)	0.80	0.69	0.52
Pitch (P)	0.11	0.10	0. 15

The cracking process converts the feed into products whose density is smaller than that of the feed, that's why the volume of outputs from this process is greater than the feed volume.

The pitch (from fractionater and catalytic cracker) can be combined with CHO (two parts of CHO to 17 parts of pitch) and sold as heavy fuel oil. DN and CN can be combined (20 parts of DN with 17 parts or greater of CN) and sold as gasoline. The quality of the gasoline improves with the proportion of CN in this blend. The following table gives the selling prices and demand for the various final products.

Final product	Selling price ($/barrel)	Daily demand
Gasoline	43.25	Up to 40,000 barrels
Heating oil	39.75	Up to 40,000 barrels
Diesel fuel	39.00	Any amount
Heavy fuel oil	30.00	Any amount

The processing cost on the fractionater is estimated to be $0.60/barrel of crude processed. On the catalytic cracker the processing costs are $1.50/barrel of fresh feed at the high severity level, and $0.95/barrel of fresh feed at the normal level. Formulate the problem of determining how much of each final product to produce daily in order to maximize daily net profit, as an LP.

2.3 The nutritionist at a food research lab is trying to develop a new type of multigrain flour. The grains that can be included have the following composition and price.

	% of Nutrient in Grain			
	1	2	3	4
Starch	30	20	40	25
Fiber	40	65	35	40
Protein	20	15	5	30
Gluten	10	0	20	5
Cost (cents/kg.)	70	40	60	80

Because of taste considerations, the percent of grain 2 in the mix cannot exceed 20, the percent of grain 3 in the mix has to be at least 30, and the percent of grain 1 in the mix has to be between 10 to 25.

The percent protein content in the flour must be at least 18, the percent gluten content has to be between 8 to 13, and the percent fiber content at most 50.

Formulate the problem of finding the least costly way of blending the grains to make the flour, subject to the constraints given, as an LP.

2.4 A company is trying to make up an optimum production - storage - distribution plan for products 1 and 2 for the winter season. Data is given in the table below.

Both products 1 and 2 manufactured in any month, can be used to meet the demand in that month itself, or stored and used to meet the demand in a later month. The demand in each month must be met exactly, and there should be no leftover product at the end of the season.

The manufacture in any month of both products put together cannot exceed 120 units. The storage cost for either product, from one month to the next is 0.5 (in 1000$/unit). There is storage space available to store at most 100 units of both products put together in each month.

It is required to determine how many units of each product to produce in each month, so as to minimize the total production + storage costs, for meeting the demand. Formulate as an LP.

	Demand in Units for Product in Month		Production Cost ($1000/unit) for Product if Produced in Month	
Month	Product 1	2	Product 1	2
Jan	12	55	4	7
Feb	35	25	5	8
Mar	55	65	5	8
Apr	90	60	4	7

2.5 An 8 year old child has 24 teeth usually. A dental researcher has estimated that on an average, 8 year old kids in a geographical region have 1.8 decayed teeth each. Give LP formulations for the problems of finding the minimum and maximum possible values for the fraction of 8 year old kids in that region who have 2 or more decayed teeth.

2.6 A village for retired persons has a population of more than 10,000 people. The residents suffer from four chronic ailments. The percentages having these ailments are 80%, 75%, 70%, 70% respectively, but there is no single person who has all four ailments. It is required to find lower and upper bounds for the percentages of this population who suffer from three and two ailments. Give LP formulations for the problems of finding these bounds.

2.7 A company makes 3 juice mixes called tropical mix (TM), pacific delight (Pa.D), and hi-nutrition (HN), using high-sugar pineapple juice (HSP), normal pineapple juice (NP), orange juice (OJ), tangerine juice (TJ), and white grape juice (GJ). Data on these mixes is given below.

Juice	Specs. on % of juice in mix			Price/unit	Available (units)
	TM	Pa.D	HN		
HSP	≥ 5	≥ 10		65	1000
NP	$10-33$	≤ 20	$30-50$	30	10,000
OJ	$50-80$	≤ 70	≥ 50	20	60,000
TJ		≥ 15		75	2000
GJ	$10-20$	$10-20$		25	40,000
Selling price/unit	100	150	80		

It is required to determine the composition of the three mixes, so as to maximize the net profit for the company while meeting the specs. Formulate this as an LP.

2.8 A company which makes two grades of industrial liquid cleaners labeled as L and H, is trying to revise their formulations. They are made by mixing 4 different chemicals C_1 to C_4. The products have to meet specifications on two different properties P_1, P_2 each measured in its own units. Assume that the linearity assumptions hold for determining the value of any of these properties for a mixture in terms of its values in the components of the mixture. Relevant data is given below. Formulate the problem of finding the best blend for L and H, to maximize net daily profit, as an LP.

Chemical	Value of P_1	Value of P_2	Daily availability	Cost
C_1	12	150	1000 brls.	55 $/brl.
C_2	15	140	900	50
C_3	9	120	1500	45
C_4	20	155	1200	35

Product	Max. for P_1	Min. for P_2	Exact demand/day	Price
L	15	142	2000 brls.	57 $/brl.
H	15	145	1300	62

2.9 A company makes two types of discrete parts called A and B. Each part has to be cast in the casting shop, machined, and then finished. Let shops 1, 2, and 3 refer to casting, machining, and finishing shops respectively. For $i = 1, 2, 3$, shop i has enough capacity to process either a_i units of A, or b_i units of B daily, or any combination of these two activities in proportions of these levels summing up to 1 where a_i, b_i are given below.

Shop no. i	Shop	Capacity for A a_i	Capacity for B b_i
1	Casting	100	70
2	Machining	80	90
3	Finishing	60	110

Up to 55 units of A, and 60 units of B can be sold daily. The net profit from unit of A and B sold is $800, 900 respectively. Ignoring the integer requirements on the units of A and B produced daily, formulate the problem of determining how many units of A and B to produce daily, to maximize total net profit.

Plot the set of feasible solutions for the problem, and determine the optimum solution geometrically.

Determine the marginal values of the RHS constants corresponding to the casting, machining, finishing shop capacity constraints.

If the casting shop capacity is fully used up in the optimum solution, determine how much the company could pay an outside supplier for a casting of A and B respectively over its production cost, and still break even.

2.10 An American textiles marketing firm buys shirts from manufacturers and sells them to clothing retailers. For the next season they are considering four styles with the total orders to be filled as given in the following table (the unit K = kilo, i.e., one thousand shirts). They are considering 3 manufacturers, M_1, M_2, M_3 who can make these styles in quantities and prices ($/shirt) as given below.

Style	Orders	M_1		M_2		M_3	
		Capacity	Price	Capacity	Price	Capacity	Price
1	200K	100K	$8	80K	$6.75	120K	$9
2	150K	80K	10	60K	10.25	100K	10.50
3	90K	75K	11	50K	11	75K	10.75
4	70K	60K	13	40K	14	50K	12.75
Capacity for all styles together		275K		150K		220K	

The manufacturing facilities of M_3 are located within US territories, so there are no limits on how many shirts can be ordered from M_3. But manufacturers M_1 and M_2 are both located in the same foreign country, and there is a limit (quota) of 350K shirts that can be imported from both of them put together. It is required to find the cheapest way to meet the demand. Ignoring the integer requirements on the number of shirts of any style purchased from any manufacturer, formulate this problem as an LP.

2.11 A company manufactures two types of cake mixes A and B using two raw materials R_1 and R_2. The following table gives the necessary data.

Raw material	Units needed to make 1 unit of		Units available
	A	B	
R_1	1	2	6000
R_2	2	1	8000
Net profit per unit made	7	5	
Maximum demand	3500	2500	

Formulate the problem of determining how many units of A and B to make, as an LP.

Solve the problem geometrically. Determine the marginal values associated with all the RHS constants in the model. Interpret them.

At this stage, how much extra profit can the company make if the supply of R_1, R_2 is increased by one unit?

A new cake mix developed by the company's kitchen needs 2 units of R_1 and two units of R_2 as input per unit. What is the minimum net profit that a unit of this new cake mix should make, if it were to be competitive with A, B?

2.12 A manufacturer of edible oils has two plants P_1, P_2 at different locations. They buy raw oils from 4 different suppliers S_1 to S_4 and process it. The processing yields two grades of oils, grades 1 and 2 and some residue. There are companies which buy the residue for use in manufacturing soap etc. Here is the data. All prices, and costs are quoted in the same money units.

Supplier	Available quantity (units/month)	Price (mu/u)	Shipping cost (mu/u) to P_1	P_2
1	1300	35.5	2.0	5.5
2	2300	34.5	6.0	5.0
3	2550	33.7	3.2	4.3
4	1850	35.0	4.0	3.7

mu/u = money units/unit

Approximately 4% of the raw oil purchased is lost in transit from the suppliers to the plants. The two plants operate with different equipment and refining procedures, and hence the yields are different, and the outputs command different market prices.

	At P_1			At P_2		
	Grade 1	Grade 2	Residue	Grade 1	Grade 2	Residue
Recovery rates	60%	30%	10%	51%	29%	20%
Open market price	200	150	29	190	145	29

The production costs at plants P_1, P_2 are 25, 24 mu/u of raw oil processed respectively. The plants at P_1, P_2 have capacities of processing 7000, 5000 units of raw oil respectively per month. The operating range of each plant is 50 to 100% of its capacity. Formulate the problem of finding a production plan that maximizes the company's net profit as an LP.

2.13 A refinery makes three grades of gasoline by blending three different crude oil distillates A, B, and C. Following data is given.

Distillate	Oc.R	Availability (brls./day)	Cost ($/brl)
A	83	20,000	26
B	88	25,000	30
C	93	15,000	34

Gasoline	Min. Oc.R	Max. % A	Min. % C	Selling price ($/brl)
Regular	87		20	33
Mid-grade	89	15	30	41
Premium	90	60	40	48

Formulate an LP model to determine the blending plan that maximizes net profit.

2.14 Faced with the difficulty of getting jobs they like near where they live, a newly married couple started making high nutrition currant jelly (CJ) and orange marmalade (OM) in a small kitchen and selling them in jars through a corner store.

It costs $0.80 and $1.50 to produce a jar of OM and CJ, respectively. With no advertising other than a few locally displayed signs, they have been able to sell about 3500 jars of OM and 5500 jars of CJ at prices of $2.20 and $4 per jar respectively, per week. They found out that if they advertise in the local newspaper and radio station, the weekly sales of OM can be increased at the rate of 2 jars/$ spent until the total sales reach 4200 jars; and that of CJ can be increased at the rate of 1.2 jars/$ spent until the total sales reach 6000 jars. They would like to spend no more than $1000 on advertising per week. It is required to determine an optimum production and advertisement plan to maximize total net profit. Ignoring the integer requirements on jars produced and sold, formulate this problem as an LP.

2.15 A regional power system has excess generating capacity in districts A, B, and C; and has contractual obligations to supply power to districts X, Y, and Z, as indicated in the following tables.

District	Excess capacity (MW)	Cost ($/MW)
A	100	500
B	75	700
C	200	400

District	Power to supply
X	55
Y	50
Z	90

Transmission lines exist between districts as indicated below. When a transmission line exists between two points, power can be transmitted between them at no cost, but there is power loss during transmission, data on which is given below.

District	Lines connect to	Transmission loss % on line
A	B, C	5%
A	X, Z	10%
B	C, A	5%
B	Z, Y	10%
C	A, B	5%
C	Y, X	10%

Formulate the problem of finding an optimum generation, distribution plan to meet the contractual obligations at X, Y, Z, at minimum cost, as an LP.

2.16 An office furniture making company makes two types of desks. They have 150 man hours of welders time, and 280 man hours of assembler time available per day. Data on the man hours needed to make a unit of 5 desks is given below.

	Type 1	Type 2
Assembler man hours/unit	1.8	2.7
Welder man hours/unit	1.9	1.2

Each unit of type 1 and 2 desks produced result in a revenue of $575 and $450, respectively. Formulate the problem of making the best utilization of available resources.

2.17 A company makes 5 different products that each need processing on three types of machines. Labor cost is $6/hour on machine types 1, 2, and $4/hour on machine type 3. Other data is given below. Formulate the problem of maximizing the net profit of this company as an LP.

	Mts. mc. time/unit of product					Mc. time available/week
	1	2	3	4	5	
Mc. type 1	12	7	8	10	7	149 hrs.
2	9	7	5	0	14	129 hrs.
3	7	8	5	4	3	1118 hrs.
Price/unit	$17	15	16	13	14	
Raw material cost/unit	$3	2	0.9	1.2	1	

2.18 A farmer has 1000 acres of crop land and 4500 man hours of labor available in the next season. The farmer can grow either corn or soybeans. Relevant data on the two crops is given below. Formulate the problem of finding the best allocation of available acreage between the crops.

	Corn	Soybean
Man hrs. labor required/acre	6	4
Cost of seed, fertilizer, insecticide/acre	$85	$35
Yield Bushels/acre	120	35
Selling price/bushel	$3.15	$6.25

2.19 Solve each of the following LPs geometrically and identify all the possible optimum solutions for each problem.

$$\text{Minimize} \quad -10x_1 - 10x_2$$
$$\text{subject to} \quad x_1 + x_2 \leqq 4$$
$$x_1 + 2x_2 \geqq 2$$
$$2x_1 + x_2 \geqq 2$$
$$x_1, x_2 \geqq 0$$

$$\text{Maximize} \quad -x_1 + 2x_2$$
$$\text{subject to} \quad x_1 + 2x_2 \geqq 2$$
$$2x_1 + x_2 \geqq 2$$
$$-x_1 + x_2 \geqq 0$$
$$x_1, x_2 \geqq 0$$

2.20 A furniture manufacturer has 3600 bd.ft. of Walnut, 4300 bd.ft. of Maple, and 6550 bd.ft. of Oak in stock. He can produce three products using this lumber, with input requirements as given below.

	Wood needed (bd. ft./unit)			Revenue/unit
	Walnut	Maple	Oak	
Table	10	50	100	$100
Desk	10	30	40	$50
Chair	80	5	0	$10

A table is always sold in combination with 4 chairs; and a desk is always sold in combination with one chair. But a chair can be sold independently. Formulate the problem of finding his product mix to maximize revenue, ignoring the integer requirement on the variables.

2.21 A farmer with 350 acres of farm land and a budget of $8000 for wages of laborers for the next season is trying to determine what he should plant. He pays laborers $4/hour. Other data is given below.

Crop	Hrs. labor/acre	Yield bushels/acre	Net profit/bushel
Corn	2	80	$0.5
Oats	2	60	0.3
Flax	5	50	0.6
Wheat	7	60	0.7
Soybean	9	55	0.9

The farmer needs to plant 45 acres of corn and oats combined (in any combination) for cattle feed for his own livestock. Also, he already made commitments to sell a customer the produce of 20 acres devoted to flax and 30 acres devoted to wheat. And his farm equipment can handle at most 100 acres of soybean. Formulate the problem of deciding how many acres to devote to each crop subject to all these constraints, to maximize the net profit.

2.22 A paper company produces three grades of paper using four types of wood pulp. Data on the monthly availability of various types of pulp, and the production requirements of various grades of paper is given below.

Paper grade	Input of pulp type tons/ton of paper				Profit ($/ton)
	1	2	3	4	
1	0	0	1.3	2.5	210
2	1.9	0.5	2.1	0	400
3	2.4	0	0.8	0	600
Monthly availability (tons)	1500	800	500	2000	

Formulate the problem of determining how many tons of each grade of paper to produce to maximize the profit derived from the available pulp.

2.23 An oil company makes two grades of gasoline by blending three types of refined oils.

The specifications are: grades 1 and 2 gasolines are required to have vapor pressure (VP) $\leqq 22, 24$; and Oc.R $\geqq 87, 92$ respectively. The company would like to produce between 100,000 to 250,000 barrels of grade 1 gasoline; and between 50,000 to 125,000 barrels of grade 2 gasoline, respectively, every week. The characteristics of the three types of refined oils are summarized below.

Type	VP	Oc.R	Weekly availability	Cost/barrel
1	26	86	185,000 barrels	$25
2	20	90	108,000	29
3	15	95	98,000	32

The company sells gasolines of grades 1 and 2 to retailers at the rate of $33 and $35/barrel, respectively. It is required to find the optimal blending plan to maximize the net weekly profit. Formulate this as an LP.

2.24 A steel company makes two types of steel sheet, each in coils of 4 tons and about 400 feet long. Each ton of sheet type 1 (2) needs 1 hour (1/2 hour) of box annealing machine time, 4 minutes (3 minutes) of cold rolling machine time, and 7 minutes (3 minutes) of strand annealing machine time. The company has 44 mc. hrs. of box annealing, 150 mc. hrs. of cold rolling, and 300 mc. hrs. of strand annealing time available/month. The demand for sheet types 1 and 2 is estimated to be about 15,000 tons and 10,000 tons, respectively. Each ton of sheet type 1 and 2 sold yield a profit of $35 and $26, respectively. Formulate the problem of finding an optimum product mix as an LP.

2.25 A company gets three byproducts A, B, and C, in quantities of 15,000, 8000, and 10,000 tons, respectively/month from operations to manufacture their main products. They can make a fertilizer consisting of 40% of A, 30% of B, and 30% of C and sell it at a profit of $5/ton. They can also make a construction material consisting of 15% of A, 25% of B, and 60% of C, and sell it at a profit of $3/ton. Any leftover quantities of A, B, and C can be given away free to a company dealing with bulk materials. Formulate the problem of using the byproducts for maximum benefit as an LP.

2.26 An automobile company is planning its fall advertising campaign to unveil the new models for the coming year. The marketing department has assembled the following data.

Medium	Cost/spot	Viewers/spot in millions	
		All viewers	Youth
TV-Prime time	100,000	6	2.5
TV-Non-prime time	78,000	4	1.5
Radio	40,000	2.5	1
Newspapers and magazines	20,000	1	0.4

The company would like to limit their TV advertising expenses to $3 million and buy at least 5 prime time spots and at least four non-prime time spots. They would like to buy a minimum of 6 radio advertising units, and at least 9 advertising units in newspapers and magazines. They also want to make sure that their message reaches at least 30 million youth viewers. It is required to devise an advertising campaign costing no more than $12 million that reaches as many viewers as possible, subject to these constraints. Ignoring any integer requirements on the variables, formulate this as an LP.

2.27 A company makes three grades of lubricating oils L, M, and H; by blending four different ingredients A, B, C, and D. There are two important specifications for lubricating oils. One is the AR rating which should be no more than 11, 10, 9 for L, M, and H, respectively; and the other is the AO rating which should be at least 95, 100, 110 for L, M, and H, respectively. The demand for L is $\geqq$ 2000 barrels, for M is exactly 1500 barrels, and for H it is at most 1000 barrels daily. The selling price of L, M, and H is \$90, \$100, \$120/barrel respectively.

The AR rating (AO rating) of the ingredients A, B, C, D are 7, 9, 12, 20 (120, 90, 100, 95), respectively. The daily availability of A, B, C, D is barrels 2000, 1000, 1500, 1700 at prices of \$85, 81, 79, 75 per barrel respectively. Formulate the problem of determining an optimum blend and production plan for each lubricating oil to maximize net profit.

2.28 A company makes a product which is in the form of a powder, and retails it under four different categories classified by particle size. The manufactured product yields material of Categories 1, 2, 3, 4 in average percentages of 20, 30, 35, 15 respectively by weight. However, Category 1 product can be ground and reentered at Classifier 2 to yield material of Categories 2, 3, 4 in average percentages of 25, 50, 25 respectively by weight. Category 2 material can also be processed on the grinder and entered at Classifier 3 to yield material of Categories 3, 4 in average percentages of 40, 60 respectively.

The market demand for Categories 1, 2, 3, 4 is estimated to be 10, 24, 22, 18 tons/day at prices of \$7, 10, 16, 25 per ton respectively. The company makes 60 tons of the main product daily. The cost of putting 1 ton of Category 1, 2 material through the grinder is \$5, 10 respectively. Material of any category which is left over each day has to be put through a palletizing unit at a cost of \$5/ton. Formulate the problem of determining a production plan that maximizes the total net profit.

2.29 A company needs a dry product that comes in several grades. It is available in several different prepared blend lots, and also pure grade lots, each sold in bags of 1000 lbs. The cost of pure grades 1 to 10 is \$183, 228, 273, 318, 363, 408, 453, 498, 543, 583 per bag respectively. Data on prepared blend lots is given below.

Blend	\multicolumn{10}{c}{Lbs. of grade i per bag of this blend}	Cost (\$/bag)
	$i = 1$	2	3	4	5	6	7	8	9	10	
A	150	200	250	150	150	100					239
B			75	200	250	200	200	75			324
C					150	200	200	125	125	200	400
D							75	50	150	725	476
E	450	300	150	100							187
F			200	200	200	200	200				306
G								250	250	500	466

A customer needs a bulk quantity of this product consisting of grades 1 to 10 in amounts of at least 0, 0, 300, 3585, 5865, 12915, 12075, 12945, 11880, 8735 lbs.

respectively. It is required to determine how many bags of prepared blend lots A to G, and lots of pure grades 1 to 10 to buy and mix to get the bulk quantity desired by the customer at minimum cost. Formulate as an LP.

How does the formulation change if the customer's demand is to be met using only the blended lots available and none of the pure grades. (C. H. White)

2.30 A small distributor buys fuel from 3 suppliers according to the following agreements, blends it and sells it to homes and small businesses (H & SB) in a small town.

Supplier	S content (% by wt.)	Cost ($/gallon)	Commitment on purchases/month Min (gallons)	Max
A	0.7	1.30	3000	4500
B	0.2	1.36	3000	3700
C	0.15	1.40	3000	4500

(i)   Formulate distributor's problem to meet demand for 4800 gallons in a month from H & SB at least cost.

(ii)   What changes take place in the model in (i) if a new constraint that the S content in the fuel sold to H & SB cannot exceed 0.2%?

(iii) The distributor has found a new industrial customer who agrees to buy 6300 gallons of fuel with S content of at most 0.6%, per month. How does this modify the model in (ii)?

(iv)   The distributor has now acquired a storage tank with a capacity of 23,000 gallons. In the month of May the tank has 16,100 gallons of fuel with S content of 0.327% and the average laid down cost of $1.36/gallon. Statistical analysis has revealed that the average monthly demand follows the pattern given below.

Customer	May to Sept. (gallons/month)	Oct. to Apr.
H & SB	2100	9800
Industrial	6000	5700

Construct the model which formulates the distributor's problem of deriving an optimal annual plan from May to April. The amount of fuel in the storage tank can fluctuate from month to month (subject to its capacity of 23,000 gallons), but it is required to end up with a stock of 16,100 gallons of fuel at the end of the planned year. ([J. R. Aronofsky, J. M. Dutton, and M. T. Tayyabkhan, 1978])

2.31 This problem arose in managing coast guard operations in Canada. They have three ships in their Pacific fleet to supply 50 manned lighthouses and service navigational aids in regions 1 to 5 (each of these regions need 10 trips per year); and to service unmanned lighthouses and navigational aids in regions 6 to 11 (each of these regions need two trips per year). Following tables present data on the costs and time needed for these tasks per trip. Here c_{ij} = cost (in $), and a_{ij} = time needed (in ship hours) to complete the tasks in region j using ship i per trip. The ships are of different weights and horse power, and hence these quantities vary with i.

	c_{ij}, a_{ij} for				
	$j = 1$	2	3	4	5
Ship $i = 1$	$c_{ij} = 5503$	3099	3298	3777	3908
	$a_{ij} = 128$	72	77	88	91
Ship 2	4002	1434	1526	1800	2056
	232	80	84	100	114
Ship 3	6144	3460	3681	4209	4363
	128	72	77	88	91

	c_{ij}, a_{ij} for					
	$j = 6$	7	8	9	10	11
Ship $i = 1$	$c_{ij} = 12652$	3291	2582	12186	17544	13873
	$a_{ij} = 404$	77	60	283	408	323
Ship 2	7273	1451	1172	6734	9540	7380
	404	81	65	374	530	410
Ship 3	14124	3674	2882	13603	19584	15487
	294	77	60	283	408	223

The available hours for the ships are 6000, 3000, 1500 respectively, in that order, per year. Let x_{ij} = fraction of annual service at region j to be performed by ship i. Using these decision variables, formulate the problem of finding a minimum cost allocation of the yearly work among the ships as an LP.

Solve this LP model by an available software package and obtain an optimum solution $(\bar{x}_{ij})$.

Start with $(\bar{x}_{ij})$, round the variables in it into integer values appropriately, or use some heuristic search; and obtain an implementable solution in which each ship makes an integer number of trips, while carrying out each regions yearly requirements subject to the constraints on ship hours available that increases the total cost over that of $(\bar{x}_{ij})$ by as small an amount as possible. ([J. G. Debanne, and J. N. Lavier, 1979])

2.32 This problem deals with planning to meet the demand for freight cars (FCs) at a railroad company. There are three types of freight cars, and the planning horizon is a period of three months. At the beginning of the planning horizon some FCs are already available, some of these are in good condition, but others need either class I or II repairs before they can be used. Also, new FCs can either be built in shop or purchased.

The total number of FCs built per day in shop cannot exceed 32 due to space limitations. Available are up to 8000 man-hours of labor/day at a cost of $10/man-hour, up to 1000 wheels/day at a cost of $200/wheel, up to 200 spring assemblies/day at a cost of $750/spring assembly, and 175 frames/day at a cost of $1000/frame.

The expected number of new breakdowns needing either class I or II repair during the planning horizon is estimated and given below. Also given are data on demand, purchase price, and inputs needed to build or repair. FCs used to meet demand during a month can be assumed to return at the end of the month; and be available for assignment towards the demand in the next month if they are in good condition, or to be repaired and then assigned towards the demand if they need repairs.

Formulate the problem of meeting the demand at minimum cost as an LP ignoring integer restrictions on variables like the number of cars built or purchased, etc.

FC type	Demand in month			Available at beginning of horizon in condition			Price of new FC ($)	Expected no. breakdowns	
	1	2	3	Good	Cl. I	Cl. II		Cl. I	Cl. II
Box	2500	3000	2000	1500	300	250	25,000	700	300
Gondola	2000	2500	2250	1700	300	100	20,000	300	200
Flat	1500	2000	3000	1200	500	100	18,000	400	100

FC type	Inputs to build new FC				Inputs for class I repair/FC				Inputs for class II repair/FC			
	M	W	S	F	M	W	S	F	M	W	S	F
Box	300	20	4	5	50	3	1	1	75	5	2	3
Gondola	250	20	4	4	40	3	1	1	60	5	2	2
Flat	225	20	4	4	40	3	1	1	50	5	2	2
M = man-hours, W = wheels, S = spring assemblies, F = frames												

2.33 A company makes two products *A* and *B* using two resources, labor and raw material. The production requirements are: 0.16 manhours of labor and 2 kg of raw material to make 1 kg of *A*, and 0.20 manhours of labor and 1 kg of raw material to make 1 kg of *B*. In selling the products, the company discriminates on the basis of price, leading to a segmentation of its markets. Following table shows the anticipated capacities and prevailing prices for each market segment.

Market segment	Capacity (kg)		Price ($/kg)	
	A	B	A	B
1	500	700	10	7
2	1000	800	9	6.5
3	1000	900	8	6
4	-	5000	-	5.5

The company has 5000 kg of raw material available weekly, but more can be purchased at a price of $2/kg. It has 15 employees who work 40 hours each per week, but if necessary, work overtime at the penalty rates shown in the following table

Type	Hrs./week	Pay rate ($/hr.)
Regular	40	10
Overtime	10	15
Double-time	10	20

It is required to find a production and selling plan that maximizes the net profit. Formulate this problem as an LP. ([R. Snyder, January 1984])

2.34 The feed for dairy cattle at a farm consists of hay (H), sugar beet (SB), grass nuts (GN), kale (K), and sileage (S). For the condition of the cow and the high quality of her milk yield, there are constraints on the content of dry matter (DM), indigestible organic matter (IOM), digestible crude protein (DCP), metabolisable energy (ME), calcium (Ca), magnisium (Mg), and sodium (Na) in the cow's daily feed. The composition of the foods and the constraints on the nutrients are stated below.

Nutrient	Content of nutrient in food (g/kg)					Daily requirements
	H	SB	GN	K	S	
DM	850	900	900	150	250	$\leqq$ 18,000 g
IOM	306	117	198	26.25	82.5	$\leqq$ 3600 g
DCP	33.15	54.9	122.4	16.5	25.5	$\geqq$ 345 g
ME	7.14	11.4	9.72	1.66	2.2	$\geqq$ 63 g
Ca	3.23	5.67	0	2.64	1.625	$\geqq$ 21 g
Na	0.595	2.7	0	0.323	0.625	$\geqq$ 9 g
Mg	1.02	1.44	0	0.293	0.5	$\geqq$ 9 g
Cost(cents/kg)	7.5	23	23	6	5	

Formulate the problem of finding a minimum cost feed for the farmer's cows and obtain an optimum solution for it using one of the available software packages. ([P. Lazarus and D. Kirkman, January 1980])

2.35 A farmer can buy two fertilizer mixes A and B. Their composition and prices are given below.

Primary nutrient	A	B
N (kg of N/ton)	240	280
Ph (kg of P_2O_5/ton)	150	100
K (kg of K_2O/ton)	220	250
Cost ($/ton)	600	800

The fertilizer requirements per acre are 120 kg of N, 50 kg of P_2O_5, and 70 kg of K_2O. It is required to determine the least cost combination of mixes A, B that will meet the requirements. Formulate as an LP.

2.36 Optimum fertilizer mixes for sugar beet Yields from irrigated fields of sugar beet grown in the western part of Spain (sown in the fall and harvested in mid-summer) can easily reach 60 tons/ha. The following upper and lower fertilizer requirement limits are recommended for high yield.

N	160 to 180 kg/ha
P_2O_5	60 to 80 kg/ha
K_2O	80 to 100 kg/ha
B	1 to 1.3 kg/ha
Mn	0.5 to 1 kg/ha
Zn	0.15 to 0.25 kg/ha

It should be emphasized that the setting of upper limits is necessary as sugar beet is sensitive to surpluses of several nutrients. The costs and composition of the available fertilizer mixtures is shown in the following table.

Fertilizer mixture	Nutrient %						Cost (pesetas/kg)
	N	P_2O_5	K_2O	B	Mn	Zn	
Urea	46	0	0	0	0	0	38.6
Super Ph. 0-45-0	0	45	0	0	0	0	32.4
8-24-8	8	24	8	0	0	0	31.3
9-18-27B	9	18	27	0.3	0	0	34.9
Laifol 0-9-18	0	9	18	0	2	1	376
Fertiluq 0-35-35	0	35	35	2	5	5	412
Utefol ficoop 6-18-27	6	18	27	2	5	5	174
Quimifole (ERT) 8-8-8	8	8	8	2.5	5	10	275

Also, the bottom four fertilizer mixtures are foliar fertilizers applied during spring. For each of the primary nutrients N, K_2O_5, and K_2O; a maximum of 35% of its requirement can come through these four mixtures. It is required to find a least cost fertilizer mixture combination that meets all these requirements and constraints. Formulate this as an LP. ([M. I. Minguez, C. Romero, and J. Domingo, January 1988])

2.37 A company estimates that it will have a cash flow of (80, 70, 0, 30) in the next four years from existing operations (the money unit is $1000). In this planning horizon it has the opportunity to invest in 9 investment projects, with cash outflows as shown below. In this table, positive outlays are investments to be made. Negative outlays are returns. In each project, the participation can be at any level between 0 to 100%.

Year	Outlay in project no. in year at 100% level								
	1	2	3	4	5	6	7	8	9
1	100	140	20						
2	50			70					
3	−20		−25		100	100			
4	−70					−109	100	150	20
5	−80			−91					
6		−150			−80				−25
7		−150			−60		−100	−25	
8							−80	−200	

There is an opportunity to borrow money in year 3 for a four year period at 9% annual interest to be paid out each year. The principal amount of this debt has to be repaid in year 7 along with the interest due in that year.

In each year any spare money can be invested in short term deposits of one year, earning 6% interest.

In every year cash outflows are limited by cash inflows. Formulate the problem of maximizing the total return at the end of year 8 as an LP.

2.38 A person who starts out with $10,000, is investigating three investments, each with a one year term, with the following expected rate of return over the next 4 year period.

Investment	Expected fraction return in year				Yearly investment
	1	2	3	4	
1	0.05	0.06	0.07	0.07	2000 - 5000
2	0.07	0.07	0.08	0.08	4000 - 7000
3	0.10	0.08	0.06	0.06	1000 - 4000

In each investment, the principal is given back with the return at the end of each year. Since there are inherent uncertainties and risks in each investment, the person would like to limit the amount allotted to each investment within the bounds specified in the table. Formulate the problem of determining an investment plan over the four year period that maximizes the total expected yield by the end of the fourth year.

2.39 A company has has two currently operating plants, plants 1 and 2, for manufacturing a product which they sell through four distribution centers. Due to increasing demand for the product, the company has decided to set up a new plant that can produce 1100 units of finished product per week. Data on the unit transportation costs and expected weekly demand is given below.

Plant	Unit transp. cost to dist. center				Weekly prod.
	1	2	3	4	capacity
1	0.4	0.6	0.3	0.5	1500
2	0.9	1.0	0.3	0.8	1800
New, to be built					1100 (planned)
Expected weekly demand	1400	1100	900	1000	

It takes 0.54 units of labor, 0.10 units of raw material R_1, 0.04 units of raw material R_2, 0.20 units of coal, to manufacture one unit of product at the new plant.

Two sites, sites 1 and 2, are available to locate the new plant 3. The cost of shipping the product from a new plant at site 1 (2) to the four distribution centers is 0.3, 0.5, 0.4, 0.2 (0.4, 0.6, 0.1, 0.3) per unit respectively.

The cost of setting up a new plant at site 1 or 2 is the same.

The cost of the various inputs at the two sites are given below.

Site	Cost per unit of			
	Labor	R_1	R_2	Coal
1	4	0.05	0.03	0.04
2	4.5	0.04	0.03	0.03

The company's plants 1 and 2 are well established and have been running for some time. It wants to determine whether to set up the new plant at site 1 or 2, in order to minimize the production costs at this plant plus the weekly shipping cost. Formulate this problem.

2.40 Optimal new vehicle inventory An automobile dealership can maximize their gross profit by concentrating on the proper model mix in inventory. Every time a customer does not buy because the dealer did not have the "right" car in stock, there is an opportunity loss for the dealership (not only lost profit from the sale, but profit that would have come in future in service and other departments within the dealership). The ordering of new vehicles operates under certain constraints, both internal and external. Internal constraints arise from limited storage space and display area, and interest expenses. External constraints arise from those imposed by vehicle manufacturer and financial institutions.

We give below data from an Oldsmobile dealership selling 6 body types resulting in 19 different models. Some models generate more gross profit than others, but the manufacturer requires the dealer to represent the entire model line as part of a franchise agreement; i.e., the dealer must keep an inventory of a certain number of slow moving models in order to get the popular models. Some of the constraints indicated in the following table are also developed based on the management's estimates of relevant boundaries (e.g. max-min sales levels, space limitations, etc.).

Model	Body type	Mean gross profit/unit	Yearly sales bounds for type	model
Cutlass-Calais	Coupe	$675		Between
	Sedan	700		360 to 400
	International series	950	$\leq$ 60	
Cutlass-Cierra	Coupe	550		Between
	Sedan	700		560 to 600
	Int. series	900	$\leq$ 60	
	Stn. wagon	875	$\leq$ 40	
Cutlass-Supreme	Base coupe	650	$\geq$ 150	At least
	SL coupe	850		300
	Int. series	950	$\leq$10	
Delta-Royal	Coupe	700		Between
	Sedan	900		500 to 600
	Brougham sedan	1000		
	Stn. wagon	900	$\geq$ 36	
98 Regency	Base sedan	1175		At least
	Brougham sedan	1350		160
	Touring sedan	1400	$\leq$ 36	
Toronado	Toronado	600		At least
	Trofeo	1150		60

Ignoring the integer requirements on the decision variables, formulate the problem of determining how many cars of each model-type the dealership should order for next year in order to maximize the profit on car sales. ([M. J. Schniederjans, N. K. Kwak, and J. S. Frueh, 1991])

2.41 A farmer has 120 acres for growing wheat or corn. The yield for wheat is 55 bushels/acre/year, and that for corn is 95 bushels/acre/year. He can devote any fraction of the available 120 acres to wheat or corn. Cultivation needs 4 manhours of labor per acre per year, plus 0.15, 0.70 manhours per bushel of wheat, corn harvested, respectively. Unlimited quantities of wheat and corn can be sold at prices of \$1.75, \$0.95 per bushel respectively to wholesalers. The farmer can also buy wheat, corn at \$2,5, \$1.5 per bushel respectively in the retail market.

The farmer also has a 10,000 sq.ft. facility for raising pigs and/or chicken. Pigs and chicken are sold at one year of age, a one year old pig sells for \$200; while one year old chicken, measured in units called coop, sell for \$200/coop. The requirements for raising pigs, chicken are given below.

	Requirements for one pig or one coop chicken		
	Grain	Floor space	Labor
Pigs	25 W or 20 C	25 sq.ft.	25 manhours
Chicken	10 W or 25 C	15	40

W = wheat, C = corn

The farmer already has 2000 manhours of labor for raising pigs and/or chicken, and he himself has 2000 hours of time that he can devote to this. Additional labor for this activity can be hired at \$4/hour, but for each hour of hired labor, he has to devote 0.1 hours of his own time for supervision.

Formulate the problem of determining how much wheat, corn, pigs, chicken he should raise to maximize his net profit. ([L. Schrage, 1987], [F. H. Murphy, E. A. Stohr, and A. Asthana, July 1992])

2.42 Food processing plant optimization This problem deals with improving the efficiency of a food processing plant. The operation of the plant involves preparing *batches*, each batch being a mixture of powdered ingredients (flour, salt, etc.) according to a given recipe. There are 17 different ingredients in all.

The powdered ingredients are held in sizeable hoppers. Several (usually 12) such hoppers are attached to a weighing machine. Ingredients are emptied from one hopper at a time into the scales of the weighing machine. After the ingredients are weighed, the contents of the scales are conveyed to a bin, where they are blended into the batch.

There are three such weighing machines working simultaneously and independently in the plant, and the whole operation is controlled by a computerized system. As soon as a batch is completed, weighing stops, the scales are cleaned, and preparations take place for the next batch.

The total weight of a batch is always $\leqq$ 1000 kg. The capacity of the scales on each of the three weighing machines is 400 kg.

Some ingredients "flow" faster than others from the hoppers. Let g_i denote the time in some time units, for one kg. of ingredient i to flow from the hopper onto the scales of a weighing machine.

Consider the recipe that requires the following quantities of ingredients I_1 to I_6. We also provide the values of g_i for each of these ingredients.

Ingredient	I_1	I_2	I_3	I_4	I_5	I_6
Kg needed for recipe	130	146	101	290	29	21
g_i	1	1.2	1.3	1.1	0.8	0.9

These ingredients are contained in the hoppers of the three weighing machines, WM1, WM2, WM3, as indicated in the following table. The * mark indicates that at least one hopper of the corresponding weighing machine contains the respective ingredient (therefore a blank entry indicates that none of the hoppers on this weighing machine contain the ingredient). An ingredient can be weighed on a weighing machine only if one of its hoppers contains that ingredient. For example, on WM1 we can weigh ingredients I_1, I_3, I_4 in the recipe, but not I_2, I_5, I_6.

	I_1	I_2	I_3	I_4	I_5	I_6
WM1	*		*	*		
WM2		*		*		
WM3	*	*			*	*

The problem is to determine a weighing schedule which specifies the quantities of the various ingredients to be weighed by each of the weighing machines, which corresponds to the shortest weighing time for this recipe.

Consider the following weighing schedule for this recipe.

	I_1	I_2	I_3	I_4	I_5	I_6
WM1	130		101	29.5		
WM2				260.5		
WM3		146			29	21

This weighing schedule is feasible because it satisfies all the constraints. It takes $(130/1) + (101/1.3) + (29.5/1.1) = 234.5$ time units on WM1, $260.5/1.1 = 236.82$ time units on WM2, and $(146/1.2) + (29/0.8) + (21/0.9) = 181.25$ time units on WM3 to complete.

Formulate the problem of finding a weighing schedule that takes the shortest total time for this recipe. Since the work on each weighing machine proceeds independently and simultaneously, the objective function in this problem is not a linear function of the times taken by the three weighing machines, so watch out.

For an extension of this problem to finding the optimum assignment of the various ingredients to hoppers on the weighing machines, see Exercise 9.23 ([R. Benveniste, May 1986]).

2.43 This exercise deals with the allocation of crude oil production to oil fields in Saudi Arabia. Crude oil coming up through an oil well contains various dissolved gas components. The GOR (gas-oil ratio) measures the cubic feet (CF) of dissolved gases per barrel of gas-oil mixture coming up through the well, this differs from field to field. From the wells, the gas-oil mixture is sent by pipelines to GOSPS (gas oil separation plants) where the gases are liberated from the mixture. The SF (shrinkage factor) is the barrels of C (crude oil) left after one barrel of gas-oil mixture is processed for gas removal in GOSPS, this also differs from field to field. From the GOSPS, C is transported to various refineries in the country or to shipping terminals for export. The liberated gases are piped from the GOSPS to gas plants where S (sulphur) and M (methane, also called fuel gas) are separated from them, and the remaining material is liquified into LNG. The LNG is then sent by pipeline to fractionation plants where it is separated into its various gas components, E (ethane), Z (propane), B (butane), T (pentane), naptha, and some others. M, E are then piped to industrial complexes in the country to be used either as fuel (in electric power generation and desalination plants), or as feed stocks in petrochemical industry. The demand for M, E from these plants has to be met, their supply however is dependent on crude oil production, which in turn depends on international market pressures and the ceilings imposed by OPEC, of which Saudi Arabia is a member.

The original problem studied dealt with a system consisting of 16 oil fields producing four types of crudes, 56 GOSPS, 3 gas plants, and 2 fractionation plants. But to keep the model size reasonable, in this exercise we will confine our attention to 3 oil fields producing two types of crudes (AL = Arabian light, and AM = Arabian medium), 10 GOSPS, 3 gas plants, and 2 fractionation plants. In one of the oil fields there are two reservoirs or producing areas, the other two oil fields have one reservoir each. The whole system layout is shown in Figure 2.4.

The GOR in the gas-oil mixture coming up through a well remains constant as long as the production is above the reservoir bubble point, which is the case in all the oil fields under consideration.

Saudi Arabia faces many conflicting objectives in setting daily production levels from the oil fields. There is tremendous political pressure from powerful oil importing countries to keep the production level high in order to keep the price those countries have to pay for C low. OPEC imposes a ceiling on the amount of C each member country can produce, in order to keep the price from being too low. Again, they are forced to maintain a certain level of production in order to obtain sufficient quantities of M, E to meet the internal demand for these products. In this exercise we will address the conflicting objectives of adhering to OPEC ceilings on C production as far as possible, while keeping the production level sufficiently high to generate enough M, E, to meet the internal demand for these gases. This may force them to divert a higher proportion of their daily production to fields with higher GOR.

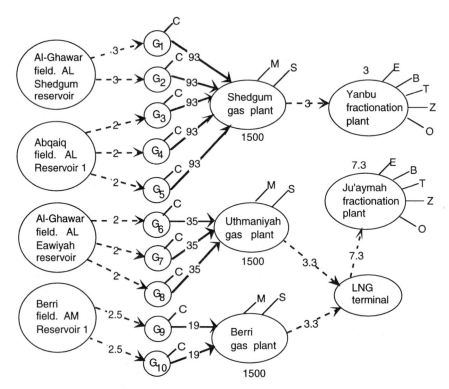

Figure 2.4 Dashed arcs are pipelines for transporting liquid material like crude or LNG, capacity on them is marked in units of 100,000 barrels/day. Solid lines represent pipelines for gases, on these, capacity is entered in units of 10^6 CF/day. Inside the node representing reservoirs, we enter the field, type of crude produced there (AL or AM), and the reservoir name or number in the field. G_1 to G_{10} are GOSPS, flows into them are of gas-oil mixture, and flows out of them are of liberated gases. Small lines with a letter at the tip represent output streams of product denoted by that letter. The capacities of gas plants measured in 10^6 CF/day, and of fractionation plants in units of 100,000 barrels of LNG/day, are entered by their side. Flows out of gas plants are of LNG.

Reservoir	Cost*	GOR	SF	Quantity/CF of gases		
				CF of M	grams of S	LNG (brl)
Shedgum	6.56	543	0.747	0.599	0.0167	0.0002
Ewiyah	6.56	400	0.794	0.606	0.0461	0.00015
Abqaiq 1	5.7	846	0.685	0.498	0.0254	0.0024
Berri 1	8.74	735	0.801	0.529	0.1105	0.0003

*Cost to bring up crude to well-head in SR (Riyals)/brl

Outputs of Components from LNG

Stream	Quantity per brl of LNG, E in CF; B, T, Z, in brl				
	E	B	T	Z	O = others
Shedgum-Yanbu	852.71	0.145	0.103	0.333	Rest
Uthmaniyah-Ju'aymah	820.47	0.153	0.102	0.341	Rest
Berri-Ju'aymah	919.78	0.148	0.089	0.310	Rest

Processing costs at gas plants and fractionation plants can be ignored. However transportation costs are to be taken into account in computing the total costs. It costs 0.0034, 0.0114, 0.0021, 0.0021, 0.0037 SR/CF to transport the liberated gases from GOSPS 1 to 5 respectively to Shedgum gas plant; 0.156, 0.0165, 0.0312 SR/CF to transport them from GOSPS 6, 7, 8 respectively to Uthmaniyah gas plant; and 0.00047 SR/CF to transport them from GOSPS 9, or 10 to the Berri gas plant. It costs 2.24 SR/brl to transport LNG from Shedgum gas plant to the Yanbu fractionation plant. It costs 0.31, 0.127 SR/brl to transport LNG from Uthmaniyah and Berri gas plants respectively to the LNG terminal. The cost of shipping LNG from the LNG terminal to the Ju'aymah fractionation plant is 1.1 SR/brl.

They would like to produce crudes AL, AM in proportion 60:40. For any deviation in the daily production of AL, AM from this proportion, a penalty of 0.0025 SR/brl has to be included in the overall cost.

The OPEC production ceilings would impose an upper bound of 4.5×10^6 per day on the daily production of crudes AL, AM put together from these three fields. They would like to adhere to these limits.

The internal demands for M, E are 1.621×10^8, 2.87×10^7 CF daily respectively, which they would like to meet as far as possible. There is a penalty for shortage of either of these products at 1000 SR/CF. Penalties are also to be assessed for overproduction (i.e., production over the daily demand for these products) at 0.0073 SR/CF of M, and 0.0293 SR/CF of E.

Formulate the problem of determining how much gas-oil mixture to produce from each of the 4 reservoirs under consideration, and the optimal routing of this production through the processes and plants described in Figure 2.4, so as to meet the demand for M, E at minimum total cost. ([S. O. Duffuaa, J. A. Al-Zayer, M. A. Al-Marhoun, and M. A. Al-Saleh, Nov. 1992]).

2.44 The health of the Egyptian economy depends critically on how well the government manages their water resources, i.e., the water releases from Aswan

high dam on the Nile river. The central issues in the operation of Aswan high dam
are the storage of river water in the reservoir, Lake Nasser, for future use; releases
of water from Lake Nasser for irrigation and hydroelectric power generation; and
preventing the occurrence of floods. The monthly inflows into Lake Nasser are
stochastic, however in this exercise we will treat them as deterministic quantities
equal to the average amounts for that month of the year computed from past data.
All the water released from Lake Nasser, whether it is for irrigation or any other
purpose, is run through the power plant, and hence generates power.

   The releasing channels have capacities, and if the actual releases exceed these
capacities, there will be scouring damage, which should be avoided.

   Some water in Lake Nasser is lost due to seepage and evaporation. These losses
depend on the climate, and so vary from month to month. If V_j is the average
volume of water in the lake (measured in units of 10^9 cubic meters, or $10^9 m^3$) in
month j of the year, then the amount of water loss due to seepage and evaporation
in that month can be estimated to be $\ell_j V_j$ where the coefficients ℓ_j are given below.
The following data is provided in the table given below.

$$
\begin{aligned}
b_j &= \text{Estimated water inflow in } 10^9 m^3 \text{ during month } j \text{ of the year} \\
a_j &= \text{Agricultural requirements for water in } 10^9 m^3 \text{ during month } j \text{ of the year} \\
\ell_j &= \text{Seepage and evaporation loss coefficient during month } j \text{ of the year} \\
p_j &= \text{Estimated return (in Egyptian pounds) per } m^3 \text{ of water released in month} \\
&\quad \text{of year for agriculture up to the requirement}
\end{aligned}
$$

Aswan high dam data

Month j	b_j	a_j	ℓ_j	p_j
January	4.221	4.0	0.007	3.05
February	3.041	4.3	0.006	3.05
March	2.584	4.3	0.005	3.14
April	2.170	4.3	0.008	3.40
May	2.033	5.6	0.010	3.54
June	2.055	6.6	0.017	3.59
July	4.847	6.9	0.024	3.59
August	18.630	3.8	0.010	3.75
September	22.317	5.1	0.009	3.75
October	15.333	6.3	0.009	3.75
November	8.235	6.3	0.008	3.05
December	5.554	4.4	0.007	3.05

   The estimated return (in Egyptian pounds) from hydroelectric power generation
per m^3 of water released anytime from Lake Nasser is 0.08. The releasing channels

capacity is $9 \times 10^9 m^3$ per month. The amount of water in Lake Nasser at the beginning of January is $50 \times 10^9 m^3$, and the lakes capacity is $168 \times 10^9 m^3$. The objective in the management of the dam is to determine the monthly water releases during the year to maximize the total benefit through the returns from agriculture and hydroelectric power generation. Formulate this as a linear program. ([S. O. Duffuaa, 1991]).

2.45 A Fertilizer Planning Model The country of Egypt is divided into 20

Region	Annual demand (in units of 1000 tons) for fertilizer in region								
	CAN 26	CAN 31	CAN 33.5	AS	U	SSP 15.5	C 1	C 2	DAP
Alex			5	3	1	8			
Beh	1		25	90	35	64	1	0.1	0.1
Ghar			17	60	28	57	1	0.2	0.1
Kes	1		10	45	22	25	2	0.2	
Dak	1		26	60	20	52	1		
Dami			2	15	8	5			
Shar	1		31	50	28	43	1	0.1	
Isma			4	6	2	4			
Suez			1			1			
Meno			24	21	30	33	2	0.1	0.1
Kalu			25	16	7	22	1		0.1
Giza			40	6	2	14	1	0.1	
Beni	1		15	1	20	13	3		
Fayo	1		20	6	20	17	1		
Mini	2	15	35	1	41	50	3	0.2	0.1
Assi	1	20	26	1	27	35	5	0.1	
Nv					1	1			
Sohag		65	3		7	20	1		
Quena		95	2		3	8			
Aswan		40				8			

Blank entries indicate that the fertilizer is not used in the region in significant quantity

separate demand regions for fertilizers. These are:

Alex = Alexandria,	Beh = Behera,	Ghar = Gharbia,	Kes = Kafr El Sheikh
Dak = Dakahria,	Dami = Damieta,	Shar = Sharkia,	Isma = Ismalia
Suez,	Meno = Menoufia,	Kalu = Kalubia,	Giza
Beni = Beni Suef,	Fayo = Fayoum,	Mini = Minia,	Assi = Assiout
Nv = New Valley,	Sohag,	Quena,	Aswan

There are 9 different types of fertilizers in common use in Egypt, these are: CAN 26 (Calcium ammonium nitrate 26 percent), CAN 31, CAN 33.5, AS (Ammonium sulfate), U (urea), SSP 15.5 (Single super-phosphate 15.5 percent), C 1 (Compound 1 : 25-5.5-0), C 2 (Compound 2 : 30-10-0), and DAP (Di-ammonium phosphate). The table given above provides the estimated annual demand for these fertilizers, in each region.

There are 5 fertilizer plants located at Aswan (AW), Helwan (H), Kafr El Zayaat (K), Assiout (AO), and Abu Zaabal (AZ). There are 9 different manufacturing processes in operation at these plants. These are:

SS	=	Sulfuric acid: Sulfur
SP	=	Sulfuric acid: Pyrites
NAO	=	Nitric acid: Ammonia oxidation
AWE	=	Ammonia: Water electrolysis
ACO	=	Ammonia: Coke oven gas
ASCO	=	Ammonium sulfate: by-product of coke oven plant
SSPD	=	Single superphosphate: Den method
CAN 31 &33.5	=	Calcium ammonium nitrate processes

The processes in use at the various plants and the respective capacities (tons/day) are given below.

Process	Production capacity at plant (tons/day)				
	AW	H	K	AO	AZ
SS			200	250	242
SP			50		227
NAO	800	282			
AWE	450				
ACO		172			
ASCO		24			
SSPD			600	600	600
CAN 31 and/or 33.5	1100	364			

Blank entry indicates that process does not exist at plant

Unless otherwise mentioned, the common unit for measuring the quantities of the various fertilizers and the following raw materials and intermediate products is ton. The various raw materials and intermediate products are:

PR (phosphate rock 30% P_2O_5),	SA (sulfuric acid)
A (ammonia),	PY (pyrites)
SU (sulfur),	NA (nitric acid)
L (limestone),	ST (pressurized steam)

The unit for measuring EP (electric power) is KWH (kilowatt hour). COG (coke-oven gas) is measured in units of 1000 cubic meters. BFG (blast furnace gas), and CW (cooling water) are measured in cubic meters. The units of major inputs needed per ton of output by the various processes are given below (if an input is not mentioned, it implies that if it is used at all in the process, it is used in insignificant quantities).

Process	Output	Major inputs needed/ton of output
SS	SA	0.334 SU, 50 EP, 20 CW
SP	SA	0.826 PY, 75 EP, 20 CW
NAO	NA	0.292 A, 231 EP, 0.6 CW
AWE	A	13960 EP, 609 BFG, 700 CW, 4 ST
ACO	A	2 COG, 19 EP, 17 CW
ASCO	AS	0.76 SA, 0.26 A, 14 EP, 6 CW
SSPD	SSP 15.5	0.62 COG, 0.41 SA, 0.6 PR
CAN 31	CAN 31	0.2 A, 0.71 NA, 0.12 L, 49 CW, 0.4 ST
CAN 33.5	CAN 33.5	0.21 A, 0.76 NA, 0.04 L, 49 CW, 0.5 ST

The prices of various raw materials (in Egyptian pounds/unit) from domestic sources, where they are available, are given below. A blank entry in the following table indicates that the raw material is not available locally near the plant.

Raw material	Price (Egyptian pound/unit) at plant location				
	H	AW	AO	K	AZ
L	1.2	1.2			
COG	16				
PR			3.5	5.0	4.0
SA	3				

The following table provides the interplant distances for shipment of raw materials or intermediate products such as sulfuric acid. The distance matrix is symmetric, so only the upper half of the matrix is given.

From	Distance in km to				
	K	AZ	H	AO	AW
K	-	85	230	479	983
AZ		-	61	420	924
H			-	400	900
AO				-	504
AW					-

Interplant shipments of raw materials and intermediate products take place by rail, the applicable rate in Egyptian pounds per ton is $0.5 + 0.03 \times$ (distance in km). The 0.5 in this cost estimate per ton, which is independent of the distance transported, represents the cost of loading, unloading, and handling.

Raw materials can also be imported from other countries. All imports enter Egypt through the seaport of Alexandria. Imported raw materials are transported by barge either directly to the destination plant, or in some cases where the canal does not go all the way, they are taken as far as possible by barge with the remaining distance being covered by lorries. Distances from Alexandria to plant locations for imports of raw materials are given below.

Plant location	Distance from Alexandria (km) by		
	Waterway	Road	Total
K	104	6	110
AZ	210	0.1	210.1
H	183		183
AO	583		583
AW	1087	10	1097

The rate schedule for transport of raw materials from Alexandria to the various plants (in Egyptian pounds/ton) is $1.0 + 0.003 \times$(distance in km) on waterway, and $0.5 + 0.0144 \times$(distance in km) for road transport.

The prices of imports of various raw materials and fertilizers (in Egyptian pounds per ton delivered at Alexandria) are given below.

Product	Average import price	Product	Average import price
PY	6.7	U	60.0
SU	22.0	AS	31.4
CAN 26	30.6	DAP	71.5
CAN 31	36.5	C 1	41.0
CAN 33.5	40.2	C 2	52.0

EP for the AWE process at AW plant is available at the very special rate of 0.002 Egyptian pounds per KWH. For all other processes and at all other plant locations, the prices (in Egyptian pounds per unit) of EP and other miscellaneous inputs are given below.

Input	EP	BFG	CW	ST
Price	0.007	0.0075	0.031	1.25

All shipments of final products from plants to the marketing regions are undertaken by lorry. Similarly all imported fertilizer is transported from the port of entry, Alexandria, to the marketing regions, by lorry. The road distances from each plant and Alexandria (for imported fertilizer) to marketing regions is given in the following table.

To region	K	AZ	H	AO	AW	Alexandria
Alex	119	210	244	607	1135	
Beh	42	50	184	547	1065	600
Ghar	20	65	122	485	1003	166
Kes	20	105	162	525	1043	161
Dak	58	138	152	515	1033	224
Dami	131	216	233	596	1114	283
Shar	78	60	110	473	991	256
Isma	241	142	173	536	1054	381
Suez	246	224	178	541	1059	386
Meno	33	154	109	472	990	157
Kalu	66	97	76	439	957	190
Giza	133	48	9	372	890	287
Beni	248	163	105	257	775	359
Fayo	230	145	88	308	826	341
Mini	372	288	230	132	650	384
Assi	504	420	362		518	616
Nv	703	619	561	199	519	815
Sohag	603	519	461	99	419	715
Quena	746	662	604	242	276	858
Aswan	1022	938	880	518		1134

Distance in km from

In Egypt, the labor force in fertilizer plants is fixed and closely monitored by the government. Only if production is halted entirely, and the plant is scrapped, can this labor cost be considered as variable. Since the plants are all going to be in operation, labor costs can be ignored.

(i)   It is required to determine the least cost supply and shipment pattern for meeting the annual demand of the 9 types of fertilizers. The model has to determine whether each fertilizer should be imported or produced domestically at the existing production facilities. Formulate this problem as an LP, and solve this model using a mathematical programming language such as AMPL or GAMS.

(ii)   The country's farmers are accustomed to use the 9 types of fertilizers mentioned earlier, in annual quantities given above. Can sizeable savings be achieved if a different combination of fertilizers is used instead? Suppose the only requirement is to meet the demands for the major nutrients nitrogen (N), and phosphorus (measured in terms of P_2O_5). Following table gives the amounts of N, and P_2O_5

contained in each of the 9 popular fertilizers. Blank entries in this table can be treated as 0.

Fertilizer	% of N	% of P_2O_5	Fertilizer	% of N	% of P_2O_5
CAN 26	26		SSP 15.5		21
CAN 31	31		C 1	25	5.5
CAN 33.5	33.5		C 2	30	10
AS	21		DAP	16	48
U	45				

From the annual fertilizer demand in each region given earlier, estimate the annual requirements for the nutrients N and P_2O_5 in each region. Formulate an LP model for determining a least cost supply (domestic fertilizer production plus imported fertilizers) and shipment pattern for fertilizers to meet the annual demands for the nutrients N and P_2O_5 in each region. Solve this model using a mathematical programming language like AMPL, or GAMS. Compare the solution obtained here with that obtained under (i). ([A. M. Choksi, A. Meeraus, and A. J. Stoutjesdijk, 1980]).

2.46 Molten pig iron mix problem A steel company makes pig iron from iron ore, which is then used to make various steel alloys in an open hearth furnace. Molten pig iron is tapped from blast furnace into torpedo cars. The torpedo cars are closed containers in which molten iron can be kept in the molten state for a few hours. From the torpedo cars, the molten pig iron is poured into an open ladle which is carried by crane and the molten pig iron in it emptied into the open hearth furnace which converts it into the steel alloy with desired chemical composition.

As the molten pig iron is being tapped into a torpedo car at the blast furnace a sample is taken for quick chemical analysis to measure the amount of manganese(Mn), silicon(Si), sulphur(S), phosphorus(P), and titanium(Ti) in it. The composition of the pig iron in each torpedo car (each of the constituent elements is measured in units of 1 out of 1000 by weight) is given below. Each ladle carries 260 tons of molten iron, which is exactly the quantity of pig iron needed for a charge (also called heat) in the open hearth furnace.

At some stage, suppose we have 4 torpedo cars containing 260 tons of molten pig iron each, with the following compositions.

Element	Content in pig iron in torpedo car no.			
	1	2	3	4
S	6	5	6	6
Si	43	42	35	47
Mn	32	37	31	36
P	100	102	103	101
Ti	61	67	60	64

It is required to prepare 4 charges using the molten pig iron in the 4 torpedo cars. The desired (or target) composition in the four charges is given below.

Element	Desired content in charge no.			
	1	2	3	4
S	5	6	5	5
Si	56	44	30	35
Mn	40	35	35	40
P	96	110	100	102
Ti	62	65	60	61

For each charge, we must mix molten pig iron from the available torpedo cars to prepare an open ladle filled with 260 tons of molten pig iron. The closer the chemical composition of the pig iron in the ladle is to the target chemical composition of the charge, the less the processing time for the charge in the open hearth furnace.

(i)   Discuss how much pig iron from each of the four torpedo cars should be used in each charge, so as to minimize the total processing time at the open hearth furnace for all the four charges put together. Formulate this as an LP.

(ii)   Torpedo cars come in pairs. Cars no. 1 and 2 form a pair, and cars no. 3 and 4 form a second pair. Suppose we have to make sure that by the time charges 1 and 2 are poured, both the torpedo cars in one of the two pairs are emptied. How does this change the formulation in (i)? (Soo Y. Chang)

2.47 A paper making machine produces master rolls of paper of width $w = 4200$ mm. The company has to fill orders for this paper in the following quantities q_i of specified widths w_i, for $i = 1$ to 9 during a planning period.

No. i	Orders to be filled (quantity in tons)								
	1	2	3	4	5	6	7	8	9
Width w_i (mm)	1115	1500	1000	1275	1430	1150	1055	575	950
Quantity q_i	105	11	9	18	5	5	12	15	16

There are many different ways of combining one or more of the required widths w_i into a master roll, these are called *cutting patterns*. A cutting pattern specifies the number of rolls n_i of width w_i to be cut out of a master roll simultaneously, for $i = 1$ to 9. For example, a cutting pattern C_1 represented by the vector (n_1 to $n_9)^T = (2, 0, 1, 0, 0, 0, 0, 0, 1)^T$ specifies that 2 rolls of width $w_1 = 1155$ mm, 1 roll of width 1000 mm, and 1 roll of width 950 mm be cut out of a master roll simultaneously. Since the width of the master roll is $w = 4200$ mm, the trim waste in implementing this cutting pattern is a strip of width $4200 - 2 \times 1115 - 1 \times 1000 - 1 \times 950 = 20$ mm for a percentage waste of $(100 \times 20)/4200 = 0.47\%$.

The company is considering the following 15 cutting patterns.

Pattern C_j	No. of rolls n_i^j in pattern, of width w_i								
	1115	1500	1000	1275	1430	1150	1055	575	950
C_1	2	0	1	0	0	0	0	0	1
C_2	1	2	0	0	0	0	0	0	0
C_3	2	0	0	1	0	0	0	1	0
C_4	0	0	0	0	2	0	1	0	0
C_5	0	0	0	3	0	0	0	0	0
C_6	0	2	0	0	0	0	1	0	0
C_7	0	0	0	0	0	0	0	0	4
C_8	0	0	0	1	2	0	0	0	0
C_9	0	0	0	0	0	3	0	1	0
C_{10}	1	0	0	0	0	0	2	0	1
C_{11}	1	0	2	0	0	0	1	0	0
C_{12}	3	0	0	0	0	0	0	1	0
C_{13}	3	0	0	0	0	0	0	0	0
C_{14}	2	0	0	0	0	0	0	0	2
C_{15}	0	0	4	0	0	0	0	0	0

Define the decision variables

$$x_j \quad = \quad \text{weight of master roll (in tons) cut according to cutting pattern } C_j$$

for $j = 1$ to 15. Formulate as an LP the problem of finding $(x_1$ to $x_{15})$ so as to fill all the orders with minimum trim loss.

How does the formulation change if the quantity to be delivered of width w_i is allowed to vary between $\pm 10\%$ of the specified q_i, for $i = 1$ to 9?

Here we discussed the problem of finding the quantities of the master roll to cut according to given cutting patterns, in order to minimize the trim loss to fill specified orders. Discuss how one might generate a good set of cutting patterns to be used as input for this process. (Case formulated by D. H. Lombardo Ferreira of UniSoma, Brazil).

2.48 UniCitrus is a major producer of frozen concentrated orange juice in Southern Brazil. Three orange varieties, Hamlin, Pera, and Valencia are used. The company buys the production from selected plantations, picks the fruit, processes it into juices, and blends them into two main products called Standard and Dairy. The Standard is primarily used by beverage industries all over the world. The slightly more expensive Dairy has to satisfy certain conditions on flavor, acidity, cell content, and color among others.

The planning horizon is a quarter. The forecasted demand over the planning horizon is given below. There is no penalty for unfulfilled demand.

Product	Demand (tons) in month			Selling Price ($/ton)
	1	2	3	
Standard	500	1500	700	1000
Dairy	200	100	100	1100

The different orange varieties have different maturation periods. The expected availability in the region, of different varieties during the planning horizon, is given below.

Month	Expected availability (in 1000s of boxes) of		
	Hamlin	Pera	Valencia
1	300	250	0
2	350	350	50
3	100	500	100
Juice (kg/box) in this variety	3.5	3.7	3.4

There are lower and/or upper limits on the percentage of juice of each variety allowed in each product. These are: Standard can have at most 25% Hamlin juice, must have at least 60% Pera juice, and at most 40% Valencia juice. Dairy can have at most 30% Hamlin juice, must have at least 50% Pera juice, and between 15 to 50% Valencia juice.

Juice made in a month can be sold in that month itself, or later. Unsold juices may be stored in refrigirated tanks at a cost of $10/ton/month for all varieties (a ton is 1000 kg).

The cost of purchasing, picking, and transporting the fruit to the juice plant is estimated to be $1/box for each variety. The cost of running the juice plant is a fixed amount of $(1/2) million/month, and the plant has adequate capacity and workforce to handle all the fruit expected to be available in each month of the planning horizon.

(i)   Juice inventories are negligible at the beginning of the planning horizon. The company would like to have no juice inventories at the end of the planning horizon. It is required to determine a production plan that maximizes the companies net profit during the planning horizon. Formulate this as an LP and solve it using an LP software package.

(ii)   In (i) we ignored possible capacity limitations of the juice plant. The equipment at the plant consists of extractors, evaporators, blenders, dryers for residues, etc. This equipment can process 500,000 boxes of Pera variety per month if this is the only variety to be processed. Hamlin variety uses 10% more, and Valencia 10% less capacity. The effective plant capacity will therefore depend on the proportion of different varieties processed. Revise your model in (i) to take into account these processing capacity constraints.

(iii) Orange growers in Northern Brazil produce a lot of Valencia oranges. A truckload of Valencia consists of 1000 boxes. Let λ denote the transportation cost for bringing a truckload of oranges from the north to the juice plant. Evaluate the effect of increasing the availability of Valencia oranges on the optimum production plan, treating λ as an unknown parameter.

(iv) The Hamlin variety is scarce in Northern Brazil. Competitors of UniCitrus located in the north may come south and buy this variety for their processing needs. Evaluate the effect on the optimum production plan, of a decrease in the availability of Hamlin Variety due to their purchases, by an analysis similar to that in (iii). (Case prepared by E. C. Marujo and M. A. Pereira of UniSoma, Brazil)

2.49 The UniPaper co. grows Eucalyptus plantations at 5 sites in Brazil. Eucalyptus is originally fron Australia, but grows exceptionally fast in Brazilian climate. The company harvests eucalyptus trees for wood from 6 to 9 years after planting. Using the harvested wood, and other eucalyptus wood purchased from independent farmers when necessary, the company makes wood pulp and a variety of paper products at its plant. This exercise addresses the problem of managing the eucalyptus plantations for maximum profit.

The work at the plantation is of the following three types.

Planting: This action consists of planting eucalyptus seedlings on land that has been cleared previously.

Reforming: This action consists of removing the stumps and roots from the soil, and preparing it for planting.

Recycling: This action consists of cutting the trees for wood, but leaving the stump and roots. The stump will bud again into a new tree.

Usually at most two recyclings (three cuts or three cycles) are allowed before replanting.

Wood is measured in units denoted by ST and called "esteros". Each ST is approximately one cubic meter in logs. The company currently needs approximately 60,000 ST annually for its plant. Eucalyptus wood can be purchased or sold at a cost of $2.5/ST in the open market.

The present condition at the 5 sites owned by the company is summarized below.

Site	Area (ha)	Present cycle no.	Age (years)
1	1500	1	4
2	1000	1	5
3	1500	2	6
4	3000	2	3
5	1500	1	4

The following table gives data on the productivity of each site.

Site	Cycle	Wood Prod. (ST/ha) if cutting age is			
		6	7	8	9
1,2,3	1	300	320	340	350
	2	270	280	290	300
	3	250	260	270	280
4,5	1	200	210	220	230
	2	180	190	200	210
	3	150	160	170	180

The costs of running a site are those of maintenance, planting and reforming. The maintenance cost is $35/ha in the first year of any rotation at any site, and $7/ha in subsequent years. The planting and reforming costs are given below.

Operation	Cost ($/ha) at site				
	1	2	3	4	5
Planting	100	200	160	260	200
Reforming	80	100	80	130	100

The cost of transporting the wood from sites 1 to 5 to the plant in $/ST are 0.2, 0.4, 0.7, 0.7, 0.2, respectively.

The plant has to be kept supplied with the necessary wood, either from the companies plantations, or purchased from outside suppliers. Any excess wood produced at the companies plantations and not used the plant, is sold in the open market.

Assume that the annual rate for money is 12%. It is required to determine an optimal cutting plan for each site (giving the alternatives of cutting ages between 6 to 9 years, and decision of reforming or recyling at the end of each cycle), that maximizes the present value of net profit. Identify the various cutting plans, and defining the decision variables to be the fraction of the area at each site following the plans, formulate the problem as an LP. Solve the model using an LP software package. (Case formulated by S. G. Abasto and M. Taube of UniSoma, Brazil).

2.50 Flat-Glass Raw Material Mix Problem The planning horizon for a glass company is a time interval of two periods. They have to produce 100,000 tons of gray glass in period 1, and an equal amoun of clear glass in period 2.

The raw materials used in the manufacture of these glasses are sand, dolomite, lime stone, soda ash, iron oxide and waste-glass. The important constituents of glass that these raw materials provide when melted down are the oxides SiO_2,

Na_2O, CaO, MgO, and Fe_2O_3. The rest of the matter in these raw materials (classified as "other matter") is either volatile, or other matter that evaporates and bubbles away during the glass making process.

The glass making oven is continuously fed with a mixture of the above raw materials. The "other matter" bubbles away, and the oxides mix and leave the oven as a thin and large glass river that is the input for plate glass manufacture.

The cooling and plate glass handling processes generate recyclable wastes. These are used as waste-glass, one of the inputs into the oven. Assume that recyclable waste-glass is produced at a rate of 10% of the oven production.

Data on the composition of the various raw materials, the two final products (gray and clear glass), and the various costs is given below.

| Raw | % by weight of | | | | | | Cost |
material	SiO_2	Na_2O	CaO	MgO	Fe_2O_3	Other matter	($/ton)
Sand	99.5	0.01	0.01	0.01	0.07	0.4	0.3
Dolomite	0.7	0.02	31.0	21.8	0.05	46.43	0.8
Limestone	0.5	0	55.3	0.4	0.05	43.75	0.4
Soda ash	0.02	58.1	0	0.01	0	41.87	3.6
Iron oxide	0	0	0	0	98.2	1.8	3.3
Gray glass	71.5	13.0	10.0	5.0	0.5	0	
Clear glass	71.91	13.0	10.0	5.0	0.09	0	

The cost of energy for melting any of the raw materials sand, dolomite, limestone, soda ash, or iron oxide, is estimated to be $ 0.4/ton for making gray glass, and $ 0.44/ton for clear glass. The cost of energy for melting waste-glass input is estimated to be $ 0.3/ton for making gray glass, and $ 0.33/ton for making clear glass.

It is required to determine the input recipe, i.e., the % by weight of each raw material in the input fed into the oven. Formulate a model for determining an input recipe for each period that minimizes   (a) the cost of raw materials,   (b) the energy costs,   (c) the total cost of raw materials and energy. How does the model change if the company decides to allow a deviation of at most 0.2 for the % of each of SiO_2, Na_2O, CaO, and MgO, from its target value for the glass? (Case formulated by J. E. C. Scheidt of UniSoma, Brazil).

2.11   References

J. R. ARONOFSKY, J. M. DUTTON, and M. T. TAYYABKHAN, 1978, *Managerial Planning with Linear Programming in Process Industry Operations*, Wiley-Interscience, NY.

R. BENVENISTE, May 1986, "An Integrated Plant Design and Scheduling Problem in the Food Industry", *Journal of the Operational Research Society*, 37, no. 5, 453-461.

A. M. CHOKSI, A. MEERAUS, and A. J. STOUTJESDIJK, 1980, *The Planning of Investment Programs in the Fertilizer Industry*, The John Hopkins University Press, Baltimore, MD, USA.

G. B. DANTZIG, 1963, *Linear Programming and Extensions*, Princeton University Press, Princeton, NJ.

J. G. DEBANNE, and J. N. LAVIER, February 1979, "Management Science in the Public Sector - The Estevan Case", *Interfaces*, 9, 2, part 2, 66-77.

S. O. DUFFUAA, 1991, "A Chance-Constrained Model for the Operation of the Aswan High Dam", *Engineering Optimization*, 17 (109-121).

S. O. DUFFUAA, J. A. AL-ZAYER, M. A. AL-MARHOUN, and M. A. AL-SALEH, Nov. 1992, "A Linear Programming Model to Evaluate Gas Availability for Vital Industries in Saudi Arabia", *Journal of the Operational Research Society*, 43, no. 11, (1035-1045).

P. LAZARUS, and D. KIRKMAN, January 1980, "An Investigation of Feedstuffs for Dairy Cattle on Liverton Farm", *Journal of the Operational Research Society*, 31, no. 1, 3-15.

M. I. MINGUEZ, C. ROMERO, and J. DOMINGO, January 1988, "Determining Optimum Fertilizer Combinations Through Goal Programming With Penalty Functions: An Application to Sugar Beet Production in Spain", *Journal of the Operational Research Society*, 39, n0. 1, 61-70.

K. G. MURTY, 1983, *Linear Programming*, Wiley, NY.

M.J. SCHNIEDERJANS, N. K. KWAK, and J. S. FRUEH, 1991, "A Stochastic Linear Programming Model to Improve Automobile Dealership Operations", *Computers and Operations Research*, 18, no. 8, 669-678.

R. D. SNYDER, January 1984, *Journal of the Operational Research Society*, 35, no. 1, 69-74.

Chapter 3

Review of Matrix Algebra and Geometry

3.1　The n-dimensional Euclidean Vector Space $\mathbb{R}^n$

$\mathbb{R}^n$ is the space of all **points** or **vectors** (column or row, depending on the context as required by the user) of the form $x = (x_1, \ldots, x_n)^T$ with n coordinates, each of which is a real number. We will treat each vector as a column vector unless specified otherwise. That's why we have written the vector x as $(x_1, \ldots, x_n)^T$. In any vector in $\mathbb{R}^n$, the first coordinate is the x_1-coordinate, the second the x_2-coordinate, etc. For example in the vector $(1, 2, 3)^T$ in $\mathbb{R}^3$, the x_1, x_2, x_3-coordinates are 1, 2, 3 respectively. This makes each point in $\mathbb{R}^n$ an **ordered vector** in the sense that if the order of the entries changes, the point changes. For example, the vectors $(1, 2)^T$ and $(2, 1)^T$ are different in $\mathbb{R}^2$ since they represent two different points. See Figure 3.1.

When all the coordinates are zero, the vector is known as the zero vector or the **origin**, and it is also denoted by the symbol "0". It will always be clear from the context whether "0" refers to the real number 0, or the zero vector in some Euclidean space.

The vector x is said to be **nonnegative** if each of its components is greater than or equal to 0. Symbolically this is written as $x \geqq 0$. Notice the two lines under the inequality sign. Obviously $0 \geqq 0$. The vector x is said to be **positive** if each of its components is strictly greater than 0. This is denoted by $x > 0$.

3.1.1　Scalar Multiples of a Vector

The word scalar means a real number. Let $x = (x_1, \ldots, x_n)^T$ and α be a scalar. Then $(\alpha x_1, \ldots, \alpha x_n)^T$ denoted by αx is the scalar multiple of x by the scalar α. In scalar multiplication of a vector, each coordinate is multiplied by that scalar. For example, $(3, -6)^T = 3(1, -2)^T$ is the scalar multiple of $(1, -2)^T$ by the scalar 3.

When $x \neq 0$ is a vector in $\mathbb{R}^n$, the set of all vectors which are scalar multiples of x, i.e., $\{\alpha x : \alpha \text{ a real number}\}$, is known as the **linear hull of** x, or **line of** x, or the **one dimensional subspace determined by** x. In Figure 3.1, the entire straight line through 0 and the vector $(1, 2)^T$ consisting of both the normal and the thick parts is the linear hull of $(1, 2)^T$.

When $x \neq 0$ is a vector in $\mathbb{R}^n$, the set of all vectors which are scalar multiples **of** x **by a nonnegative scalar, i.e.,** $\{\alpha x : \alpha \geq 0\}$**, is called the half-line or the ray of** x. It is the part of the line of x beginning at the origin, going towards x and continuing in that direction indefinitely. In Figure 3.1, the thick part of the line of $(1, 2)^T$ is the half-line or the ray of $(1, 2)^T$.

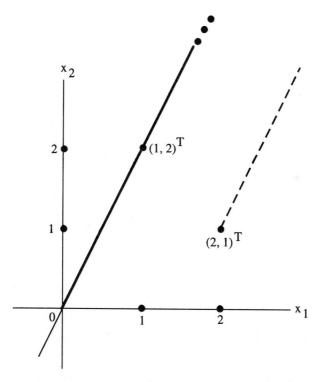

Figure 3.1 The line (both thin and thick parts) and the half-line or the ray (thick part) of $(1, 2)^T$ in $\mathbb{R}^2$. The dashed half-line through $(2, 1)^T$ is the half-line through $(2, 1)^T$ parallel to the ray of $(1, 2)^T$.

Exercise

3.1 Let $x^1 = (-3, 0, 1, -1, 2)^T$. Check which of the following vectors are scalar multiples of x^1: $(6, 0, -2, 2, -4)^T$, $(-12, 0, 4, -4, 8)^T$, 0, $(-6, 2, 2, -2, 4)^T$. Also check which of these vectors are on the ray of x^1, and which are on the line of x^1 but not on the ray.

3.1.2   Linear Combinations of Vectors, Linear Hulls

The sum or a linear combination of two vectors is only defined if either both vectors are row vectors or both are column vectors in the same dimension space. For any vector $x \in \mathbb{R}^n$, $x + 0 = x$. If $x = (x_1, \ldots, x_n)^T, y = (y_1, \ldots, y_n)^T$ are two nonzero vectors in $\mathbb{R}^n$, then the vector $(x_1 + y_1, \ldots, x_n + y_n)^T$ obtained by adding corresponding coordinates in x and y is their sum $x + y$. If y is not a scalar multiple of x, then $x + y$ is the fourth vertex in the unique parallelogram whose other three vertices are 0, x, and y. This geometric interpretation of the sum of two vectors is known as the **parallelogram law of addition of vectors**. See Figure 3.2.

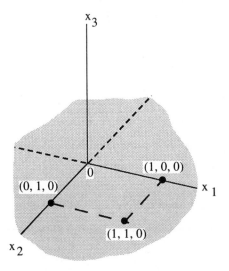

Figure 3.2 The parallelogram law. The sum $(1, 0, 0)^T + (0, 1, 0)^T = (1, 1, 0)^T$ is the fourth vertex of the unique parallelogram defined by 0, $(1, 0, 0)^T$ and $(0, 1, 0)^T$. The dotted x_1, x_2-coordinate plane is the linear hull of $(1, 0, 0)^T, (0, 1, 0)^T$.

Let $x^1 = (x_1^1, \ldots, x_n^1)^T$, $x^2 = (x_1^2, \ldots, x_n^2)^T$, and let α_1, α_2 be two scalars. Then the sum $\alpha_1 x^1 + \alpha_2 x^2 = (\alpha_1 x_1^1 + \alpha_2 x_1^2, \ldots, \alpha_1 x_n^1 + \alpha_2 x_n^2)^T$ is known as a **linear combination of** x^1, x^2 **with coefficients** α_1, α_2. For example, if $x^1 = (1, 0, -2)^T$, $x^2 = (-3, 4, 16)^T$, any point of the form $\alpha_1 x^1 + \alpha_2 x^2 = (\alpha_1 - 3\alpha_2, 4\alpha_2, -2\alpha_1 + 16\alpha_2)^T$ where α_1, α_2 are any two real numbers, is a linear combination of x^1, x^2. Taking $\alpha_1 = 2, \alpha_2 = -1$ gives the specific point $(5, -4, -20)^T$.

The set of all linear combinations of x^1, x^2, $\{\alpha_1 x^1 + \alpha_2 x^2 : \alpha_1, \alpha_2$ take all possible real values$\}$ is known as the **linear hull** or the **linear span of** $\{x^1, x^2\}$. It is the smallest dimensional **subspace** of $\mathbb{R}^n$ containing x^1 and x^2. For example, in Figure 3.2, the dotted x_1, x_2-coordinate plane is the linear hull of $\{(1, 0, 0)^T, (0, 1, 0)^T\}$.

In general, let $\{x^1, \ldots, x^k\}$ be a finite set of points in $\mathbb{R}^n$, where $x^r = (x_1^r, \ldots, x_n^r)^T$ for $r = 1$ to k. A linear combination of these points with coefficients $\alpha_1, \ldots, \alpha_k$ is the point $x = \alpha_1 x^1 + \ldots + \alpha_k x^k = (\alpha_1 x_1^1 + \ldots + \alpha_k x_1^k, \ldots, \alpha_1 x_n^1 + \ldots + \alpha_k x_n^k)^T$. The set of all linear combinations of $x^1, \ldots, x^k$, i.e., $\{\alpha_1 x^1 + \ldots + \alpha_k x^k : \alpha_1, \ldots, \alpha_k$ take all possible real values$\}$ is the subspace of $\mathbb{R}^n$ of smallest dimension containing $x^1, \ldots, x^k$; it is called the **linear hull** or the **linear span of** $\{x^1, \ldots, x^k\}$.

Spanning Property

Let $\Gamma = \{x^1, \ldots, x^k\}$, $\Omega = \{y^1, \ldots, y^p\}$ be two sets of vectors from $\mathbb{R}^n$. The set Γ is said to span Ω if every vector in Ω can be expressed as a linear combination of vectors in Γ, i.e., if Ω is a subset of the linear hull of Γ.

How to Check Whether a Given Vector is a Linear Combination of a Set of Vectors

Let $\Gamma = \{A_1, \ldots, A_k\}$ be a set of vectors from $\mathbb{R}^n$ and b another vector from $\mathbb{R}^n$. Either all these vectors are row vectors, or all of them are column vectors. To check whether b is a linear combination of vectors from Γ (i.e., whether b is in the linear hull of Γ), we need to solve the following system of n linear equations in k variables, $\alpha_1, \ldots, \alpha_k$.

$$\alpha_1 A_1 + \ldots + \alpha_k A_k = b \tag{3.1}$$

If $(\bar{\alpha}_1, \ldots, \bar{\alpha}_k)$ is a solution for (3.1), then b is a linear combination of $A_1, \ldots, A_k$ with coefficients $\bar{\alpha}_1, \ldots, \bar{\alpha}_k$. If (3.1) has no solution, b is not in the linear hull of Γ.

If k and n are small, (3.1) can be solved by some trial and error method. Pivotal methods which are guaranteed to handle systems of linear equations are discussed in Section 3.4; they can be used when k or n is not small.

As an example consider the column vectors from $\mathbb{R}^4$ given in the following Tableau 3.1.

Tableau 3.1

A_1	A_2	A_3	b^1	b^2	b^3	b^4
1	0	1	3	2	2	3/4
0	1	−1	−1	0	−4	−1/4
0	0	1	2	3	3	1/2
0	0	−1	−2	−4	−3	−1/2

Suppose we wish to check whether b^1 is in the linear hull of $\{A_1, A_2, A_3\}$. For this we need to solve the system of linear equations in detached coefficient form given below on the left. Using the method discussed in Section 3.4 we find the solution $(\bar{\alpha}_1, \bar{\alpha}_2, \bar{\alpha}_3) = (1,\ 1,\ 2)$ for this system on the left. So, b^1 is a linear combination of A_1, A_2, A_3 with coefficients 1, 1, 2 (i.e., $b^1 = A_1 + A_2 + 2A_3$).

α_1	α_2	α_3	b^1
1	0	1	3
0	1	−1	−1
0	0	1	2
0	0	−1	−2

α_1	α_2	α_3	b^2
1	0	1	2
0	1	−1	0
0	0	1	3
0	0	−1	−4

In the same way to check whether b^2 is in the linear hull of $\{A_1, A_2, A_3\}$, we need to solve the system of linear equations in detached coefficient form given above on the right. It can be verified that this system has no solution (the bottom two equations contradict each other). So, b^2 is not in the linear hull of $\{A_1, A_2, A_3\}$.

Importance of Linear Hulls or Subspaces

Linear hulls or subspaces are very important in the study of systems of homogeneous linear equations, i.e., equations of the form

$$Ax = 0 \tag{3.2}$$

In (3.2), x is the vector of variables, and A is the coefficient matrix. This system is called a homogeneous system since the RHS constants vector is 0. $x = 0$ is always a solution to a homogeneous system. If x^1, x^2 are two solutions to (3.2), then $Ax^1 = 0$, $Ax^2 = 0$, so, $A(\alpha_1 x^1 + \alpha_2 x^2) = \alpha_1 Ax^1 + \alpha_2 Ax^2 = 0$, for any scalars α_1, α_2; so every linear combination of any two solutions of (3.2) is also a solution. Hence the set of solutions of (3.2) is a subspace or a linear hull. So, linear hulls play an important role in the study of homogeneous systems of linear equations.

Half-Line through a Parallel to the Ray of b

Let a, b be two vectors in $\mathbb{R}^n$ with $b \neq 0$. Then the half-line through a parallel to the ray of b is the translate of the ray of b to a; i.e., it is the set $\{a + \alpha b : \alpha \geq 0\}$. For example in Figure 3.1, the dashed half-line is the half-line through $(2,1)^T$ parallel to the ray of $(1,2)^T$.

Given three points a, b, c in $\mathbb{R}^n$ with $a \neq 0$, suppose we want to check whether c is on the half-line through b parallel to the ray of a. The answer is yes iff there exists a scalar α such that $c = b + \alpha a$, i.e., iff $c - b$ is a scalar multiple of a. For example, let $a = (1,2)^T, b = (2,1)^T, c = (12,21)^T, d = (17,36)^T$. $c - b = (10,20)^T = 10a$, so c is on the half-line through b parallel to the ray of a, $d - b = (15,35)^T$ is not a scalar multiple of a, so, d is not on the half-line through b parallel to the ray of a.

3.1.3 Affine Combinations, Affine Hulls or Affine Spaces

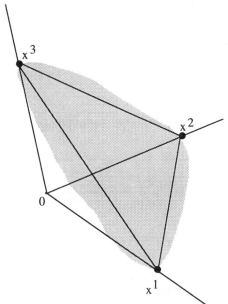

Figure 3.3 The affine hull of $\{x^1, x^2, x^3\}$ in $\mathbb{R}^3$ is the dotted two dimensional hyperplane containing x^1, x^2, and x^3.

Let x^1, x^2 be two distinct points in $\mathbb{R}^n$. The linear combination $\alpha_1 x^1 + \alpha_2 x^2$ is said to be an **affine combination** of x^1, x^2 if the coefficients satisfy the condition $\alpha_1 + \alpha_2 = 1$. See Figure 3.4. In the same way, the linear combination $\alpha_1 x^1 + \ldots \alpha_k x^k$ is said to be an **affine combination** of $x^1, \ldots, x^k$ if the coefficients satisfy the condition $\alpha_1 + \ldots + \alpha_k = 1$. The set of all affine combinations of $x^1, \ldots, x^k$; $\{\alpha_1 x^1 + \ldots \alpha_k x^k : \alpha_1, \ldots, \alpha_k$ take all possible real values subject to $\alpha_1 + \ldots + \alpha_k = 1\}$ is known as the **affine hull** or **affine space** or **flat of** $\{x^1, \ldots, x^k\}$. The affine hull of a set of points is a subset of its linear hull. See Figure 3.3.

When $\alpha_1 + \ldots + \alpha_k = 1$, $\alpha_1 x^1 + \ldots \alpha_k x^k = (1 - \alpha_2 - \alpha_3 - \ldots - \alpha_k) x^1 + \alpha_2 x^2 + \ldots + \alpha_k x^k = x^1 + [\alpha_2(x^2 - x^1) + \ldots + \alpha_k(x^k - x^1)]$. If $y = \alpha_2(x^2 - x^1) + \ldots + \alpha_k(x^k - x^1)$, it is a linear combination of $x^2 - x^1, \ldots, x^k - x^1$. As $\alpha_2, \ldots, \alpha_k$ assume all possible real values, y varies over the linear hull of $\{x^2 - x^1, \ldots, x^k - x^1\}$. Therefore the affine hull of $\{x^1, \ldots, x^k\}$ is the translate of the linear hull of $\{x^2 - x^1, \ldots, x^k - x^1\}$ to x^1.

The Straight Line Joining Two Distinct Points in $\mathbb{R}^n$

Let a, b be any two distinct points in $\mathbb{R}^n$ for any n. Then there is a unique straight line joining a and b, it is their affine hull $\{\alpha a + (1 - \alpha)b : \alpha \text{ takes all real values}\}$. See Figure 3.4.

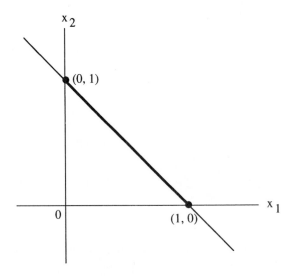

Figure 3.4 The straight line joining (0, 1), (1, 0) is their affine hull. The thick portion is the line segment joining (0, 1), (1, 0), it is their convex hull.

Given three points a, b, c in $\mathbb{R}^n$ with $a \neq b$, suppose we need to check whether c is on the straight line joining a and b. The answer is yes iff there exists a real number α such that $c = \alpha a + (1 - \alpha)b$, i.e., iff $c - b$ is a scalar multiple of $a - b$. For example, let $a = (2, -1, 1)^T$, $b = (1, 2, -1)^T$, $c = (-2, 11, -7)^T$, $d = (2, -1, 0)^T$. $c - b = (-3, 9, -6)^T = -3(1, -3, 2)^T = -3(a - b)$, so c is on the straight line joining a and b. $d - b = (1, -3, 1)^T$ is not a scalar multiple of $a - b$, so d is not on the straight line joining a and b.

How to Check Whether a Given Vector is an Affine Combination of a Set of Vectors

To check whether b in $\mathbb{R}^n$ is an affine combination of the set of vectors $\Gamma = \{A_1, \ldots, A_k\}$ in $\mathbb{R}^n$, we need to solve the following system of $n + 1$ equations in k variables $\alpha_1, \ldots, \alpha_k$.

$$\alpha_1 A_1 + \ldots + \alpha_k A_k = b \tag{3.3}$$
$$\alpha_1 + \ldots + \alpha_k = 1$$

If $(\bar{\alpha}_1, \ldots, \bar{\alpha}_k)$ is a solution for (3.3), then b is an affine combination of $A_1, \ldots, A_k$ with coefficients $\bar{\alpha}_1, \ldots, \bar{\alpha}_k$. If (3.3) has no solution, b is not in the affine hull of Γ.

As an example, to check whether the vector b^1 in Tableau 3.1 is in the affine hull of $\{A_1, A_2, A_3\}$ there, we need to solve the system of equations in detached coefficient form given below on the left. It can be verified that this system on the left has no solution, so b^1 is not in the affine hull of $\{A_1, A_2, A_3\}$ of Tableau 3.1.

α_1	α_2	α_3	b^1
1	0	1	3
0	1	-1	-1
0	0	1	2
0	0	-1	-2
1	1	1	1

α_1	α_2	α_3	b^3
1	0	1	2
0	1	-1	-4
0	0	1	3
0	0	-1	-3
1	1	1	1

To check whether b^3 in Tableau 3.1 is an affine combination of $\{A_1, A_2, A_3\}$ there, we need to solve the system of linear equations given above on the right. This system has a solution $(\alpha_1, \alpha_2, \alpha_3) = (-1, -1, 3)$. So, $b^3 = -A_1 - A_2 + 3A_3$ this expresses b^3 as an affine combination of $\{A_1, A_2, A_3\}$.

Importance of Affine Spaces

Affine spaces are very important in the study of systems of linear equations

$$Ax = b \tag{3.4}$$

in which A, b are the data, and x is the vector of variables. When $b \neq 0$, the system (3.4) is known as a nonhomogeneous system of linear equations. If x^1, x^2 are two solutions of (3.4), then $Ax^1 = b = Ax^2$, so $A(\alpha_1 x^1 + \alpha_2 x^2) = \alpha_1 Ax^1 + \alpha_2 Ax^2 = (\alpha_1 + \alpha_2)b = b$ if $\alpha_1 + \alpha_2 = 1$; so every affine combination of two solutions of (3.4) is also a solution. Hence the set of solutions of (3.4) is an affine space. Therefore affine spaces play an important role in the study of systems of linear equations.

3.1.4   Convex Combinations, Convex Hull

Let x^1, x^2 be two distinct points in $\mathbb{R}^n$. The linear combination $\alpha_1 x^1 + \alpha_2 x^2$ is said to be a **convex combination of** x^1, x^2 if the coefficients satisfy the conditions $\alpha_1 + \alpha_2 = 1$ and $\alpha_1, \alpha_2 \geqq 0$. The set of all convex combinations of x^1, x^2, $\{\alpha_1 x^1 + \alpha_2 x^2 : \alpha_1, \alpha_2$ take all real values satisfying $\alpha_1 + \alpha_2 = 1$ and $\alpha_1, \alpha_2 \geqq 0\}$, called *the convex hull of* $\{x^1, x^2\}$, is the **line segment** joining x^1, x^2. It is the portion of the straight line joining x^1, x^2 between these two points. See Figure 3.4. In the same way, the linear combination $\alpha_1 x^1 + \ldots \alpha_k x^k$ is said to be a **convex combination of** $x^1, \ldots, x^k$ if the coefficients satisfy the conditions $\alpha_1 + \ldots + \alpha_k = 1$ and $\alpha_1, \ldots, \alpha_k \geqq 0$. The set of all convex combinations of $x^1, \ldots, x^k$; i.e. the set $\{\alpha_1 x^1 + \ldots \alpha_k x^k : \alpha_1, \ldots, \alpha_k$ take all possible real values subject to $\alpha_1 + \ldots + \alpha_k = 1$ and $\alpha_1, \ldots, \alpha_k \geqq 0\}$ is known as the **convex hull of** $\{x^1, \ldots, x^k\}$. The convex hull of a set of points is a subset of its affine hull. See Figure 3.5.

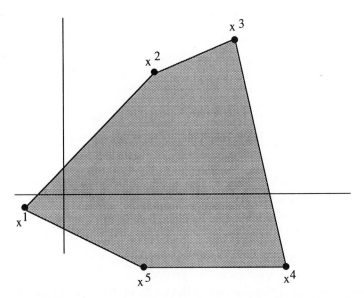

Figure 3.5 Convex hull of $\{x^1, x^2, x^3, x^4, x^5\}$ in $\mathbb{R}^2$ is the shaded polytope.

How to Check Whether a Given Point is in the Convex Hull of a Set of Points

To check whether b in $\mathbb{R}^n$ is a convex combination of the set of vectors $\Gamma = \{A_1, \ldots, A_k\}$ in $\mathbb{R}^n$, we need to solve the following system of $n + 1$ equations in k nonnegative variables $\alpha_1, \ldots, \alpha_k$.

$$
\begin{aligned}
\alpha_1 A_1 + \ldots + \alpha_k A_k &= b \\
\alpha_1 + \ldots + \alpha_k &= 1 \\
\alpha_1, \quad \ldots, \alpha_k &\geqq 0
\end{aligned}
\tag{3.5}
$$

Since the variables in (3.5) are constrained to be nonnegative, it may not be possible to solve it by methods for solving systems of linear equations only. If k and n are small, we can try to solve (3.5) by some trial and error method. A method that is guaranteed to handle systems of the form (3.5) is Phase I of the simplex method for LP discussed in Chapter 7. If $(\bar{\alpha}_1, \ldots, \bar{\alpha}_k)$ is a solution for (3.5), then b is a convex combination of $A_1, \ldots, A_k$ with coefficients $\bar{\alpha}_1, \ldots, \bar{\alpha}_k$. If (3.5) has no feasible solution, b is not in the convex hull of Γ.

As an example, to check whether the vector b^3 in Tableau 3.1 is in the convex hull of $\{A_1, A_2, A_3\}$ there, we need to solve the system of equations in nonnegative variables given below on the left. It can be verified that this system has no solution, so b^3 is not in the convex hull of $\{A_1, A_2, A_3\}$ of Tableau 3.1.

α_1	α_2	α_3	b^3
1	0	1	2
0	1	-1	-4
0	0	1	3
0	0	-1	-3
1	1	1	1

$$\alpha_1, \alpha_2, \alpha_3 \geqq 0$$

α_1	α_2	α_3	b^4
1	0	1	$3/4$
0	1	-1	$-1/4$
0	0	1	$1/2$
0	0	-1	$-1/2$
1	1	1	1

$$\alpha_1, \alpha_2, \alpha_3 \geqq 0$$

To check whether b^4 in Tableau 3.1 is a convex combination of $\{A_1, A_2, A_3\}$ there, we need to solve the system given above on the right. This system has a feasible solution $(\alpha_1, \alpha_2, \alpha_3) = (1/4, 1/4, 1/2)$. So, $b^4 = (1/4)A_1 + (1/4)A_2 + (1/2)A_3$. This expresses b^4 as a convex combination of $\{A_1, A_2, A_3\}$.

Importance of Convex Hulls

Convex hulls are very important in the study of linear constraints involving some linear inequalities. Consider the system

$$
Ax \geqq b
\tag{3.6}
$$

in which A, b are the data, and x is the vector of variables. If x^1, x^2 are two feasible solutions of (3.6), then $Ax^1 \geqq b, Ax^2 \geqq b$, so $A(\alpha_1 x^1 + \alpha_2 x^2) = \alpha_1 Ax^1 + \alpha_2 Ax^2 \geqq (\alpha_1 + \alpha_2)b = b$ if $\alpha_1, \alpha_2 \geqq 0$, and $\alpha_1 + \alpha_2 = 1$; so every convex combination of two feasible solutions of (3.6) is also a feasible solution. Therefore convex hulls play an important role in the study of systems of linear constraints involving some inequalities.

3.1.5   Nonnegative Linear Combinations, Pos Cones

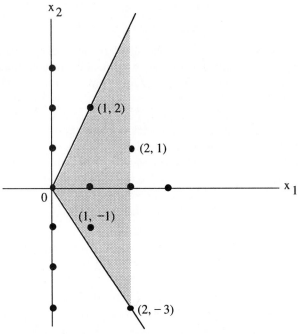

Figure 3.6 The cone Pos{(1, 2), (2, 1), (1, −1),
(2, −3)} in $\mathbb{R}^2$. It contains the rays of each of
these points, and is the smallest angle at the origin
containing all these rays.

The linear combination $\alpha_1 x^1 + \ldots \alpha_k x^k$ is said to be a **nonnegative combination of** $x^1, \ldots, x^k$ if the coefficients $\alpha_1, \ldots, \alpha_k$ are all $\geqq 0$. The set of all
nonnegative combinations of $x^1, \ldots, x^k$; i.e., the set $\{\alpha_1 x^1 + \ldots \alpha_k x^k : \alpha_1, \ldots, \alpha_k$
are all $\geqq 0\}$ is known as the **nonnegative hull** or the **pos cone of** $\{x^1, \ldots, x^k\}$
and is denoted by $\text{Pos}\{x^1, \ldots, x^k\}$. It is a cone; see Figures 3.6, 3.7.

To check whether a vector $b \in \mathbb{R}^n$ is in the cone $\text{Pos}\{A_1, \ldots, A_k\}$ in $\mathbb{R}^n$, we
need to solve the following system of n linear equations in k nonnegative variables
$\alpha_1, \ldots, \alpha_k$.

$$
\begin{aligned}
\alpha_1 A_1 + \ldots + \alpha_k A_k &= b \\
\alpha_1, \quad \ldots, \alpha_k &\geqq 0
\end{aligned}
\tag{3.7}
$$

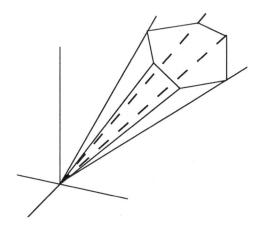

Figure 3.7 A pos cone in $\mathbb{R}^3$.

Since the variables in (3.7) are constrained to be nonnegative, it may not be possible to solve it by methods for solving systems of linear equations only. It can be solved by applying the simplex algorithm discussed in Chapter 7 on a Phase I formulation of the problem.

Consider a homogeneous system of linear inequalities, $Ax \overset{\geq}{=} 0$. If x^1, x^2 are two feasible solutions of this system, then $A(\alpha_1 x^1 + \alpha_2 x^2) = \alpha_1 Ax^1 + \alpha_2 Ax^2 \overset{\geq}{=} 0$ if $\alpha_1, \alpha_2 \overset{\geq}{=} 0$. So, a nonnegative combination of feasible solutions of this system is also a feasible solution. Thus the set of feasible solutions of this system is a cone. Hence pos cones play an important role in the study of homogeneous systems of linear inequalities.

We summarize the facts about various types of combinations of vectors in a set $\Gamma = \{x^1, \ldots, x^k\} \subset \mathbb{R}^n$ in the table given at the top of the next page.

3.1.6 Hyperplanes, Half-Spaces, Convex Sets

A nontrivial linear equation in n variables $x_1, \ldots, x_n$ is one of the form $a_1 x_1 + \ldots + a_n x_n = b$ where at least one of the coefficients $a_1, \ldots, a_n$ is nonzero (the RHS constant b may be either 0 or nonzero; it does not matter). A **hyperplane** in $\mathbb{R}^n$ is the set of all points x in $\mathbb{R}^n$ satisfying a single nontrivial linear equation. In the special case of $\mathbb{R}^2$, a hyperplane is a straight line and vice versa. See Figure 3.8.

Each hyperplane partitions the space into two **half-spaces**. The hyperplane represented by $a_1 x_1 + \ldots + a_n x_n = b$ in $\mathbb{R}^n$ partitions $\mathbb{R}^n$ into one half-space

Type	Definition	Name for set of combinations
Linear combination of $\boldsymbol{\Gamma}$	Any point of form $\alpha_1 x^1 + \ldots + \alpha_k x^k$, where $\alpha_1, \ldots, \alpha_k$ are all real	Linear hull or linear span of $\boldsymbol{\Gamma}$; it is a subspace
Affine combination of $\boldsymbol{\Gamma}$	Any point of form $\alpha_1 x^1 + \ldots + \alpha_k x^k$, where $\alpha_1, \ldots, \alpha_k$ are all real satisfying $\alpha_1 + \ldots + \alpha_k = 1$	Affine hull or affine space of $\boldsymbol{\Gamma}$; it is a translate of a subspace
Convex combination of $\boldsymbol{\Gamma}$	Any point of form $\alpha_1 x^1 + \ldots + \alpha_k x^k$, where $\alpha_1, \ldots, \alpha_k$ are all real satisfying $\alpha_1 + \ldots + \alpha_k = 1$ and $\alpha_1 \geqq 0, \ldots, \alpha_k \geqq 0$	Convex hull of $\boldsymbol{\Gamma}$
Nonnegative combination of $\boldsymbol{\Gamma}$	Any point of form $\alpha_1 x^1 + \ldots + \alpha_k x^k$, where $\alpha_1, \ldots, \alpha_k$ are all $\geqq 0$	Nonnegative hull or Pos cone of $\boldsymbol{\Gamma}$; or Pos($\boldsymbol{\Gamma}$)

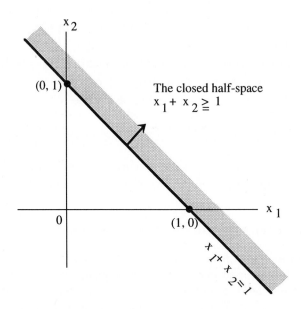

Figure 3.8 The hyperplane in $\mathbb{R}^2$ represented by $x_1 + x_2 = 1$ is the thick straight line. The inequality $x_1 + x_2 \geqq 1$ represents the half-space of all points in $\mathbb{R}^2$ lying above this hyperplane (i.e., on the dotted side).

represented by the inequality $a_1 x_1 + \ldots + a_n x_n \geqq b$, and the other represented by the inequality $a_1 x_1 + \ldots + a_n x_n \leqq b$. The intersection of these two half-spaces is the hyperplane itself. Thus a half-space is the set of all points satisfying a single nontrivial linear inequality. The half-space (or **closed half-space** to be mathematically precise) represented by $a_1 x_1 + \ldots + a_n x_n \geqq$ (or $\leqq$) b is the set of all points in $\mathbb{R}^n$ lying on one side of the hyperplane represented by $a_1 x_1 + \ldots + a_n x_n = b$. In Figure 3.8, the set of all points above the hyperplane represented by $x_1 + x_2 = 1$ is the half-space represented by $x_1 + x_2 \geqq 1$; and the set of all points below this hyperplane is the half-space represented by $x_1 + x_2 \leqq 1$ (notice that the origin 0 satisfies this inequality strictly, and is in this half-space).

The single equation

$$a_1 x_1 + \ldots + a_n x_n = b \qquad (3.8)$$

is equivalent to the pair of inequality constraints

$$
\begin{aligned}
a_1 x_1 + \ldots + a_n x_n &\geqq b \\
a_1 x_1 + \ldots + a_n x_n &\leqq b
\end{aligned}
\qquad (3.9)
$$

This explains the fact that a hyperplane in $\mathbb{R}^n$ is the intersection of the two closed half-spaces it defines.

In a general LP there may be equality constraints, inequality constraints, and sign restrictions on the variables. Each sign restriction is an inequality constraint. Each equality constraint is equivalent to a pair of inequality constraints. Hence all the constraints in an LP may be expressed as linear inequality constraints. The set of points satisfying a linear inequality constraint is a closed half-space. Every feasible solution for the LP must satisfy all the constraints, and hence it must be in each of the corresponding half-spaces. *Hence, for any LP, the set of feasible solutions is the intersection of a finite number of half-spaces.*

Convex Sets

A subset $\mathbf{K} \subset \mathbb{R}^n$ is said to be a **convex set** if every convex combination of any pair of points in $\mathbf{K}$ is also in $\mathbf{K}$; i.e., if x^1, x^2 are both in $\mathbf{K}$, then $\alpha x^1 + (1 - \alpha) x^2$ is in $\mathbf{K}$ for all $0 \leqq \alpha \leqq 1$. Hence a set is convex if for every pair of points in it, it contains the entire line segment joining those points. Examples of convex sets in the plane are given in Figure 3.9. Figure 3.10 has examples of nonconvex sets. The intersection of a finite number of half-spaces is a special type of convex set called a **convex polyhedron** or a **convex polyhedral set**. See Figure 3.11. So, from earlier discussion, the set of feasible solutions of an LP is a convex polyhedron. A bounded convex polyhedron is called a **convex polytope** .

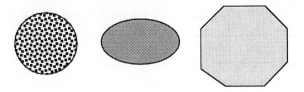

Figure 3.9 Convex sets. (Left) All points inside or on the circle. (Middle) All points inside or on the ellipse. (Right) All points inside or on the convex polygon which is a convex polytope since it is bounded.

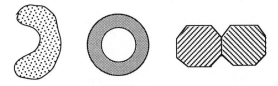

Figure 3.10 Nonconvex sets. (Left) All points inside or on the cashew nut. (Middle) All points on or between two circles. (Right) The union of two convex polygons.

Verify that if we add the additional constraint $x_1 + x_2 \overset{\le}{=} 3$ to those already in Figure 3.11, the resulting set of feasible solutions becomes bounded, and hence a convex polytope.

Given a convex polyhedron represented by a system of linear constraints, it is easy to check whether a given point is contained in it. If the point satisfies all the constraints, it is in the polyhedron. If it violates at least one of the constraints, it is not in the polyhedron. For example, the point $(1,1)^T$ satisfies all the constraints given in Figure 3.11, so it is in the convex polyhedron in this figure (verify this in Figure 3.11). Also verify that $(1,3)^T$ is not in the convex polyhedron in Figure 3.11, since it violates the constraint $x_1 - x_2 \overset{\ge}{=} -1$.

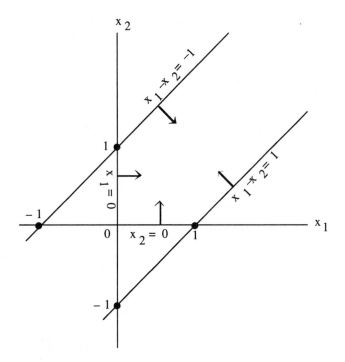

Figure 3.11 The convex polyhedron that is the intersection of 4 half-spaces represented by $x_1 \geqq 0$, $x_2 \geqq 0$, $x_1 - x_2 \geqq -1$, $x_1 - x_2 \leqq 1$ in $\mathbb{R}^2$. Since this set is unbounded it is a convex polyhedron and not a polytope.

Exercises

3.2 Column vectors called $A_{.1}$ to $A_{.7}$ in $\mathbb{R}^2$ are given in the following tableau.

$A_{.1}$	$A_{.2}$	$A_{.3}$	$A_{.4}$	$A_{.5}$	$A_{.6}$	$A_{.7}$
1	1	−1	−1	0	3	1
1	−2	−1	−4	−2	−1	−4

Answer the following questions: (1) Is $A_{.4}$ on the half-line through $A_{.2}$ parallel to the ray of $A_{.3}$? What about $A_{.5}$? (2) Is $A_{.3}$ on the line of $A_{.1}$? Is it on the ray of $A_{.1}$? Is $A_{.4}$ on the line of $A_{.1}$? (3) Is $A_{.2}$ in the linear hull of $A_{.1}$ and $A_{.5}$? (4) Is $A_{.6}$ in Pos$\{A_{.1}, A_{.2}, A_{.4}, A_{.5}\}$? What about $A_{.3}$? (5) Is $A_{.7}$ an affine combination of $A_{.1}$ and $A_{.2}$? Is it a convex combination? (6) Is $A_{.2}$ a convex combination of $A_{.1}$ and $A_{.7}$? (7) Draw Pos$\{A_{.1}, A_{.2}, A_{.4}, A_{.5}\}$ and verify the answer in (4). (8) Draw the half-line through $A_{.2}$ parallel to the ray of $A_{.3}$ and verify the answer in (1). (9) Draw the convex hull of $\{A_{.1},\ldots, A_{.7}\}$. (10) Is $A_{.1}$ contained in the polyhedron represented by $x_1 \geqq 0$, $x_2 \geqq 0$, $-x_1 + 2x_2 \leqq 2$, $3x_1 - x_2 \leqq 3$? What about $A_{.5}$? Draw this convex polyhedron and verify the answers geometrically. Is this polyhedron a polytope?

3.3 Column vectors called $A_{.1}$ to $A_{.8}$ in $\mathbb{R}^4$ are given in the following tableau.

$A_{.1}$	$A_{.2}$	$A_{.3}$	$A_{.4}$	$A_{.5}$	$A_{.6}$	$A_{.7}$	$A_{.8}$
1	0	0	2	-1	3	$1/2$	2
0	1	0	1	-1	2	0	0
1	-1	2	-1	6	3	$3/2$	4
-1	2	1	-1	2	4	0	-1

Answer the following questions: (1) Is $A_{.4}$ in the linear hull of $\{A_{.1}, A_{.2}, A_{.3}\}$? If so, is it in the convex or affine hull of the same set? What about $A_{.5}$, $A_{.6}$? (2) Is $A_{.7}$ in the convex hull of $\{A_{.1}, A_{.2}, A_{.3}\}$? (3) Is $A_{.8}$ on the straight line joining $A_{.1}$ and $A_{.3}$? Is it on the half-line through $A_{.3}$ parallel to the ray of $A_{.1}$?

3.2 Matrices, Row Operations, and Pivot Steps

A **matrix** is a rectangular array of numbers. Here is a matrix of order $m \times n$. It has m rows and n columns

$$A = \begin{pmatrix} a_{11} & \cdots & a_{1j} & \cdots & a_{1n} \\ \vdots & & \vdots & & \vdots \\ a_{i1} & \cdots & a_{ij} & \cdots & a_{in} \\ \vdots & & \vdots & & \vdots \\ a_{m1} & \cdots & a_{mj} & \cdots & a_{mn} \end{pmatrix}$$

The entry in the ith row and the jth column of A, a_{ij} is known as the (i, j)**th entry in** A. When the order of A is understood, we will also denote A by (a_{ij}). The coefficients of variables in a system of linear equations is a matrix. For example $A = (a_{ij})$ is the matrix of coefficients of the variables in the system

$$
\begin{aligned}
a_{11}x_1 &+ \ldots + a_{1n}x_n = b_1 \\
&\vdots \qquad\qquad \vdots \\
a_{m1}x_1 &+ \ldots + a_{mn}x_n = b_m
\end{aligned}
\tag{3.10}
$$

If α is a real number, αA is the matrix obtained by multiplying each entry in A by α. Thus the (i, j)th entry in αA is αa_{ij}.

Two matrices $A = (a_{ij})$ and $B = (b_{ij})$ can only be added if they are of the same order. In this case $A + B = C$ where C is of the same order as A and B, and the (i,j)th entry in C is $c_{ij} = a_{ij} + b_{ij}$. Thus matrix addition consists of adding the entries in corresponding positions.

The **transpose** of the matrix A, denoted by A^T, is obtained by recording each row vector of A as a column. Thus if A is of order $m \times n$, A^T will be of order $n \times m$.

$$A^T = \begin{pmatrix} a_{11} & \cdots & a_{i1} & \cdots & a_{m1} \\ \vdots & & \vdots & & \vdots \\ a_{1j} & \cdots & a_{ij} & \cdots & a_{mj} \\ \vdots & & \vdots & & \vdots \\ a_{1n} & \cdots & a_{in} & \cdots & a_{mn} \end{pmatrix}$$

A row vector in $\mathbb{R}^n$ can be treated as a matrix of order $1 \times n$. Likewise a column vector in $\mathbb{R}^n$ can be treated as a matrix of order $n \times 1$.

A square matrix is a matrix that has the same number of rows and columns. The matrices

$$D = \begin{pmatrix} d_{11} & \cdots & d_{1m} \\ \vdots & & \vdots \\ d_{m1} & \cdots & d_{mm} \end{pmatrix}, I = \begin{pmatrix} 1 & 0 & 0 \\ 0 & 1 & 0 \\ 0 & 0 & 1 \end{pmatrix}$$

are square matrices. D is said to be of order m. The entries $d_{11}, d_{22}, \ldots, d_{mm}$ are the **diagonal entries** in D. They constitute its **principal diagonal**. All the other entries in D are its **off-diagonal entries**. The matrix I is the unit matrix of order 3. A **unit matrix** or **identity matrix** is a square matrix in which all diagonal entries are equal to 1, and all off-diagonal entries are 0.

If $A = (a_{ij})$ is a matrix of order $m \times n$, we will denote by $A_{i.}$ the ith row vector of A, and by $A_{.j}$ the jth column vector of A. Thus

$$A_{i.} = (a_{i1}, \ldots, a_{in}), \qquad A_{.j} = (a_{1j}, \ldots, a_{mj})^T$$

The rules for multiplication among matrices have evolved out of the study of linear transformations on systems of linear equations. Matrix multiplication is not commutative, i.e., the order in which matrices are multiplied is very important. If A is a matrix of order $m \times n$, and B is a matrix of order $r \times s$, the product AB in this order is defined only if $n = r$. In this case the product AB is a matrix $C = (c_{ij})$ of order $m \times s$, where the (i,j)th entry in C, $c_{ij} = \sum_{t=1}^{n} a_{it} b_{tj}$. Hence the (i,j)th entry in AB is the sum of the products of the entries in the i th row in A, by the corresponding entry in the jth column of B. Therefore matrix multiplication is **row by column multiplication**, the (i,j)th entry in the matrix product AB is the dot product of the ith row vector of A and the jth column vector of B. Here is an example.

$$A = \begin{pmatrix} 1 & -1 & -5 \\ 2 & -3 & -6 \end{pmatrix}, \quad B = \begin{pmatrix} -2 & -4 \\ -6 & -4 \\ 3 & -2 \end{pmatrix}, \quad AB = \begin{pmatrix} -11 & 10 \\ -4 & 16 \end{pmatrix}$$

When the product AB exists, the product BA may not even be defined, and even if it is defined, AB and BA may not be equal and may be of different orders. So, when writing a matrix product one should specify the order of the entries very carefully.

If I is the unit matrix of order n, and A, D are matrices or vectors such that the products IA and DI are defined, then it can be verified that $IA = A$, and $DI = D$. So, in the space of matrices, the unit matrices play the same role as the number one does among real numbers.

If $x = (x_i), y = (y_j)$ are two column vectors in $\mathbb{R}^n$, $x^T y = \sum_{i=1}^n x_i y_i$ is the dot product of the two vectors x and y, it is a real number. However the matrix product yx^T is the matrix $(x_i y_j)$ of order $n \times n$.

If A is a matrix of order $m \times n$, and π, x are vectors, the products πA, Ax exist only if π is a row vector in $\mathbb{R}^m$ and x is a column vector in $\mathbb{R}^n$.

3.2.1   Row Operations on a Matrix or on a Tableau

A **row operation** (or sometimes called an **elementary row operation**) is a very fundamental tool in computations involving matrices. To perform a row operation on a matrix, we put the matrix in a tableau, and identify each row (column) of the matrix by the corresponding row (column) of the tableau. Often each row and each column is also given an identifying name or label to keep track of the changes that occur due to operations carried out.

There are two types of row operations. One is to select a row of the tableau and multiply every element in it by the same nonzero constant. The second is to add a scalar multiple of a row to another. This operation is performed element by element using corresponding elements in the respective rows. The following tableau with an added RHS constants column illustrates how a row operation is performed using two rows, rows 1 and 2. The third row in this tableau is $\alpha(\text{row } 1) + \text{row } 2$, where α is a nonzero scalar. The fourth row in this tableau is $\beta(\text{row } 1)$ where β is a nonzero scalar. This tableau is followed by a numerical example.

Row operations

Row 1	a_{11}	$\ldots$	a_{1j}	$\ldots$	a_{1n}	b_1
Row 2	a_{21}	$\ldots$	a_{2j}	$\ldots$	a_{2n}	b_2
Row 2 + α(Row 1)	$a_{21} + \alpha a_{11}$	$\ldots$	$a_{2j} + \alpha a_{1j}$	$\ldots$	$a_{2n} + \alpha a_{1n}$	$b_2 + \alpha b_1$
β(Row 1)	βa_{11}	$\ldots$	βa_{1j}	$\ldots$	βa_{1n}	βb_1

$A_{1.}$	1	0	3	-5	3
$A_{2.}$	0	1	2	4	1
$A_{2.} - 5A_{1.}$	-5	1	-13	29	-14
$(1/2)A_{1.}$	$1/2$	0	$3/2$	$-5/2$	$3/2$

3.2.2   Pivot Operations

A **pivot operation** or **pivot step** on a matrix, or on a tableau representing a matrix, or on a system of equality constraints in detached coefficient form; consists of a series of row operations; and is specified by designating a row as the **pivot row** and a column as the **pivot column**, the element that lies in both the pivot row and the pivot column is called the **pivot element** for the pivot step. The pivot element will always be nonzero (it cannot be zero, because the pivot operation requires division by the pivot element). The pivot operation consists of a series of elementary row operations to transform the pivot column into one containing an entry of $+1$ in the pivot row, and an entry of 0 in all the other rows. It requires the following.

(i) For each row i other than the pivot row, if the element in this row and the pivot column is nonzero, subtract a suitable multiple of the pivot row from row i so that the entry in the pivot column and row i becomes 0.

(ii) Divide the pivot row by the pivot element.

Designating the pivot row and pivot column for a pivot operation (pivot element is boxed) in a tableau

a_{11}	a_{12}	$\cdots$	a_{1s}	$\cdots$	a_{1n}	b_1	
$\vdots$	$\vdots$		$\vdots$		$\vdots$	$\vdots$	
a_{r1}	a_{r2}	$\cdots$	$\boxed{a_{rs}}$	$\cdots$	a_{rn}	b_r	Pivot row
$\vdots$	$\vdots$		$\vdots$		$\vdots$	$\vdots$	
a_{m1}	a_{m2}	$\cdots$	a_{ms}	$\cdots$	a_{mn}	b_m	
			Pivot col.				

Tableau after the pivot step

$\bar{a}_{11}$	$\bar{a}_{12}$	$\cdots$	0	$\cdots$	$\bar{a}_{1n}$	b_1
$\vdots$	$\vdots$		$\vdots$		$\vdots$	$\vdots$
$\bar{a}_{r1}$	$\bar{a}_{r2}$	$\cdots$	1	$\cdots$	$\bar{a}_{rn}$	$\bar{b}_r$
$\vdots$	$\vdots$		$\vdots$		$\vdots$	$\vdots$
$\bar{a}_{m1}$	$\bar{a}_{m2}$	$\cdots$	0	$\cdots$	$\bar{a}_{mn}$	$\bar{b}_m$

Here for $j = 1$ to n, $\bar{a}_{rj} = a_{rj}/a_{rs}$, $\bar{a}_{ij} = a_{ij} - a_{is}a_{rj}/a_{rs}$ for $i \neq r$; $\bar{b}_r = b_r/a_{rs}$, $\bar{b}_i = b_i - a_{is}b_r/a_{rs}$ for $i \neq r$. Here is a numerical illustration.

			Pivot col.				
Tableau	1	0	1	−1	2	1	
before	6	4	2	−4	0	6	Pivot row
pivot	3	−2	0	1	−1	4	
step	1	0	−1	3	4	10	
Tableau	−2	−2	0	1	2	−2	
after	3	2	1	−2	0	3	
pivot	3	−2	0	1	−1	4	
step	4	2	0	1	4	13	

In a tableau representing a system of linear equations in detached coefficient tabular form, each row of the tableau represents an equation in the system. If a pivot operation is carried out on this tableau, it leads to the detached coefficient tableau representation of an equivalent system of equations.

An algorithm based on pivot steps is called a **pivotal algorithm.**

Why Row Operations Involving Two Rows Are Not Valid on a System of Inequalities

When row operations are performed on a system of linear equations, or on the detached coefficient tableau representing it, we get another system or tableau that is equivalent to the original one. For example, consider the tableau on the left given below representing a system of two equations in two unknowns. We perform the row operation [5(row 1) + row 2] on this tableau, leading to the one on the right, representing a system that is equivalent to the original one. It can be verified that both systems have the unique solution $(x_1, x_2)^T = (6, 15)^T$.

x_1	x_2	
1	0	6
−2	1	3

x_1	x_2	
1	0	6
3	1	33

Similar row operations are not valid on a system of linear inequalities, or on the tableau representing them. For example, consider the simple system of linear inequalities on the left given below; its set of feasible solutions is the nonnegative orthant in $\mathbb{R}^2$ part of which is dotted in Figure 3.12 left. Performing the row operation [2(row 1) + row 2] leads to the system on the right, the set of feasible solutions of which is partly shaded in Figure 3.12 right; different from the set of feasible solutions of the system on the left.

x_1	x_2	
1	0	$\geqq 0$
0	1	$\geqq 0$

x_1	x_2	
1	0	$\geqq 0$
2	1	$\geqq 0$

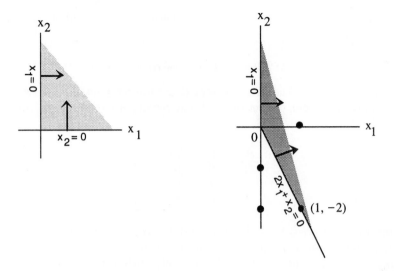

Figure 3.12 (Left) Nonnegative orthant, the set of feasible solutions of $x_1 \geqq 0, x_2 \geqq 0$. (Right) The set of feasible solutions of $x_1 \geqq 0, 2x_1 + x_2 \geqq 0$.

So row operations are not valid on a system of linear inequalities, because these operations change the set of feasible solutions.

Row operations, or pivot operations are perfectly valid on a system of equations, since they lead to an equivalent system with the same set of solutions.

When a system of linear inequalities is converted into a system of equations by introducing the appropriate slack variables, it is again perfectly valid to perform any row operations on these equations. For example consider the system: $x_1 + x_2 \geqq 1$, $x_1 - x_2 \leqq 0$. After introducing the slack variables s_1, s_2, this system is the one on the left given below, with each row in the tableau representing an equation. We perform the row operation [2(row 1) + row 2] on these equations, leading to the system on the right which is equivalent to the original system.

x_1	x_2	s_1	s_2	
1	1	−1	0	1
1	−1	0	1	0
	$s_1, s_2 \geqq 0$			

x_1	x_2	s_1	s_2	
1	0	−1	0	1
3	1	−2	1	2
	$s_1, s_2 \geqq 0$			

3.3 Linear Dependence, Independence, Rank, Bases

A set of vectors $\{A_1, \ldots, A_k\}$ in $\mathbb{R}^n$ is said to be **linearly dependent** iff it is possible to express the zero vector as a linear combination of vectors in the set, with at least one nonzero coefficient; i.e., iff there exist real numbers $\alpha_1, \ldots, \alpha_k$ not all zero such that

$$0 = \alpha_1 A_1 + \ldots + \alpha_k A_k \qquad (3.11)$$

In this case, (3.11) is known as a **linear dependence relation** for $\{A_1, \ldots, A_k\}$. As an example, consider the set of 4 column vectors $A_{.1}$ to $A_{.4}$ in $\mathbb{R}^3$ given in the following tableau.

$A_{.1}$	$A_{.2}$	$A_{.3}$	$A_{.4}$
1	0	0	5
0	1	0	−6
0	0	1	0

Since $0 = -5A_{.1} + 6A_{.2} + A_{.4}$, the set $\{A_{.1}$ to $A_{.4}\}$ is linearly dependent, and this equation is a linear dependence relation for this set.

A set of vectors in $\mathbb{R}^n$ is said to be **linearly independent** if it is not linearly dependent, i.e., if the only linear combination of vectors in the set which is equal to zero has all zero coefficients. For example, the set of vectors $\{A_{.1}, A_{.2}, A_{.3}\}$ given above is linearly independent since the only solution to the system of equations

α_1	α_2	α_3	
1	0	0	0
0	1	0	0
0	0	1	0

is $(\alpha_1, \alpha_2, \alpha_3) = (0, 0, 0)$.

The empty set containing no vectors is linearly independent, since there is no way we can express the 0-vector as a linear combination of vectors from the empty set.

The set $\{a\}$ containing a single vector in $\mathbb{R}^n$ is linearly dependent if $a = 0$, linearly independent otherwise.

3.3.1   An Algorithm for Testing Linear Independence

Let $\mathbf{\Gamma} = \{A_1, \ldots, A_k\}$ be a set of vectors in $\mathbb{R}^n$, either all row vectors, or all column vectors, with $a_{t1}, \ldots, a_{tn}$ as the coordinates of A_t for $t = 1$ to k. We will now describe an algorithm based on row operations, to check whether this set $\mathbf{\Gamma}$ is linearly independent.

First write each vector in the set as a row vector in a tableau (even though all of them may be column vectors in the data given), and enter its name in a column on the left hand side of the tableau. The entries in the tableau will be altered in the course of the algorithm. However, as only row operations are carried out, at every step each row in the current tableau will be a nonzero linear combination of vectors in the original set $\mathbf{\Gamma}$. In each row, this expression will be indicated in the column set up on the left-hand side of the tableau. A convenient method for keeping track of these expressions is to store them in detached coefficient form by storing the coefficients of $A_1, \ldots, A_k$ in the expression for each row. In this form, the original tableau is recorded as below.

Original Tableau

A_1	A_2	$\ldots$	A_k			
1	0	$\ldots$	0	a_{11}	$\ldots\ldots$	a_{n1}
0	1	$\ldots$	0	a_{12}	$\ldots\ldots$	a_{n2}
$\vdots$	$\vdots$		$\vdots$	$\vdots$		$\vdots$
0	0	$\ldots$	1	a_{1k}	$\ldots\ldots$	a_{nk}

The matrix on the left-hand side of this original tableau is the unit matrix of order k. This indicates that the first row in the original tableau is $1A_1 + 0A_2 + \ldots + 0A_k = A_1$; the second row in the same way is A_2, etc.

Whenever row operations are carried out, they are also performed on the left-hand portion of the tableau. At any stage, the ith row in the left-hand portion of the current tableau contains the coefficients of $A_1, A_2, \ldots, A_k$ in the expression for the vector in this row on the right-hand portion; i.e., if the entries in this row are

β_{i1}	β_{i2}	$\ldots$	β_{ik}	$\bar{a}_{1i}$	$\ldots\ldots$	$\bar{a}_{ni}$

then the vector $(\bar{a}_{1i}, \ldots, \bar{a}_{ni})$ is equal to $\sum_{j=1}^{k} \beta_{ij} A_j$. So, the matrix on the left-hand portion of the tableau helps us to remember what each row in the tableau is as a linear combination of vectors in the original set Γ, hence this matrix is often called the **memory matrix**.

The algorithm is initiated with the original tableau as recorded above, and consists of the following steps.

Step 1: Select the first row of the original tableau to examine and go to Step 2.

Step 2   Examining the selected row in the current tableau: If all entries are zero in the right-hand portion of the current tableau in the row being examined, the left-hand portion in this row provides the coefficients in an expression for the zero vector as a nonzero linear combination of vectors in the original set Γ. This expression provides a linear dependence relation for Γ, and Γ is linearly dependent, terminate.

If there are some nonzero entries in the right-hand portion of the row being examined in the current tableau, select one of them as the pivot element. The row being examined is the pivot row, and the column in the current tableau containing the pivot element is the pivot column. Go to Step 3.

Step 3   Row operations: If the pivot row is the last row, conclude that the original set Γ is linearly independent, and terminate.

Suppose the pivot row is not the last row. For each row in the current tableau below the pivot row, subtract a suitable multiple of the pivot row from it to convert the element in this row in the pivot column to 0. When this has been completed for each row below the pivot row, we have the new tableau. In this new tableau select the row just below the pivot row to examine, and go back to Step 2.

This algorithm terminates after examining a subset of rows, and if Γ is linearly dependent, it provides a linear dependence relation for it at termination. The row operations carried out in Step 3 only convert every element in the pivot column below the pivot row to 0, so they require much less work than a full pivot step.

As an example consider the set $\Gamma = \{A_1, A_2, A_3, A_4\}$ of column vectors given below.

A_1	A_2	A_3	A_4
6	1	1	14
2	0	1	3
4	1	1	10
0	1	1	2
2	0	1	3

To apply the algorithm for checking whether Γ is linearly independent, the original tableau is given in Tableau 3.2.

Tableau 3.2

— Coefficient of				Pivot col.					
A_1	A_2	A_3	A_4						
1	0	0	0	6	2	4	0	2	Pivot row
0	1	0	0	1	0	1	1	0	
0	0	1	0	1	1	1	1	1	
0	0	0	1	14	3	10	2	3	

Tableau 3.3

Coefficient of				Pivot col.					
A_1	A_2	A_3	A_4						
1	0	0	0	6	2	4	0	2	
0	1	0	0	1	0	1	1	0	Pivot row
$-1/2$	0	1	0	-2	0	-1	1	0	
$-3/2$	0	0	1	5	0	4	2	0	

We examine the first row in Tableau 3.2 and select the nonzero entry of 2 in column 2 in it as the pivot element. Since row 2 already has a 0 in the pivot column, we don't have to do anything in row 2. Row 3 has a 1 in the pivot column, to convert that element to 0 we perform the operation [Row 3 − (1/2)Row 1]. Row 4 has a 3 in the pivot column, to convert that element to 0, we perform the row operation [Row 4 − (3/2)Row 1]. This leads to the next tableau, Tableau 3.3

In Tableau 3.3, row 2 is examined and the element of 1 in it in column 1 is selected as the pivot element. We need to convert the element of −2 in row 3, and 5 in row 4 in the pivot column to zero. This needs row operations [Row 3 + 2(Row 2)], [Row 4 −5(Row 2)] in Tableau 3.3, these lead to Tableau 3.4.

Tableau 3.4

Coefficient of						Pivot			
A_1	A_2	A_3	A_4			col.			
1	0	0	0	6	2	4	0	2	
0	1	0	0	1	0	1	1	0	
−1/2	2	1	0	0	0	1	3	0	Pivot row
−3/2	−5	0	1	0	0	−1	−3	0	

We need to examine row 3 in Tableau 3.4, and we select the element of 3 in column 4 in it as the pivot element. We need to convert the element of −3 in row 4 in the pivot column to 0, this needs the row operation [Row 4 + Row 3]. This leads to the next Tableau 3.5.

Tableau 3.5

Coefficient of									
A_1	A_2	A_3	A_4						
1	0	0	0	6	2	4	0	2	
0	1	0	0	1	0	1	1	0	
−1/2	2	1	0	0	0	1	3	0	
−2	−3	1	1	0	0	0	0	0	To examine

In Tableau 3.5, row 4 is to be examined, and there is no nonzero entry in it in the right hand portion of the tableau. So, the original set Γ is linearly dependent. From the memory portion in the row being examined we get the linear dependence relation for Γ: $-2A_1 - 3A_2 + A_3 + A_4 = 0$.

In this example, it is clear that the set of column vectors $\{A_1, A_2, A_3\}$ is linearly independent.

3.3.2 A Maximal Linearly Independent Subset of a Set of Vectors

From the definition it is clear that any subset of a linearly independent set of vectors is linearly independent. So, a linearly independent set of vectors remains linearly independent when vectors are deleted from it, but it may become linearly dependent if new vectors are included in it. For example, the linearly independent set

$\{(1, 0, 0)^T, (0, 1, 0)^T, (0, 0, 1)^T\}$ becomes linearly dependent when any other vector from $\mathbb{R}^3$ is included in it.

Let $\mathbf{\Gamma} \subset \mathbb{R}^n$ be a set of vectors. A subset $\mathbf{E} \subset \mathbf{\Gamma}$ is said to be a **maximal linearly independent subset of $\mathbf{\Gamma}$** if $\mathbf{E}$ is linearly independent, and either $\mathbf{E} = \mathbf{\Gamma}$ or for every vector a in $\mathbf{\Gamma}$ not in $\mathbf{E}$, $\mathbf{E} \cup \{a\}$ is linearly dependent.

The definition itself suggests a simple augmentation scheme for finding a maximal linearly independent subset of $\mathbf{\Gamma}$. If 0 is the only vector in $\mathbf{\Gamma}$, the only maximal linearly independent subset of $\mathbf{\Gamma}$ is the empty set, terminate. If $\mathbf{\Gamma}$ contains some nonzero vectors, start the scheme with a singleton subset consisting of exactly one nonzero vector from $\mathbf{\Gamma}$. Maintaining the subset linearly independent always, augment it by one new element from $\mathbf{\Gamma}$ at a time, making sure that linear independence holds, until no more new vectors from $\mathbf{\Gamma}$ can be included; at which stage the subset becomes a maximal linearly independent subset of $\mathbf{\Gamma}$. But this scheme is computationally expensive. We will discuss an efficient method for finding a maximal linearly independent subset based on row operations in Section 3.3.5.

Given a set of vectors $\mathbf{\Gamma} \subset \mathbb{R}^n$, there may be several maximal linearly independent subsets of it. A fundamental theorem in linear algebra states that all maximal linearly independent subsets of $\mathbf{\Gamma}$ have the same cardinality, and this number is known as the **rank** of the set $\mathbf{\Gamma}$.

3.3.3   A Minimal Spanning Subset of a Set of Vectors

Let $\mathbf{\Gamma} \subset \mathbb{R}^n$ and let $\mathbf{F} \subset \mathbf{\Gamma}$ satisfy the property that every vector in $\mathbf{\Gamma}$ can be expressed as a linear combination of vectors in $\mathbf{F}$. Then $\mathbf{F}$ is said to be a **spanning subset of $\mathbf{\Gamma}$** (or to have the property of spanning $\mathbf{\Gamma}$). As an example let $\mathbf{\Gamma}_1 = \{(1, 0, 0), (-2, 0, 0), (1, 1, 0), (-3, -4, 0), (1, 1, -1)\}$. The subset $\mathbf{F}_1 = \{(1, 0, 0), (1, 1, 0), (1, 1, -1)\}$ is a spanning subset of $\mathbf{\Gamma}_1$, but $\mathbf{F}_2 = \{(1, 0, 0), (1, 1, 0), (-3, -4, 0)\}$ is not a spanning subset of $\mathbf{\Gamma}_1$ since the vector $(1, 1, -1)$ in $\mathbf{\Gamma}_1$ cannot be expressed as a linear combination of vectors in $\mathbf{F}_2$.

Let $\mathbf{\Gamma} \subset \mathbb{R}^n$ be a set of vectors. From the definition it is clear that a spanning subset of $\mathbf{\Gamma}$ retains its spanning property when more vectors are included in it, but may loose the spanning property if vectors are deleted from it. A subset $\mathbf{F} \subset \mathbf{\Gamma}$ is said to be a **minimal spanning subset of $\mathbf{\Gamma}$** if $\mathbf{F}$ is a spanning subset of $\mathbf{\Gamma}$, but no proper subset of $\mathbf{F}$ has the spanning property for $\mathbf{\Gamma}$.

The definition itself suggests a simple deletion scheme for finding a minimal spanning subset of $\mathbf{\Gamma}$. The scheme maintains a subset of $\mathbf{\Gamma}$, initially that subset is $\mathbf{\Gamma}$ itself. In each step, delete one vector from the subset making sure that the remaining subset always has the spanning property for $\mathbf{\Gamma}$ (for this it is enough if the vector to be deleted is in the linear hull of the remaining elements in the subset at that stage). When you reach a stage where no more vectors can be deleted from the subset without losing the spanning property for $\mathbf{\Gamma}$ (this will happen if the subset becomes linearly independent), you have a minimal spanning subset of $\mathbf{\Gamma}$. Again this scheme is computationally expensive. We will discuss an efficient method for finding a minimal spanning subset based on row operations in Section 3.3.5.

3.3.4 The Rank of a Set of Vectors, Bases

A theorem in linear algebra states that every maximal linearly independent subset of a set $\Gamma \subset \mathbb{R}^n$, is a minimal spanning subset for Γ and vice versa. Hence all minimal spanning subsets of Γ have the same cardinality; it is the same as the cardinality of any maximal linearly independent subset of Γ; this number is called the **rank** of the set Γ. It is also the dimension of the linear hull of Γ. Every maximal linearly independent subset of Γ or a minimal spanning subset of Γ is called a **basis** for Γ. Thus a basis for a set of vectors Γ in $\mathbb{R}^n$ is a linearly independent subset satisfying the property that all vectors in the set Γ not in the basis can be represented as linear combinations of vectors in the basis.

When referring to a basis **B** for a set Γ, vectors in **B** are called **basic vectors**, and vectors in Γ not in **B** are called **nonbasic vectors**. In most applications, the basic vectors themselves are arranged in some order and called the **first basic vector**, **second basic vector**, etc. Then the word basis refers to this ordered set with the basic vectors arranged in this order.

Let I be the unit matrix of order n. Its set of column vectors $\{I_{.1}, I_{.2}, \ldots, I_{.n}\}$ is a basis for $\mathbb{R}^n$ called the **unit basis**. For example, $\{(1,0,0)^T, (0,1,0)^T, (0,0,1)^T\}$ is the unit basis for $\mathbb{R}^3$. So, the rank of $\mathbb{R}^n$ is n. Every basis for $\mathbb{R}^n$ always consists of n vectors. And therefore any set of $n+1$ or more vectors from $\mathbb{R}^n$ is always linearly dependent.

By definition, every nonbasic vector can be expressed as a linear combination of basic vectors. We will now show that this expression is always unique. Let the basis **B** $= \{A_1, \ldots, A_r\}$ and let A_{r+1} be a nonbasic vector. Suppose A_{r+1} can be expressed as a linear combination of vectors in **B** in two different ways, say as $\sum_{t=1}^{r} \beta_t A_t$ and as $\sum_{t=1}^{r} \gamma_t A_t$ where $(\beta_1, \ldots, \beta_r) \neq (\gamma_1, \ldots, \gamma_r)$. So, if $\alpha = (\alpha_1, \ldots, \alpha_r) = (\beta_1 - \gamma_1, \ldots, \beta_r - \gamma_r)$ then $\alpha \neq 0$. Since $A_{r+1} = \sum_{t=1}^{r} \beta_t A_t = \sum_{t=1}^{r} \gamma_t A_t$, we have

$$0 = \sum_{t=1}^{r} \beta_t A_t - \sum_{t=1}^{r} \gamma_t A_t = \sum_{t=1}^{r} (\beta_t - \gamma_t) A_t = \sum_{t=1}^{r} \alpha_t A_t \qquad (3.12)$$

and since $\alpha = (\alpha_1, \ldots, \alpha_r) \neq 0$, (3.12) is a linear dependence relation for **B** $= \{A_1, \ldots, A_r\}$, a contradiction since **B** is a basis. So, the expression for the nonbasic vector A_{r+1} as a linear combination of basic vectors must be unique.

The vector of coefficients in the expression of a nonbasic vector as a linear combination of basic vectors, arranged in the same order in which the basic vectors are arranged in the basis, is called **the representation of this nonbasic vector in terms of this basis**. Thus if the basis **B** $= \{A_1, \ldots, A_r\}$ with the vectors arranged in this order, and the nonbasic vector $A_{r+1} = \sum_{t=1}^{r} \beta_t A_t$, then $(\beta_1, \ldots, \beta_r)$ is the representation of A_{r+1} in terms of the basis **B**. As an example, $\mathbf{F}_1$ is a basis for the set Γ_1 discussed in Section 3.3.3. The nonbasic vector $(-3, -4, 0) = (1, 0, 0) - 4(1, 1, 0)$, and hence the representation of $(-3, -4, 0)$ in terms of the basis $\mathbf{F}_1$ for Γ_1 is $(1, -4, 0)$.

3.3.5 An Algorithm to Find the Rank, a Basis, and the Representation of Each Nonbasic Vector

Let $\Gamma = \{A_1, \ldots, A_k\} \subset \mathbb{R}^n$ be the given set of vectors. The algorithm discussed in Section 3.3.1 to check whether Γ is linearly independent, will find a basis for Γ and a representation for each nonbasic vector, if all the vectors are examined instead of terminating when a linear dependence relation for Γ is found. For the sake of completeness we state the algorithm below.

Step 1: Enter the vectors $A_1, \ldots, A_k$ in Γ as row vectors in a tableau in that order, and set up the memory matrix (initially unit matrix of order k) with its columns labeled $A_1, \ldots, A_k$ in that order.

Step 2: Select the first row of the original tableau to examine and go to Step 3.

Step 3   Examining the selected row in the current tableau: Suppose all entries are zero in the right-hand portion of the current tableau in the row being examined. If this is the last row in the tableau go to Step 5. Otherwise move to the next row and start this Step 3 to examine it.

If there are some nonzero entries in the right-hand portion of the row being examined in the current tableau, select one of them as the pivot element. The row being examined is the pivot row, and the column in the current tableau containing the pivot element is the pivot column. Go to Step 4.

Step 4   Row operations: If the pivot row is the last row in the current tableau, go to Step 5.

Suppose the pivot row is not the last row. For each row in the current tableau below the pivot row, subtract a suitable multiple of the pivot row from it to convert the element in this row in the pivot column to 0. When this has been completed for each row below the pivot row, we have the new tableau. In this new tableau select the row just below the pivot row to examine, and go back to Step 3.

Step 5: Identify the rows in the final tableau that have nonzero entries in the right-hand portion. Suppose these are rows $i_1, \ldots, i_r$. Then $\mathbf{B} = \{A_{i_1}, \ldots, A_{i_r}\}$ is a basis for the original set Γ, and the rank of Γ is r.

For each t such that row t has all zero entries in the right-hand portion in the final tableau, let $\beta_{t1}, \ldots, \beta_{tk}$ be the coefficients in the memory part (the left-hand part) in this row. β_{tt} will be 1, A_t is a nonbasic vector outside the basis $\mathbf{B}$, and since $\sum_{j=1}^{k} \beta_{tj} A_{tj} = 0$, we have $A_t = \sum(-\beta_{tj} A_j : \text{over } j = 1 \text{ to } k, j \neq t)$. Hence after removing the zero entries corresponding to nonbasic vectors from $(-\beta_{t1}, \ldots, -\beta_{t,t-1}, 0, -\beta_{t,t+1}, \ldots, -\beta_{tk})$, it becomes the representation of the nonbasic vector A_t in terms of the basis $\mathbf{B}$. Terminate.

So, this algorithm terminates after examining every row. As an example consider the set $\Gamma = \{A_1, \text{ to } A_5\}$ of column vectors in $\mathbb{R}^6$ given below.

Tableau 3.6

A_1	A_2	A_3	A_4	A_5
1	2	0	2	1
2	4	0	4	-1
-1	-2	3	1	-1
1	2	1	3	1
0	0	2	2	-2
-2	-4	2	-2	1

We enter each of the vectors A_1 to A_5 as a row in a tableau, set up the memory matrix and apply the algorithm. The various tableaus obtained are given below. The nonzero rows in the right-hand portion of the final tableau are rows 1, 3, 5. Hence $\{A_1, A_3, A_5\}$ is a basis for the original set Γ in this example, and its rank is 3. From rows 2 and 4 in the final tableau we have $-2A_1 + A_2 = 0$ and $-2A_1 - A_3 + A_4 = 0$. So, $A_2 = 2A_1$ and $A_4 = 2A_1 + A_3$. So, the representations of the nonbasic vectors A_2, A_4 in terms of the basis $\{A_1, A_3, A_5\}$ are $(2, 0, 0)$ and $(2, 1, 0)$ respectively.

	Coefficient of										
A_1	A_2	A_3	A_4	A_5	PC						
1	0	0	0	0	1	2	-1	1	0	-2	PR
0	1	0	0	0	2	4	-2	2	0	-4	
0	0	1	0	0	0	0	3	1	2	2	
0	0	0	1	0	2	4	1	3	2	-2	
0	0	0	0	1	1	-1	-1	1	-2	1	
								PC			
1	0	0	0	0	1	2	-1	1	0	-2	
-2	1	0	0	0	0	0	0	0	0	0	
0	0	1	0	0	0	0	3	1	2	2	PR
-2	0	0	1	0	0	0	3	1	2	2	
-1	0	0	0	1	0	-3	0	0	-2	3	
1	0	0	0	0	1	2	-1	1	0	-2	
-2	1	0	0	0	0	0	0	0	0	0	
0	0	1	0	0	0	0	3	1	2	2	
-2	0	-1	1	0	0	0	0	0	0	0	
-1	0	0	0	1	0	-3	0	0	-2	3	To examine

PR = pivot row, PC = pivot col.

Which basic vectors can a nonbasic vector replace?

Let $\Gamma = \{A_1, \ldots, A_k\} \subset \mathbb{R}^n$, and suppose $\mathbf{B} = \{A_1, \ldots, A_r\}$ is a basis for Γ. Consider the nonbasic vector A_{r+1}. Can we get a new basic vector for Γ by replacing a basic vector in $\mathbf{B}$ with A_{r+1}? If so, which basic vectors can A_{r+1} replace? The answer to these questions is completely contained in the representation of A_{r+1} in terms of **B**. Let $(\alpha_1, \ldots, \alpha_r)$ be this representation. Then for $j = 1$ to r, A_{r+1} can replace the basic vector A_j in **B** to lead to a new basis for Γ iff $\alpha_j \neq 0$.

As an example, let $\Gamma = \{A_1 \text{ to } A_5\}$, the set of column vectors in $\mathbb{R}^6$ in Tableau 3.6. We have seen above that $\mathbf{B} = \{A_1, A_3, A_5\}$ is a basis for this set Γ, and that the representation of A_4 in terms of **B** is $(2, 1, 0)$. So, replacing A_1 or A_3 in **B** by A_4 leads to other bases for Γ. Hence $\{A_4, A_3, A_5\}, \{A_1, A_4, A_5\}$ are both bases for Γ. However, since the coefficient of A_5 in the representation for A_4 in terms of **B** is 0, $\{A_1, A_3, A_4\}$ is not a basis for Γ, in fact this set is linearly dependent.

3.3.6   The Rank of a Matrix

Let $A = (a_{ij})$ be a matrix of order $m \times n$. The column rank of A is the rank of the set of column vectors of A. The row rank of A is the rank of the set of row vectors of A. A standard theorem in matrix algebra states that the column and row ranks of a matrix are always equal, and their common value is called the **rank** of the matrix A. So, if rank of A is r, then $r \overset{\leq}{=} m, r \overset{\leq}{=} n$. The rank of the matrix A can be found by applying the algorithm discussed in Section 3.3.5 to find the rank of the set of row (or column) vectors of A.

3.3.7   The Inverse of a Nonsingular Square Matrix

A square matrix that can be transformed into the unit matrix by rearranging its row vectors is known as a **permutation matrix**. The matrix in (2.4) is a permutation matrix of order 5. When the rows of this matrix are rearranged in the order (row 2, row 4, row 5, row 3, row 1) it becomes the unit matrix of order 5.

If P is a permutation matrix of order n, and A is any matrix of order $n \times r$, the product PA is a matrix that can be obtained from A by rearranging its rows. As an example let P be the permutation matrix of order 5 in (2.4), and A the matrix of order 5×3 given below; verify that PA is a matrix that is obtained by permuting the rows of A.

$$A = \begin{pmatrix} 1 & -1 & 3 \\ 7 & 6 & -4 \\ 5 & -8 & 2 \\ -4 & 0 & 1 \\ -9 & 1 & 2 \end{pmatrix}, \quad PA = \begin{pmatrix} -9 & 1 & 2 \\ 1 & -1 & 3 \\ -4 & 0 & 1 \\ 7 & 6 & -4 \\ 5 & -8 & 2 \end{pmatrix}$$

This is the reason for calling P a permutation matrix. In the same way if P is a permutation matrix of order n and D is a matrix of order $r \times n$, the matrix DP can be obtained from D by permuting its columns.

A square matrix of order n is said to be **nonsingular** if its rank is n, **singular** if its rank is $\overset{\leq}{=} n - 1$.

Inverse matrices are only defined for square matrices. If A is a square matrix of order n, its inverse, if it exists, is another square matrix D of order n satisfying $DA = AD = I$, the unit matrix of order n. This definition is analogous to the definition of the inverse of a nonzero real number γ, as the number δ satisfying $\gamma\delta = 1$, since the unit matrix plays the role of the number 1 among square matrices. If it exists, the inverse of a square matrix A is denoted by A^{-1}. A^{-1} exists iff A is nonsingular.

The inverse of a unit matrix I is itself. The inverse of any permutation matrix P is its transpose P^T.

Let B be a given square matrix of order n, and b a given column vector in $\mathbb{R}^n$. Here we present a pivotal algorithm for checking whether B is nonsingular, and obtain its inverse if it is. This algorithm can also be used to compute the representation of b in terms of the set of column vectors of B if B is nonsingular.

Let I be the unit matrix of order n. The algorithm starts with the following initial tableau.

Initial tableau

| I | B | b |

We will refer to columns 1 to n in the tableau that contain the unit matrix in the initial tableau as the I-columns; and columns $n + 1$ to $2n$ that contain the matrix B in the initial tableau as the B-columns. We now describe the algorithm.

Step 1 Pick the first row in the initial tableau to examine and go to Step 2.

Step 2 **Examining the selected row in the current tableau** If the row being examined in the current tableau consists of 0 entries in all the B-columns, the matrix B is singular and hence B^{-1} does not exist, terminate. In this case the entries in the row vector being examined in the I-columns, are the coefficients of a linear dependence relation for the set of row vectors of B.

If the row being examined in the current tableau contains some nonzero entries among the B-columns, select one of these nonzero elements as the pivot element. In the current tableau, the row being examined is the pivot row, and the column containing the pivot element is the pivot column. Perform the pivot step in the current tableau with this pivot row and pivot column. Go to Step 3 with the new tableau obtained.

Step 3 If the row examined was not the last row in the tableau, select the next row in the current tableau to examine and go to Step 2.

If all the rows have been examined, the pivot operations performed so far would have transformed the matrix contained among the B-columns into a permutation matrix in the current tableau. Let this matrix be P. Rearrange the rows of the current tableau so that P becomes the unit matrix. This operation is equivalent to multiplying every column in the current tableau on the left by P^T. The square matrix contained among the I-columns in the final tableau is B^{-1}.

The last column in the final tableau is called the **updated column of b wrt**
the basis B; it is the representation of the original b in terms of the columns
of B. So, if the entries in the final tableau in this column are $\bar{b}_1, \ldots, \bar{b}_n$, then
$b = \sum_{j=1}^{n} \bar{b}_j B_{.j}$, and $\bar{b} = (\bar{b}_1, \ldots, \bar{b}_n)^T = B^{-1}b$.

As an example, consider the following matrix B and the column vectors b, d.

$$B = \begin{pmatrix} 0 & -2 & -1 \\ 1 & 0 & -1 \\ 1 & 1 & 0 \end{pmatrix}, \quad b = \begin{pmatrix} 1 \\ 1 \\ -2 \end{pmatrix}, \quad d = \begin{pmatrix} 2 \\ 0 \\ 1 \end{pmatrix} \tag{3.13}$$

Suppose it is required to compute B^{-1} and the representations of b, d in terms
of the columns of B. We apply the algorithm discussed above. The pivot element
in each step is enclosed in a box.

I-columns			B-columns			b	d
1	0	0	0	$\boxed{-2}$	−1	1	2
0	1	0	1	0	−1	1	0
0	0	1	1	1	0	−2	1
−1/2	0	0	0	1	1/2	−1/2	−1
0	1	0	$\boxed{1}$	0	−1	1	0
1/2	0	1	1	0	−1/2	−3/2	2
−1/2	0	0	0	1	1/2	−1/2	−1
0	1	0	1	0	−1	1	0
1/2	−1	1	0	0	$\boxed{1/2}$	−5/2	2
−1	1	−1	0	1	0	2	−3
1	−1	2	1	0	0	−4	4
1	−2	2	0	0	1	−5	4
1	−1	2	1	0	0	−4	4
−1	1	−1	0	1	0	2	−3
1	−2	2	0	0	1	−5	4

Hence

$$B^{-1} = \begin{pmatrix} 1 & -1 & 2 \\ -1 & 1 & -1 \\ 1 & -2 & 2 \end{pmatrix} \tag{3.14}$$

It can be verified that $BB^{-1} = B^{-1}B = I$, the unit matrix of order 3. Also, we
see that the representation of b is $\bar{b} = (-4, 2, -5)^T$, i.e., $b = -4B_{.1} + 2B_{.2} - 5B_{.3}$;
verify this and that $\bar{b} = B^{-1}b$. Similarly the representation of d is $\bar{d} = (4, -3, 4)^T = B^{-1}d$.

As another example, suppose we want to find the inverse of the following matrix
$\hat{B}$. The algorithm is applied below, boxing the pivot element in each step.

$$\hat{B} = \begin{pmatrix} 1 & 0 & -2 \\ -1 & 1 & 1 \\ 0 & 1 & -1 \end{pmatrix}$$

I-columns			B-columns		
1	0	0	$\boxed{1}$	0	-2
0	1	0	-1	1	1
0	0	1	0	1	-1
1	0	0	1	0	-2
1	1	0	0	$\boxed{1}$	-1
0	0	1	0	1	-1
1	0	0	1	0	-2
1	1	0	0	1	-1
-1	-1	1	0	0	0

Since the last row has all zero entries among the B-columns, we conclude that $\hat{B}$ is singular and hence has no inverse. From the I-part of the last row we obtain the linear dependence relation for the set of row vectors of $\hat{B}$, $-\hat{B}_{1.} - \hat{B}_{2.} + \hat{B}_{3.} = 0$.

3.3.8 How to Update the Inverse of a Matrix When One of Its Column Vectors Changes

Let B be a nonsingular square matrix of order n, whose inverse B^{-1} is known. Suppose we want to find the inverse of another matrix, $\tilde{B}$, which differs from B in just one column. Here we show that if it exists, $\tilde{B}^{-1}$ can be obtained by performing a single pivot step.

Suppose $\tilde{B}$ is obtained by replacing $B_{.j}$, the jth column of B, by the column vector $d \in \mathbb{R}^n$. Find the representation, $\bar{d} = (\bar{d}_1, \ldots, \bar{d}_n)^T = B^{-1}d$ of d in terms of the columns of B. $\bar{d}$ is known as the **updated column** corresponding to d wrt the basis B. From Section 3.3.5, the matrix $\tilde{B}$ obtained by replacing the column $B_{.j}$ in B by d is nonsingular iff $\bar{d}_j \neq 0$. So, if $\bar{d}_j = 0$, $\tilde{B}$ is singular and $\tilde{B}^{-1}$ does not exist. If $\bar{d}_j \neq 0$, set up the following tableau.

	$\bar{d}_1$
	$\vdots$
B^{-1}	$\boxed{\bar{d}_j}$ Pivot row
	$\vdots$
	$\bar{d}_n$
	Pivot col.

In this tableau perform a pivot step with $\bar{d}$ as the pivot column, jth row as the pivot row, and the element $\bar{d}_j$ in $\bar{d}$ as the pivot element. This pivot step transforms the tableau into the following ($I_{.j}$ is the jth column vector of the unit matrix I of order n), with the matrix on the left hand side becoming $\tilde{B}^{-1}$.

$$
\begin{array}{c|c}
\tilde{B}^{-1} & I_{.j} \\
\end{array}
$$

As an example, let B, d be the matrix and column vector given in (3.13), with B^{-1} given in (3.14). The updated column corresponding to d wrt the basis B is $\bar{d} = B^{-1}d = (4, -3, 4)^T$. Let $\tilde{B}$ be the matrix obtained by replacing $B_{.2}$ in B by d, it is given below. Since $\bar{d}_2 = -3 \neq 0$, $\tilde{B}$ is nonsingular. To obtain $\tilde{B}^{-1}$, we set up the tableau as given below and perform the pivot step.

$$
\tilde{B} = \begin{pmatrix} 0 & 2 & -1 \\ 1 & 0 & -1 \\ 1 & 1 & 0 \end{pmatrix}
$$

			Pivot col.	
1	-1	2	4	
-1	1	-1	$\boxed{-3}$	Pivot row
1	-2	2	4	
$-1/3$	$1/3$	$2/3$	0	
$1/3$	$-1/3$	$1/3$	1	
$-1/3$	$-2/3$	$2/3$	0	

So

$$
\tilde{B}^{-1} = \begin{pmatrix} -1/3 & 1/3 & 2/3 \\ 1/3 & -1/3 & 1/3 \\ -1/3 & -2/3 & 2/3 \end{pmatrix}
$$

It can be verified that $\tilde{B}\tilde{B}^{-1} = \tilde{B}^{-1}\tilde{B} = I$.

3.3.9   Inverses of Special Structured Matrices in LP

Let I_n be the unit matrix of order n; c and d be row vectors in $\mathbb{R}^n$; and 0 the zero column vector in $\mathbb{R}^n$. Then the matrices of order $(n + 1) \times (n + 1)$, and $(n + 2) \times (n + 2)$ given below on the left appear at the beginning of applying the simplex method on an LP, and their inverses are given.

$$
\begin{pmatrix} I_n & \vdots & 0 \\ \cdots & & \\ c & & 1 \end{pmatrix}^{-1} = \begin{pmatrix} I_n & \vdots & 0 \\ \cdots & & \\ -c & & 1 \end{pmatrix}
$$

$$
\begin{pmatrix}
I_n & \vdots & 0 & \vdots & 0 \\
 & & \cdots & & \\
c & & 1 & & 0 \\
d & & 0 & & 1
\end{pmatrix}^{-1}
=
\begin{pmatrix}
I_n & \vdots & 0 & \vdots & 0 \\
 & & \cdots & & \\
-c & & 1 & & 0 \\
-d & & 0 & & 1
\end{pmatrix}
$$

For example, let $n = 2$, $c = (-2, 7)$, $d = (1, 1)$. Then

$$
\begin{pmatrix}
1 & 0 & 0 \\
0 & 1 & 0 \\
-2 & 7 & 1
\end{pmatrix}^{-1}
=
\begin{pmatrix}
1 & 0 & 0 \\
0 & 1 & 0 \\
2 & -7 & 1
\end{pmatrix},
\begin{pmatrix}
1 & 0 & 0 & 0 \\
0 & 1 & 0 & 0 \\
-2 & 7 & 1 & 0 \\
1 & 1 & 0 & 1
\end{pmatrix}^{-1}
=
\begin{pmatrix}
1 & 0 & 0 & 0 \\
0 & 1 & 0 & 0 \\
2 & -7 & 1 & 0 \\
-1 & -1 & 0 & 1
\end{pmatrix}
$$

3.4 Systems of Simultaneous Linear Equations

Consider the system of m linear equations in n variables, (3.10), represented in matrix form as

$$ Ax = b \tag{3.15} $$

in which the coefficient matrix $A = (a_{ij})$ of order $m \times n$, and the RHS constants vector $b = (b_i) \in \mathbb{R}^n$ are the data, and $x = (x_1, \ldots, x_n)^T$ is the vector of variables. The ith equation in this system is $A_i.x = b_i$, hence this system can be written in an equivalent manner as: $A_i.x = b_i$, $i = 1$ to m. If $\bar{x} = (\bar{x}_1, \ldots, \bar{x}_n)$ is a solution of this system, then

$$ A\bar{x} = \sum_{j=1}^{n} A_{.j}\bar{x}_j = b \tag{3.16} $$

So, any solution x of (3.15) is the vector of coefficients in an expression for b as a linear combination of column vectors of the matrix A. Thus (3.15) has a solution iff the RHS constants vector b is in the linear hull of the set of column vectors of A. Equivalently, (3.15) has no solution iff it is impossible to express b as a linear combination of column vectors of A. Because of this property, the space of the column vectors of A plays a major role in the study of systems like (3.15).

It is convenient to represent the system (3.10) or (3.15) in the form of a detached coefficient tableau as given below.

Tableau 3.7: Detached coefficient representation

x_1	$\cdots$	x_j	$\cdots$	x_n	
a_{11}	$\cdots$	a_{1j}	$\cdots$	a_{1n}	b_1
$\vdots$		$\vdots$		$\vdots$	$\vdots$
a_{i1}	$\cdots$	a_{ij}	$\cdots$	a_{in}	b_i
$\vdots$		$\vdots$		$\vdots$	$\vdots$
a_{m1}	$\cdots$	a_{mj}	$\cdots$	a_{mn}	b_m

Each row of this tableau corresponds to a constraint and vice versa. The last column in the tableau, $(b_1, \ldots, b_m)^T$ is the RHS constants column. The column $(a_{1j}, \ldots, a_{ij}, \ldots, a_{mj})^T$ in this tableau, called the **column of** x_j, is the coefficient vector of the variable x_j in the various constraints in the system. So each column in the tableau other than the RHS constants column, corresponds to a variable. The equality symbol is normally omitted in tabular representation, and it is understood that each row represents an equality constraint.

Let $(A\vdots b)$ denote the $m \times (n+1)$ matrix obtained by including b as an additional $(n + 1)$th column to A. We now present some results on the conditions for the existence of solutions and uniqueness.

Result 3.1   *The system (3.15) has a solution iff rank$(A\vdots b)$ = rank(A). If rank$(A\vdots b)$ = 1 + rank(A), (3.15) has no solution.*

PROOF   Let **E** denote a maximal linearly independent subset of column vectors of A. So, rank$(A) = |\mathbf{E}|$, and every column vector in A can be expressed as a linear combination of the vectors in **E**.

Suppose (3.15) has no solution. By the above discussion, this holds iff b cannot be expressed as a linear combination of columns of A; i.e., iff $\mathbf{E} \cup \{b\}$ is a linearly independent set; which holds iff $\mathbf{E} \cup \{b\}$ is a maximal linearly independent subset of columns of $(A\vdots b)$; i.e., iff rank$(A\vdots b)$ = 1 + rank(A).

Suppose (3.15) has a solution. This holds iff b can be expressed as a linear combination of columns of A; i.e., iff b can be expressed as a linear combination of columns in the set **E**; i.e., iff **E** is also a maximal linearly independent subset of columns of $(A\vdots b)$; or iff rank$(A\vdots b)$ = rank(A). ∎

Result 3.2   *If (3.15) has a solution, that solution is unique iff the set of column vectors of A is linearly independent, i.e., rank$(A) = n$.*

PROOF   Suppose $\bar{x}$ is a solution for (3.15). So, $A\bar{x} = b$. If the set of column vectors of A is linearly dependent, there exists a vector $y \in \mathbb{R}^n, y \neq 0$ such that $Ay = 0$. From $A\bar{x} = b, Ay = 0$ we conclude that $A(\bar{x} + \theta y) = b$, i.e., $(\bar{x} + \theta y)$ is also a solution of (3.15) for all θ real. And since $y \neq 0$, $\bar{x} + \theta y$ determines the straight line through $\bar{x}$ parallel to the line of y as θ takes all real values, and every point on this line is a solution of (3.15). So, if (3.15) has a solution, and the set of column vectors of A is linearly dependent, (3.15) has many distinct solutions.

Now, suppose (3.15) has two distinct solutions, $\bar{x}$ and $\hat{x}$. So, $\bar{x} - \hat{x} \neq 0$. Also, $A\bar{x} = b$ and $A\hat{x} = b$. So, $A(\bar{x} - \hat{x}) = 0$, this is a linear dependence relation for the set of column vectors of A. So, if the solution to (3.15) is not unique, the set of column vectors of A must be linearly dependent. ∎

As an example consider the following system.

x_1	x_2	x_3	
1	−1	0	−1
1	1	2	25
1	−1	0	−1

$\bar{x} = (5, 6, 7)^T$ can be verified to be a solution to this system. The set of column vectors of the coefficient matrix A is linearly dependent, in fact if $y = (1, 1, -1)^T$, then $Ay = 0$. $\bar{x} + \theta y = (5 + \theta, 6 + \theta, 7 - \theta)^T$ can be verified to be a solution of the system for all real values of θ.

Redundant Equality Constraints

An equality constraint in (3.15) can be considered to be a redundant constraint and deleted from the system without changing the set of solutions, if it can be obtained as a linear combination of the remaining constraints. As an example, consider the following system

$$
\begin{array}{rrrrcl}
x_1 & -x_2 & +2x_3 & +x_4 & = & 1 \\
-x_1 & & +x_3 & -2x_4 & = & 2 \\
-x_1 & -2x_2 & +7x_3 & -4x_4 & = & 8
\end{array}
$$

Each equation in this system can be obtained as a linear combination of the others, for example, the third equation equals two times the first plus three times the second. So, any one of the equations in this system can be considered as a redundant equation and deleted without any change in the set of solutions. The same process can be repeated again in the remaining system if it has redundant equations.

If (3.15) has a solution, and $\text{rank}(A) = r < m$, exactly $m - r$ redundant equations can be deleted from (3.15), one at a time. The Gaussian elimination method for solving linear equations discussed later identifies redundant equations in this case, that can be deleted one after the other. If (3.15) has a solution, after the deletion of redundant equality constraints, it will be reduced into a system in which the number of constraints is equal to the rank of A.

Elimination Interpretation of a Pivot Step

A standard technique for solving systems of linear equations is based on **elimination**. This technique selects an equation, and a variable, x_1 say, with a nonzero coefficient in that equation. It uses the equation to find an expression for that variable x_1 in terms of the other variables. By substituting that expression for x_1 in all the other equations, it **eliminates** x_1 from them. After this elimination the remaining system has one equation and one variable less, and it is handled in the same way. As an example, consider the following system.

Tableau 3.8

x_1	x_2	x_3	x_4	
-1	0	2	0	5
1	-1	1	2	10
0	1	-1	3	25
1	0	1	-1	20

The equation corresponding to the second row in this tableau is $x_1 - x_2 + x_3 + 2x_4 = 10$. To eliminate the variable x_1 using this equation, we have $x_1 = 10 + x_2 - x_3 - 2x_4$ from it. Substituting this expression for x_1 in the other equations and rearranging terms, we are lead to the following reduced system.

$$
\begin{array}{rrrcl}
-x_2 & +3x_3 & +2x_4 & = & 15 \\
x_2 & -x_3 & +3x_4 & = & 25 \\
x_2 & & -3x_4 & = & 10
\end{array}
\qquad (3.17)
$$

This reduced system has only 3 equations in 3 unknowns. Once a solution to this system is obtained, we can substitute this solution in $10 + x_2 - x_3 - 2x_4$ and get the corresponding value for x_1. This reduced system can again be reduced further by eliminating a variable from it using the same technique, and the process can be continued until we reach a system with only one equation which can be solved trivially. This is the main strategy used by methods based on elimination.

In Tableau 3.8 if we perform a pivot step with row 2 (the row used above to get an expression for x_1 in terms of the other variables) as the pivot row, and the column of x_1 as the pivot column, we are lead to the following equivalent system.

x_1	x_2	x_3	x_4	
0	-1	3	2	15
1	-1	1	2	10
0	1	-1	3	25
0	1	0	-3	10

We verify that rows 1, 3, 4 in this tableau define exactly the reduced system in (3.17). This shows that a pivot step on the tableau representing a system of linear equations is exactly the same as eliminating the variable associated with the pivot column, using the equation corresponding to the pivot row, from all other equations. The second row in this tableau (the one which was the pivot row for the pivot step) can be thought of as providing an equation for finding the corresponding value for the variable x_1 (the one associated with the pivot column in that pivot step) once the values of the remaining variables are determined. For this reason the variable x_1 is known as the **dependent variable** in this row. In linear programming literature it is commonly referred to as the **basic variable** in this row.

Special Systems

Consider the system of linear equations (3.10) or (3.15) with $m = n$, i.e., the number of equations in the system is the same as the number of unknowns. In this case A is a square matrix of order n.

If $A = I$, the unit matrix of order n, the system is very special; it is $Ix = x = b$. So, the unique solution is $x = b$.

If A is nonsingular and A^{-1} is known, then (3.15) can be converted into a system in which the coefficient matrix is I, by multiplying both sides of it on the left by the known A^{-1}. This leads to $A^{-1}Ax = A^{-1}b$, or $Ix = x = A^{-1}b$. So, in this case the unique solution of the system is $x = A^{-1}b$. As an example consider the following system

x_1	x_2	x_3	
0	-2	-1	1
1	0	-1	1
1	1	0	-2

The coefficient matrix, RHS constants vector are B, b from (3.13). We already computed B^{-1} in Section 3.3.7, it is given in (3.14). So, the unique solution of this system is $x = B^{-1}b = (-4, 2, -5)^T$.

When A is a nonsingular square matrix, one method for solving the system $Ax = b$, is to compute A^{-1} and then use the formula $x = A^{-1}b$ to get the solution. However, this method is computationally expensive, the Gaussian elimination method discussed later on is more efficient if A^{-1} is not known already.

Another special system that can be solved very efficiently by a direct method called **back substitution method** is one associated with a coefficient matrix which is upper-triangular. A square matrix is said to be **upper-triangular** if all its diagonal entries are nonzero; but all the elements below the diagonal are zero. A **lower-triangular** matrix can be defined in a similar way. In the following matrices, D_1 is upper-triangular, but D_2 is neither upper- nor lower-triangular.

$$D_1 = \begin{pmatrix} 2 & 0 & -2 & 1 \\ 0 & 1 & 2 & 0 \\ 0 & 0 & 3 & 6 \\ 0 & 0 & 0 & -2 \end{pmatrix}, \quad D_2 = \begin{pmatrix} 1 & 0 & 0 & 0 \\ 0 & 1 & -1 & 0 \\ 0 & -1 & 1 & 0 \\ 0 & 0 & 0 & 1 \end{pmatrix}$$

If $A = (a_{ij})$ is an upper-triangular matrix of order n, the system $Ax = b$ can be solved very efficiently by the following procedure. Since $a_{n1}, \ldots, a_{n,n-1}$ are all 0, and $a_{nn} \neq 0$ by the upper-triangularity property, from the last equation of the system we have $x_n = b_n/a_{nn}$. Again, since $a_{n-1,1}, \ldots, a_{n-1,n-2}$ are all 0 and $a_{n-1,n-1} \neq 0$, by substituting the value of x_n computed above into the $(n-1)$th equation in the system, the value of x_{n-1} can be computed directly (it is $(b_{n-1} - a_{n-1,n}x_n)/a_{n-1,n-1}$). In this manner, the values of the variables $x_n, x_{n-1}, \ldots, x_1$ can be computed one by one in this order. At the stage that the values of $x_n, \ldots, x_{r+1}$ are already computed, the value of x_r is computed from the rth equation as $x_r = (b_r - a_{r,r+1}x_{r+1} - \ldots - a_{rn}x_n)/a_{rr}$ by substituting the known values of $x_n, \ldots, x_{r+1}$; for $r = n-1, \ldots, 1$. For this reason this procedure is known as the **back substitution method.**

As an example, consider the following system of equations.

x_1	x_2	x_3	x_4	
2	0	-2	1	-1
0	1	2	0	9
0	0	3	6	30
0	0	0	-2	-6

The coefficient matrix of this system is the upper triangular matrix D_1 given above. From the last equation we have $x_4 = 3$. Substituting $x_4 = 3$ in the third equation leads to $x_3 = 4$. Substituting $x_4 = 3, x_3 = 4$ in the second equation leads to $x_2 = 1$. Substituting $x_4 = 3, x_3 = 4, x_2 = 1$ in the first equation leads to $x_1 = 2$. Hence the solution of this system is $(x_1, x_2, x_3, x_4)^T = (2, 1, 4, 3)^T$, found by back substitution.

The Gaussian Elimination Method

This method, based on row operations, is an efficient method for solving systems of linear equations. The number of equations in the system may be $<, =$, or $>$ the number of variables. If redundant equality constraints exist, they are identified one by one and deleted from the system as they are identified. Whenever row operations are performed with a row as the pivot row, the method selects a basic variable in that row, and performs row operations with its column as the pivot column. These row operations convert all the entries in the pivot column, below the pivot row, into zeros. If the system has no solution, at some stage the fundamental inconsistent equation "$0 = 1$" will appear as a linear combination of the equations in the original system, and the method will terminate then. If the system has a solution, it will be converted at the end into one in which the number of constraints is equal to the rank of the original coefficient matrix (this will clearly be $\leqq$ the number of variables); and in the final tableau the matrix of columns of the basic variables in their proper order will be an upper-triangular matrix. From the final tableau, one can find a solution for the system by back substitution after fixing all the nonbasic variables at 0; or if there are multiple solutions, a formula for the general solution of the system expressing each basic variable as a function of the nonbasic variables which can assume arbitrary values. We now describe the method for solving the general system (3.10) or (3.15).

Step 1 Express the system in detached coefficient tableau form as in Tableau 3.7. Select the the first row in this tableau to examine and go to Step 2.

Step 2   **Examining the selected row in the current tableau:** Suppose all entries are zero in the left-hand portion of the current tableau in the row being examined. If the updated RHS constant entry in this row is nonzero, this row represents an inconsistent equation, $0 =$ a nonzero number, so the original system has no solution, terminate. If the updated RHS constant entry in this row is zero, this row represents a redundant equation in the system, delete it from the tableau and go to Step 4 if this was the last row, or move to the next row and start this Step 2 to examine it.

If there are some nonzero entries in the left-hand portion of the row being examined in the current tableau, select one of them as the pivot element. The row being examined is the pivot row, and the column in the current tableau containing the pivot element is the pivot column. Record the variable associated with the pivot column as the basic variable in this pivot row. Go to Step 3.

Step 3   **Row operations:** If the pivot row is the last row in the current tableau, go to Step 4.

Suppose the pivot row is not the last row. For each row in the current tableau below the pivot row, subtract a suitable multiple of the pivot row from it to convert the element in this row in the pivot column to 0. When this has been completed for each row below the pivot row, we have the new tableau. In this new tableau select the row just below the pivot row to examine, and go back to Step 2.

Step 4   We reach this step after all the rows are examined if the system is not inconsistent. The number of rows remaining in the final tableau is equal to the rank of the original coefficient matrix; number these remaining rows serially, and number the basic variables in them also serially. When the columns of the basic variables in the final tableau are rearranged in proper serial order, they form an upper-triangular matrix. Variables which are not basic are called **nonbasic** (or **independent variables**).

If there are no nonbasic variables, the system has a unique solution; this can be found from the final tableau by back substitution by evaluating the basic variables one by one in reverse serial order.

If there are some nonbasic variables, the system has alternate solutions. In each row transfer the terms involving the nonbasic variables to the RHS constants column with a negative sign. Now you can give arbitrary values (for example 0) to each nonbasic variable, and then compute the corresponding values of the basic variables in a solution for the system in reverse serial order by back substitution. Or, you can obtain expressions for basic variables in terms of the nonbasic variables, again by back substitution; this will lead to an expression for the general solution of the system.

As an example, consider the original system in the first tableau given below. Applying the Gaussian elimination method on this problem leads to the following tableaus. In each tableau, PR indicates the pivot row, and PC the pivot column.

Basic variable	x_1	x_2	x_3	x_4	x_5	x_6	b	
		PC						
	1	-2	1	1	1	-1	-4	PR
	2	2	0	1	-1	-1	8	
	2	-4	-2	0	-2	-2	-7	
	0	0	-1	4	1	1	5	
				PC				
x_2	1	-2	1	1	1	-1	-4	
	3	0	1	2	0	-2	4	PR
	0	0	-4	-2	-4	0	1	
	0	0	-1	4	1	1	5	
	PC							
x_2	1	-2	1	1	1	-1	-4	
x_4	3	0	1	2	0	-2	4	
	3	0	-3	0	-4	-2	5	PR
	-6	0	-3	0	1	5	-3	
					PC			
x_2	1	-2	1	1	1	-1	-4	
x_4	3	0	1	2	0	-2	4	
x_1	3	0	-3	0	-4	-2	5	
x_5	0	0	-9	0	-7	1	7	PR

Hence the system of equations from the final tableau, after transforming the nonbasic variable terms to the right-hand side, and rearranging the basic variables in serial order, is

x_2	x_4	x_1	x_5				
-2	1	1	1	-4	$-($	x_3	$-x_6)$
0	2	3	0	4	$-($	x_3	$-2x_6)$
0	0	3	-4	5	$-($	$-3x_3$	$-2x_6)$
0	0	0	-7	7	$-($	$-9x_3$	$+x_6)$

Giving both the nonbasic variables x_3, x_6 the value zero, leads to the solution $(x_1, \ldots, x_6)^T = (1/3, 29/12, 0, 3/2, -1, 0)^T$. The general solution to the system in which the basic variables are expressed as functions of the independent variables x_3, x_6 can be obtained by solving the above system by back substitution.

An example illustrating the inconsistency termination is given below. The original system is the one in the top tableau. In the final tableau at the bottom, row 3 is being examined, all entries in the left-hand portion are 0, and the updated RHS constant is $-2 \neq 0$ (this row represents the constraint "$0 = -2$" which is inconsistent). So, this system has no solution.

Basic variable	x_1	x_2	x_3	x_4	x_5	b	
	PC						
	1	−2	2	−1	1	−8	PR
	−1	0	4	−7	7	16	
	0	−2	6	−8	8	6	
	PC						
x_1	1	−2	2	−1	1	−8	
	0	−2	6	−8	8	8	PR
	0	−2	6	−8	8	6	
x_1	1	−2	2	−1	1	−8	
x_2	0	−2	6	−8	8	8	
	0	0	0	0	0	−2	To examine

A third example illustrating the discovery and deletion of a redundant equality constraint is given below.

Basic variable	x_1	x_2	x_3	x_4	x_5	x_6	b	
	PC							
	0	1	−1	2	−2	1	−7	PR
	−1	−1	−2	3	−1	−1	4	
	−1	0	−3	5	−3	0	−3	
	1	0	2	−1	2	−2	8	
	PC							
x_2	0	1	−1	2	−2	1	−7	
	−1	0	−3	5	−3	0	−3	PR
	−1	0	−3	5	−3	0	−3	
	1	0	2	−1	2	−2	8	
		PC						
x_2	0	1	−1	2	−2	1	−7	
x_1	−1	0	−3	5	−3	0	−3	
	0	0	0	0	0	0	0	
	0	0	−1	4	−1	−2	5	PR
x_2	0	1	−1	2	−2	1	−7	
x_1	−1	0	−3	5	−3	0	−3	
x_3	0	0	−1	4	−1	−2	5	

In the third tableau, the third equation became a redundant equation, and was deleted. Applying back substitution on the final tableau we find the general solution for this system to be

$$\begin{aligned}
x_1 &= 18 - (\quad 7x_4 \quad\quad\quad -6x_6) \\
x_2 &= -12 - (-2x_4 \quad -x_5 \quad +3x_6) \\
x_3 &= -5 - (-4x_4 \quad +x_5 \quad +2x_6) \\
x_4, x_5, x_6 & \quad \text{arbitrary}
\end{aligned}$$

In particular, giving the value 0 to all the nonbasic variables x_4, x_5, x_6 leads to the solution $x = (18, -12, -5, 0, 0, 0)^T$.

When inconsistency termination occurs, we know that a linear combination of the constraints in the original system leads to the inconsistent equation. It may be useful to know the coefficients in such a linear combination. This coefficient vector could be obtained by including a memory matrix in the computations, as in the algorithms discussed in Sections 3.3.1, 3.3.5. If a linear combination of a system of constraints leads to an inconsistent equation, one of the constraints with a nonzero coefficient in the combination must either be removed, or its RHS constant modified appropriately, to make the system feasible. This information is useful in exploring ways to modify the system to make it feasible.

3.4.1 Bases and Basic Vectors for a System of Linear Equations

From the results discussed above we know that if a system of linear equations has a solution, then any redundant constraints in the system can be removed one by one until the number of constraints in the system becomes equal to the rank of the coefficient matrix in the original system. So, without any loss of generality, we can restrict our study to systems of the form

$$Ax = b \qquad\qquad (3.18)$$

where A is a matrix of order $m \times n$ and rank m. In such a system, the number of variables n will be $\geqq$ the number of constraints m. If $n = m$, the system is a **square system** with a unique solution. If $n > m$, the system has many solutions.

A **basis for this system** is a square nonsingular submatrix of A of order m, consisting of a linearly independent set of m columns of A. The vector of variables associated with these columns is the corresponding **basic vector for the system**. When referring to a basic vector, variables that are not in it are called **nonbasic variables**, and those in it are called **basic variables**. Every basic vector for (3.18) consists of exactly m basic variables. Each basis or the corresponding basic vector leads to a unique **basic solution** to the system. This basic solution is obtained by setting all the nonbasic variables equal to 0, and then solving the remaining square system for the values of the basic variables. Let $x_B = (x_1, \ldots, x_m)$ denote a basic vector for (3.18), and x_D denote the corresponding vector of nonbasic variables. Let B denote the basis consisting of columns in A of the basic variables in x_B, and D denote the submatrix consisting of the columns in A associated with the nonbasic variables. Then after rearranging the variables into the basic and nonbasic groups, (3.18) can be written in the following form

x_B	x_D	
B	D	b

So, the basic solution of (3.18) corresponding to the basis B, or the basic vector x_B is

$$\begin{pmatrix} x_D \\ x_B \end{pmatrix} = \begin{pmatrix} 0 \\ B^{-1}b \end{pmatrix}$$

As an example, consider the following system.

x_1	x_2	x_3	x_4	x_5	x_6	b
1	3	1	−1	1	4	6
0	0	1	−3	1	1	5
0	0	0	0	−1	−1	−3

The vector (x_1, x_2, x_6) is not a basic vector for this system since the set of column vectors associated with these variables is linearly dependent.

It can be verified that $x_B = (x_1, x_3, x_5)$ is a basic vector for this system, associated with the basis

$$B = \begin{pmatrix} 1 & 1 & 1 \\ 0 & 1 & 1 \\ 0 & 0 & -1 \end{pmatrix}$$

The basic solution corresponding to this basic vector is $x = (x_1, \text{ to } x_6) = (1, 0, 2, 0, 3, 0)^T$.

3.5 Exercises

3.4 Check whether the set of column vectors in the following tableau is linearly dependent. If it is, find a linear dependence relation for it. Also, find the rank and a basis for this set, and the representation of each nonbasic vector in terms of this basis. Select the leftmost vector in the tableau that is nonbasic, and explain which basic vectors it can replace in this basis, to lead to other bases for the set.

$A_{.1}$	$A_{.2}$	$A_{.3}$	$A_{.4}$	$A_{.5}$	$A_{.6}$
1	0	1	0	1	0
0	1	1	0	1	0
1	−1	0	1	2	0
−1	2	1	1	3	−1
1	1	2	1	4	1

3.5 Compute the inverses of the following matrices. If the inverse does not exist, find a linear dependence for the set of row vectors of the matrix, and its rank.

$$A = \begin{pmatrix} 0 & 1 & -2 \\ 1 & 0 & -1 \\ 2 & 2 & -6 \end{pmatrix}, \quad B = \begin{pmatrix} 0 & 1 & -1 \\ 2 & 0 & 1 \\ 1 & 1 & 0 \end{pmatrix}.$$

3.6 A is a matrix of order $m \times n$. Show that the set of column (row) vectors of A is linearly dependent iff there exists a column (row) vector $y \neq 0$ ($\pi \neq 0$) such that $Ay = 0$ ($\pi A = 0$).

3.7 Let $A_{1.} = (-1, 1, 2), A_{2.} = (1, 2, -3), A_{3.} = (-1, 4, \alpha)$. Find the set of values of α for which the set $\{A_{1.}, A_{2.}, A_{3.}\}$ is linearly independent. Is this set convex? Explain.

3.8 Let

$$D = \begin{pmatrix} 1 & 5 & -1 & -5 \\ 2 & 3 & 2 & 0 \\ -3 & -5 & -4 & 2 \end{pmatrix}, \quad A = \begin{pmatrix} 1 & 2 & 3 \\ -1 & 2 & 0 \\ 0 & 1 & -1 \end{pmatrix}, \quad b = \begin{pmatrix} -4 \\ 1 \\ 6 \end{pmatrix}$$

Express the straight line L_1 joining $D_{.1}$ and $D_{.2}$ algebraically.
Express the line segment L_2 joining $D_{.3}$ and $D_{.4}$ algebraically.
Do L_1 and L_2 intersect? If so, find the point of intersection.
Is $D_{.1}$ contained in the convex polyhedron defined by $Ax \overset{\leq}{=} b$?
Are $D_{.3}$ and $D_{.4}$ contained on opposite sides of the hyperplane $\mathbf{H}$ defined by $x_1 + 5x_2 + x_3 = -2$?

3.9 Let

$$D = \begin{pmatrix} 1 & -1 & 0 & 1 \\ -1 & 2 & 1 & -1 \\ 0 & 1 & 2 & -1 \\ 2 & -1 & 2 & 2 \end{pmatrix}, \quad b = \begin{pmatrix} 2 \\ 1 \\ 2 \\ \lambda \end{pmatrix}$$

Compute D^{-1} and the representation of b in terms of the columns of D. Use this to find the range of values of λ for which b is in the affine hull of the columns of D.

3.10 Prove that if the set of vectors $\{A_{.1}, A_{.2}, A_{.3}\}$ from $\mathbb{R}^n$ is linearly independent, then so is $\{A_{.2} + A_{.3}, A_{.3} + A_{.1}, A_{.1} + A_{.2}\}$. Is the converse also correct?

Assume that $\{A_{.1}, A_{.2}, A_{.3}\}$ is linearly independent. Let $y^1 = A_{.1} + A_{.2} + A_{.3}, y^2 = A_{.1} + \alpha A_{.2}, y^3 = A_{.2} + \beta A_{.3}$. Find a condition on α, β to guarantee that $\{y^1, y^2, y^3\}$ is also linearly independent.

3.11 $\mathbf{S} = \{A_{.1}, \ldots, A_{.m}\}$ is a linearly independent set of vectors in $\mathbb{R}^n$. $A_{.m+1} \in \mathbb{R}^n$ is not in the linear hull of $\mathbf{S}$. Then prove that $\{A_{.1}, \ldots, A_{.m}, A_{.m+1}\}$ is linearly independent.

3.12 Let

$$\Gamma_2 = \left\{ \begin{pmatrix} 1 \\ -1 \\ 1 \end{pmatrix}, \begin{pmatrix} 0 \\ 1 \\ -1 \end{pmatrix}, \begin{pmatrix} -1 \\ 0 \\ 1 \end{pmatrix} \right\}, \quad a^2 = \begin{pmatrix} -18 \\ 2 \\ 10 \end{pmatrix}$$

Is a^2 a linear, affine, or convex combination of Γ_2?

Is there an affine combination of Γ_2 on the ray of a^2?

Is $(-2, 8, -5)$ on the line segment joining $(4, 2, 3)$ and $(1, 5, -1)$?

Is $(1, 8, 7)$ on the half-line through $(4, 7, 8)$ parallel to the ray of $(-6, 2, -2)$?

3.13 Let

$$\Gamma_1 = \left\{ \begin{pmatrix} 1 \\ -1 \\ 1 \end{pmatrix}, \begin{pmatrix} 0 \\ 1 \\ -1 \end{pmatrix}, \begin{pmatrix} -1 \\ 0 \\ 1 \end{pmatrix}, \begin{pmatrix} 1 \\ 1 \\ 1 \end{pmatrix} \right\}, \quad a^1 = \begin{pmatrix} 0 \\ 0 \\ 1 \end{pmatrix}$$

Is a^1 a linear, affine, or convex combination of Γ_1?

3.14 Find the rank of the following matrix, and $\mathbf{E}$, a maximal linearly independent subset of row vectors in it. Also, write down the representation in terms of $\mathbf{E}$ for each row vector of this matrix not in $\mathbf{E}$.

$$\begin{pmatrix} 1 & 1 & -1 & 2 & 1 \\ 2 & -1 & 2 & 1 & 2 \\ 4 & 1 & 0 & 5 & 4 \\ 0 & 0 & 1 & 0 & 0 \\ 7 & 1 & 2 & 8 & 7 \end{pmatrix}$$

3.15 Solve the following system by back substitution.

x_1	x_2	x_3	x_4	
1	2	0	1	-2
0	-1	-2	1	-1
0	0	-3	1	-5
0	0	0	2	2

3.16 Solve the following systems of equations. In each case check for the uniqueness of the solution, and derive an expression for the general solution otherwise.

(a)

x_1	x_2	x_3	x_4	x_5	
1	-1	2	1	1	3
0	-1	1	2	1	0
1	-1	1	1	2	1

(b)

x_1	x_2	x_3	x_4	x_5	
1	2	-1	1	-1	1
-1	1	-2	2	1	2
0	3	-3	3	0	3
1	1	-1	-2	1	-2

(c)

x_1	x_2	x_3	x_4	x_5	
1	−1	1	1	1	1
2	−2	−1	−1	2	2
4	4	1	1	4	5

3.17 Consider the system of linear equations $Ax = b$, where A is a matrix of order $m \times n$.

If it is known that this system has a unique solution, what should rank$(A \vdots b)$ be?

Let rank$(A) = r$. If it is known that the system has no solution, what should rank$(A \vdots b)$ be?

If it is known that the system has more than one solution, what relationship exists between m, n, rank(A), rank$(A \vdots b)$?

What relationship exists between the following three numbers: rank(A), the number of vectors in a maximal linearly independent subset of row vectors of A, the number of vectors in a maximal linearly independent subset of column vectors of A.

3.18 Consider a square in the two dimensional Cartesian plane with sides of length 1. At each corner draw a circular arc with that corner as center, and 1 as radius, joining the two adjacent corners. These arcs divide the square into various regions as shown in Figure 3.13. It is required to find the area of the region marked with "C" in the center bounded by the four circular arcs. Use your knowledge of the properties of circles and triangles, and derive a system of three linearly independent equations in three unknowns, whose solution contains the value of the area of "C". Find the area of "C" by solving this system of equations (communicated by S. N. Kabadi).

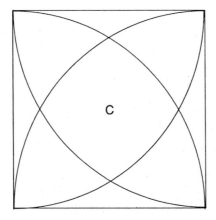

Figure 3.13

3.19 Find the general solution of the following system of linear equations. What kind of a set is the set of all solutions?

x_1	x_2	x_3	x_4	x_5	
1	0	0	0	2	3
1	−1	0	0	2	1
−1	0	1	0	3	−2
0	1	−1	1	5	−1
1	0	0	1	12	1

3.20 Determine the conditions that b_1, b_2, b_3 have to satisfy (in the form of an equation) so that the following system has at least one solution.

x_1	x_2	x_3	b
1	2	−3	b_1
2	6	−11	b_2
1	−2	7	b_3

When the above condition is satisfied, show that the set of solutions of the above system is a straight line in $\mathbb{R}^3$. Write down the coordinates of a general point on this straight line.

3.6   References

G. GOLUB, and C. VAN LOAN, 1983, *Matrix Computations*, John Hopkins Press.
B. NOBLE, and J. W. DANIEL, 1977, *Applied Linear Algebra*, 2nd. ed., Prentice Hall, Englewood Cliffs, NJ.
D. I. STEINBERG, 1974, *Computational Matrix Algebra*, McGraw Hill, NY.
G. STRANG, 1980, *Linear Algebra and Its Applications*, Academic Press, 2nd ed.
G. STRANG, 1986, *Introduction to Applied Mathematics*, Wellesley-Cambridge Press, Wellesley, MA-02181.

Chapter 4

Duality and Optimality Conditions in LP

4.1 The Dual Problem, Complementary Pairs, and Complementary Slackness Optimality Conditions

Associated with every linear programming problem, there is another linear program called its **dual**, involving a different set of variables, but sharing the same data. When referring to the dual problem of an LP, the original LP is called **the primal** or **the primal problem**.

The dual of an LP arises from economic considerations that come up in marginal analysis. In LP's, each constraint usually comes from the requirement that the total amount of some item utilized should be $\leqq$ (or =) the total amount of this item available, or that the total number of units of some item produced should be $\geqq$ (or =) the known requirement for this item. In the dual problem there will be a dual variable associated with this primal constraint, and it can be interpreted as the **marginal value of that item** or **its primal constraint**. The variables in the dual problem can also be interpreted as the **Lagrange multipliers associated with the constraints in the primal problem**.

We provide an example of the economic arguments behind duality by deriving the dual of the fertilizer manufacturer's problem (2.1) discussed in Chapter 2, with data given in Table 2.1. The fertilizer manufacturer has a daily supply of 1500 tons of RM 1, 1200 tons of RM 2, and 500 tons of RM 3 from the company's quarries; presently these supplies can be used to manufacture Hi-ph or Lo-ph fertilizers to make profit. There is a detergent company in the area that needs RM 1, 2, 3. The detergent manufacturer wants to persuade the fertilizer manufacturer to give up fertilizer making, and instead sell the supplies of RM 1, 2, 3 to the detergent company. Being very profit conscious, the fertilizer manufacturer will not agree to this deal unless the prices offered by the detergent manufacturer for the raw materials fetch as much income as each of the options in the fertilizer making business.

Let π_i = price (\$/ton) offered by the detergent manufacturer for RM i, i = 1, 2, 3. Obviously these prices have to be $\geqq 0$. Now consider the Hi-ph fertilizer. Manufacturing one ton of it yields a profit of \$15, and uses up 2 tons RM 1, 1 ton RM 2, and 1 ton RM 3. The same basket of raw materials fetches a price of $2\pi_1 + \pi_2 + \pi_3$ from the detergent manufacturer. So, the fertilizer manufacturer will not find the price vector $\pi = (\pi_1, \pi_2, \pi_3)$ acceptable unless $2\pi_1 + \pi_2 + \pi_3 \geqq 15$. Similar economic analysis with the Lo-ph fertilizer leads to the constraint $\pi_1 + \pi_2 \geqq 10$. With the price vector π, the cost to the detergent company of acquiring the daily raw material supply is $1500\pi_1 + 1200\pi_2 + 500\pi_3$, and the detergent manufacturer would clearly like to see this minimized. Thus the price vector $\pi = (\pi_1, \pi_2, \pi_3)$ that the detergent manufacturer offers for the supplies of RM 1, 2, 3, should minimize $v(\pi) = 1500\pi_1 + 1200\pi_2 + 500\pi_3$, subject to the constraints $2\pi_1 + \pi_2 + \pi_3 \geqq 15$, $\pi_1 + \pi_2 \geqq 10$, $\pi_1, \pi_2, \pi_3 \geqq 0$, to make it acceptable to the fertilizer manufacturer. Thus the detergent manufacturer's problem, that of determining the best price vector acceptable to the fertilizer manufacturer, is

$$
\begin{array}{llllll}
\text{Minimize } v(\pi) = & 1500\pi_1 + & 1200\pi_2 + & 500\pi_3 & & \\
\text{Subject to} & 2\pi_1 + & \pi_2 + & \pi_3 & \geqq & 15 \qquad (4.1)\\
& \pi_1 + & \pi_2 & & \geqq & 10 \\
& \pi_1, & \pi_2, & \pi_3 & \geqq & 0
\end{array}
$$

(4.1) is the dual of (2.1) and vice versa. This pair of problems is a primal-dual pair of LPs. When considering the primal (2.1), the variables in its dual (4.1) are called the **dual variables**, and the slacks in (4.1) corresponding to the inequality constraints in it are called the **dual slack variables**.

Since the first constraint in (4.1) comes from the economic analysis of the Hi-ph manufacturing process, this dual constraint is said to correspond to the Hi-ph primal variable x_1. Likewise, the second dual constraint in (4.1) corresponds to the primal variable x_2. In the same way, the dual variable π_1, the detergent manufacturer's price for the item RM 1, is associated with the RM 1 (first) primal constraint in (2.1). Similarly the dual variables π_2, π_3 are associated with the second (RM 2), and third (RM 3) primal constraints in (2.1), respectively. Thus there is a dual variable associated with each primal constraint, and a dual constraint corresponding to each primal variable. Also, verify the following facts.

1. The coefficient matrix in the detergent manufacturer's problem (4.1) is just the transpose of the coefficient matrix in the fertilizer manufacturer's problem (2.1) and vice versa.

2. The RHS constants in (4.1) are the objective coefficients in (2.1) and vice versa.

3. Each variable in (2.1) leads to a constraint in (4.1) and vice versa.

4. (2.1) is a maximization problem in which the constraints are $\leqq$ type; and (4.1) is a minimization problem in which the constraints are $\geqq$ type.

In the detergent manufacturer's problem, the nonnegative RM 1 price π_1 is associated with the nonnegative primal (RM 1) slack variable $1500 - 2x_1 - x_2$, and together they form a pair $(1500 - 2x_1 - x_2, \pi_1)$ that is known as a **complementary pair** in these primal, dual problems. Similarly the pair $(1200 - x_1 - x_2, \pi_2)$, $(500 - x_1, \pi_3)$ are two other complementary pairs. Likewise the nonnegative primal variable x_1 (amount of Hi-ph manufactured) in the fertilizer manufacturer's problem, is associated with the nonnegative dual slack variable $15 - 2\pi_1 - \pi_2 - \pi_3$ in the detergent manufacturer's problem, and $(x_1, 15 - 2\pi_1 - \pi_2 - \pi_3)$ forms another complementary pair. In a similar way $(x_2, 10 - \pi_1 - \pi_2)$ is another complementary pair.

The marginal value of RM i in the fertilizer manufacturer's problem is the rate of change in the maximum profit per unit change in the availability of RM i from its present value; thus it is the net worth of one additional unit of RM i over the present supply, for $i = 1, 2, 3$, to the fertilizer manufacturer. Hence, if the detergent manufacturer offered to buy RM i at a price $\geqq$ its marginal value, for $i = 1, 2, 3$, the fertilizer manufacturer would find the deal acceptable. Being cost conscious, the detergent manufacturer wants to make the price offered for any raw material to be the smallest value that will be acceptable to the fertilizer manufacturer. Hence, in an optimum solution of (4.1), the π_i will be the marginal value of RM i, for $i = 1, 2, 3$, in (2.1). Thus the dual variables are the marginal values of the items associated with the constraints in the primal problem. These marginal values depend on the data, and may change if the data does.

The fertilizer manufacturer's problem is a maximization problem, all constraints in it are "$\leqq$" inequalities, and all the decision variables in it are nonnegative. In its dual, which is a minimization problem, all constraints came out to be "$\geqq$" inequalities, and all the dual variables are also nonnegative. In a general LP there may be equality, "$\leqq$", and/or "$\geqq$" inequality constraints and variables that are unrestricted (i.e., they can be 0, or take negative or positive values), or restricted to be $\geqq 0$, or $\leqq 0$. In writing the dual of such a general LP, the dual variables associated with equality constraints are always unrestricted variables in the dual problem. The dual variable associated with a "$\geqq$" primal inequality is restricted to be nonnegative (nonpositive) in the dual problem, if the primal is a minimization (maximization) problem. Similarly, the dual variable associated with a "$\leqq$" primal inequality is restricted to be nonpositive (nonnegative) in the dual problem if the primal is a minimization (maximization) problem. And correspondingly, the dual constraints associated with unrestricted primal variables are always equality constraints, and those associated with primal variables restricted to be $\overset{\geq}{=} 0$ ($\overset{\leq}{=} 0$) or "$\overset{\geq}{=}$" ("$\overset{\leq}{=}$") inequalities if the dual is a minimization problem; or "$\overset{\leq}{=}$" ("$\overset{\geq}{=}$") inequalities if the dual is a maximization problem. For the economic and mathematical justification for these rules, see [D. Gale, 1960] or [K. G. Murty, 1983 of Chapter 2].

An LP is said to be in **standard form** if all the constraints in it are linear equations in nonnegative variables, and the objective function is in minimization form. Here is an LP in standard form in matrix notation.

$$
\begin{aligned}
\text{Minimize} \quad & z(x) = cx \\
\text{subject to} \quad & Ax = b \\
& x \overset{\geq}{=} 0
\end{aligned}
\tag{4.2}
$$

where the coefficient matrix A is of order $m \times n$, the cost vector $c = (c_j : j = 1$ to $n)$ is a row vector, and the RHS constants vector $b = (b_i : i = 1$ to $m)$ is a column vector, and $x = (x_j : j = 1$ to $n)$ is the column vector of decision variables in the problem. Every LP can be transformed into standard form, and in fact LPs are put in this form before solving them by the simplex method. Hence for practical applications the standard form of LP is the most important. To write the dual of the LP (4.2), associate a dual variable, π_i say, to the ith constraint in it, $i = 1$ to m; and let $\pi = (\pi_1, \ldots, \pi_m)$ be the row vector of dual variables. Then, the dual of (4.2), derived using the same economic arguments as above, is:

$$
\begin{aligned}
\text{Maximize} \quad & v(\pi) = \pi b \\
\text{subject to} \quad & \pi A \overset{\leq}{=} c
\end{aligned}
\tag{4.3}
$$

As mentioned above, the dual variables are unrestricted in (4.3) since the associated primal constraints in (4.2) are all equality constraints. And the dual constraints in (4.3) are "$\overset{\leq}{=}$" inequalities since the corresponding primal variables are nonnegative.

In matrix notation, $\pi A \overset{\leq}{=} c$, is a system of n inequality constraints; these are $\pi A_{.j} \overset{\leq}{=} c_j$, $j = 1$ to n. There is one dual constraint corresponding to each primal variable. $\pi A_{.j} \overset{\leq}{=} c_j$ is the one corresponding to x_j, $j = 1$ to n; $c_j - \pi A_{.j}$ denoted by $\bar{c}_j$ is the dual slack variable for it. It is called the **reduced** or **relative cost coefficient** of x_j wrt the dual vector π.

Every complementary pair in a primal, dual pair of LPs always consists of a variable restricted to be $\overset{\geq}{=} 0$ in one problem, and the nonnegative slack variable of the corresponding constraint in the other problem. There are no complementary pairs associated with equality constraints in one problem, and the corresponding unrestricted variables in the other problem. In the primal, dual pair (4.2), (4.3), the complementary pairs are $(x_j, \bar{c}_j = c_j - \pi A_{.j})$ for $j = 1$ to n.

As an example, consider the following LP in standard form. It is expressed in detached coefficient form for clarity. It has 3 constraints in 6 nonnegative variables. The last row in the tableau gives the objective function. In a column on the left hand side, we listed the dual variables associated with the primal constraints. We tabulate the dual problem just after the primal.

Tableau 4.1: Primal problem

Associated dual var.	x_1	x_2	x_3	x_4	x_5	x_6	b
π_1	1	2	3	-2	1	16	17
π_2	0	1	-4	1	1	1	2
π_3	0	0	1	-2	1	0	1
Primal obj. row	3	11	-15	10	4	57	$= z$, minimize

$$x_j \geqq 0 \text{ for all } j.$$

Tableau 4.2: Dual problem

π_1	π_2	π_3			Primal var. corresponding to dual constraint
1	0	0	$\leqq$	3	x_1
2	1	0	$\leqq$	11	x_2
3	-4	1	$\leqq$	-15	x_3
-2	1	-2	$\leqq$	10	x_4
1	1	1	$\leqq$	4	x_5
16	1	0	$\leqq$	57	x_6
17	2	1	$= v(\pi)$, maximize		

Introducing the dual slack variables $\bar{c}_1$ to $\bar{c}_6$, the dual can be written with its constraints as equality constraints as in Tableau 4.3.

Tableau 4.3: Dual problem

π_1	π_2	π_3	$\bar{c}_1$	$\bar{c}_2$	$\bar{c}_3$	$\bar{c}_4$	$\bar{c}_5$	$\bar{c}_6$	
1	0	0	1	0	0	0	0	0	3
2	1	0	0	1	0	0	0	0	11
3	-4	1	0	0	1	0	0	0	-15
-2	1	-2	0	0	0	1	0	0	10
1	1	1	0	0	0	0	1	0	4
16	1	0	0	0	0	0	0	1	57
17	2	1	0	0	0	0	0	0	$= v(\pi)$, maximize

$$\bar{c}_j \geqq 0, j = 1 \text{ to } 6$$

$\bar{c}_j$ here is the relative cost coefficient of x_j, for $j = 1$ to 6. The complementary pairs in these primal, dual problems are $(x_1, \bar{c}_1 = 3 - \pi_1)$, $(x_2, \bar{c}_2 = 11 - (2\pi_1 + \pi_2))$, $(x_3, \bar{c}_3 = -15 - (3\pi_1 - 4\pi_2 + \pi_3))$, $(x_4, \bar{c}_4 = 10 - (-2\pi_1 + \pi_2 - 2\pi_3))$, $(x_5, \bar{c}_5 = 4 - (\pi_1 + \pi_2 + \pi_3))$, $(x_6, \bar{c}_6 = 57 - (16\pi_1 + \pi_2))$.

We now state a fundamental result in LP theory that serves as the basis for designing algorithms to solve LPs, and for checking when an algorithm has reached an optimum solution.

Theorem 4.1 *In a primal, dual pair of LPs, let x be the vector of primal variables, and π the vector of dual variables. A primal vector $\bar{x}$ is an optimum solution for the primal problem iff it satisfies the following condition (i), and there exists a dual vector $\bar{\pi}$ satisfying (ii), which together with $\bar{x}$ also satisfies (iii).*

(i) **Primal feasibility***: The vector $\bar{x}$ must satisfy all the constraints and bound restrictions in the primal problem.*

(ii) **Dual feasibility***: The vector $\bar{\pi}$ must satisfy all the constraints in the dual problem.*

(iii) **Complementary slackness optimality conditions***: In every complementary pair for these primal, dual problems, at least one of the two quantities in the pair is zero at the solutions $(\bar{x}, \bar{\pi})$. Or, equivalently, the product of the two quantities in every complementary pair is zero.*

If all three conditions are satisfied, $\bar{x}$ is an optimum solution for the primal problem, and $\bar{\pi}$ is an optimum solution of the dual problem, and the optimum objective values in the two problems are equal.

For a proof of this theorem, see [D. Gale, 1960] or [K. G. Murty, 1983 of Chapter 2]. We will use the result in this theorem repeatedly in algorithms for the assignment and transportation problems and general LPs discussed in later chapters.

We now explain what the complementary slackness conditions are for the primal, dual problems (4.2), (4.3). If x, π are primal and dual solutions for (4.2), (4.3), and $(\bar{c}_j) = (c_j - \pi A_{.j})$, since the complementary pairs in these problems are $(x_j, \bar{c}_j)$ for $j = 1$ to n; the complementary slackness conditions for these problems can be stated in one of two ways: At least one quantity in each pair $(x_j, \bar{c}_j)$ is zero; or equivalently, $x_j \bar{c}_j = 0$ for all j.

As an example, consider the fertilizer manufacturer's problem (2.1). The vector $\bar{x} = (300, 900)^T$ can be verified to be feasible to (2.1), so it satisfies the primal feasibility criterion. The dual vector $\bar{\pi} = (5, 5, 0)$ can be verified to be feasible to its dual (4.1). So, $\bar{\pi}$ satisfies the dual feasibility condition. And the values of the various complementary pairs in these primal, dual problems at $\bar{x}, \bar{\pi}$ are: $(\bar{x}_1, 15 - 2\bar{\pi}_1 - \bar{\pi}_2 - \bar{\pi}_3) = (300, 0)$, $(\bar{x}_2, 10 - \bar{\pi}_1 - \bar{\pi}_2) = (900, 0)$, $(\bar{\pi}_1, 1500 - 2\bar{x}_1 -$

$\bar{x}_2) = (5, 0)$, $(\bar{\pi}_2, 1200 - \bar{x}_1 - \bar{x}_2) = (5, 0)$, $(\bar{\pi}_3, 500 - \bar{x}_1) = (0, 200)$. In every complementary pair, one of the two quantities is zero, so $\bar{x}, \bar{\pi}$ together satisfy the complementary slackness optimality conditions. So, $\bar{x}$ is an optimum solution for the fertilizer problem (2.1), and $\bar{\pi}$ is an optimum solution for its dual (4.1). We already verified that $\bar{x}$ is optimal to (2.1) by the geometric method in Section 2.1. Also, the maximum profit in (2.1) is \$13,500, which is also the minimum cost in its dual (4.1), so the optimum objective values in the primal and dual problems are equal.

As another example, consider the LP in Tableau 4.1. Consider the primal vector $\bar{x} = (2, 6, 1, 0, 0, 0)^T$. It satisfies all the constraints and sign restrictions in the primal problem, so it is primal feasible. Consider the dual vector $\bar{\pi} = (3, 5, -4)$, which can be verified to be dual feasible. The dual slack vector corresponding to $\bar{\pi}$ is $\bar{c} = (\bar{c}_1, \bar{c}_2, \bar{c}_3, \bar{c}_4, \bar{c}_5, \bar{c}_6) = (0, 0, 0, 3, 0, 4)$. So the values of the various complementary pairs at $\bar{x}, \bar{\pi}, (\bar{x}_j, \bar{c}_j); j = 1$ to 6 are: (2, 0), (6, 0), (1, 0), (0, 3), (0, 0), (0, 4). At least one quantity in each pair is zero. So $\bar{x}, \bar{\pi}$ satisfy all the complementary slackness optimality conditions. Hence, by Theorem 4.1, $\bar{x}$ is an optimum solution of the LP in Tableau 4.1, $\bar{\pi}$ is an optimum solution of its dual in Tableau 4.2. Both optimum objective values can be verified to be equal to 57.

In these two examples we used Theorem 4.1 to check whether a given solution is optimal to an LP. Later on we will see how Theorem 4.1 also provides a guiding light for designing algorithms to try to construct solutions which satisfy the conditions there, and thereby solve both the primal and dual problems together.

We will use the abbreviation "c.s." for "complementary slackness".

The Dual of the Balanced Transportation Problem

Consider the balanced transportation problem for shipping iron ore from mines 1, 2 to plants 1, 2, 3 at minimum cost, formulated in Chapter 2. In this problem, the primal variable x_{ij} = ore (in tons) shipped from mine i to plant j; $i = 1, 2$; $j = 1, 2, 3$. Here is the problem in detached coefficient form. In a column on the left hand side we list the dual variables that we associate with the primal constraints for writing the dual problem.

Ore shipping problem

Associated dual var.	x_{11}	x_{12}	x_{13}	x_{21}	x_{22}	x_{23}		Item
u_1	1	1	1	0	0	0	800	Ore/mine 1
u_2	0	0	0	1	1	1	300	Ore/mine 2
v_1	1	0	0	1	0	0	400	Ore/plant 1
v_2	0	1	0	0	1	0	500	Ore/plant 2
v_3	0	0	1	0	0	1	200	Ore/plant 3
	11	8	2	7	5	4	$= z$, minimize	

$$x_{ij} \geq 0 \text{ for all } i, j.$$

So, the dual of this problem is the following.

Maximize $\qquad 800u_1 + 300u_2 + 400v_1 + 500v_2 + 200v_3$

Associated
primal var.

subject to $\qquad$
$$
\begin{array}{rcl}
u_1 + v_1 & \stackrel{\leq}{=} & 11 \qquad x_{11} \\
u_1 + v_2 & \stackrel{\leq}{=} & 8 \qquad x_{12} \\
u_1 + v_3 & \stackrel{\leq}{=} & 2 \qquad x_{13} \\
u_2 + v_1 & \stackrel{\leq}{=} & 7 \qquad x_{21} \\
u_2 + v_2 & \stackrel{\leq}{=} & 5 \qquad x_{22} \\
u_2 + v_3 & \stackrel{\leq}{=} & 4 \qquad x_{23}
\end{array}
$$

Here u_i is the dual variable associated with source i (mines 1, 2 in this problem), and v_j is the dual variable associated with demand center j (plants 1, 2, 3 in this problem). If c_{ij} is the original cost coefficient of the primal variable x_{ij} in this problem, the corresponding dual constraint is $u_i + v_j \stackrel{\leq}{=} c_{ij}$; its dual slack or reduced cost coefficient is $\bar{c}_{ij} = c_{ij} - u_i - v_j$. The pairs $(x_{ij}, \bar{c}_{ij} = c_{ij} - u_i - v_j)$ for various values of i, j are the complementary pairs in these primal, dual problems.

In Chapter 2 we mentioned that all the constraints and the decision variables or their values in a particular solution in a balanced transportation problem can be displayed very conveniently in the form of a two dimensional transportation array. In this array representation we can also include the dual variables u_i associated with the rows of the array (representing sources in the problem) in a right hand column, and the dual variables v_j associated with the columns of the array (representing demand centers in the problem) in a bottom row. With these things, the array representation of this iron ore shipping problem is given below.

Array Representation of the Ore Shipping Problem

	Steel Plant			Availability (tons)	Dual var.
	1	2	3		
Mine 1	$\bar{c}_{11}$ x_{11} $\quad$ 11	$\bar{c}_{12}$ x_{12} $\quad$ 8	$\bar{c}_{13}$ x_{13} $\quad$ 2	$= 800$	u_1
Mine 2	$\bar{c}_{21}$ x_{21} $\quad$ 7	$\bar{c}_{22}$ x_{22} $\quad$ 5	$\bar{c}_{23}$ x_{23} $\quad$ 4	$= 300$	u_2
Requirement (tons)	$= 400$	$= 500$	$= 200$		
Dual var.	v_1	v_2	v_3		

$x_{ij} \stackrel{\geq}{=} 0$ for all i, j. Minimize cost.

In this array representation it is very convenient to check whether the given dual vector $(u = (u_i), v = (v_j))$ is dual feasible. It is dual feasible if, $\bar{c}_{ij} = c_{ij} - u_i - v_j$ is $\geqq 0$ for all i, j. For this it is convenient to compute $\bar{c}_{ij}$ and enter it in the top left corner of the cell (i, j) for all i, j. When both $x_{ij}, \bar{c}_{ij}$ are entered this way in each cell of the array, it is easy to check whether the complementary slackness optimality conditions hold (at least one of $x_{ij}, \bar{c}_{ij}$ have to be zero for each (i, j), or equivalently $x_{ij}\bar{c}_{ij} = 0$ for every (i, j)).

The array form of the balanced transportation problem is very convenient for displaying the current primal and dual solutions and the relative cost coefficients.

In a general balanced transportation problem, there may be m sources, and n demand centers with the following data

$$
\begin{aligned}
a_i &= \text{material (in units) available at source } i,\ i = 1 \text{ to } m\\
b_j &= \text{material required at demand center } j,\ j = 1 \text{ to } n\\
c_{ij} &= \text{cost (\$/unit) to ship from source } i, \text{ to demand center } j,\ i\\
&= 1 \text{ to } m,\ j = 1 \text{ to } n
\end{aligned}
$$

The problem is a balanced transportation problem if the data satisfies

$$
\sum_{i=1}^{m} a_i = \sum_{j=1}^{n} b_j \tag{4.4}
$$

i.e., the total amount of material required at all the demand centers is equal to the total amount of material available at all the sources. We assume that this condition holds.

The primal variables are: $x_{ij} = $ units shipped from source i to demand center j, $i = 1$ to m, $j = 1$ to n. Associate the dual variable u_i with the primal constraint of source i, and the dual variable v_j with the primal constraint of demand center j. Then the balanced transportation problem with this data is (4.5), and its dual is (4.6).

$$
\begin{aligned}
\text{Minimize} \quad & z(x) = \sum_{i=1}^{m}\sum_{j=1}^{n} c_{ij} x_{ij}\\[2mm]
\text{subject to} \quad & \sum_{j=1}^{n} x_{ij} = a_i, i = 1 \text{ to } m\\[2mm]
& \sum_{i=1}^{m} x_{ij} = b_j, j = 1 \text{ to } n\\[2mm]
& x_{ij} \geqq 0,\ \text{for all } i, j
\end{aligned} \tag{4.5}
$$

$$\text{Maximize} \quad w(u,v) = \sum_{i=1}^{m} a_i u_i \; + \; \sum_{j=1}^{n} b_j v_j$$

$$\text{subject to} \quad u_i + v_j \; \overset{\leq}{=} \; c_{ij}, \text{ for all } i,j \qquad (4.6)$$

$\bar{c}_{ij} = c_{ij} - u_i - v_j$ is the relative cost coefficient of x_{ij}, i.e., the dual slack associated with it. The various $(x_{ij}, \bar{c}_{ij})$ are the complementary pairs in these primal, dual problems.

The Dual of the Assignment Problem

As discussed in Chapter 2, an assignment problem of order n is a balanced transportation problem involving n sources with one unit supply each, and n demand centers with one unit requirement each. Typical applications of the assignment model are in problems of assigning candidates to positions (see Chapter 1), jobs to machines, etc. The decision variable x_{ij} in the model is defined by

$$x_{ij} = \begin{cases} 1 & \text{if source } i \text{ ships a unit to demand center } j \text{ (or candidate } i \\ & \text{is assigned to position } j, \text{ etc., depending on the context of} \\ & \text{the problem)} \\ 0 & \text{otherwise} \end{cases}$$

Even though the problem is one involving continuous variables, it has an integral (i.e., 0–1) optimum solution by Theorem 2.1, and only such an integral solution is desired. Since the supply at each source and the requirement at each demand center is always 1 in an assignment problem, this information is understood and not recorded in an array representation of the assignment problem. Let c_{ij} be the cost coefficient associated with x_{ij}. Then the cost matrix $C = (c_{ij})$ of order $n \times n$ is the data for the assignment problem of order n. With this data, the assignment problem is to find an integral (i.e., 0–1) solution to

$$\text{Minimize} \quad z(x) = \sum_{i=1}^{n}\sum_{j=1}^{n} c_{ij} x_{ij}$$

$$\text{subject to} \quad \sum_{j=1}^{n} x_{ij} \; = \; 1, i = 1 \text{ to } m$$

$$\sum_{i=1}^{n} x_{ij} \; = \; 1, j = 1 \text{ to } n \qquad (4.7)$$

$$x_{ij} \; \overset{\geq}{=} \; 0, \text{ for all } i,j$$

Its dual is

$$\text{Maximize} \quad w(u, v) = \sum_{i=1}^{n} u_i \; + \; \sum_{j=1}^{n} v_j$$

$$\text{subject to} \quad u_i + v_j \overset{\leq}{=} c_{ij}, \text{ for all } i, j \tag{4.8}$$

And from Theorem 4.1, if $\bar{x} = (\bar{x}_{ij}), (\bar{u} = (\bar{u}_i), \bar{v} = (\bar{v}_j)$ are optimal to (4.7), (4.8), then we will have

$$
\begin{aligned}
\bar{c}_{ij} &= c_{ij} - \bar{u}_i' - \bar{v}_j \overset{\geq}{=} 0 \text{ for all } i, j \text{ (dual feasibility)} \\
\bar{x}_{ij} &= 0 \text{ or } 1 \text{ for all } i, j \text{ (primal feasibility)} \\
\bar{c}_{ij} &= 0 \text{ whenever } \bar{x}_{ij} = 1 \text{ (complementary slackness property)}
\end{aligned} \tag{4.9}
$$

The Hungarian method discussed in Chapter 5 is an efficient method for the assignment problem. It maintains both primal and dual vectors from the beginning, and uses the dual solution in every step in an effort to generate primal and dual solutions which together satisfy all the above optimality conditions.

4.2   Using C. S. Optimality Conditions to Check Optimality of a Given Feasible Solution

Given an LP, and a feasible solution $\hat{x}$ for it, sometimes we can check whether $\hat{x}$ is an optimum solution by solving the system of linear equations in the dual variables identified by the c.s. optimality conditions. We will illustrate this with several examples.

EXAMPLE 4.1

Consider the following LP in standard form. It involves 5 nonnegative variables subject to 3 equality constraints. The objective function $z(x) = -6x_1 + 9x_2 - 5x_3 + 10x_4 - 25x_5$ is required to be minimized subject to these constraints and nonnegativity restrictions on the variables.

x_1	x_2	x_3	x_4	x_5	b
1	-2	2	1	3	3
0	1	-1	-2	4	3
0	0	2	5	-1	9
-6	9	-5	10	-25	$= z(x)$, minimize

$$x_1 \text{ to } x_5 \overset{\geq}{=} 0$$

We want to check whether $\hat{x} = (12, 7, 2, 1, 0)^T$ is an optimum solution for this problem.

The first thing to do is to check whether $\hat{x}$ is feasible to the problem. It is ≥ 0, and we verify that it satisfies all the three equality constraints. So, $\hat{x}$ is primal feasible, i.e., it satisfies condition (i) of Theotrem 4.1.

Now we write the dual problem. Associating the dual variables π_1, π_2, π_3 to the three primal constraints in that order, we find the dual problem to be

Maximize	$v(\pi) =$			$3\pi_1$	$+3\pi_2$	$+9\pi_3$		Associated primal var.
subject to	$\bar{c}_1 =$	-6	$-(\pi_1$			$)$	≥ 0	x_1
	$\bar{c}_2 =$	9	$-(-2\pi_1$	$+\pi_2$		$)$	≥ 0	x_2
	$\bar{c}_3 =$	-5	$-(2\pi_1$	$-\pi_2$	$+2\pi_3)$		≥ 0	x_3
	$\bar{c}_4 =$	10	$-(\pi_1$	$-2\pi_2$	$+5\pi_3)$		≥ 0	x_4
	$\bar{c}_5 =$	-25	$-(3\pi_1$	$+4\pi_2$	$-\pi_3)$		≥ 0	x_5

Since the primal constraints are equations, there are no sign restrictions on the dual variables π_1, π_2, π_3. The complementary pairs in this primal, dual pair of LPs are $(x_j, \bar{c}_j)$, $j = 1$ to 5. And the c.s. optimality conditions for this pair are

$$x_j \bar{c}_j = 0, \quad \text{for } j = 1 \text{ to } 5$$

To check whether $\hat{x}$ is an optimum solution, by Theorem 4.1 we need to check whether there exists a dual feasible solution π which satisfies these c.s. optimality conditions together with $\hat{x}$.

Since $\hat{x}_1 = 12$, $\hat{x}_2 = 7$, $\hat{x}_3 = 2$, $\hat{x}_4 = 1$ are all positive; if there is a dual solution π which satisfies the c.s. optimality conditions together with $\hat{x}$, then it must satisfy $\bar{c}_1 = \bar{c}_2 = \bar{c}_3 = \bar{c}_4 = 0$; i.e., it must satisfy

$$\begin{array}{rlll}
-6 & -(\pi_1 & &) & = 0 \\
9 & -(-2\pi_1 & +\pi_2 &) & = 0 \\
-5 & -(2\pi_1 & -\pi_2 & +2\pi_3) & = 0 \\
10 & -(\pi_1 & -2\pi_2 & +5\pi_3) & = 0
\end{array}$$

This is a system of 4 equations in 3 unknowns, and we verify that it has the unique solution $\hat{\pi} = (-6, -3, 2)$. Also, we verify that $\bar{c}_5$ for $\hat{\pi}$ is $-25 - (3\hat{\pi}_1 + 4\hat{\pi}_2 - \hat{\pi}_3)$ $= -25 - (-18 - 12 - 2) = 7 > 0$. Hence $\hat{\pi}$ satisfies all the constraints in the dual problem, and is therefore dual feasible.

Since $\hat{x}$ is primal feasible, $\hat{\pi}$ is dual feasible, and they together satisfy the c.s. optimality conditions, we conclude by Theorem 4.1 that $\hat{x}$ is in fact an optimum solution for the original LP, and that $\hat{\pi}$ is a dual optimum solution. We also verify that $z(\hat{x}) = -9 = v(\hat{\pi})$; i.e., the primal and dual optimum objective values are equal.

EXAMPLE 4.2

Consider the following LP in standard form, involving 5 nonnegative variables subject to two equality constraints.

x_1	x_2	x_3	x_4	x_5	b
2	1	−1	2	1	25
2	0	2	−1	3	20
8	6	−10	20	−2	$= z(x)$, minimize

$$x_1 \text{ to } x_5 \geqq 0$$

We want to check whether $\tilde{x} = (10, 5, 0, 0, 0)^T$ is an optimum solution for this problem. We verify that $\tilde{x}$ is primal feasible. Associating the dual variables π_1, π_2 to the two constraints, the dual problem is

Maximize $v(\pi) = \quad\quad\quad 25\pi_1 \quad +20\pi_2$ Associated primal var.

subject to
$$\bar{c}_1 = \quad 8 \quad -(2\pi_1 \quad +2\pi_2) \quad \geqq 0 \quad\quad x_1$$
$$\bar{c}_2 = \quad 6 \quad -(\pi_1 \quad\quad\quad) \quad \geqq 0 \quad\quad x_2$$
$$\bar{c}_3 = \quad -10 \quad -(-\pi_1 \quad +2\pi_2) \quad \geqq 0 \quad\quad x_3$$
$$\bar{c}_4 = \quad 20 \quad -(2\pi_1 \quad -\pi_2) \quad \geqq 0 \quad\quad x_4$$
$$\bar{c}_5 = \quad -2 \quad -(\pi_1 \quad +3\pi_2) \quad \geqq 0 \quad\quad x_5$$

As before, the c.s. optimality conditions for this problem are: $x_j \bar{c}_j = 0$, $j = 1$ to 5. Since $\tilde{x}_1 = 10, \tilde{x}_2 = 5$ are > 0, any dual solution π that satisfies the c.s. optimality conditions together with $\tilde{x}$ must satisfy $\bar{c}_1 = 0$ and $\bar{c}_2 = 0$, i.e.,

$$8 \quad -(2\pi_1 \quad +2\pi_2) \quad = 0$$
$$6 \quad -(\pi_1 \quad\quad\quad) \quad = 0$$

This system of equations has the unique solution $\tilde{\pi} = (6, -2)$. For $\tilde{\pi}$, we compute $\bar{c}_3 = -10 - (-\tilde{\pi}_1 + 2\tilde{\pi}_2) = 0$, $\bar{c}_4 = 20 - (2\tilde{\pi}_1 - \tilde{\pi}_2) = 4$, $\bar{c}_5 = -2 - (\tilde{\pi}_1 + 3\tilde{\pi}_2) = -2$. Since $\bar{c}_5 < 0$, $\tilde{\pi}$ is not dual feasible (it violates the dual constraint associated with the primal variable x_5). So, there is no dual feasible solution which satisfies c.s. optimality conditions together with $\tilde{x}$. Hence, by Theorem 4.1, $\tilde{x}$ is not an optimum solution of the primal problem.

EXAMPLE 4.3

Consider the following LP in standard form.

x_1	x_2	x_3	x_4	x_5	b
1	0	1	−2	1	10
0	1	−1	2	2	15
1	1	1	1	−1	25
1	4	−2	4	20	$= z(x)$, minimize

$$x_1 \text{ to } x_5 \geqq 0$$

We want to check whether $\hat{x} = (10, 15, 0, 0, 0)^T$ is an optimum solution for this problem. We verify that $\hat{x}$ is primal feasible. Associating the dual variables π_1, π_2, π_3 to the three primal equality constraints, the dual problem is

						Associated primal var.
Maximize	$v(\pi) =$		$10\pi_1$	$+15\pi_2$	$+25\pi_3$	
subject to	$\bar{c}_1 =$	1	$-(\pi_1$		$+\pi_3)$	$\geqq 0$	x_1
	$\bar{c}_2 =$	4	$-($	$+\pi_2$	$+\pi_3)$	$\geqq 0$	x_2
	$\bar{c}_3 =$	-2	$-(\pi_1$	$-\pi_2$	$+\pi_3)$	$\geqq 0$	x_3
	$\bar{c}_4 =$	4	$-(-2\pi_1$	$+2\pi_2$	$+\pi_3)$	$\geqq 0$	x_4
	$\bar{c}_5 =$	20	$-(\pi_1$	$+2\pi_2$	$-\pi_3)$	$\geqq 0$	x_5

As before, the c.s. optimality conditions for this problem are: $x_j \bar{c}_j = 0$, $j = 1$ to 5. Since $\hat{x}_1 = 10$, $\hat{x}_2 = 15$ are > 0, any dual solution that satisfies the c.s. optimality conditions together with $\hat{x}$ must satisfy $\bar{c}_1 = 0$ and $\bar{c}_2 = 0$; i.e.,

$$
\begin{array}{llll}
1 & -(\pi_1 & +\pi_3) & = 0 \\
4 & -(\quad +\pi_2 & +\pi_3) & = 0
\end{array}
$$

In this case the c.s. optimality conditions have only identified two equations in the three dual variables π_1, π_2, π_3; not enough to identify the dual solution uniquely. So, in this example we are unable to check whether the given primal feasible solution $\hat{x}$ is optimal, from the c.s. optimality conditions, using only methods for solving linear equations. In this example, Theorem 4.1 says that $\hat{x}$ is an optimum solution iff there is a $\pi = (\pi_1, \pi_2, \pi_3)$ satisfying: $\bar{c}_1 = \bar{c}_2 = 0$ and $\bar{c}_3 \geqq 0$, $\bar{c}_4 \geqq 0$, and $\bar{c}_5 \geqq 0$; i.e.,

$$
\begin{array}{rrrrrr}
1 & -(\pi_1 &   & +\pi_3) & = 0 \\
4 & -(& +\pi_2   & +\pi_3) & = 0 \\
-2 & -(\pi_1 & -\pi_2   & +\pi_3) & \geq 0 \\
4 & -(-2\pi_1 & +2\pi_2 & +\pi_3) & \geq 0 \\
20 & -(\pi_1 & +2\pi_2 & -\pi_3) & \geq 0 \\
\end{array}
$$

Since this system involves linear inequalities, it cannot be solved directly by the Gaussian elimination method discussed in Chapter 3. This problem can ofcourse be solved by the primal simplex method discussed in Chapter 7.

EXAMPLE 4.4

Consider the following balanced transportation problem.

	Market 1		Market 2		Market 3		Availability
Source 1	x_{11}	11	x_{12}	13	x_{13}	20	= 300
Source 2	x_{21}	6	x_{22}	6	x_{23}	8	= 300
	= 100		= 400		= 100		

$x_{ij} \geq 0$ for all i, j. Minimize cost.

Supplies, requirements in large size numbers

x_{ij} is the amount shipped from source i to market j, $i = 1, 2$; $j = 1, 2, 3$. The unit shipping cost on this route, c_{ij}, is entered in the lower right corner of cell (i, j) in the array. The problem is to find an (x_{ij}) that meets all the requirements at minimum cost.

Associate the dual variables u_1, u_2 to the source availability constraints; and v_1, v_2, v_3 to the market requirement constraints. As described earlier, the dual of this problem is

Maximize $\quad 300u_1 + 300u_2 + 100v_1 + 400v_2 + 100v_3 \quad$ Associated primal var.

subject to

$$\bar{c}_{11} = 11 - u_1 - v_1 \overset{\geq}{=} 0 \qquad x_{11}$$
$$\bar{c}_{12} = 13 - u_1 - v_2 \overset{\geq}{=} 0 \qquad x_{12}$$
$$\bar{c}_{13} = 20 - u_1 - v_3 \overset{\geq}{=} 0 \qquad x_{13}$$
$$\bar{c}_{21} = 6 - u_2 - v_1 \overset{\geq}{=} 0 \qquad x_{21}$$
$$\bar{c}_{22} = 6 - u_2 - v_2 \overset{\geq}{=} 0 \qquad x_{22}$$
$$\bar{c}_{23} = 8 - u_2 - v_3 \overset{\geq}{=} 0 \qquad x_{23}$$

Suppose we are given that $\hat{u} = (3, -4), \hat{v} = (8, 10, 12)$ is an optimum solution of the dual problem. Using this information find an optimum solution of the primal transportation problem.

We verify that $\hat{u}, \hat{v}$ is actually dual feasible. For this primal, dual pair the c.s. optimality conditions are

$$x_{ij}\bar{c}_{ij} = 0, \quad i = 1, 2; \, j = 1, 2, 3$$

For $\hat{u}, \hat{v}$ we compute $(\bar{c}_{11}, \bar{c}_{12}, \bar{c}_{13}, \bar{c}_{21}, \bar{c}_{22}, \bar{c}_{23}) = (0, 0, 5, 2, 0, 0)$. Since $\hat{u}, \hat{v}$ is dual optimum, from the c.s. optimality conditions we conclude that $x_{13} = x_{21} = 0$ (since $\bar{c}_{13}$ and $\bar{c}_{21}$ are > 0) in every primal optimum solution. Substituting $x_{13} = x_{21} = 0$ in the system of primal constraints, it becomes

$$
\begin{array}{llll}
x_{11} & +x_{12} & & = 300 \\
 & & x_{22} \quad +x_{23} & = 300 \\
x_{11} & & & = 100 \\
 & x_{12} \quad +x_{22} & & = 400 \\
 & & x_{23} & = 100
\end{array}
$$

which by back substitution yields the solution $\hat{x} = (\hat{x}_{11}, \hat{x}_{12}, \hat{x}_{13}, \hat{x}_{21}, \hat{x}_{22}, \hat{x}_{23}) = (100, 200, 0, 0, 200, 100)$. Since $\hat{x}, (\hat{u}, \hat{v})$ satisfy all the conditions stated in Theorem 4.1, we verify that $(\hat{u}, \hat{v})$ is in fact dual optimum as stated in the hypothesis, and that $\hat{x}$ is an optimum solution of the primal transportation problem.

4.3 Primal and Dual Basic Solutions Associated With a Basis

Consider the LP (4.2) in standard form where A is a matrix of order $m \times n$ and rank m. A basis B for (4.2) is a nonsingular square submatrix of A of order m, and x_B, the corresponding basic vector, is the vector of variables in (4.2) associated with the columns in the basis B. Every basic vector for (4.2) is a vector of m variables the column vectors associated with which form a linearly independent set, and hence a basis.

When considering the basis B, and the basic vector x_B, variables not in x_B are called **nonbasic variables**, the columns in (4.2) associated with them are called **nonbasic columns**. Let x_D denote the vector of nonbasic variables in some order, and D the submatrix of A consisting of the columns of A associated with these nonbasic variables. Let c_B be the row vector of basic cost coefficients, and c_D the row vector of nonbasic cost coefficients. Rearranging the variables in (4.2) into basic and nonbasic parts, (4.2) can be written as

Tableau 4.4

x_B	x_D	
B	D	b
c_B	c_D	$= z$, minimize

$x_B,\ x_D \gtreqless 0.$

The basic vector x_B defines a unique **basic solution** for the system of equality constraints "$Ax = b$", which may not be feasible in the sense it may not satisfy the sign restrictions "$x \gtreqless 0$". It is obtained by setting all the nonbasic variables equal to zero ($x_D = 0$) and then solving the remaining system for the values of the basic variables in the solution. This remaining system is $Bx_B = b$, and its solution is $x_B = B^{-1}b$. So, the primal basic solution of (4.2) associated with the basic vector x_B, or the corresponding basis B is

$$\text{vector of nonbasic variables,} \quad x_D = 0$$
$$\text{basic vector,} \quad x_B = B^{-1}b \qquad (4.10)$$

If $B^{-1}b \gtreqless 0$, the solution in (4.10) is feasible to (4.2) since it satisfies all the constraints and sign restrictions, in this case it is called a **basic feasible solution (BFS) of** (4.2); and the basic vector x_B and the basis B are said to the **primal feasible basic vector** and **primal feasible basis**, respectively.

If $B^{-1}b \ngtreqless 0$, the solution in (4.10) satisfies the constraints "$Ax = b$" but not the sign restrictions "$x \gtreqless 0$". In this case the solution in (4.10) is infeasible to (4.2), and the basic vector x_B and basis B are called **primal infeasible basic vector**, **primal infeasible basis** respectively.

The primal basic solution in (4.10) is said to be **primal degenerate** if at least one of the basic variables has value zero in it; i.e., if at least one of the components of $B^{-1}b$ is zero. In this case the basic vector x_B and basis B are said to be **primal degenerate**.

If all the components of $B^{-1}b$ are nonzero (i.e., every basic variable has a nonzero value) then the basic solution in (4.10) is said to be a **primal nondegenerate basic solution of** (4.2). In this case the basic vector x_B and the basis B are said to be **primal nondegenerate**.

Rearranging the constraints in the dual problem (4.3) in order of the primal variables corresponding to them as arranged in Tableau 4.4, they are

$$\pi B \quad \overset{\leq}{\underset{=}{}} \quad c_B$$
$$\pi D \quad \overset{\leq}{\underset{=}{}} \quad c_D. \tag{4.11}$$

The first line in (4.11) contains the dual constraints corresponding to the m basic variables in x_B, and the second line contains those corresponding to the nonbasic variables in x_D. Denote the row vectors of dual slacks variables in these sets by $\bar{c}_B$, $\bar{c}_D$. Introducing these slack variables, (4.11) becomes

$$\pi B + \bar{c}_B \quad = \quad c_B$$
$$\pi D + \bar{c}_D \quad = \quad c_D \tag{4.12}$$
$$\bar{c}_B, \; \bar{c}_D \quad \overset{\geq}{\underset{=}{}} \quad 0.$$

The definition of the **dual basic solution** corresponding to the basis B is tailored to make sure that it satisfies the complementary slackness conditions together with the primal basic solution associated with the basis B given in (4.10). In the primal basic solution in (4.10), only basic variables in x_B can have nonzero values, and the complements of these variables are the dual slacks in the vector $\bar{c}_B$. So, for the dual basic solution to satisfy the complementary slackness conditions with the primal basic solution in (4.10), it is enough if we make sure that $\bar{c}_B = 0$, from (4.12); this defines the dual basic solution associated with B to be the unique solution of

$$\pi B = c_B \tag{4.13}$$

or $\tilde{\pi} = c_B B^{-1}$. So, the dual solution corresponding to the basic vector x_B of (4.2) is the unique solution of the system of dual constraints corresponding to the basic variables in x_B, each treated as an equation (this is (4.13)).

This dual basic solution $\tilde{\pi}$, is feasible to the dual problem if it satisfies all the dual constraints (those in (4.3) or (4.12)). It satisfies the dual constraints corresponding to the basic variables in x_B as equations. So to be dual feasible it has to satisfy the dual constraints associated with nonbasic variables; i.e., the relative cost coefficients of all the nonbasic variables, $\bar{c}_j = c_j - \tilde{\pi} A_{.j}$ have to be $\overset{\geq}{\underset{=}{}} 0$. If this happens, the basis B, and the basic vector x_B are said to be a **dual feasible basis** and **dual feasible basic vector** for (4.2), respectively. If at least one of the nonbasic relative cost coefficients $\bar{c}_j = c_j - \tilde{\pi}A_{.j}$ is < 0, $\tilde{\pi}$ is dual infeasible; in this case the basis B, and the basic vector x_B are said to be **dual infeasible** for (4.2).

To summarize, let x_B be a basic vector for (4.2) associated with the basis B, nonbasic vector x_D, basic cost (row) vector c_B, nonbasic cost (row) vector c_D. The primal basic solution corresponding to x_B is obtained by the system on the left of (4.14); and the dual basic solution corresponding to x_B is obtained by the system on the right in (4.14). x_B is primal feasible if $B^{-1}b \geqq 0$; it is dual feasible if $c_j - (c_B B^{-1})A_{.j} \geqq 0$ for all nonbasic x_j.

$$Bx_B = b \qquad\qquad \pi B = c_B \qquad\qquad (4.14)$$
$$x_D = 0.$$

As an example, consider the vector $x_B = (x_1, x_2, x_3)$ for the LP in standard form in Tableau 4.1. The corresponding coefficient submatrix B is the 3×3 coefficient matrix for the system on the left given below; it is nonsingular, and hence a basis and so x_B is a basic vector. The primal basic solution corresponding to it is obtained from the system of equations on the left given below. It is $\tilde{x} = (2, 6, 1, 0, 0, 0)^T$. So, this basic vector x_B is primal feasible, and it is primal nondegenerate since all the basic variables x_1, x_2, x_3 are nonzero in the basic solution. The dual basic solution corresponding to x_B is the solution of the system of equations on the right given below. It is $\tilde{\pi} = (3, 5, -4)$. By substituting this solution in the dual constraints given in Tableau 4.3, we find that the vector of dual slacks at $\tilde{\pi}$ are $\bar{c} = (0, 0, 0, 3, 0, 4)$, since $\bar{c} \geqq 0$, it is dual feasible. So for this problem, x_B is both a primal and dual feasible basic vector. Also, verify that $\tilde{x}, \tilde{\pi}$ satisfy the complementary slackness conditions "$x_j \bar{c}_j = 0$" for all j (this automatically follows from the manner in which the dual basic solution corresponding to a basic vector is defined).

x_1	x_2	x_3	
1	2	3	17
0	1	−4	2
0	0	1	1
$x_4 = x_5 = x_6 = 0$.			

π_1	π_2	π_3	
1	0	0	3
2	1	0	11
3	−4	1	−15

As another example, consider the vector $x_{B_2} = (x_4, x_5, x_6)$ for the LP in standard form in Tableau 4.1. The corresponding coefficient submatrix B_2 is the 3×3 coefficient matrix for the system on the left given below, it is also nonsingular and hence a basis. The primal basic solution, obtained from the system on the left given below, is $\hat{x} = (0, 0, 0, 0, 1, 1)^T$; it is primal feasible, but since the basic variable x_4 is zero in it, it is primal degenerate. So x_{B_2} is a degenerate primal feasible basic vector for this problem. The dual basic solution corresponding to x_{B_2}, obtained from the system on the right given below is $\hat{\pi} = (51/16, 6, -83/16)$. By substituting $\hat{\pi}$

in the dual constraints given in Tableau 4.3, we find that the vector of dual slacks at $\hat{\pi}$ is $\bar{c} = (-3/16, -11/8, 37/8, 0, 0, 0)$. Since the first two components in this vector are < 0, $\hat{\pi}$ is dual infeasible; so x_{B_2} is a dual infeasible basic vector for the LP in Tableau 4.1.

x_4	x_5	x_6	
-2	1	16	17
1	1	1	2
-2	1	0	1
$x_1 = x_2 = x_3 = 0$.			

π_1	π_2	π_3	
-2	1	-2	10
1	1	1	4
16	1	0	57

Suppose the basic vector x_B associated with the basis B for (4.2) is both primal and dual feasible. Let $\tilde{x}, \tilde{\pi}$ be the corresponding primal and dual basic solutions. Then by their definition $\tilde{x}, \tilde{\pi}$ satisfy all three conditions for optimality (primal and dual feasibility, and complementary slackness conditions) stated in Theorem 4.1. So, $\tilde{x}$ is optimal to (4.2), and $\tilde{\pi}$ is optimal to its dual (4.3). Hence the BFS associated with a basic vector for (4.2) which is both primal and dual feasible is always optimal. For this reason a basic vector for (4.2) which is both primal and dual feasible, is called an **optimal basic vector**.

We now state an important result for LP theory that is used in the simplex algorithm.

Theorem 4.2 *If the LP in standard form (4.2) has an optimum solution, then it has an optimum solution which is the BFS associated with an optimum basic vector for (4.2).*

See [D. Gale, 1960] or [K. G. Murty, 1983 of Chapter 2] for a proof of this theorem. The number of possible basic vectors for (4.2) is at most $\binom{n}{m}$, a finite number, and Theorem 4.2 states that it is sufficient to search among the finite number of BFSs for an optimum solution. This makes it possible to solve the continuous problem (4.2) by a finite search. The simplex algorithm for LP discussed later is based on this idea; it searches among the BFSs in an efficient way moving from one BFS to a better one in each step until an optimum solution is found.

4.3.1 How to Compute the Basic Solutions

Consider the LP in standard form, (4.2), where A is a matrix of order $m \times n$ and rank m. To understand the computations to be carried out when solving this problem by the simplex method, or to actually carry out these computations on a numerical instance by hand computation, it is convenient to express this problem in the form of a detached coefficient tableau as given below.

Tableau 4.5: Original tableau

x	$-z$	
A	0	b
c	1	0

$$x_j \geqq 0 \qquad \text{for all } j, \text{ minimize } z$$

It is called the original tableau because it contains the data from the original statement of the problem. There are $m + 1$ rows in this tableau. Each of the first m rows represents an equality constraint in (4.2). The $(m + 1)$th (last) row in this tableau just represents the equation $cx - z = 0$, so it defines the objective function, and is called the objective row.

Suppose $x_B = (x_1, \ldots, x_m)$ is a basic vector for this problem. The **associated basis** B is the submatrix of A consisting of the columns of A associated with the basic variables in x_B, it is $(A_{.1} \vdots \cdots \vdots A_{.m})$. The row vector of basic cost coefficients is $c_B = (c_1, \ldots, c_m)$.

Each of the variables in x_B will be a basic variable in one of the constraint rows in Tableau 4.5. The last row in Tableau 4.5 is not a constraint row, it just defines the objective function in the problem. We extend the basic vector to all the rows in Tableau 4.5 by selecting $-z$ as the basic variable in row $m+1$. The **augmented basis**, consisting of the column vectors of all the basic variables including $-z$ in their proper order is the submatrix of order $(m + 1) \times (m + 1)$ given below on the left.

$$\begin{pmatrix} B & \vdots & 0 \\ \cdot & \cdot & \cdot \\ c_B & \vdots & 1 \end{pmatrix}, \qquad T = \begin{pmatrix} B & \vdots & 0 \\ \cdot & \cdot & \cdot \\ c_B & \vdots & 1 \end{pmatrix}^{-1} = \begin{pmatrix} B^{-1} & \vdots & 0 \\ \cdot & \cdot & \cdot \\ -c_B B^{-1} & \vdots & 1 \end{pmatrix} \qquad (4.15)$$

The inverse of the augmented basis is the matrix T given above on the right. The first m entries in the last row of T constitute $-c_B B^{-1} = -\pi$, which is the negative of the dual basic solution associated with the basic vector x_B. Verify that

$$T \begin{pmatrix} b \\ \ldots \\ 0 \end{pmatrix} = \begin{pmatrix} B^{-1} b \\ \ldots \\ -c_B B^{-1} b \end{pmatrix} = \begin{pmatrix} \bar{b} \\ \ldots \\ -\bar{z} \end{pmatrix} \qquad (4.16)$$

where $\bar{b}$ is the vector of values of the basic variables in the primal basic solution corresponding to x_B. Since all the nonbasic variables are zero in this basic solution, we get the primal basic solution completely from this information. Also $-\bar{z}$ in (4.16) can be verified to be the negative of the value of z at this primal basic solution. The column vector on the right hand side of (4.16) is known as the **updated RHS constants vector** wrt x_B. The following tableau consisting of T and the updated RHS constants vector is known as the **inverse tableau** wrt the basic vector x_B.

Inverse Tableau

Basic var.	T		Basic values
x_B	B^{-1}	0	$\bar{b}$
$-z$	$-\pi$	1	$-\bar{z}$

As mentioned above, the basic values column in the inverse tableau gives the values of the basic variables and the negative objective value at the primal basic solution wrt x_B. From the last row of T we can read off the dual basic solution associated with x_B. Also, the relative cost coefficient of any nonbasic variable x_j

wrt the basic vector x_B is $c_j - \pi A_{.j} = (-\pi, 1) \begin{pmatrix} A_{.j} \\ \dots \\ c_j \end{pmatrix}$, so it is the dot product of

the last row vector of T with the column vector of x_j in the original tableau, and hence can be computed very efficiently.

So, given any basic vector for (4.2), the primal and dual basic solutions associated with it can be computed by constructing the inverse tableau wrt x_B using the formulae in (4.15) and (4.16).

As an example, consider the basic vector $x_B = (x_1, x_2, x_3)^T$ for the LP in Tableau 4.1. We have the following for this basic vector.

$$\text{Basis} = B = \begin{pmatrix} 1 & 2 & 3 \\ 0 & 1 & -4 \\ 0 & 0 & 1 \end{pmatrix}, \text{Augmented basis} = \begin{pmatrix} 1 & 2 & 3 & 0 \\ 0 & 1 & -4 & 0 \\ 0 & 0 & 1 & 0 \\ 3 & 11 & -15 & 1 \end{pmatrix}$$

The inverse tableau corresponding to x_B is given below

Inverse tableau wrt x_B

Basic var.	T				basic values
x_1	1	-2	-11	0	2
x_2	0	1	4	0	6
x_3	0	0	1	0	1
$-z$	-3	-5	4	1	-57

Here T is the inverse of the augmented basis. From its last row we verify that the dual basic solution is $\pi = (3, 5, -4)$ computed earlier. The basic values column is obtained by multiplying the original RHS constants column $(17, 2, 1, 0)^T$ by T on the left. From it we read out that $x_1 = 2, x_2 = 6, x_3 = 1$ in the primal basic solution. x_4, x_5, x_6 being nonbasics are 0 in this solution. And the objective value at this solution is 57.

The simplex algorithm moves from one basic vector to an adjacent basic vector, maintaining the inverse tableau by updating it with a single pivot step as discussed in Section 3.3.8.

4.4   When is an Optimum Dual Solution the Marginal Value Vector?

Consider the LP in standard form (4.2). The marginal value of b_i in this problem has been defined to be the rate of change in the optimum objective value per unit change b_i from its current value, when this rate exists. Select a b_i, say b_1. Suppose we keep all the other data in the problem fixed at their current value, except b_1. Then as b_1 varies, the optimum objective value in the problem is a function of b_1 which we denote by $f(b_1)$. It can be shown that $f(b_1)$ is a continuous piecewise linear function with monotonically increasing slope (see Figure 4.1) in the range where it is defined and finite. For a proof of this result see [K. G. Murty, 1983 of Chapter 2].

In Figure 4.1 we called the values of b_1 where the slope of $f(b_1)$ changes, β_1, β_2, etc. These are known as **breakpoints** since the slope of $f(b_1)$ changes at them. If the current value of b_1 is not at a breakpoint, it must lie in an interval between

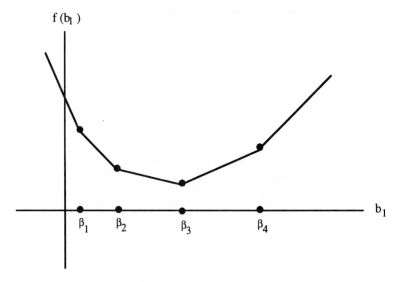

Figure 4.1 Optimum objective value as a function of the RHS constant b_1 is piecewise linear with monotonically increasing slope.

two consecutive break points, say $\beta_t < b_1 < \beta_{t+1}$. Then in this interval $f(b_1)$ is linear, the marginal value of $f(b_1)$ at such a point b_1 is well defined and it is the slope of $f(b_1)$ in that interval, and this value remains unchanged as long as b_1 remains within this interval. If the current value of b_1 is at a breakpoint β_1 say, then the marginal value at β_1 is not well defined since the slope of $f(b_1)$ changes at β_1. However, at β_1 there is a **positive marginal value** which is the rate of change in $f(b_1)$ as b_1 increases from β_1 (the **right slope** of $f(b_1)$ at β_1), and a **negative marginal value** which is the rate of change in $f(b_1)$ as b_1 decreases from β_1 (the **left slope** of $f(b_1)$ at β_1).

So marginal values wrt all b_i are well defined if none of the current values of any b_i are at a breakpoint for it. We have the following theorem which makes it possible to conclude if this is the case, and derive these marginal values in terms of the optimum dual solution.

Theorem 4.3 *Consider the LP (4.2). If an optimum BFS for the problem obtained is primal nondegenerate (i.e., all basic variables are > 0 in that BFS), then for each i the current value of b_i is not at a breakpoint for it. In this case, there is a unique optimum dual solution, and it is the vector of marginal values.*

For a proof of this theorem see [K. G. Murty, 1983 of Chapter 2]. It says that if the primal optimal BFS obtained is nondegenerate, it is perfectly valid to interpret the optimum dual solution obtained, $\pi = (\pi_i)$ as the marginal value vector. In fact in Chapter 7 we discuss techniques for determining the range of values of the RHS constant b_i around its present value, for which the marginal value wrt b_i remains equal to this π_i, for any i.

And we have the following result that covers the other possibility.

Theorem 4.4 *Let $\bar{x}, \bar{\pi} = (\bar{\pi}_i)$ be optimal primal and dual basic solutions associated with (4.2) obtained by an algorithm. If $\bar{x}$ is a degenerate BFS, $\bar{\pi}$ may not be the unique optimum dual solution. In this case it is possible that for some i, the current value of b_i may be at a break point for it, and so marginal values may not be well defined. However, in this case, for each i,*

$$\text{negative marginal value wrt } b_i \overset{\le}{=} \bar{\pi}_i \overset{\le}{=} \text{positive marginal value wrt } b_i.$$

So the rate of change in the optimum objective value per unit increase in b_i from present value is $\overset{\ge}{=} \bar{\pi}_i$; and the same rate per unit decrease in b_i from present value is $\overset{\le}{=} \bar{\pi}_i$.

For a proof of this theorem again see [K. G. Murty, 1983 of Chapter 2]. In the case covered by Theorem 4.4, the positive and negative marginal values wrt each b_i can be computed by means of efficient algorithms involving a further analysis of the dual problem. These algorithms are discussed in [K. G. Murty, 1983 of Chapter 2]. But for some reason, these positive and negative marginal values have not come into popular use, and none of the currently existing commercial LP software packages have subroutines for evaluating them when the optimal primal solution obtained is degenerate.

4.5   Exercises

4.1 Consider the following LP in standard form.

x_1	x_2	x_3	x_4	x_5	
1	0	1	0	-1	20
0	1	0	1	1	10
0	0	1	$-3/2$	0	5
1	1	1	2	0	$= z$, minimize

$$x_j \overset{\geq}{=} 0 \text{ for all } j$$

Find the primal and dual basic solutions associated with the basic vectors $x_{B_1} = (x_1, x_2, x_3)$, $x_{B_2} = (x_2, x_3, x_4)$ for this problem. Mention whether each of these basic vectors is primal feasible or not, primal nondegenerate or not, and dual feasible or not. Compute the inverse tableau for this problem wrt x_{B_1}.

4.2 Consider the following LP

$$
\begin{array}{rlllllll}
\text{Minimize } z(x) = & 12x_1 & -x_2 & +3x_3 & +5x_4 & & -8x_6 & \\
\text{subject to} & x_1 & -2x_2 & +x_3 & -x_4 & +x_5 & -x_6 & +x_7 & = -4 \\
& x_1 & +x_2 & & +2x_4 & -x_5 & -x_6 & -2x_7 & = 9 \\
& -2x_1 & +x_2 & & & -x_5 & +x_6 & +x_7 & = 5 \\
\end{array}
$$
$$x_j \overset{\geq}{=} 0, \ j = 1 \text{ to } 7.$$

Write the dual of this problem and all the complementary pairs in these primal, dual problems.

Is $\bar{\pi} = (3, 4, -1)$ dual feasible in this problem? Compute all the relative cost coefficients wrt $\bar{\pi}$.

Mention all the primal variables which have to be equal to 0 in every optimum solution of the primal LP if it is known that $\bar{\pi}$ is an optimum dual solution.

In the system of *equality constraints* in the primal problem, substitute 0 for each primal variable determined above, and obtain the general solution for the remaining system. What kind of geometric object is the set of all solutions to the remaining system?

Identify nonnegative solutions of the remaining system. Are these optimum solutions of the primal LP? Why?

4.3 Check whether the basic vector (x_4, x_2, x_6) is primal feasible for the following LP. Is it dual feasible? Is it optimal?

x_1	x_2	x_3	x_4	x_5	x_6	b
1	1	1	1	1	1	9
1	1	0	1	1	0	5
1	0	0	1	0	0	2
3	−2	1	−5	4	−2	$= z$, minimize

$$x_j \geqq 0 \text{ for all } j.$$

4.4 Write the dual of the following LP, and the complementary slackness optimality conditions for this primal, dual pair. Using them check whether $\pi = (-2, -3, 1)$ is an optimum solution of the dual problem. If it is, derive an optimum solution of the LP using it and the optimality conditions.

x_1	x_2	x_3	x_4	x_5	x_6	x_7	b
1	2	−2	1	0	3	1	−5
−2	1	−1	2	1	1	2	1
1	0	2	2	2	2	−2	8
10	0	9	−2	−1	−1	−10	$= z$, minimize

$$x_j \geqq 0 \text{ for all } j.$$

4.5 Ice cream is a mixture of 4 ingredients, and its flavor is measured in units of mmm. Each gallon of a certain brand of ice cream is required to have at least one mmm. Data on the cost and mmm content of the ingredients is given below.

Ingredient	1	2	3	4
cost (\$/gallon)	2	0.7	1	1.5
mmm/gallon	3	0	0.5	1

Formulate the problem of determining a minimum cost composition for acceptable ice cream.

Write the dual of this problem and solve this dual geometrically. Find the optimum composition from the dual optimum using the complementary slackness optimality conditions.

4.6 For the following LP check whether $\hat{x} = (5, 2, 3, 1, 0, 0, 0)^T$ is an optimum solution using the c.s. optimality conditions.

x_1	x_2	x_3	x_4	x_5	x_6	x_7	b
1	−1	1	0	1	0	2	6
0	2	1	−1	−1	2	0	6
0	0	2	1	0	1	2	7
0	0	0	2	1	−1	1	2
−2	4	6	3	−1	10	5	$= z(x)$, minimize

$$x_1 \text{ to } x_7 \geqq 0$$

4.7 For the following LP check whether $\tilde{x} = (5, 6, 4, 0, 0)^T$ is an optimum solution using the c.s. optimality conditions.

x_1	x_2	x_3	x_4	x_5	b
1	2	4	2	0	28
2	−1	3	−4	2	16
−1	1	−1	3	5	−3
6	3	10	4	6	$= z(x)$, minimize

$$x_1 \text{ to } x_5 \overset{\geq}{=} 0$$

4.8 Consider the balanced transportion problem with the following data, and dual solution $(\tilde{u}, \tilde{v})$.

Source i	Unit Shipping cost to mkt. j				Availability	$\tilde{u}_i$
	1	2	3	4		
1	11	10	8	5	200	5
2	6	8	5	2	100	2
3	11	5	6	5	150	3
Requirement	50	50	200	150		
$\tilde{v}_j$	4	2	3	0		

Verify that $(\tilde{u}, \tilde{v})$ is dual feasible. Assuming that $(\tilde{u}, \tilde{v})$ is an optimum dual solution, find an optimum solution of the primal transportation problem.

4.9 Consider the problem of allocating candidates to marketing director positions in zones discussed in Section 1.4, with annual sales data given in Table 1.1. Using $\tilde{u}$ = (0, 2, 4, 9, 16, 24), $\tilde{v}$ = (1, 2, 5, 7, 9, 13) as the dual solution, show that the greedy solution obtained in Section 1.4, $\{(C_1, Z_1), (C_2, Z_2), (C_3, Z_3), (C_4, Z_4), (C_5, Z_5), (C_6, Z_6)\}$ actually minimizes the total sales volume as claimed in Section 1.4, instead of maximizing it.

4.6　References

D. GALE, 1960, *Theory of Linear Economic Models*, McGraw Hill, NY.

Chapter 5

Hungarian Method: A Primal-Dual Algorithm for the Assignment Problem

Primal-dual algorithms are a class of methods for LPs with the following characteristic features:

1. They maintain a dual feasible solution, and a primal vector (this vector is primal infeasible until termination) that together satisfy all the complementary slackness optimality conditions for the original problem throughout the algorithm. This primal vector is usually feasible to a relaxation of the primal problem (typically this is obtained by changing the equality constraints in the problem to " $\leq$ " inequalities).

2. In each step, the algorithm either performs (a) below, or (b), if this is not possible.

   (a) Keeps the dual solution fixed, and tries to alter the primal vector to bring it closer to primal feasibility while continuing to satisfy the complementary slackness conditions together with the present dual solution.

   (b) Keeps the primal vector fixed, and changes the dual feasible solution. The aim of this is to get a new dual feasible solution satisfying two conditions. The first is that the new dual feasible solution satisfies the complementary slackness conditions together with the present primal vector. The second is that it makes it possible to get a new primal vector closer to primal feasibility when the algorithm continues.

3. As the algorithm progresses, the primal vector moves closer and closer to primal feasibility. In other words, there is a measure of primal infeasibility which improves monotonically during the algorithm.

There are two possible conclusions at termination. One occurs if the primal vector being maintained becomes primal feasible at some stage; then it is an optimum solution. The second occurs if a primal infeasibility criterion is satisfied at some stage.

For the assignment problem it is easy to obtain an initial dual feasible solution. And the task in (a) above is a maximum value flow problem on a subnetwork known as the **admissible** or **equality subnetwork** wrt the present dual feasible solution which can be solved very efficiently by a simple labeling routine. These facts make the primal-dual method particularly attractive to solve the assignment problem. The primal-dual approach can be used to solve a general LP, although, for these general problems it seems to offer no particular advantage over the primal simplex algorithm discussed in Chapter 7.

5.1   The Hungarian Method for the Assignment Problem

The data in an assignment problem of order n is the cost matrix $c = (c_{ij})$ of order $n \times n$. Given c, the problem is to find $x = (x_{ij})$ of order $n \times n$ to

$$\text{Minimize } z_c(x) \quad = \quad \sum_{i=1}^{n} \sum_{j=1}^{n} c_{ij} x_{ij}$$

$$\text{Subject to } \sum_{j=1}^{n} x_{ij} \quad = \quad 1 \quad \text{for } i = 1 \text{ to } n \qquad (5.1)$$

$$\sum_{i=1}^{n} x_{ij} \quad = \quad 1 \quad \text{for } j = 1 \text{ to } n$$

$$x_{ij} \quad \geqq \quad 0 \quad \text{for all } i, j$$

$$\text{and } x_{ij} \quad = \quad 0 \text{ or } 1 \quad \text{for all } i, j \qquad (5.2)$$

Every feasible solution of (5.1) and (5.2) is an assignment of order n and vice versa. See (2.4) for an assignment of order 5. Here is an assignment $\bar{x}$ of order 4.

$$\begin{pmatrix} 0 & 1 & 0 & 0 \\ 0 & 0 & 0 & 1 \\ 1 & 0 & 0 & 0 \\ 0 & 0 & 1 & 0 \end{pmatrix} \qquad (5.3)$$

Verify that each row and each column of an assignment always contains a single nonzero entry of "1", so it satisfies the constraints in (5.1). So, every assignment is a permutation matrix and vice versa. The Hungarian method does not use basic vectors for (5.1), but it maintains (5.2) throughout.

In an assignment x, if a particular $x_{ij} = 1$, the cell (i, j) is said to have an **allocation** (in this case row i is said to be **allocated to**, or **matched with**, column j in x). The assignment in (2.4) has allocations in cells (1, 5), (2, 1), (3, 4), (4, 2), (5, 3) because $x_{ij} = 1$ in all these cells in this assignment; all other cells have no allocation in them since $x_{ij} = 0$ in them.

Let j_r be the column which has the unique "1" entry in row $r, r = 1$ to n, in an assignment x of order n. Then the only nonzero variables in x are x_{r,j_r}, for $r = 1$ to n. In this case we will denote the assignment x by the set of its unit cells, namely $\{(1, j_1), \ldots, (n, j_n)\}$. In this notation, the assignment $\bar{x}$ in (5.3) will be denoted by $\{(1, 2), (2, 4), (3, 1), (4, 3)\}$.

Assignments are useful to model situations in which there are two distinct sets of objects of equal numbers, and we need to form them into pairs, each pair consisting of one object of each set. For example, $\mathcal{N}_1$ might be a set of n boys, and $\mathcal{N}_2$ a set of an equal number of girls. If each boy in $\mathcal{N}_1$ marries a girl in $\mathcal{N}_2$, the resulting set of couples will be an assignment of order n. Sometimes we will take both the sets $\mathcal{N}_1, \mathcal{N}_2$, to be $\{1, \ldots, n\}$. It should be understood that they refer to the serial numbers of distinct sets of objects.

A **partial assignment** of order n is a 0-1 square matrix of order n which contains at most one nonzero entry of 1 in each row and column. Here is a partial assignment of order 3, only row 1 has an allocation in it, rows 2, 3 have no allocations.

$$\begin{pmatrix} 0 & 1 & 0 \\ 0 & 0 & 0 \\ 0 & 0 & 0 \end{pmatrix}$$

Clearly, a partial assignment is a feasible solution for a relaxed version of (5.1) and (5.2) in which the equality constraints in (5.1) are replaced by the corresponding "$\leq$" inequalities. The Hungarian method moves among partial assignments in which the number of allocations keeps increasing as the algorithm progresses. For the sake of distinguishing between them, we will refer to assignments which are not partial assignments as **full** or **complete assignments**.

As discussed in Chapter 4, the dual of (5.1) is: find $u = (u_1, \ldots, u_n), v = (v_1, \ldots, v_n)$ to

$$\text{Maximize } w(u, v) = \sum_{i=1}^{n} u_i + \sum_{j=1}^{n} v_j$$

$$\text{Subject to } u_i + v_j \overset{\leq}{=} c_{ij}, \ i, j = 1 \text{ to } n \tag{5.4}$$

The dual constraints can be written in an equivalent manner as

$$\bar{c}_{ij} = c_{ij} - u_i - v_j \overset{\geq}{=} 0, \quad i, j, = 1 \text{ to } n \tag{5.5}$$

The conditions (5.5) are the **dual feasibility conditions** for (u, v). The $\bar{c}_{ij}$ are known as the **reduced** or **relative cost coefficients** wrt (u, v), and the matrix $\bar{c} = (\bar{c}_{ij})$ is known as the reduced cost matrix wrt (u, v). So, the vectors (u, v) are dual feasible iff the reduced cost matrix wrt them, $\bar{c}$ is $\overset{\geq}{=} 0$.

Given the vectors $(u = (u_i), v = (v_j))$, to compute the reduced cost coefficients, we subtract u_i from each entry in the ith row of the original cost matrix c, for $i = 1$ to n. Then in the resulting matrix we subtract v_j from each entry in the jth column, for $j = 1$ to n. This leads to the reduced cost matrix $\bar{c}$ wrt (u, v). The total subtracted from the rows and columns is $\sum_{i=1}^{n} u_i + \sum_{j=1}^{n} v_j$, this quantity is called the **total reduction**, it is the dual objective function $w(u, v)$.

Given a dual feasible solution (u, v), the cell (i, j) in the assignment array is classified as follows wrt (u, v)

$$\text{if} \quad \bar{c}_{ij} = c_{ij} - u_i - v_j \left\{ \begin{array}{ll} = 0, & \text{it is an \textbf{admissible cell}} \\ > 0, & \text{it is an \textbf{inadmissible cell}} \end{array} \right.$$

Admissible cells are also called **equality cells** in the literature; this name refers to the fact that the dual constraint corresponding to this cell (which is $u_i + v_j \overset{\leq}{=} c_{ij}$) holds as an equation if the cell is an equality cell.

From Chapter 4 we know that an assignment x and a dual feasible solution (u, v) are optimal to the respective problems if

$$x_{ij}(c_{ij} - u_i - v_j) = x_{ij}\bar{c}_{ij} = 0, \qquad \text{for all } i, j = 1 \text{ to } n \qquad (5.6)$$

i.e., if all the allocations in x occur in admissible cells wrt (u, v). (5.6) are the **complementary slackness optimality conditions** for the assignment problem and its dual. These conditions are the active principle of the Hungarian method. It begins with a dual feasible solution and tries to find an assignment with allocations only among the admissible cells wrt it.

As an example, consider the assignment problem of order 3 with cost matrix given in the following array. In all the arrays in this chapter we enter the original cost coefficient c_{ij} in the lower right corner of cell (i, j). We record the dual vector u in the right most column, and the dual vector v in the bottommost row of the array. The relative cost coefficient $\bar{c}_{ij} = c_{ij} - u_i - v_j$ is entered in the upper left corner of cell (i, j). Since there are already so many numbers in cell (i, j), if we also record the value of x_{ij} in the assignment in that cell, it will create confusion. To avoid confusion, we record the assignment by putting a little square " $\square$ " in the center of each allocated cell.

$j =$	1	2	3	u_i
$i = 1$	$3 = \bar{c}_{11}$ $c_{11} = 10$	0 9	0 $\square$ 10	3
2	19 22	0 $\square$ 5	6 12	-1
3	0 $\square$ 9	9 20	3 15	5
v_j	4	6	7	

Since all the $\bar{c}_{ij}$ are $\geqq 0$, the (u, v) given on the array is dual feasible. The total reduction is $\sum_{i=1}^{n} u_i + \sum_{j=1}^{n} v_j = 24$ in this example. The admissible cells wrt (u, v) in this example are $(1, 2)$, $(1, 3)$, $(2, 2)$, $(3, 1)$; these are the cells with $\bar{c}_{ij} = 0$. All other cells are inadmissible. To find an optimum assignment in this example, we try to find an assignment with allocations among admissible cells only. Such an assignment $x = \{(1, 3), (2, 2), (3, 1)\}$ has been entered in the array. As a matrix this assignment x is the following.

$$x = \begin{pmatrix} 0 & 0 & 1 \\ 0 & 1 & 0 \\ 1 & 0 & 0 \end{pmatrix}$$

Since, x, (u, v) satisfy the complementary slackness optimality conditions (5.6), x is an optimum assignment in this example, and (u, v) is an optimum dual solution. The optimum objective value is $c_{13} + c_{22} + c_{31} = 10 + 5 + 9 = 24$, is the total reduction wrt (u, v).

For practice, see whether you can find an optimum assignment in the problems with cost matrices given in the following arrays, using the dual solution entered on the array, by the same procedure. In each case check whether the dual solution given is dual feasible or not.

$j =$	1	2	3	u_i
$i = 1$	$c_{11} = 3$	9	10	10
2	4	5	19	6
3	6	8	8	3
v_j	5	8	4	

$j =$	1	2	3	u_i
$i = 1$	c_{11} $= 1$	6	8	1
2	12	33	45	12
3	14	84	93	14
v_j	0	5	7	

The Total Reduction wrt a Dual Feasible Solution as a Lower Bound for the Optimum Objective Value

Consider an optimization problem in which an objective function $f(x)$ in decision variables x is to be minimized subject to some constraints on x. Let f^* denote the unknown minimum value of $f(x)$ in the problem. If there is no way to eyeball the value of f^* from the original data directly, an extremely valuable thing to have is an efficient technique that can compute a good lower bound for it. Such techniques are called **lower bounding strategies**. The output of a lower bounding strategy for this problem is therefore a number ℓ which is guaranteed to satisfy $\ell \stackrel{\leq}{=} f^*$. The quality of this lower bound ℓ becomes higher as the difference $f^* - \ell$ decreases closer to 0. A lower bounding strategy that is guaranteed to produce high quality lower bounds with very little computational effort is very highly useful to get a handle on the original minimization problem. Many approaches for solving minimization problems are based on judicious use of very efficient lower bounding strategies. For the assignment problem, we have a very nice lower bounding strategy based on its dual, and the Hungarian method makes use of it. Lower bounding strategies are particularly important in algorithms for solving integer and combinatorial optimization problems; in fact lower bounding strategies are a basic building block in the branch and bound approach for solving these problems (see Chapter 10).

We will now state the lower bounding result for the assignment problem in the form of a theorem.

Theorem 5.1 *Consider the assignment problem (5.1) of order n with c as the cost matrix. Let (u, v) be any dual feasible solution with total reduction $w = \sum_{i=1}^{n} u_i + \sum_{j=1}^{n} v_j$. Then w is a lower bound for the minimum value of $z_c(x)$ in the problem, i.e.,*

$$z_c(x) \stackrel{\geq}{=} w, \qquad \text{for all assignments} \quad x \tag{5.7}$$

Proof $c = (c_{ij})$ is the original cost matrix, and $z_c(x)$ is the original objective function to be minimized in (5.1). Suppose the matrix $c^1 = (c_{ij}^1)$ is obtained by selecting any row or column of c, and subtracting a number α from each element in it. For example, if we selected row 1 for this operation

$$c^1 = \begin{pmatrix} c_{11} - \alpha & c_{12} - \alpha & \cdots & c_{1n} - \alpha \\ c_{21} & c_{22} & \cdots & c_{2n} \\ \vdots & \vdots & & \vdots \\ c_{n1} & c_{n2} & \cdots & c_{nn} \end{pmatrix}$$

Let x be any assignment of order n; and let $z_c(x), z_{c^1}(x)$ be the objective values of any assignment x with c, c^1 as cost matrices, respectively. The fact that each assignment contains exactly one allocation in any row or column of the array implies that

$$z_{c^1}(x) = z_c(x) - \alpha, \qquad \text{for all assignments } x$$

or $z_c(x) = z_{c^1}(x) + \alpha$.

Let $\bar{c} = (\bar{c}_{ij} = c_{ij} - u_i - v_j)$ be the reduced cost matrix wrt the dual feasible solution (u, v). Since $\bar{c}$ is obtained by subtracting u_i from each entry in the ith row of the original cost matrix c, for $i = 1$ to n; and then subtracting v_j from each entry in the jth column of the resulting resulting matrix , for $j = 1$ to n; by applying the above result repeatedly we conclude that $z_{\bar{c}}(x)$, the cost any assignment x with $\bar{c}$ as the cost matrix, satisfies

$$z_c(x) = z_{\bar{c}}(x) + \sum_{i=1}^n u_i + \sum_{j=1}^n v_j = z_{\bar{c}}(x) + w \tag{5.8}$$

Since (u, v) is dual feasible, $\bar{c} = (\bar{c}_{ij}) \geqq 0$. And $\bar{c} \geqq 0, x \geqq 0$ implies that $z_{\bar{c}}(x) = \sum_{i=1}^n \sum_{j=1}^n \bar{c}_{ij} x_{ij} \geqq 0$. So from (5.9) we conclude that $z_c(x) \geqq w$, i.e., (5.7) holds, or, the total reduction w is a lower bound for the minimum objective value in the assignment problem. ∎

Theorem 5.2 *Let (u, v) be a dual feasible solution corresponding to the assignment problem (5.1). Let $\bar{c} = (\bar{c}_{ij} = c_{ij} - u_i - v_j)$ be the reduced cost matrix wrt (u, v). If $\bar{x} = (\bar{x}_{ij})$ is an assignment of order n which has allocations only among admissible cells wrt (u, v), then $\bar{x}$ is an optimum assignment for (5.1).*

Proof Verify the hypothesis in the theorem that the assignment $\bar{x}$ has allocations only among admissible cells wrt (u, v) is the same as the requirement that $\bar{x}, (u, v)$ together satisfy the complementary slackness conditions (5.6). This hypothesis implies that $\bar{x}_{ij} = 1$ (i.e., cell (i, j) has an allocation) only if $\bar{c}_{ij} = 0$ (i.e., cell (i, j) is an admissible cell wrt (u, v)). So, $z_{\bar{c}}(\bar{x}) = \sum_{i=1}^n \sum_{j=1}^n \bar{c}_{ij} \bar{x}_{ij} = 0$. By using (5.8) proved in Theorem 5.1 on $\bar{x}$ we get $z_c(\bar{x}) = w = \sum_{i=1}^n u_i + \sum_{j=1}^n v_j = \text{total}$

reduction. We already proved in Theorem 5.1 that $z_c(x) \geqq w$ for every assignment x. So, $z_c(x) \geqq w = z_{\bar{c}}(x)$ for all assignments x, i.e., $\bar{x}$ is a minimum cost assignment wrt c as the cost matrix. ∎

Strategy Followed By the Hungarian Method

Of the three conditions needed for optimality stated in Theorem 4.1 (namely primal feasibility, dual feasibility, and complementary slackness property) the Hungarian method maintains dual feasibility and the complementary slackness property throughout, and in each step strives to move closer to primal feasibility.

For the assignment problem, it is very easy to find an initial dual feasible solution (u, v) (Step 1 in the Hungarian method finds it). Then the method finds a partial assignment x with maximum number of allocations (called a **maximum partial assignment**) subject to the condition that allocations can only occur among admissible cells, using a labeling routine. Several basic mathematical results on the maximum partial assignment were proved earlier by Hungarian mathematicians J. Egerváry and D. König, hence the name Hungarian method for this algorithm. If the maximum partial assignment is a complete assignment, it is an optimum assignment, and the method terminates. Otherwise the method goes through a dual solution change step and repeats the whole process with the new dual feasible solution. The method maintains the partial assignment x, and dual feasible solution (u, v) throughout. (u, v) will always be dual feasible, x will always be a 0-1 matrix which is a partial assignment, and together they will always satisfy the complementary slackness conditions (5.6). The number of allocations in x keeps on increasing as the method progresses (in fact once a row or a column of the array receives an allocation, it will continue to have an allocation in the sequel), when the number of allocations becomes equal to n, x becomes a complete assignment which is optimal; and the method terminates.

Forbidden Cells

In many applications usually a set of **forbidden cells** in the array is specified with the condition that no allocations can be made among cells in it. For example, in the application of assigning candidates to jobs, if candidate 1 (the one corresponding to row 1 of the array) is not qualified to do job 1 (the one corresponding to column 1 of the array) then we cannot have an allocation in cell $(1, 1)$ and it will be a forbidden cell. Let **F** denote the set of all forbidden cells. For each $(i, j) \in$ **F** we need to make sure that $x_{ij} = 0$ under this condition.

In this case an assignment x is said to be a **feasible assignment** if it has no allocations in any forbidden cells, **infeasible assignment** otherwise. In this case our aim is to find a minimum cost feasible assignment, or conclusively establish that no feasible assignment exists.

In the minimization problem (5.1) one way to force a variable x_{ij} to be zero in the optimum solution is to make its cost coefficient $c_{ij} = +\infty$. So, we make

$c_{ij} = +\infty$ for all forbidden cells $(i,j) \in \mathbf{F}$. With this change all forbidden cells have cost coefficient $+\infty$ and vice versa. If no forbidden cells are specified, all entries in the cost matrix will be finite.

If all cells in any row or column are forbidden, clearly there can be no feasible assignment for the problem. So, we assume that each row and column of the array has at least one non-forbidden cell (i.e., one that can have an allocation).

Since the dual solution will always be finite, the relative cost coefficient in cell (i,j) wrt (u,v), namely $\bar{c}_{ij} = c_{ij} - u_i - v_j$, will be ∞ iff cell (i,j) is a forbidden cell.

During labeling routines of the Hungarian method, each row and column of the array may be in three possible states: **unlabeled, labeled and unscanned, labeled and scanned**. The **list** is always the set of current labeled and unscanned rows and columns.

THE HUNGARIAN METHOD

Step 1   Initialization   If all cells in any row or column are forbidden, there is no feasible assignment, terminate. Otherwise, make $c_{ij} = +\infty$ in all forbidden cells (i,j).

Substep 1.1   The initial dual feasible solution   If some dual feasible solution is available, use it as the initial one; otherwise for $i = 1$ to n define $u_i^0 = \min\{c_{i1}, \ldots, c_{in}\}$. Enter the u_i^0s in a right hand column of the assignment array. Subtract u_i^0 from each original cost coefficient in the ith row.

Define $v_j^0 = \min\{c_{1j} - u_1^0, \ldots, c_{nj} - u_n^0\}$, $j = 1$ to n. Enter the v_j^0s in a bottom row of the array. Subtract v_j^0 from each present cost entry in the jth column.

$(u^0 = (u_i^0), v^0 = (v_j^0))$ is the initial dual solution, and these operations transform the original cost matrix into the reduced cost matrix wrt it. Since the minimum in each row and then the column has been subtracted from all elements in that row or column, the reduced cost matrix obtained will be ≥ 0, so $(u^0 = (u_i^0), v^0 = (v_j^0))$ is dual feasible.

Identify all admissible cells. By the manner in which the initial dual feasible solution is obtained, it is clear that there is at least one admissible cell in each row and column of the array. Go to Substep 1.2.

Substep 1.2   The Initial Partial Assignment   The aim of this substep is to get a partial assignment with as many allocations as possible among present admissible cells. In the beginning all admissible cells are **uncrossed**. As allocations are made, all un-allocated admissible cells in the row or column of an allocated cell will be crossed out, to make sure that no row or column receives more than one allocation.

Look for a row or a column of the array with a single uncrossed admissible cell. If you find one, make an allocation in the admissible cell in it, and **cross out** other admissible cells in its column, row.

If each row and column without an allocation has 2 or more uncrossed admissible cells in it, select one of these admissible cells, make an allocation in it, cross out the other admissible cells in its row and column, and repeat.

Continue this work as far as possible, until either there are no rows or columns without an allocation, or every admissible cell without an allocation has been crossed out. Now remove all the crossings at this stage.

If each row and column has an allocation, we have a full assignment, it is an optimum assignment, terminate. Otherwise go to Step 2.

Step 2   Labeling Routine to Find a Path for Putting a New Allocation
The aim of this routine is to check if the present partial assignment is a maximum partial assignment among admissible cells, or show how to place one more allocation among admissible cells.

Substep 2.1   Label Rows Without an Allocation   For each i such that row i of the array has no allocation at this stage, label it $(s, +)$. This label just means that this row needs an allocation. You can write these labels in a column to the right of the u-column. Each of these rows is now labeled but not scanned; include them in the list.

Substep 2.2   Select a Labeled But Unscanned Row or Column to Scan
If the list of labeled but unscanned nodes is $= \emptyset$, we say that the labeling routine has terminated with a **nonbreakthrough**. This outcome implies that the present partial assignment is a maximum partial assignment among present admissible cells. In this case go to Step 4. Otherwise, select a row or column from the list for scanning and delete it from the list.

Substep 2.3   Scanning   If you are scanning a row, say row i, go to forward labeling given below. If you are scanning a column, say column j, go to reverse labeling given below.

Forward labeling   Scanning row i consists of labeling each unlabeled column j for which (i, j) is an admissible cell, with the label (row i, +).

You can enter the labels on the columns in a bottom row just below the v-row. The label of $(i, +)$ on column j just means that an allocation can be placed in cell (i, j) while revising the allocations, since cell (i, j) is an admissible cell.

During forward labeling if a column without an allocation is labeled, we say that the labeling routine has reached a **breakthrough**. This outcome means that a way for finding a new partial assignment with one more allocation among the admissible cells has been discovered. This will need canceling the present allocations in some allocated cells and putting new allocations along a path indicated by the present labels on rows and columns. So, while forward labeling, the moment the first column without an allocation receives a label, we terminate the labeling routine and go to Step 3.

Reverse labeling Suppose column j is to be scanned. By the way the computation is organized in the method, it will have an allocation. If the row in which that allocation occurs is unlabeled so far, label it with (column j, $-$).

If breakthrough has not occurred, include all newly labeled rows and columns in the list of labeled and unscanned nodes, and go to Substep 2.2.

Step 3   Allocation change routine Suppose column j, which does not have an allocation, has been labeled. Let the label on it be (row i, $+$). The $+$ sign inside the label indicates that a new allocation should be made in the cell (i, j). Next look at the label on row i. Suppose it is (column j_1, $-$). This implies that the present allocation in cell (i, j_1) should be deleted. Then look at the label on column j_1. Continue along the path guided by the labels, adding a new allocation and deleting a present allocation in this manner, until a row with the label $(s, +)$ is reached. At this stage the allocation change routine stops. If all columns have allocations now, these allocations define an optimum assignment, terminate the method. Otherwise erase the labels on all the rows and columns and go back to Step 2 afresh with the new partial assignment and the present dual feasible solution.

Step 4   Dual solution change routine At this stage the labeled rows and the labeled columns may be scattered all across the array. To understand the current position, it is convenient to visualize all labeled rows as constituting a block of labeled rows as in Array 5.1 given below (even though they may not all be together in the array at this stage), and all unlabeled rows as a block, etc.

At this stage all labeled rows and columns have been scanned. So, if there is an admissible cell, (i, j) say, with row i labeled and column j unlabeled; then column j would have been labeled while scanning row i. However, since column j is unlabeled, it implies that cell (i, j) could not be an admissible cell. Thus all cells in the block of labeled rows and unlabeled columns must be inadmissible cells at this stage; i.e., their present relative cost coefficients must be > 0.

Also, at this stage if there is an allocation in a cell (i, j), with row i unlabeled and column j labeled, then when column j was scanned row i would have been labeled. Since row i is now unlabeled, this implies that there are no allocations among unlabeled rows and labeled columns.

Compute δ, the minimum value of reduced cost coefficient among cells in labeled rows and unlabeled columns. δ will be > 0 by the above argument. If $\delta = +\infty$, this can only happen if forbidden cells are specified in the problem; there is no feasible assignment, terminate the method. If δ is finite, add it to the value of u_i in all labeled rows and subtract it from the value of v_j in all labeled columns; this yields the new dual feasible solution. Compute the new reduced cost coefficient in each cell. From the formula for the new dual feasible solution it follows that it can be obtained as given below.

New reduced cost coefficient in cell $(i,j) =$

$$\begin{cases} \text{Present reduced cost coefficient, if both} \\ \text{row } i \text{ and column } j \text{ are either labeled, or} \\ \text{both are unlabeled} \\ \\ -\delta + \text{ present reduced cost coefficient, if} \\ \text{row } i \text{ is labeled, and column } j \text{ is unlabeled} \\ \\ \delta + \text{ present reduced cost coefficient, if row} \\ i \text{ is unlabeled, and column } j \text{ is ulabeled} \end{cases}$$

From the definition of δ it can be verified that all the new reduced cost coefficients are $\geqq 0$, so the new dual solution is dual feasible. In fact the new reduced cost coefficient is 0 in all cells (i,j) with row i labeled, column j unlabeled, and the old reduced cost coefficient in $(i,j) = \delta$; i.e., all cells which tie for δ in its definition above. Hence, new admissible cells are created in the labeled row-unlabeled column block as a result of this dual solution change.

Also, all admissible cells in the unlabeled row-labeled column block become inadmissible wrt the new dual feasible solution. However, since there are no allocations in this block as explained above, the complementary slackness conditions (5.6) continue to hold in the present partial assignment and the new dual feasible solution.

A simple procedure to compute δ and the new reduced cost coefficients when solving small problems by hand computation is the following:

Draw a line through all the present reduced cost coefficients in each unlabeled row and each labeled column. When these lines are drawn, the following properties will hold

a) The lines cover all admissible cells.

b) The number of lines drawn will be equal to the number of allocations in the present partial assignment.

c) The lines drawn in a) are the smallest number of lines through the rows and columns of the array, needed to cover all the admissible cells. The present partial assignment contains the maximum number allocations that can be put among admissible cells, subject to the condition that no more than one allocation can be put in any row or column.

Results b), c) follow from a theorem proved by J. Egerváry and D. König.

Then $\delta =$ minimum reduced cost coefficient among cells not covered by any line. $\delta > 0$ by (a). Subtract δ from all reduced cost coefficients not covered by any line. Add δ to all reduced cost coefficients covered by two lines. Leave others unchanged. This gives the new reduced cost coefficients.

Retain the present labels on all the labeled rows and columns, but include all the labeled rows in the list of labeled and unscanned nodes, and resume the labeling process from where it was left off previously, by going to Substep 2.2 in Step 2.

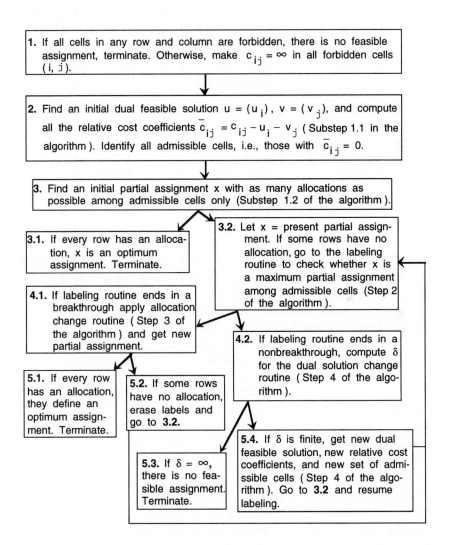

Figure 5.1 A flow chart for the Hungarian Method for the assignment problem.

Discussion

Array 5.1 Summary of Position When Method Reaches Step 4.

$(\tilde{u}, \tilde{v}, \tilde{\tilde{c}} = (\tilde{\tilde{c}}_{ij} = c_{ij} - \tilde{u}_i - \tilde{v}_j))$ are the dual solution and reduced cost matrix before the change. $(\hat{u}, \hat{v}, \hat{\tilde{c}} = (\hat{\tilde{c}}_{ij} = c_{ij} - \hat{u}_i - \hat{v}_j))$ are the corresponding things after the change.

	Block of labeled cols.	Block of unlabeled cols.	Allocations	St. lines	Dual change
Block of labeled rows	Each col. here has an allocation among labeled rows (there is breakthrough otherwise). $\hat{\tilde{c}}_{ij} = \tilde{\tilde{c}}_{ij}$ here, so, admissibility pattern remains unchanged.	No admissible cells here (one col. would be labeled otherwise). $\tilde{\tilde{c}}_{ij} > 0$, here. $\delta = \text{Min}.\{\tilde{\tilde{c}}_{ij}: (i,j) \text{ here}\} > 0$. $\hat{\tilde{c}}_{ij} = \tilde{\tilde{c}}_{ij} - \delta$ here. New admissible cells created here, their cols. will be labeled next.	Some rows have no allocation.		$\hat{u}_i = \tilde{u}_i + \delta$ for i here.
Block of unlabeled rows	No allocation here (otherwise a row here could be labeled). $\hat{\tilde{c}}_{ij} = \tilde{\tilde{c}}_{ij} + \delta$ here. All cells here become inadmissible.	Each row here contains an allocation in these cols. $\hat{\tilde{c}}_{ij} = \tilde{\tilde{c}}_{ij}$ here, so admissibility pattern remains unchanged here.	Each row has allocation.	Draw through each row.	$\hat{u}_i = \tilde{u}_i$ for i here.
Allocations	Each col. has allocation.	Some have no allocation.			
St. lines	Draw through each col.				
Dual change	$\hat{v}_j = \tilde{v}_j - \delta$ for j here	$\hat{v}_j = \tilde{v}_j$ for j here.			

Cells that have allocations at present remain admissible in the new reduced cost matrix, and hence (5.6) continues to hold. New admissible cells are created among labeled rows and unlabeled columns; so, when the list is made equal to the set of all labeled rows, and labeling is resumed, at least one new column will be labeled.

Suppose δ came out to be $+\infty$ in Step 4 at some stage. Let **J** be the set of all cells (i, j) with row i labeled, and column j unlabeled at this stage. Since $\delta = \min\{\bar{c}_{ij} : (i, j) \in \mathbf{J}\}$, $\delta = \infty$ iff the present reduced cost coefficient $\bar{c}_{ij} = +\infty$ for all cells $(i, j) \in \mathbf{J}$ (i.e., all cells $(i, j) \in \mathbf{J}$ are forbidden cells). Even if all cells not in **J** are made admissible, no more labeling can be carried out, and the current nonbreakthrough continues to hold. This implies that the present partial assignment contains the maximum number of allocations possible under the constraint that x_{ij} must be 0 for all $(i, j) \in \mathbf{J}$, hence there is no feasible assignment in the problem.

EXAMPLE 5.1 : Illustration of the allocation change routine

In this example $n = 6$ and Array 5.2 contains all the relevant information. Admissible cells are those with a zero in the upper left corner. Allocations are marked with a $\square$ in the cell. For simplicity all other information is omitted, since it is not needed for this illustration.

We will now explain the labeling sequence. The labeling routine begins by labeling rows 4, 5 without allocations with the label $(s, +)$. Now list = {rows 4, 5}.

We select row 4 to scan and delete it from the list. Unlabeled columns 3, 4 with admissible cells in this row get labeled with the label (Row 4, +). Now list = {row 5, cols. 3, 4}.

We select row 5 to scan and delete it from the list. There are no unlabeled columns with admissible cells in row 5. So, when row 5 is scanned, no new columns are labeled. List = {cols. 3, 4}.

We select column 3 to scan and delete it from the list. Row 2 which has an allocation in column 3, gets labeled with the label (Col. 3, $-$). List = {col. 4, row 2}.

We select column 4 to scan and delete it from the list. Row 3 which has an allocation in column 4, gets labeled with the label (Col. 4, $-$). List = {rows 2, 3}.

We select row 3 to scan and delete it from the list. Unlabeled column 1 with admissible cell in this row get labeled with the label (Row 3, +). Now list = {row 2, col. 1}.

We select column 1 to scan and delete it from the list. Row 6 which has an allocation in column 1, gets labeled with the label (Col. 1, $-$). List = {rows 2, 6}.

We select row 6 to scan and delete it from the list. Unlabeled column 6 with admissible cell in this row get labeled with the label (Row 6, +), and since column 6 has no allocation at this stage we have a breakthrough. We terminate the labeling routine and go to the allocation change step. Notice that the scanning of row 6 is not completed here since column 5 can be labeled also, but we terminate the labeling routine with the breakthrough declaration the moment one column without an allocation is labeled.

When column 6 without an allocation is labeled (Row 6, +) we had a breakthrough, so we put a new allocation in the cell (6, 6). Now look at the label on row 6, which is (Col 1, −). Thus we delete the allocation in cell (6, 1). Continuing this way using the labels, we put a new allocation in (3, 1), delete the one in (3, 4), add on allocation in (4, 4), and reach row 4 labeled $(s, +)$, implying that the allocation change routine is complete. The allocation change path indicated by the labels on Array 5.2 is shown in Figure 5.2. The wavy edges in Figure 5.2 correspond to allocated cells in Array 5.2, on this allocation change path. This path is clearly an **alternating path** (nodes in it correspond alternatively to unallocated cells and allocated cells in Array 5.2). It is called the alternating predecessor path of column 6 traced by the labels.

Array 5.2

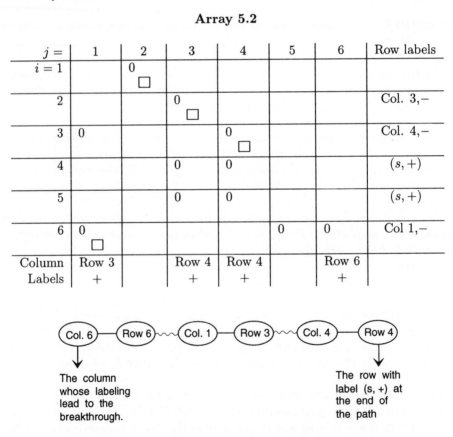

$j =$	1	2	3	4	5	6	Row labels
$i = 1$		0 □					
2			0 □				Col. 3,−
3	0			0 □			Col. 4,−
4			0	0			$(s, +)$
5			0	0			$(s, +)$
6	0 □				0	0	Col 1,−
Column Labels	Row 3 +		Row 4 +	Row 4 +		Row 6 +	

Figure 5.2 Allocation change path.

The allocation change routine reverses the roles of unallocated and allocated cells along this path. It has the effect of increasing the number of allocations by 1. Hence a path like this is called an **augmenting path** in the admissible network. An augmenting path in the admissible network wrt the present allocations is an alternating path of unallocated and allocated arcs, joining a column node and a row node both of which have no allocated cells in them. The labeling routine discussed above is an efficient scheme to look for such an augmenting path. When a break-through occurs, it is an indication that an augmenting path has been identified; it is the predecessor path of the column node without an allocation, whose labeling lead to the breakthrough. If a nonbreakthrough occurs, it is an indication that no augmenting path exists in the admissible network wrt the present allocations, and that the present partial assignment is a maximum partial assignment in the admissible network.

The new allocations after the allocation change are shown in Array 5.3. The new partial assignment has five allocations, one more than the previous. Now the labels on all the rows and columns are erased, and the algorithm begins another labeling cycle to see whether yet another allocation can be squeezed in among the present admissible cells.

Array 5.3

$j =$	1	2	3	4	5	6
$i = 1$		0 $\square$				
2			0 $\square$			
3	0 $\square$			0		
4			0	0 $\square$		
5			0	0		
6	0				0	0 $\square$

EXAMPLE 5.2

Here we solve the assignment problem of order 4 with the original cost matrix $c = (c_{ij})$ given below.

	c_{ij}				Initial		$c_{ij} - u_i$		
$j =$	1	2	3	4	u_i	$j = 1$	2	3	4
$i = 1$	15	22	13	4	4	11	18	9	0
2	12	21	15	7	7	5	14	8	0
3	16	20	22	6	6	10	14	16	0
4	6	11	8	5	5	1	6	3	0
					Initial				
					$v_j \rightarrow$	1	6	3	0

Total Reduction $= 32$

First reduced cost matrix with initial allocations

$j =$	1	2	3	4	u_i	Row labels
$i = 1$	$10 = \bar{c}_{11}$	12	6	0		$s, +$
					4	
2	4	8	5	0 ☐	7	Col. 4, $-$
3	9	8	13	0	6	$s, +$
4	0	0	0 ☐	0	5	
v_j	1	6	3	0		
Col. labels				Row 3, $+$		

Nonbreakthrough
$\delta = 4$
Labeling resumed below

Second reduced cost matrix

$j =$	1	2	3	4	u_i	Row labels
$i = 1$	6	8	2	0	8	$s, +$
2	0	4	1	0 ☐	11	Col. 4, $-$
3	5	4	9	0	10	$s, +$
4	0	0	0 ☐	4	5	
v_j	1	6	3	-4		
Col. labels	Row 2 $+$			Row 3, $+$		

Breakthrough
since col. 1 is labeled
Total reduction $= 40$

Second reduced cost matrix with new allocations and new labels

$j =$	1	2	3	4	u_i	Row labels
$i = 1$	6	8	2	0		$s, +$
					8	
2	0 □	4	1	0	11	
3	5	4	9	0 □	10	Col. 4, −
4	0	0	0 □	4	5	
v_j	1	6	3	−4		
Col. labels				Row 1, +		

Nonbreakthrough
$\delta = 2$
Labeling resumed below

Third reduced cost matrix

$j =$	1	2	3	4	u_i	Row labels
$i = 1$	4	6	0	0	10	$s, +$
2	0 □	4	1	2	11	
3	3	2	7	0 □	12	Col. 4, −
4	0	0	0 □	6	5	Col. 3, −
v_j	1	6	3	−6		
Col. labels		Row 4 +	Row 1 +	Row 1, +		

Breakthrough
since col. 2 labeled.
Total reduction = 42

**Third reduced cost matrix
with new allocations**

$j =$	1	2	3	4	u_i
$i = 1$	4	6	0 ☐	0	10
2	0 ☐	4	1	2	11
3	3	2	7	0 ☐	12
4	0	0 ☐	0	6	5
v_j	1	6	3	-6	

Now we have a full assignment, $\{(1, 3), (2, 1), (3, 4), (4, 2)\}$ among the admissible cells, which is an optimum assignment for the problem. Its cost (wrt the original cost matrix) can be verified to be 42, which is also the total reduction at this stage. An optimum dual solution is the (u, v) from the final reduced cost matrix.

Computational Complexity of the Hungarian Method

Measures of the **largeness** of a problem instance are provided by its parameters such as its **dimension** (the total number of data elements in it), or its **size** when all the data is rational (the total number of digits in the data when it is encoded in binary form).

The data in an assignment problem of order n consists of the n^2 entries in the cost matrix. The parameter n (the order of the problem instance) is normally used as the largeness measure of an instance of the assignment problem.

We will measure the **computational effort** of an algorithm to solve a problem instance by the number of basic operations (additions, subtractions, multiplications, divisions, comparisons, lookups, etc.) that it takes when it is applied to solve that instance. We would expect the relative difficulty of problem instances to increase in general with their largeness measure, so the computational effort of an algorithm should be expressed in terms of this largeness measure. But, even among problem instances with the same largeness measure, the computational effort of an algorithm may vary considerably depending on the actual values for the input data. A measure of the efficiency of an algorithm, known as the **worst case computational complexity** or just **computational complexity**, is a mathematical upperbound (i.e., the maximum) for the computational effort needed by the algorithm, expressed as a function of the largeness measure of the instance, as the values of the input data elements take all possible values in their range. When n is some measure of how large a problem instance is (either the size, or the dimension), an algorithm for solving it is said to be of order n^r or $O(n^r)$ if its worst case computational effort grows as αn^r, where α and r are numbers that are independent of the largeness measure and the data in the instance.

Consider the Hungarian method applied to solve an assignment problem of order n. Whenever Step 3 is carried out, the number of allocations increases by 1. Thus Step 3 is carried out at most n times in the algorithm. From Array 5.1 we see that at least one new column gets labeled when labeling is resumed after a dual solution change step. Thus Step 4 can occur at most n times between two consecutive occurrences of Step 3. The effort needed to carry out Step 4 (updating all the reduced cost coefficients) and the following labeling, before going to Step 3 or 4 again is at most $O(n^2)$. Thus the effort between two consecutive occurrences of Step 3 is $O(n^3)$, and therefore the entire method takes at most $O(n^4)$.

The Hungarian method can actually be implemented so that we only update some of the reduced cost coefficients as needed, after each dual solution change between two consecutive allocation change steps. This implementation maintains another piece of data called an **index** on each unlabeled column. Under this implementation, the computational effort between two consecutive occurrences of Step 3 is at most $O(n^2)$, and since Step 3 occurs at most n times, the overall computational complexity of the Hungarian method with this implementation is at most $O(n^3)$. See [K. G. Murty, 1992] for details on this efficient implementation of the Hungarian method.

Comment 5.1 The Hungarian method for the assignment problem is due to [H. W. Kuhn, 1955]. The name for the method recognizes the work of the Hungarian mathematicians J. Egerváry and D. König which is the basis for the method.

5.2 Exercises

5.1 Solve the assignment problems with the following cost matrices, to minimize cost.

$$
\begin{pmatrix}
10 & 11 & 10 & 5 & 6 & 4 & 3 \\
5 & 26 & 14 & 18 & 15 & 10 & 10 \\
6 & 22 & 18 & 17 & 15 & 8 & 8 \\
2 & 14 & 16 & 16 & 24 & 25 & 12 \\
4 & 15 & 19 & 10 & 8 & 14 & 11 \\
10 & 22 & 22 & 15 & 28 & 24 & 12 \\
8 & 18 & 21 & 18 & 18 & 18 & 14
\end{pmatrix}
,
\begin{pmatrix}
121 & 6 & 17 & 9 & 8 & 10 \\
8 & 33 & 45 & 15 & 20 & 31 \\
5 & 19 & 30 & 16 & 14 & 22 \\
7 & 22 & 35 & 25 & 27 & 26 \\
2 & 10 & 24 & 18 & 31 & 14 \\
4 & 12 & 31 & 17 & 18 & 18
\end{pmatrix}
$$

5.2 The coach of a swim team needs to assign swimmers to a 200-yard medley relay team. The "best times" (in seconds for 50 yards) achieved by his five swimmers in each of the strokes are given below. Which swimmer should the coach assign to each of the four strokes?

Stroke	Carl	Chris	Ram	Tony	Ken
Backstroke	37.7	32.9	33.8	37.0	35.4
Breast Stroke	43.4	33.1	42.2	34.7	41.8
Butterfly	33.3	28.5	38.9	30.4	33.6
Freestyle	29.2	26.4	29.6	28.5	31.1

5.3 A colonel has five positions to fill and five eligible candidates to fill them. The number of years of experience of each candidate in each field is given in the following table. How should the candidates be assigned to positions to give the greatest total years of experience for all jobs?

Candidate	Adjutant	Intelli.	Position Operations	Supply	Training
1	3	5	6	2	2
2	2	3	5	3	2
3	3	0	4	2	2
4	3	0	3	2	2
5	0	3	0	1	0

5.4 Find a minimum cost assignment wrt the cost matrix given below. No allocations are allowed in cells with a dot in them.

$$
\begin{pmatrix}
. & 12 & . & 6 & 9 & 6 & . & . \\
. & 8 & . & 4 & 3 & 8 & . & . \\
. & 3 & . & 18 & 3 & 19 & . & . \\
. & 1 & . & 6 & 5 & 11 & . & . \\
5 & 1 & 13 & 4 & 5 & 6 & 1 & 2 \\
. & 13 & . & 12 & 3 & 1 & . & . \\
3 & 12 & 3 & 7 & 13 & 6 & 8 & 3 \\
13 & 4 & 1 & 5 & 5 & 5 & 4 & 9
\end{pmatrix}
$$

5.5 Solve the assignment problem with the following cost matrix, where the entries are the costs of assigning a worker to a job, to find an assignment of one worker per job which minimizes the total cost.

$$
\begin{pmatrix}
5 & 3 & 7 & 3 & 4 \\
5 & 6 & 12 & 7 & 8 \\
2 & 8 & 3 & 4 & 5 \\
9 & 6 & 10 & 5 & 6 \\
3 & 2 & 1 & 4 & 5
\end{pmatrix}
$$

5.6 A truck rental company has an extra truck at each of locations 6 to 10, and is short of a truck at each of locations 1 to 5. The distances between these locations are given below.

	to 6	7	8	9	10
From 1	20	7	13	9	6
2	7	12	15	16	16
3	3	14	23	7	4
4	6	14	14	17	18
5	4	15	12	18	13

Determine how the trucks should be transferred from the excess locations where they are now, to the shortage locations where they are needed, so as to minimize the total distance travelled by all the trucks.

5.7 A marketing firm operates in 7 small regions. They hired 7 salesmen, each one to handle one region. The work involves driving on official duty, and the company is required to reimburse at the rate of 29 cents/mile for the estimated average distance the salesman is expected to drive on office duty daily. This average distance, estimated based on the location of each salesman's home and the region, is given below.

	Average daily mileage of salesman if assigned to						
	region 1	2	3	4	5	6	7
Salesman 1	100	101	108	110	109	113	105
2	101	100	103	111	103	107	109
3	100	100	100	106	100	102	102
4	100	100	100	100	100	100	100
5	100	100	100	107	100	102	102
6	100	100	112	118	115	111	113
7	100	100	109	114	117	114	110

Find the assignment of salesmen to regions, which minimizes the company's travel reimbursement bill.

5.8 The swimming coach has to select four swimmers to compete in the 400-meter medley relay in which the swimmers successively swim 100 meters of backstroke, breaststroke, butterfly, and freestyle. The coach can choose from 7 fastest swimmers whose times in seconds for 100 meters in individual events are given below.

Event	Time (in seconds), of swimmer						
	1	2	3	4	5	6	7
Backstroke	66	67	66	64	70	68	64
Breaststroke	71	72	70	69	72	72	73
Butterfly	65	67	71	74	65	64	64
Freestyle	59	59	55	59	54	54	56

The coach needs to select one swimmer for each event, so that the sum of their times in minimized. Formulate the coach's problem and obtain an optimum solution for it.

5.9 Consider the problem of allocating candidates to marketing director positions in zones discussed in Section 1.4, with annual sales data given in Table 1.1. Find the assignment of candidates to zones that actually maximizes the sales volume, using the Hungarian method. Compare this assignment with that obtained by the greedy method in Section 1.4.

5.3   References

H. W. KUHN, 1955, "The Hungarian Method for the Assignment Problem", *Naval Research Logistics Quarterly*, 2 (83-97).

K. G. MURTY, 1992, *Network Programming*, Prentice Hall, Englewood Cliffs, NJ.

Chapter 6

Primal Algorithm for the Transportation Problem

6.1 The Balanced Transportation Problem

We consider the transportation problem with the following data

m = number of sources where material is available

n = number of sinks or demand centers where material is required

a_i = units of material available at source i, $a_i > 0$, $i = 1$ to m

b_j = units of material required at sink j, $b_j > 0$, $j = 1$ to n

c_{ij} = unit shipping cost (\$/unit) from source i to sink j, $i = 1$ to m, $j = 1$ to n

The transportation problem with this data is said to satisfy the **balance condition** if it satisfies

$$\sum_{i=1}^{m} a_i = \sum_{j=1}^{n} b_j \tag{6.1}$$

If (6.1) holds, the problem is known as a **balanced transportation problem**. Letting x_{ij} denote the amount of material transported from source i to sink j, $i = 1$ to m, $j = 1$ to n, the problem is (6.2) given in the next page. It is known as an **uncapacitated balanced transportation problem** (uncapacitated because there are no specified upper bounds on the decision variables x_{ij}s). It is a **transportation problem of order** $m \times n$ (m is the number of sources, and n is the number of sinks here). Let $x = (x_{ij})$ be a feasible solution for (6.2). Summing the set of first m constraints, and the set of last n constraints in (6.2) separately, we see that $\sum_{i=1}^{m} a_i = \sum_{i=1}^{m} \sum_{j=1}^{n} x_{ij} = \sum_{j=1}^{n} b_j$.

$$\text{Minimize} \quad z(x) = \sum_{i=1}^{m} \sum_{j=1}^{n} c_{ij} x_{ij}$$

$$\text{subject to} \quad \sum_{j=1}^{n} x_{ij} \ = \ a_i, \quad i = 1 \text{ to } m$$

$$\sum_{i=1}^{m} x_{ij} \ = \ b_j, \quad j = 1 \text{ to } n \qquad (6.2)$$

$$x_{ij} \ \gtreqless \ 0, \quad \text{for all } i, j$$

So, we see that (6.1) is a **necessary condition for the feasibility of** (6.2). We assume that the data satisfies (6.1).

Redundancy in the constraints

Add the first m constraints in (6.2), and from the sum subtract the sum of the last n constraints. By (6.1), this leads to the equation "$0 = 0$". Hence there is a redundant constraint among the equality constraints in (6.2), and any one of the equality constraints in (6.2) can be treated as a redundant constraint and deleted from the system without affecting the set of feasible solutions. We treat the constraint corresponding to sink n

$$\sum_{i=1}^{m} x_{in} = b_n \qquad (6.3)$$

as the redundant constraint to eliminate from (6.2) (one could have chosen any of the other constraint as being redundant instead of this one). After the constraint in (6.3) is deleted from (6.2), we obtain the following problem in which all the equality constraints are nonredundant

$$\text{Minimize} \quad z(x) = \sum_{i=1}^{m} \sum_{j=1}^{n} c_{ij} x_{ij}$$

$$\text{subject to} \quad \sum_{j=1}^{n} x_{ij} \ = \ a_i, i = 1 \text{ to } m$$

$$\sum_{i=1}^{m} x_{ij} \ = \ b_j, j = 1 \text{ to } n - 1 \qquad (6.4)$$

$$x_{ij} \ \gtreqless \ 0, \text{ for all } i, j$$

The coefficient matrix of the system of equality constraints in (6.4) is of order $(m+n-1) \times mn$ and its rank is $(m+n-1)$. So, every basic vector for the balanced transportation problem of order $m \times n$ consists of $(m + n - 1)$ basic variables.

The Dual Problem

Associating the dual variable u_i to the constraint corresponding to source i, $i = 1$ to m; and the dual variable v_j to the constraint corresponding to sink j, $j = 1$ to n; from Section 4.1 we know that the dual of (6.2) is (4.6). Deleting the constraint (6.3) corresponding to $j = n$ in (6.2) has the effect of setting $v_n = 0$ in the dual problem. So, the dual problem is

$$\text{Maximize} \quad w(u,v) = \sum_{i=1}^{m} a_i u_i \;+\; \sum_{j=1}^{n} b_j v_j$$

$$\text{subject to} \quad u_i + v_j \;\overset{\leq}{=}\; c_{ij}, \text{ for all } i,j \qquad (6.5)$$

$$v_n \;=\; 0$$

Given the dual solution (u,v), the relative cost coefficient of x_{ij} wrt it, i.e., the dual slack variable associated with it, is $\bar{c}_{ij} = c_{ij} - u_i - v_j$, for $i = 1$ to m, $j = 1$ to n. The various pairs $(x_{ij}, \bar{c}_{ij})$ are the complementary pairs in (6.4) and its dual (6.5). And from Chapter 4, we know that the complementary slackness conditions for optimality for a primal feasible solution $x = (x_{ij})$ and dual feasible solution $(u = (u_i), v = (v_j))$ to be optimal to the respective problems, are

$$x_{ij}\bar{c}_{ij} = x_{ij}(c_{ij} - u_i - v_j) = 0 \quad \text{for all } i, j \qquad (6.6)$$

The Algorithm that We Will Discuss

The Hungarian method for the assignment problem discussed in Chapter 5 can be extended into a primal-dual algorithm for the balanced transportation problem. However, here we will discuss a new algorithm based on a different approach, the **primal simplex algorithm**, for the balanced transportation problem. Extensive computational testing has shown that the primal simplex algorithm is superior to a primal-dual approach for solving balanced transportation problems. The primal simplex algorithm begins with a primal feasible basic vector obtained by a special initialization routine. The corresponding dual basic solution is then computed. If it is dual feasible, i.e., if all the relative cost coefficients $\bar{c}_{ij}$ wrt it are ≥ 0, the present solutions are optimal to the respective problems and the algorithm terminates. If some $\bar{c}_{ij} < 0$, the present basic vector is dual infeasible. In this case the algorithm selects exactly one nonbasic variable x_{ij} corresponding to a negative $\bar{c}_{ij}$, and brings it into the basic vector; thus generating a new primal feasible basic vector with which the whole process is repeated. Since the algorithm moves only among basic vectors, the complementary slackness optimality conditions (6.6) hold automatically throughout the algorithm because in a BFS only basic primal variables can be nonzero, and basic relative cost coefficients are always zero.

Thus the primal simplex algorithm maintains primal feasibility and complementary slackness property throughout, and tries to attain dual feasibility in each step.

Because of its special structure, we can implement the primal simplex algorithm for solving the balanced transportation problem without using the inverse tableaus, but doing all the computations on transportation arrays instead. We discuss this simpler implementation.

Forbidden Cells

In most applications involving a large number of sources and sinks, a source may not be able to transport material to all the sinks. Some of the sinks may be too far away from it, or there may be no direct route from it to all the sinks. In such applications, a set of **forbidden cells** in the transportation array is specified with the condition that there should be no transportation among cells in it. Let **F** denote the set of all forbidden cells. For each $(i, j) \in \mathbf{F}$ we need to make sure that $x_{ij} = 0$ under this condition. In the minimization problem (6.2) one way to force a variable x_{ij} to be zero in the optimum solution is to make its cost coefficient $c_{ij} = +\infty$, or a very large positive number α (taking $\alpha > (\sum_{i=1}^{m} a_i)(\max\{|c_{ij}| : i = 1 \text{ to } m, j = 1 \text{ to } n\})$ would do). So, we make $c_{ij} = \alpha$ for all forbidden cells $(i, j) \in \mathbf{F}$. With this change all forbidden cells have cost coefficient α and vice versa. If no forbidden cells are specified, all entries in the cost matrix will be as specified in the original data.

6.2   Notation Used to Display the Data

Computer implementations of the primal simplex algorithm for the transportation problem are usually based on the representation of the problem as a minimum cost flow problem on a bipartite network. But for hand computation on small problems, it is convenient to work with transportation arrays discussed in Section 4.1. Each row in the array corresponds to a source, and each column corresponds to a sink. The variable x_{ij} in the problem is associated with cell (i, j) in the array. Forbidden cells (these correspond to variables x_{ij} which are required to be 0) have very large positive cost coefficients, and they are essentially crossed out and ignored in the algorithm (i.e., the values of the variables in them remain zero) once they become nonbasic.

The original cost coefficient in cell (i, j) will be entered in the lower right corner of the cell using small size numerals. The relative cost coefficients, $\bar{c}_{ij}$, will be entered in the upper left corner of the cells, also using small size numerals. The relative cost coefficient in every nonbasic forbidden cell will always be $+\infty$ if c_{ij} was defined to be $+\infty$, or some large positive number if c_{ij} was defined to be a large positive number.

Basic cells will have a small square in their center, with the value of the corresponding variable in the present BFS entered inside the square in normal size numerals. So, after an initial basic vector is selected, the basic vector at any stage consists of the set of cells with little squares in their center.

The availabilities at the sources and the requirements at the sinks are typeset using larger size numerals to distinguish them from the cost data. These are maintained on the array until a BFS to the problem is obtained.

The dual solution $((u_i), (v_j))$ is entered on the array again using smaller size numerals.

6.3 Routine for Finding an Initial Feasible Basic Vector and its BFS

This special routine for finding a feasible basic vector for a balanced transportation problem selects one basic cell per step, and hence needs $(m + n - 1)$ steps on a problem of order $m \times n$. Initially, all cells in the transportation array are open for selection as basic cells. In each step, all the remaining cells in either a row or a column of the basic cell selected in that step will be crossed out from selection in subsequent steps. Also, the row and column totals will be modified after each step. The current row and column totals will be denoted by a_i', b_j' respectively; these will always be $\geqq 0$, and they represent the remaining quantity of material still to be shipped from a source, or unfulfilled demand at a sink, at that stage. A row or column will always have an uncrossed cell not yet selected as a basic cell, that is open for selection as a basic cell, as long as the current total in it is > 0. Initially, $a_i' = $ original a_i, $b_j' = $ original b_j, for all i, j.

Initialization   All cells in the $m \times n$ transportation array are open for selection as basic cells initially, and $a_i' = $ original a_i, $b_j' = $ original b_j, for all i, j. With these go to first step. We describe the general step.

General Step   If all the remaining cells open for selection as basic cells, are all in a single row (column), select each of them as a basic cell; and make the value of the basic variable in each of them equal to the modified column (row) total at this stage. Terminate.

If the remaining cells open for being selected as basic cells are in two or more rows and two or more columns of the array at this stage, select one of them as a basic cell. Two popular rules for making this selection are given below. If (r, s) is the selected cell, make $x_{rs} = \min\{a_r', b_s'\} = \beta$ say. It is possible for β to be zero. Subtract β from both a_r' and b_s'; this updates them. If new $a_r' = 0 < b_s'$ $(b_s' = 0 < a_r')$ cross out all remaining cells in row r (column s) from being selected as basic cells in subsequent steps. If new $a_r' = $ new $b_s' = 0$, cross out all remaining cells in either row r or column s, but not both, from being selected as basic cells in subsequent steps. Go to the next step.

Rules for Selecting an Open Cell as a Basic Cell

Here we discuss two rules that are commonly used for making this selection.

The Greedy Choice Rule   Under this rule, the cell (r, s) selected as the basic cell is one which has the smallest cost coefficient among all cells open for selection at that stage.

Vogel's Choice Rule   Let line refer to a row or column of the array that contains some cells open for selection at this stage. In each line compute the **cost difference**, which is the second minimum cost coefficient − minimum cost coefficient, among all open cells in this line. Identify the line that has the maximum cost difference at this stage, and select a least cost open cell in it as the basic cell in this step. The rationale for this selection is the following. If that cell is not selected, any remaining supply or demand in this line has to be shipped using an open cell with the second minimum cost or higher cost in that line, and hence results in the highest increase in unit cost at this stage.

If forbidden cells are specified in the problem, it is possible that some of them may be selected as basic cells in this routine, and the basic variables corresponding to them may have positive values in the initial BFS. If the original problem has a feasible solution in which all the forbidden variables are zero, when the simplex algorithm is applied to solve the problem beginning with the initial BFS, the forbidden basic variables will become 0 in the BFS before the algorithm terminates with an optimum solution.

EXAMPLE 6.1

Array 6.1

	Plant			a'_i
	1	2	3	
Mine 1			200	600
	11	8	2	
Mine 2			CR	300
	7	5	4	
b'_j	400	500		

Consider the iron ore shipping problem discussed in Chapter 2. The array for this problem containing all the data is given above. The smallest cost coefficient in the entire array is $2 = c_{13}$, so we select $(1, 3)$ as the first basic cell and make $x_{13} = \min\{200, 800\} = 200$. We change the amount still to be shipped from mine 1 to $800 - 200 = 600$, and cross out cell $(2, 3)$ in column 3 from being selected as a basic cell, and enter CR in it to indicate this fact. The array at this stage is given above.

The least cost cell among the remaining open cells is $(2, 2)$ with cost coefficient 5, which is selected as the next basic cell, and we make $x_{22} = \min\{500, 300\} = 300$. We change the remaining requirement at plant 2 to $500 - 300 = 200$, cross out the remaining cell $(2, 1)$ in row 1 from being selected, and get the situation in the next array.

	Plant 1	Plant 2	Plant 3	a'_i
Mine 1	11	8	200 2	**600**
Mine 2	CR 7	300 5	CR 4	
b'_j	**400**	**200**		

Now the remaining open cells are $(1, 1)$, $(1, 2)$, both in row 1, so we select both of them as basic cells and make $x_{11} = 400$, and $x_{12} = 200$, and obtain the basic vector and associated BFS marked in the following array. The transportation cost in this BFS is $\sum\sum c_{ij}x_{ij} = 11 \times 400 + 8 \times 200 + 5 \times 300 + 2 \times 200 = \$7900$.

Array 6.2: The basic vector and BFS

	Plant 1	Plant 2	Plant 3
Mine 1	400 11	200 8	200 2
Mine 2	7	300 5	4

EXAMPLE 6.2 : Finding an initial basic vector using Vogel's choice rule

Here we will find an initial feasible basic vector for the iron ore transportation problem using Vogel's rule for selecting basic cells in each step. In row 1, the smallest and second smallest cost coefficients are 2, 8, and hence the cost difference in row 1 is $8 - 2 = 6$. In the same way, the cost differences for all the rows and columns in the array are computed and given below.

Line	Cost difference	
Row 1	$8 - 2$	$= 6$
Row 2	$5 - 4$	$= 1$
Col. 1	$11 - 7$	$= 4$
Col. 2	$8 - 5$	$= 3$
Col. 3	$4 - 2$	$= 2$

The highest cost difference occurs in row 1, and hence we select the least cost cell $(1, 3)$ in it as the first basic cell, and get the same situation as in Array 6.1 given above. Now column 3 is done, and we recompute the cost difference for the remaining lines using only the data from the remaining open cells. These are given below.

Line	Cost difference	
Row 1	$11 - 8$	$= 3$
Row 2	$7 - 5$	$= 2$
Col. 1	$11 - 7$	$= 4$
Col. 2	$8 - 5$	$= 3$

The highest cost difference occurs in column 1, and hence we select the least cost open cell in it, $(2, 1)$, as the next basic cell and make $x_{21} = \min\{400, 300\} = 300$. This leads to the array given below.

	Plant			a_i'
	1	2	3	
Mine 1			$\boxed{200}$	600
	11	8	2	
Mine 2	$\boxed{300}$	CR	CR	
	7	5	4	
b_j'	100	500		

Now the only remaining open cells, $(1, 1)$, $(1, 2)$, are in row 1, so we select both of them as basic cells and make $x_{11} = 100$, $x_{12} = 500$, leading to the basic vector given in the following array. The transportation cost in this BFS is $11 \times 100 + 8 \times 500 + 7 \times 300 + 2 \times 200 = \$7600$. Verify that this BFS is better than the BFS obtained with the greedy choice rule in Example 6.1.

Array 6.3

	Plant		
	1	2	3
Mine 1	100	500	200
	11	8	2
Mine 2	300		
	7	5	4

The computation of cost differences, and finding the maximum among them, imposes additional work in each step when using Vogel's selection rule. The effort needed to do this additional work is very worthwhile, as Vogel's rule usually produces a much better BFS than the simple greedy selection rule. Unfortunately, neither rule can guarantee that the BFS produced will be optimal, hence it is necessary to check the BFS for its optimality. Empirical tests show that the BFS produced by Vogel's rule is usually near optimal. So, some practitioners do not bother to obtain a true optimum solution to the problem, instead they implement the initial BFS obtained by using Vogel's selection rule. When used this way, the method is called **Vogel's approximation method** (or **VAM** in short) for the balanced transportation problem.

Nondegenerate, Degenerate BFSs

As discussed in Chapter 4, a BFS corresponding to a feasible basic vector for the uncapacitated balanced transportation problem is primal nondegenerate if all primal basic variables are > 0 in it, primal degenerate otherwise. In both the BFSs obtained in Examples 6.1 and 6.2 for the iron ore transportation problem, all the 4 basic variables are > 0, hence they are both primal nondegenerate for that problem. We will now consider an example which leads to a primal degenerate BFS.

EXAMPLE 6.3 : Example of a primal degenerate BFS

Consider the following balanced transportation problem with data given below.

	Sink 1	Sink 2	Sink 3	a_i
Source 1	1	2	3	15
2	4	4	10	19
3	6	5	15	11
b_j	7	8	30	

We will use the greedy selection rule for selecting basic cells in each step to get an initial BFS. The least cost cell (1, 1) is selected as the first basic cell, and $x_{11} = \min\{7, 15\} = 7$. So, all other cells in column 1 are crossed out from being selected. The position at this stage is indicated in the following array.

	Sink 1	Sink 2	Sink 3	a_i'
Source 1	[7]	2	3	8
	1			
2	CR	4	10	19
	4			
3	CR	5	15	11
	6			
b_j'		8	30	

The least cost cell among open cells now, (1, 2), with a cost coefficient of 2, is selected as the next basic cell, and we make $x_{12} = \min\{8, 8\} = 8$. At this stage we modify the totals in both row 1 and column 2 to 0, and have to cross out all remaining cells in one of them from being selected in subsequent stages. Suppose we select row 1 for this. This leads to the next array.

	Sink			a_i'
	1	2	3	
Source 1	$\boxed{7}$ ⟨1⟩	$\boxed{8}$ ⟨2⟩	CR ⟨3⟩	
2	CR ⟨4⟩	⟨4⟩	⟨10⟩	19
3	CR ⟨6⟩	⟨5⟩	⟨15⟩	11
b_j'		0	30	

Next we select $(2, 2)$ as a basic cell, and make $x_{22} = \min\{0, 19\} = 0$, and cross out the remaining cell in column 2. The remaining open cells are both in column 3, so we select both of them as basic cells. This leads to the BFS in Array 6.4.

Array 6.4

	Sink		
	1	2	3
Source 1	$\boxed{7}$ ⟨1⟩	$\boxed{8}$ ⟨2⟩	⟨3⟩
2	⟨4⟩	$\boxed{0}$ ⟨4⟩	$\boxed{19}$ ⟨10⟩
3	⟨6⟩	⟨5⟩	$\boxed{11}$ ⟨15⟩

In this BFS the basic variable $x_{22} = 0$, hence it is primal degenerate. It is necessary to record the zero valued basic variables clearly so as to distinguish them from nonbasic variables which are always 0 in every BFS. For the $m \times n$ balanced transportation problem, every basic vector must have exactly $(m + n - 1)$ basic variables or cells.

6.4 How to Compute the Dual Basic Solution and Check Optimality

As discussed in Chapter 4, given a feasible basic vector $\mathcal{B}$ for (6.4), the dual basic solution associated with it can be computed by solving the system of equations

(6.7). This system is obtained by treating all the dual constraints in the dual (6.5) corresponding to basic variables in $\mathcal{B}$ as equations. The last equation $v_n = 0$ is associated with the constraint corresponding to sink n which we have treated as a redundant constraint in (6.2) and eliminated, to get (6.4).

$$u_i + v_j \;=\; c_{ij} \text{ for each basic cell } (i,j) \in \mathcal{B} \qquad (6.7)$$
$$v_n \;=\; 0$$

For the $m \times n$ transportation problem, there are $m + n$ dual variables, and $m + n - 1$ basic variables in every basic vector. So, (6.7) is a system of $m + n$ equations in $m + n$ unknowns, a square nonsingular system of equations with a unique solution. This is the reason for requiring that all the zero valued basic variables be recorded carefully, without them (6.7) will not be a square system for computing the dual basic solution uniquely.

The special structure of the transportation problem makes it possible to solve (6.7) very easily by back substitution beginning with $v_n = 0$. Since we know $v_n = 0$, from the equations in (6.7) corresponding to basic cells in column n of the array, we can get the values of u_i for rows i of these basic cells. Now column n is processed. Knowing the values of these u_i, again from the equations in (6.7) corresponding to basic cells in the remaining columns in these rows, we can get the values of v_j for columns j of these basic cells. Now the rows of basic cells in column n are processed, and we continue the method with the columns of the newly computed v_j in the same way, until all the u_i and v_j are computed.

Having obtained the dual basic solution (u, v) corresponding to $\mathcal{B}$, we compute the relative cost coefficients $\bar{c}_{ij} = c_{ij} - u_i - v_j$ in all nonbasic cells (i, j). The optimality criterion is

$$\textbf{Optimality criterion} \quad \bar{c}_{ij} \overset{\geq}{=} 0 \quad \text{for all nonbasic } (i,j) \qquad (6.8)$$

If (6.8) is satisfied, then (u, v) is dual feasible and hence $\mathcal{B}$ is a dual feasible basic vector. Since $\mathcal{B}$ is also primal feasible, by the results discussed in Chapter 4 it is an optimal basic vector, and the primal and dual basic solutions associated with it are optimal to the respective problems.

EXAMPLE 6.4

Consider the basic vector in Array 6.2 for the iron ore shipping problem. To compute the dual basic solution, we start with $v_3 = 0$. Since $(1, 3)$ is a basic cell in this basic vector, we have $u_1 + v_3 = c_{13} = 2$, so $u_1 = 2$, and the processing of column 3 is done. As $(1, 1)$, $(1, 2)$ are basic cells, we have $u_1 + v_1 = c_{11} = 11$, $u_1 + v_2 = c_{12} = 8$, and since $u_1 = 2$ these equations yield $v_1 = 9$, $v_2 = 6$; and the processing of row 1 is done. As $(2, 2)$ is a basic cell, we have $u_2 + v_2 = c_{22} = 5$, and from $v_2 = 6$ this yields $u_2 = -1$. Now we have the complete dual solution; it is entered in the following array.

Array 6.5

	Plant			
	1	2	3	u_i
Mine 1	400 11	200 8	200 2	2
Mine 2	−1 7	300 5	5 4	−1
v_j	9	6	0	

The relative cost coefficients of the nonbasic cells (2, 1), (2, 3) are $\bar{c}_{21} = c_{21} - u_2 - v_1 = 7 - (-1) - 9 = -1$, $\bar{c}_{23} = c_{23} - u_2 - v_3 = 4 - (-1) - 0 = 5$. These are entered in the upper left corners of these cells. Since $\bar{c}_{21} < 0$, the optimality criterion (6.8) is not satisfied in this basic vector.

6.5 A Pivot Step: Moving to an Improved Adjacent Basic Vector

When we have a feasible basic vector $\mathcal{B}$ associated with the BFS $\bar{x} = (\bar{x}_{ij})$, which does not satisfy the optimality criterion (6.8), the primal simplex algorithm obtains a better solution by moving to an adjacent basic vector by replacing exactly one basic variable with a judiciously selected nonbasic variable. It can be shown that a better solution will be obtained if the nonbasic variable that is introduced into the present basic vector has its relative cost coefficient $\bar{c}_{ij} < 0$, i.e., for which the optimality criterion (6.8) is violated. That's why the set of nonbasic cells $\mathbf{E} = \{(i, j) : \bar{c}_{ij} < 0\}$, is called **the set of cells eligible to enter the present basic set**.

The method selects exactly one of these eligible nonbasic cells as the **entering cell**. This selection can be made arbitrarily, but a couple of rules that are used most commonly for solving small size problems by hand computation are the following.

First eligible cell encountered When computing the relative cost coefficients, the moment the first negative one turns up, select the corresponding cell as the entering cell. You don't even have to compute the relative cost coefficients in the remaining nonbasic cells.

Most negative $\bar{c}_{ij}$ rule Under this rule, you compute the relative cost coefficients in all the nonbasic cells, and if (6.8) is not satisfied, select as the entering cell the eligible cell (i, j) with the most negative $\bar{c}_{ij}$ (break any ties arbitrarily). This rule is also called **the minimum $\bar{c}_{ij}$ rule**, or **Dantzig's rule**.

Since every basic vector for the $m \times n$ transportation problem has exactly $m + n - 1$ basic cells, when an entering cell is brought into the basic vector, some present basic cell should be dropped from the basic vector, this cell is called **the dropping basic cell**, and the variable corresponding to it is called **the dropping basic variable**.

To determine the dropping basic variable and the new BFS, the following procedure is used. All the nonbasic variables other than the entering variable are fixed at the present value of 0, and the value of the entering variable is changed from 0 (its present value) to a value denoted by θ. So, if the entering cell is (p, q), the procedure changes the value of x_{pq} from its present 0 (since it is a nonbasic variable) to θ. Now to make sure that exactly a_p units are shipped out of source p, and b_q units are shipped to sink q, we have to add a $-\theta$ to one of the basic values in row p, and another $-\theta$ to one of the basic values in column q. These subsequent adjustments have to be made among basic values only, because every nonbasic variable other than the entering variable is fixed at its present value of 0. There is a unique way of continuing these adjustments among basic values, adding alternately a $-\theta$ to the basic value in one basic cell, and then a balancing $+\theta$ to the basic value in another basic cell; until all the adjustments cancel each other in every row and column, so that the new solution again satisfies all the equality constraints in the problem. All the cells which have the value in them modified by the adjustment process belong to a loop called the θ-**loop**. It satisfies the following properties.

(i)   Every row and column of the array either has no cells in the θ-loop; or has exactly two cells, one with a $+\theta$ adjustment, and the other with a $-\theta$ adjustment.

(ii)   All the cells in the θ-loop other than the entering cell are present basic cells.

(iii)   No proper subset of a θ-loop satisfies property (i).

This set of cells is called the θ-loop in $\mathcal{B} \cup \{(p, q)\}$. Cells in the θ-loop with a $+\theta$ adjustment are called the **recipient cells**. The only nonbasic recipient cell is the entering cell. Cells with a $-\theta$ adjustment are called the **donor cells**. All donor cells are basic cells.

So, the new solution obtained by fixing all nonbasic variables other than the entering variable at their present value of 0, changing the value of the entering variable x_{pq} from its present 0 to θ, and then reevaluating the values of the basic variables so as to satisfy all the equality constraints in the problem, is $x(\theta) = (x_{ij}(\theta))$ where

$$x_{ij}(\theta) = \begin{cases} \bar{x}_{ij} & \text{the value of the basic variable in the present BFS,} \\ & \text{if } (i,j) \text{ is not in the } \theta\text{-loop} \\ \\ \bar{x}_{ij} + \theta & \text{if } (i,j) \text{ is a recipient cell in the } \theta\text{-loop} \\ \\ \bar{x}_{ij} - \theta & \text{if } (i,j) \text{ is a donor cell in the } \theta\text{-loop} \end{cases} \qquad (6.9)$$

Since the shipments in all the recipient cells have to be increased, and in all the donor cells have to be decreased, the net cost of making a unit adjustment along the θ-loop is $\mathcal{B} \cup \{(p,q)\}$ is

$$\sum \left(c_{ij} : \begin{array}{c} \text{over recipient cells} \\ (i,j) \text{ in the } \theta\text{-loop} \end{array} \right) - \sum \left(c_{ij} : \begin{array}{c} \text{over donor cells} \\ (i,j) \text{ in the } \theta\text{-loop} \end{array} \right) \quad (6.10)$$

and this will always be equal to the relative cost coefficient $\bar{c}_{pq}$ of the nonbasic entering cell (p,q) in the θ-loop. We state this fact in the following theorem.

THEOREM 6.1   *Let $\mathcal{B}$ be a basic set of cells for the $m \times n$ transportation problem (6.4)* wrt *which the relative cost coefficients for nonbasic cells are $(\bar{c}_{ij})$. Let (p,q) be a nonbasic cell. Then the set of cells $\mathcal{B} \cup \{(p,q)\}$ contains a unique θ-loop which can be obtained by putting a $+\theta$ entry in the nonbasic cell (p,q), and alternately entries of $-\theta$ and $+\theta$ among basic cells as described above, until these adjustments cancel out in each row and column. This θ-loop satisfies conditions (i), (ii), (iii) stated above. And the net cost of this θ-loop as defined in (6.10) is $\bar{c}_{pq}$, the relative cost coefficient in (p,q)* wrt $\mathcal{B}$.

For a proof of this theorem see [K. G. Murty, 1983 of Chapter 2]. Now considering the present BFS $\bar{x}$ wrt the basic set $\mathcal{B}$, and the new solution $x(\theta)$ obtained as above, with the nonbasic cell (p,q) as the entering cell, we find from Theorem 6.1 that the objective value of $x(\theta)$ is $z(x(\theta)) = z(\bar{x}) + \theta(\text{net cost of the } \theta\text{-loop in } \mathcal{B} \cup \{(p,q)\}) = z(\bar{x}) + \theta\bar{c}_{pq}$.

Thus the relative cost coefficient $\bar{c}_{pq}$ in the nonbasic cell (p,q) is the rate of change in the objective value, per unit change in the value of the nonbasic variable x_{pq} from its present value of 0, while all the other nonbasic variables stay fixed at their present value of 0.

This is the reason for selecting the entering cell to be one with a negative relative cost coefficient, since it can lead to an improved solution with reduced objective value. If the relative cost coefficient of the entering cell is $0(> 0)$, you get a solution with the same (higher) objective value. This also explains the rationale behind the optimality criterion. If all nonbasic relative cost coefficients are $\overset{\geq}{=} 0$, there is no way we can get a new feasible solution with a strictly better objective value by increasing the values of any nonbasic variables from their present values of 0.

Since $z(x(\theta)) = z(\bar{x}) + \theta\bar{c}_{pq}$, and $\bar{c}_{pq} < 0$, as θ increases, the objective value of $x(\theta)$ decreases. To get the best solution in this step, we should give θ the maximum value it can have. As θ increases, the value of $x_{ij}(\theta)$ decreases in donor cells (i, j). So, for $x(\theta)$ to remain feasible to the problem, we need $\bar{x}_{ij} - \theta \geqq 0$ for all donor cells (i, j) in the θ-loop. Thus the maximum value that θ can have while keeping $x(\theta)$ feasible is

$$\theta = \min\{\bar{x}_{ij} : (i, j) \text{ a donor cell in the } \theta\text{-loop}\} \qquad (6.11)$$

The value of θ defined in (6.11) is called **the minimum ratio** for the operation of bringing the entering cell (p, q) into the present basic vector $\mathcal{B}$.

All donor cells (i, j) which tie for the minimum in (6.11) are said to be **eligible to drop from the present basic vector when** (p, q) **enters it**. When θ is made equal to the minimum ratio defined in (6.11) in $x(\theta)$, $x_{ij}(\theta)$ becomes zero in all these cells (i, j). One of these eligible to drop basic cells is selected as the dropping cell to be replaced by the entering cell, leading to the new feasible basic vector. $x(\theta)$ with the value of θ defined by (6.11) is the BFS associated with it; its objective value is $z(x(\theta)) = z(\bar{x}) + \theta\bar{c}_{pq} \leqq z(\bar{x})$ since $\bar{c}_{pq} < 0$ and $\theta \geqq 0$. Since $x_{ij}(\theta) = 0$ for the dropping cell (i, j), it becomes a nonbasic cell now. Any other donor cells which tied for the minimum in (6.11) will stay as basic cells with the value of the basic variable in them zero in the new BFS.

After selecting the entering cell (p, q) into the basic vector $\mathcal{B}$, the work involved in finding the θ-loop in $\mathcal{B} \cup \{(p, q)\}$, finding the minimum ratio, the dropping cell, and the new basic vector and its BFS is called a **pivot step** in the basic vector $\mathcal{B}$. In a pivot step the basic vector changes by exactly one variable.

EXAMPLE 6.5 : Example of a pivot step

Consider the feasible basic vector and the BFS displayed in Array 6.5 for the problem discussed in Example 6.4. The nonbasic cell $(2, 1)$ with relative cost coefficient $\bar{c}_{21} = -1$ is the only cell eligible to enter this basic vector. $x(\theta)$ is marked with $+\theta$, $-\theta$ entries in the following array.

	Plant			
	1	2	3	u_i
Mine 1	$400 - \theta$	$200 + \theta$	200	
	11	8	2	2
Mine 2	-1 $\quad$ θ	$300 - \theta$	5	
	7	5	4	-1
v_j	9	6	0	

The recipient cells in this θ-loop are $(2, 1)$, $(1, 2)$; and the donor cells are $(1, 1)$, $(2, 2)$. The net cost of making a unit adjustment along this θ-loop is

$c_{21} + c_{12} - c_{11} - c_{22} = 7 + 8 - 11 - 5 = -1 = \bar{c}_{21}$, verifying the statement in Theorem 6.1. For $x(\theta)$ to be feasible, we need $400 - \theta \geqq 0, 300 - \theta \geqq 0$, i.e., the maximum value that θ can have is $\min\{400, 300\} = 300$ which is the minimum ratio. When $\theta = 300$, $x_{22}(\theta)$ becomes zero, it is the only basic cell with this property, so it is the dropping basic cell. So, we put $\theta = 300$ and replace the basic cell (2, 2) by the entering cell (2, 1) leading to the next basic vector given in Array 6.6. Since $\bar{c}_{21} = -1$, and minimum ratio $\theta = 300$, the change in the objective value in this pivot step is $-1 \times 300 = -300$ (it drops from \$7900 to \$7600).

Array 6.6

	Plant			
	1	2	3	u_i
Mine 1	100	500	200	
	11	8	2	2
		1	6	
Mine 2	300			
	7	5	4	−2
v_j	9	6	0	

We computed the dual basic solution and the nonbasic relative cost coefficients wrt the basic vector in Array 6.6 and entered them. Since all the nonbasic relative cost coefficients are > 0, the optimality criterion (6.8) holds in Array 6.6, hence the BFS there is an optimum solution to the problem and its cost is \$7600. This solution requires shipping

100 tons of ore from mine 1 to plant 1
500 tons of ore from mine 1 to plant 2
200 tons of ore from mine 1 to plant 3
300 tons of ore from mine 2 to plant 1

EXAMPLE 6.6 : A trial and error method for finding the θ-loop

---

Here we give another example of using the trial and error method to find the θ-loop in $\mathcal{B} \cup \{(p, q)\}$ where $\mathcal{B}$ is a given feasible basic set for a balanced transportation problem, and (p, q) is the selected nonbasic entering cell. In the following array, the basic vector consists of all the cells with a square in the middle, with the value of the corresponding basic variable in the BFS entered inside this square.

Array 6.7

	1	2	3	4	5	u_i
1	1 ⟶ 6	[18 − θ] ⟶ 3	[20 + θ] ⟶ 4	7 ⟶ 5	−6 ⟶ 5	11
2	[27 − θ] ⟶ 10	7 ⟶ 15	6 ⟶ 15	11 ⟶ 14	[35 + θ] ⟶ 16	16
3	[0 + θ] ⟶ 8	−1 ⟶ 5	[27 − θ] ⟶ 7	10 ⟶ 11	−5 ⟶ 9	14
4	−6 ⟶ 13	−13, θ ⟶ 4	−2 ⟶ 16	[25] ⟶ 12	[65 − θ] ⟶ 25	25
v_j	−6	−8	−7	−13	0	

Relative cost coefficients in nonbasic cells are entered in the upper left corner of the cell as usual. Every cell with a negative cost coefficient is eligible to enter this basic vector; of these we selected the cell (4, 2) as the entering cell.

We make the value of $x_{42} = \theta$ by putting a θ in the center of cell (4, 2). All other nonbasic variables remain fixed at their present value of 0. To continue to satisfy the equality constraints in the problem, we need to add a $-\theta$ to the value in one of the basic cells in row 4, i.e., in cells (4, 4) or (4, 5). If we add $-\theta$ to $\bar{x}_{44}$, since this is the only basic cell in column 4, we cannot make the next balancing correction of adding a $+\theta$ in another basic cell in this column. So, the basic cell (4, 4) is the wrong cell to make the $-\theta$ adjustment in row 4, hence this adjustment must be made in the basic cell (4, 5). This is the trial and error feature of this procedure. Continuing in this manner, we get the entire θ-loop in this example, marked in the above array.

Exercise

6.1 (i)   Write the donor, recipient cells in the θ-loop in Array 6.7. (ii)   Verify that the net cost of making a unit adjustment along the θ-loop in Array 6.7 is $-13 = \bar{c}_{42}$, the relative cost coefficient of the entering cell. (iii)   Compute the cost of the present BFS (remember that $\theta = 0$ in it). (iv)   Find the minimum ratio, and select a dropping basic cell when (4, 2) enters this basic vector. Draw another array, and mark the new basic vector and the new BFS in it. (v)   Compute the cost of the new BFS and verify that it is = cost of the old BFS + $\theta\bar{c}_{42}$. (vi)   Is the new basic vector optimal? Why? (vii)   If the new basic vector is not optimal, continue the process by selecting an entering variable into it and performing a pivot step. Repeat until you get an optimum solution to the problem.

How to Find the θ-loop in a Pivot Step

We only discussed a trial and error procedure for finding the θ-loop in a pivot step. This trial and error procedure is fine for solving small problems by hand computation, but it is very inefficient for solving large real world problems on a computer. Efficient methods for finding θ-loops are based on predecessor labeling schemes for storing tree structures; see [K. G. Murty, 1983 of Chapter 2] for details on them. Using these efficient schemes, large scale transportation problems with thousands of sources and sinks can be solved in a matter of seconds on modern digital computers.

Nondegenerate, Degenerate Pivot Steps

A BFS for (6.2) is said to be **nondegenerate** if all the $m + n - 1$ basic variables are strictly > 0 in it; otherwise it is said to be **degenerate**. The BFSs found in Examples 6.1 and 6.2 are all nondegenerate, while the BFS found in Example 6.3 is degenerate.

Let $\mathcal{B}$ be a feasible basic vector for the problem, associated with the BFS $\bar{x} = (\bar{x}_{ij})$. If the nonbasic cell (p, q) is selected as the entering cell into $\mathcal{B}$, the ensuing pivot step is said to be a **nondegenerate pivot step** if the minimum ratio in it, θ is > 0; **degenerate pivot step** if this minimum ratio is 0.

If $\bar{x}$ is a nondegenerate BFS, since the minimum ratio in this step is the minimum of $\bar{x}_{ij}$ over all donor cells (i, j) in the θ-loop, all of which are basic cells, it is strictly > 0, and hence this pivot step will be a nondegenerate pivot step. So, a pivot step in a basic vector can only be degenerate if that basic vector is primal degenerate. Even if the BFS $\bar{x}$ is degenerate, if all the donor cells (i, j) in the θ-loop satisfy $\bar{x}_{ij} > 0$, the pivot step will be nondegenerate.

If $\bar{c}_{pq}$ is the relative cost coefficient of the entering cell, we have seen that the objective value of the new BFS obtained at the end of this pivot step is = objective value of the old BFS $+ \theta \bar{c}_{pq}$. Since $\bar{c}_{pq} < 0$ and $\theta > 0$ in a nondegenerate pivot step, the change in the objective value in it, $\theta \bar{c}_{pq}$, is < 0. Thus after a nondegenerate pivot step, we will obtain a new BFS with a strictly better objective value. In a degenerate pivot step, $\theta = 0$, and hence the BFS and its objective value do not change, but we get a new basic vector corresponding to the same old BFS with a different set of zero valued basic variables in it.

The pivot steps discussed in Examples 6.5 and 6.6 are nondegenerate pivot steps, since the minimum ratios are > 0 in them.

EXAMPLE 6.7 : Example of a degenerate pivot step

Consider the degenerate BFS associated with the basic vector in Array 6.4 derived in Example 6.3. The dual basic solution and the relative cost coefficients wrt this basic vector are given below.

Array 6.8

	Sink 1	Sink 2	Sink 3	u_i
Source 1	$\boxed{7-\theta}$ 1	$\boxed{8+\theta}$ 2	-5 3	8
2	1 4	$\boxed{0-\theta}$ 4	$\boxed{19+\theta}$ 10	10
3	-2 θ 6	-4 5	$\boxed{11-\theta}$ 15	15
v_j	-7	-6	0	

The nonbasic cell (3, 1) with relative cost coefficient -2 is selected as the entering cell. The θ-loop is entered on the array. The minimum ratio $= \min\{7, 0, 11\} = 0$, hence this is a degenerate pivot step. The entering cell (3, 1) replaces the basic cell (2, 2) as the new zero valued basic cell, leading to the new basic vector given below.

Array 6.9

	Sink 1	Sink 2	Sink 3
Source 1	$\boxed{7}$ 1	$\boxed{8}$ 2	3
2	4	4	$\boxed{19}$ 10
3	$\boxed{0}$ 6	5	$\boxed{11}$ 15

Even though the basic vector is different, the BFS and the objective value corresponding to it are exactly the same as before. This is what happens in a degenerate pivot step. Thus in a degenerate pivot step there is no change in the primal solution or objective value; but in every pivot step, whether degenerate or not, the basic vector changes by one variable.

Verify that in the basic vector in Array 6.8, if we had selected the entering cell to be (1, 3) instead of (3, 1), it would have resulted in a nondegenerate pivot step with a strict decrease in the objective value.

We now state the steps in the primal simplex algorithm for the balanced transportation problem.

6.6 The Primal Simplex Algorithm for the Balanced Transportation Problem

Initialization   Obtain an initial primal feasible basic vector for the problem and the BFS associated with it, as discussed in Section 6.3. With this basic vector go to the first iteration.

General Iteration   Find the dual basic solution and the nonbasic relative cost coefficients corresponding to the present basic vector, as discussed in Section 6.4. If all the nonbasic relative cost coefficients are $\geqq 0$, the optimality criterion (6.8) is satisfied, and the present primal and dual solutions are optimal to the respective problems. When (6.8) is satisfied, if some forbidden cells are still in the final basic vector with positive values for the basic variables in them in the BFS, it is an indication that there is no feasible solution for the original problem with $x_{ij} = 0$ for all forbidden cells $(i,j) \in \mathbf{F}$. On the other hand, if all forbidden variables are zero in the final BFS when (6.8) is satisfied, that BFS is an optimum feasible solution for the original problem with the constraints that all forbidden variables be zero. Terminate.

If (6.8) is not satisfied, select a nonbasic cell with a negative relative cost coefficient as the entering cell, and perform the pivot step as in Section 6.5. With the new basic vector and the BFS associated with it, go to the next iteration.

A flow chart of the primal simplex algorithm is given in the next page.

EXAMPLE 6.8

Consider the following balanced problem faced by a truck rental agency. They have some free trucks available at Detroit, Washinton DC, and Denver; and need additional trucks in Orlando, Dallas, and Seattle. Data on the cost of transportation per truck (c_{ij} in coded units of money) and other data is given below.

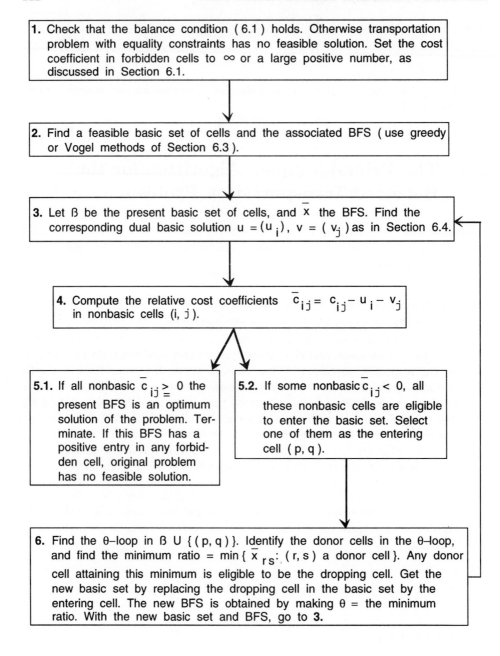

1. Check that the balance condition (6.1) holds. Otherwise transportation problem with equality constraints has no feasible solution. Set the cost coefficient in forbidden cells to ∞ or a large positive number, as discussed in Section 6.1.

2. Find a feasible basic set of cells and the associated BFS (use greedy or Vogel methods of Section 6.3).

3. Let ß be the present basic set of cells, and $\bar{x}$ the BFS. Find the corresponding dual basic solution $u = (u_i)$, $v = (v_j)$ as in Section 6.4.

4. Compute the relative cost coefficients $\bar{c}_{ij} = c_{ij} - u_i - v_j$ in nonbasic cells (i, j).

5.1. If all nonbasic $\bar{c}_{ij} \geq 0$ the present BFS is an optimum solution of the problem. Terminate. If this BFS has a positive entry in any forbidden cell, original problem has no feasible solution.

5.2. If some nonbasic $\bar{c}_{ij} < 0$, all these nonbasic cells are eligible to enter the basic set. Select one of them as the entering cell (p, q).

6. Find the θ–loop in ß U { (p, q) }. Identify the donor cells in the θ–loop, and find the minimum ratio = min { $\bar{x}_{rs}$: (r, s) a donor cell }. Any donor cell attaining this minimum is eligible to be the dropping cell. Get the new basic set by replacing the dropping cell in the basic set by the entering cell. The new BFS is obtained by making θ = the minimum ratio. With the new basic set and BFS, go to **3.**

Figure 6.1 A flow chart for the primal simplex algorithm for the balanced transportation problem.

Source city i	Sink city j			No. trucks available, a_i
	1	2	3	
Detroit 1				6
	9	6	8	
Washington DC 2				11
	10	5	12	
Denver 3				4
	11	13	20	
No. trucks needed, b_j	3	4	14	

Let x_{ij} denote the number of trucks sent from source city i to sink city j; i, j = 1 to 3. The transportation cost is $z(x) = \sum_{i=1}^{3} \sum_{j=1}^{3} c_{ij} x_{ij}$. The objective is to find an x that meets the requirements at minimal cost.

To solve this problem we determine an initial primal feasible basic vector and the associated BFS as discussed in Section 6.3 using the greedy rule to select basic cells in each step. We show this basic vector in the following array. The associated basic solution and the relative cost coefficients of nonbasic variables, computed as shown in Section 6.4, are also entered in the array.

	Sink			
	1	2	3	u_i
Source 1	3	5	$\boxed{6}$	
	9	6	8	8
2	$\boxed{3 - \theta}$	$\boxed{4}$	$\boxed{4 + \theta}$	
	10	5	12	12
3	-7 θ	0	$\boxed{4 - \theta}$	
	11	13	20	20
v_j	-2	-7	0	

The optimality criterion (6.8) is violated since $\bar{c}_{31} = -7 < 0$. (3, 1) is selected as the entering cell, and the θ-loop is already entered on the array. The minimum ratio is min$\{3, 4\} = 3$. So, this is a nondegenerate pivot step, and the basic cell (2, 1) is the dropping cell. The next BFS is given in the following array.

Array 6.10

Source	Sink 1	Sink 2	Sink 3	u_i
Source 1	10 9	5 6	[6] 8	8
2	7 10	[4] 5	[7] 12	12
3	[3] 11	0 13	[1] 20	20
v_j	−9	−7	0	

Now the optimality criterion (6.8) is satisfied, so the present BFS is an optimum solution. It requires the following shipments and has the minimum cost of 205 units of money.

6 trucks   from Detroit to Seattle
4 trucks   from Washington DC to Dallas
7 trucks   from Washington DC to Seattle
3 trucks   from Denver to Orlando
1 truck   from Denver to Seattle

If a degenerate basic vector appears in the course of this algorithm, it is mathematically possible for it to get into a never ending sequence of degenerate pivot steps. If this happens, the BFS and the objective value never change in the entire sequence, and the optimality criterion (6.8) will never be satisfied. In the simplex algorithm this phenomenon is called **cycling under degeneracy**. Even though cycling under degeneracy is a mathematical possibility, it does not seem to occur while solving real world LP models. Special, simple techniques, to guarantee that cycling under degeneracy will not occur, have been developed, these are called **techniques for resolving cycling under degeneracy**. There are simple technique for resolving cycling under degeneracy in the transportation problem, see [K. G. Murty, 1983 of Chapter 2]. These resolution techniques add only a bit more work, but they seem to be unnecessary for solving real world problems, hence we will not discuss them here.

Initiating the Primal Simplex Algorithm With a Given Primal Feasible Basic Vector

Consider a balanced transportation problem for which a primal feasible basic vector $\mathcal{B}$ is provided. We can initiate the primal simplex algorithm with $\mathcal{B}$. First we have to compute the primal basic solution corresponding to $\mathcal{B}$. All variables x_{ij} not contained in $\mathcal{B}$ are nonbasic variables, they are fixed at 0 in this basic solution. In the system of equality constraints in (6.2), when all these nonbasic variables are fixed at 0 and deleted, the remaining system can be solved by back substitution for the values of the basic variables; leading to the BFS corresponding to $\mathcal{B}$. Once this BFS is computed, the primal simplex algorithm can be applied beginning with it.

As an example, consider the iron ore shipping problem discussed in Section 2.4. Consider the basic vector $\mathcal{B} = (x_{11}, x_{12}, x_{22}, x_{23})$ for this problem. Fixing the nonbasic variables $x_{13} = x_{12} = 0$, the system of equality constraints in this problem becomes

$$
\begin{array}{lllll}
x_{11} & +x_{12} & & & = 800 \\
& & x_{22} & +x_{23} & = 300 \\
x_{11} & & & & = 400 \\
& x_{12} & +x_{22} & & = 500 \\
& & & x_{23} & = 200
\end{array}
$$

When this system is solved by back substitution, it leads to the BFS corresponding to $\mathcal{B}$ given below.

	Plant		
	1	2	3
Mine 1	400	400	
	11	8	2
Mine 2		100	200
	7	5	4

6.7 Finding Alternate Optimum Solutions for the Balanced Transportation Problem

When dealing with an optimization model, it is always helpful to know if there are alternate optimum solutions for the model, and have techniques to find them efficiently. These provide choices for the decision maker in case a particular optimum solution obtained by the algorithm turns out to be difficult to implement for reasons that have not been taken into consideration while constructing the model.

If alternate optimum solutions exist for a balanced transportation model, they can be obtained very efficiently. Given an optimum basic vector for the problem, look at the nonbasic relative cost coefficients, $\bar{c}_{ij}$ wrt it. If all these $\bar{c}_{ij}$ are > 0, the present BFS is the unique optimum solution to the problem. On the other hand, if there are some nonbasic cells (i, j) with $\bar{c}_{ij} = 0$, we can get alternate optimum BFSs by bringing these cells into the basic vector. For example, consider the optimum BFS in Array 6.10 for the problem solved in Example 6.8. The nonbasic cell (3, 2) has a relative cost coefficient of 0. By bringing this cell into the basic vector in Array 6.10, we get the following alternate optimum BFS for this problem.

Array 6.11

	Sink			
	1	2	3	u_i
Source 1	10 9	5 6	[6] 8	8
2	7 10	[3] 5	[8] 12	12
3	[3] 11	0 [1] 13	20	20
v_j	-9	-7	0	

In the same manner, if there are other nonbasic cells with zero relative cost coefficients in an optimum array, by bringing them into the basic vector we can generate alternate optimum BFSs. The convex hull of all the optimum BFSs generated is the set of optimum solutions for the problem.

6.8   Marginal Analysis in the Balanced Transportation Problem

Marginal analysis deals with the rate of change in the optimum objective value, per unit change in the RHS constants (i.e., the availabilities and requirements, a_i and b_j) in the problem. In the balanced transportation model (6.2), the condition (6.1) is necessary for feasibility. Since (6.1) holds originally, if only one quantity among $a_1, \ldots, a_m; b_1, \ldots, b_n$ changes while all the others remain fixed, the modified problem will be infeasible. So, if changes occur, at least two quantities among $a_1, \ldots, a_m; b_1, \ldots, b_n$ must change, and the changes must be such that the modified data also satisfies (6.1).

We will consider three fundamental types of changes in the availability and requirement data: (i) increased demand at sink j and a balancing increase in availability at source i (i.e., same increase in both an a_i and a b_j), (ii) increase a_p and decrease a_i by the same amount (this shifts the supply from source i to source p), and (iii) increase b_q and decrease b_j by the same amount. In each of these categories, all the other data in the problem is assumed to remain fixed at present values. The marginal value of each type is the rate of change in the optimum objective value, per unit change of this type.

Let $\bar{x} = (\bar{x}_{ij})$ be an optimum BFS of (6.4) and $(\bar{u}, \bar{v})$ an optimum dual solution. Assume that $\bar{x}$ is a nondegenerate BFS. Then by the results in Section 4.3, $(\bar{u}, \bar{v})$ is the unique optimum dual solution, and the marginal values associated with the three types of changes discussed above are as given below.

Change	Marginal value
(i) b_j and a_i increase by the same amount	$v_j + u_i$
(ii) a_p increases and a_i decreases by the same amount	$u_p - u_i$
(iii) b_q increases and b_j decreases by the same amount	$v_q - v_j$

EXAMPLE 6.9

Consider the balanced transportation problem with the following data, and an optimum BFS for it marked in the following array.

Array 6.12

	1	2	3	4	a_i	u_i
1	10 25	 36	15 20	20 10	45	10
2	 47	40 40	 30	30 20	70	20
3	50 20	 33	 15	 14	50	5
b_j	60	40	15	50		
v_j	15	20	10	0		

The optimum transportation cost in this problem is \$3950.

What will the rate of change in the optimum objective value be if b_2 were to increase from its current value of 40, and a corresponding change were made in a_3 to keep the problem balanced? It is $v_2 + u_3 = 20 + 5 = \$25$ per unit change.

From answers to the above questions, it is clear that if demand were to increase at any demand center, the best place to create additional supplies to satisfy that additional demand, purely from a transportation cost point of view, is source 3 (it is the source with the smallest u_i). This results in the smallest growth in the total transportation cost to meet the additional demand.

How much can the company save in transportation dollars by shifting supply from source 2 to source 3? The rate of change in the optimum transportation cost per unit shift is $u_3 - u_2 = 5 - 20 = -\$15$, or a saving of \$15.

Thus using this marginal analysis, we can determine if the transportation costs can be reduced by shifting production from existing centers to different places. However this analysis has not taken into account any differences in production costs between centers. To determine the net savings in shifting supplies, one has to take into account the differences in production costs between places too.

6.9   Sensitivity Analysis in the Balanced Transportation Problem

Given an optimum basic vector for a balanced transportation problem, sensitivity analysis has techniques for determining the **optimality range** of each cost coefficient. This is the interval within which that element can vary, while all the other data remain fixed at their current values, while keeping the present solution or basic vector optimal. It also has efficient techniques for finding a new optimum solution beginning with the current one, if simple changes occur in the model.

6.9.1   Cost Ranging of a Nonbasic Cost Coefficient

Let $\mathcal{B}$ be an optimum basic vector for (6.4) associated with the primal and dual optimum solutions $\bar{x}$ and $(\bar{u}, \bar{v})$. Suppose the original cost coefficient c_{pq} in a nonbasic cell (p, q) is likely to change to its new value γ_{pq} while all the other data in the problem remain unchanged. What is the range of values of γ_{pq} for which $\bar{x}$ remains optimal to the problem?

Since the only thing that changes in the array is c_{pq}, the optimality criterion (6.8) will continue to be satisfied if the new relative cost coefficient in cell (p, q), which is $= \gamma_{pq} - \bar{u}_p - \bar{v}_q$ is $\geqq 0$, i.e., if $\gamma_{pq} \geqq \bar{u}_p - \bar{v}_q$. So, $\bar{x}$ remains an optimal BFS to the problem if $\gamma_{pq} \geqq \bar{u}_p - \bar{v}_q$. Thus the optimality range for γ_{pq} is $[\bar{u}_p + \bar{v}_q, +\infty]$.

If $\gamma_{pq} < \bar{u}_p + \bar{v}_q$, the only thing that changes in the current optimal array is the relative cost coefficient in cell (p, q) to $\gamma_{pq} - \bar{u}_p - \bar{v}_q < 0$. So, (p, q) becomes eligible to enter the basic vector. By selecting (p, q) as the entering cell into $\mathcal{B}$ and continuing the primal simplex algorithm until the optimality criterion (6.8) is satisfied again, the new optimum solution can be found.

EXAMPLE 6.10

Consider the optimum basic vector given in Array 6.6 in Example 6.5. Suppose c_{22} is likely to change from its present value of 5 to γ_{22}. With this change, the relative cost coefficient in this cell becomes $\gamma_{22} - \bar{u}_2 - \bar{v}_2 = \gamma_{22} - 4$. So, the present BFS remains optimal to the problem as long as $\gamma_{22} \stackrel{\geq}{=} 4$.

If $\gamma_{22} = 2$, the relative cost coefficient in this cell becomes -2. In this case bring this cell into the basic vector to get the new optimum solution.

6.9.2 Cost Ranging of a Basic Cost Coefficient

Let $\mathcal{B}$ be an optimum basic vector for (6.4) associated with the primal and dual optimum solutions $\bar{x}$ and $(\bar{u}, \bar{v})$. Suppose c_{ij}, the cost coefficient in a basic cell (i, j) in $\mathcal{B}$ is changing to a new value γ_{ij}, but all the other data remains unchanged. What is the range of values of γ_{ij} for which $\bar{x}$ remains optimal to the problem?

We cannot answer this question by the technique used above to find the optimality range for a nonbasic cost coefficient, since basic relative cost coefficients are always zero. Also, when a basic cost coefficient changes, the dual solution changes. So, to answer this question, we need to find the dual basic solution treating γ_{ij} as a parameter, while all other cost coefficients are at their present numerical values. In the new dual solution (u, v), u_i, v_j may be linear functions of γ_{ij}. Using the new dual solution, compute the relative cost coefficients of all the nonbasic variables, these will also be linear functions of γ_{ij}. Express the condition that each of these relative cost coefficients have to be $\stackrel{\geq}{=} 0$. This will lead to lower and upper bounds for the parameter γ_{ij} which define its optimality range. As long as γ_{ij} is in the interval defined by these bounds, the present BFS remains optimal.

If the value of γ_{ij} is outside its optimality range, plug in this value and determine the new relative cost coefficients of all the nonbasic variables. At least one of them must be negative. Select one of the cells with a negative relative cost coefficient as the entering cell, and continue the application of the primal simplex method until the optimality criterion is satisfied again. At that stage you have an optimum solution for the problem with the modified data.

EXAMPLE 6.11

As an example consider the optimum BFS in Array 6.10 for the problem solved in Example 6.8. Suppose the value of c_{33} is likely to change from the present 20 to γ_{33}. We enter the data in the array given below. We compute the dual basic solution corresponding to the present basic vector treating γ_{33} as a parameter, and then the relative cost coefficients in all the nonbasic cells. All these are given in the following array. For the present basic vector to remain optimal, all these relative cost coefficients have to be $\stackrel{\geq}{=} 0$. This leads to

$$\gamma_{33} - 10 \stackrel{\geq}{=} 0, \quad \text{or } \gamma_{33} \stackrel{\geq}{=} 10$$
$$\gamma_{33} - 13 \stackrel{\geq}{=} 0, \quad \text{or } \gamma_{33} \stackrel{\geq}{=} 13$$
$$20 - \gamma_{33} \stackrel{\geq}{=} 0, \quad \text{or } \gamma_{33} \stackrel{\leq}{=} 20$$

Thus the optimality range for γ_{33} is $13 \leqq \gamma_{33} \leqq 20$.

Array 6.13

	Sink 1	Sink 2	Sink 3	u_i
	Sink			
Source 1	$\gamma_{33} - 10$ 9	5 6	6̲ 8	8
2	$\gamma_{33} - 13$ 10	4̲ 5	7̲ 12	12
3	3̲ 11	$20 - \gamma_{33}$ 13	1̲ γ_{33}	γ_{33}
v_j	$11 - \gamma_{33}$	-7	0	

If γ_{33} is outside its optimality range, say $\gamma_{33} = 12$, we plug in $\gamma_{33} = 12$ in Array 6.13. Now $\bar{c}_{21} < 0$. To get the new optimum solution, bring $(2, 1)$ into the basic vector and continue the application of the primal simplex algorithm until the optimality criterion is satisfied again.

6.9.3   Changes in Material Requirements

Here we discuss how to find the modified primal basic solution wrt the present optimum basic vector $\mathcal{B}$ for (6.4) efficiently when changes occur in the a, b-vectors. Let $\bar{x}$ denote the present BFS corresponding to $\mathcal{B}$. We consider three types of changes.

> Type 1 a_p, b_q change to $a_p + \delta, b_q + \delta$ respectively, for specified p, q. All other data remains unchanged.
>
> Type 2 a_t, a_s change to $a_t + \delta, a_s - \delta$ respectively, for specified t, s. All other data remains unchanged.
>
> Type 3 b_f, b_g change to $b_f + \delta, b_g - \delta$ respectively, for specified f, g. All other data remains unchanged.

To find the modified primal basic solution corresponding to $\mathcal{B}$ as a function of δ use the following procedures.

Type 1   Find the unique δ-path among basic cells, beginning with a $+\delta$-adjustment to the basic value in a basic cell in row p of the array, and alternately making $-\delta, +\delta$-adjustments to basic values in basic cells, until it terminates with a $+\delta$-adjustment to the basic value in a basic cell in column q.

Type 2   Find the unique δ-path among basic cells as above, beginning with a $+\delta$-adjustment to the basic value in a basic cell in row t, and terminating with a $-\delta$-adjustment to the basic value in a basic cell in row s.

Type 3   Find the unique δ-path among basic cells as above, beginning with a $+\delta$-adjustment to the basic value in a basic cell in column f and terminating with a $-\delta$-adjustment to the basic value in a basic cell in column g.

A δ-path as described above always exists and is unique and easily found. The basic values with these adjustments along the δ-path, define the basic solution corresponding to $\mathcal{B}$, as a function of δ, for the problem with modified data. Its optimality range is the range of values of δ which keeps all the basic values $\geqq 0$.

As an example consider the optimum basic vector in Array 6.12 for the transportation problem discussed in Example 6.9. In this problem, suppose a_3, b_2 change to $50 + \delta$, $40 + \delta$ respectively. This is a change of type 1 discussed above. The primal optimum BFS is nondegenerate, and as discussed in Section 6.8, the marginal value associated with this change is $u_3 + v_2 = 25$. The δ-path among basic cells from row 3 in the array to column 2, and the basic solution wrt the modified data is given in the following array.

	1		2		3		4		a_i	u_i
1	$\boxed{10 - \delta}$	25		36	$\boxed{15}$	20	$\boxed{20 + \delta}$	10	45	10
2		47	$\boxed{40 + \delta}$	40		30	$\boxed{30 - \delta}$	20	70	20
3	$\boxed{50 + \delta}$	20		33		15		14	$50 + \delta$	5
b_j	60		$40 + \delta$		15		50			
v_j		15		20		10		0		

This primal solution remains nonnegative for all δ satisfying $-20 \leqq \delta \leqq 10$. This is the optimality range for δ. For all δ in this range, the BFS given in the above array as a function of δ, is an optimum solution to the problem with the modified data. It can be verified that the objective value for this optimum solution is \$3950 + 25 δ. This shows that the rate of change in the optimum objective value per unit change in the value of δ in its optimality range is 25, which is exactly the marginal value for this type of change, computed as discussed in Section 6.8.

Changes of types 2, 3 are handled in a similar way.

6.10   What to do if There is Excess Supply or Demand

The transportation problem (6.2) in which all the constraints are equations has a feasible solution iff the total supply $\sum a_i$ is equal to the total demand $\sum b_j$.

Suppose we have a situation in which the total supply $\sum a_i$ strictly exceeds the total demand $\sum b_j$. In this case, after all the demand is met, an amount $\Delta = \sum a_i - \sum b_j$ will be left unused at the sources. So, to solve this problem, we open a new $(n+1)$th column in the array. In row i, the cell $(i, n+1)$ represents the material left unused at source i, i.e., not shipped out of source i. Since there is no cost for not shipping the material, we make the cost coefficients for all the cells in the new column $(n+1)$ equal to zero. Make $b_{n+1} = \Delta$, the total amount of material that will be left unused at all the sources. Solve this modified $m \times (n+1)$ problem as a balanced transportation problem. In the optimum solution of this modified problem, basic values in the cells of column $(n+1)$ represent unused material at the sources.

As an example, consider an oil company which has three refineries producing gasoline with daily capacities as shown in the following array. Right now the company has contracts to sell gasoline to four wholesalers to meet their daily requirements as shown below.

	Transportation cost from refinery to wholesaler(money/unit)				Daily availability (unit = 10^6 gal.)
Wholesaler	1	2	3	4	
Refinery 1	4	7	9	10	8
2	6	4	3	6	10
3	9	6	4	8	6
Daily requirement	5	3	8	4	

Here the total availability $= \sum_{i=1}^{3} a_i = 24$ units, while the total requirement is $\sum_{j=1}^{4} b_j = 20$ units. So, there is an excess supply of $\Delta = 4$ units daily. We open up a fifth column in the array corresponding to a dummy sink with a demand for 4 units (this represents unused supply at the refineries) and get the following balanced transportation problem.

	1	2	3	4	Dummy 5	a_i
1	$\boxed{5}$ 4	7	9	10	$\boxed{3}$ 0	8
2	6	4	$\boxed{8}$ 3	$\boxed{2}$ 6	0	10
3	9	$\boxed{3}$ 6	4	$\boxed{2}$ 8	$\boxed{1}$ 0	6
b_j	5	3	8	4	4	

An optimum solution for this problem is also entered in the array. From the basic values in the dummy column, we find that in order to meet the existing demand at minimum transportation cost, it is best to cut down production at refinery 1 to $8 - 3 = 5$ units/day, and that at refinery 3 to also $6 - 1 = 5$ units/day, while operating refinery 2 at its full capacity of 10 units/day.

Consider the other situation where total demand $\sum b_j$ exceeds total supply $\sum a_i$. In this case there is a shortage of $d = \sum b_j - \sum a_i$, and there is no way we can meet all the demand with the existing supply only.

To meet all of the existing demand, we need to identify a new source that can supply d units. In this case, if it is only required to find how to distribute the existing supply to meet as much of the demand as possible at minimum transportation cost, we open a dummy source row (the $(m + 1)$th), cells in which represent unfulfilled demand at the sinks. Make the cost coefficients of all the cells in this dummy row equal to zero (since they represent demand not fulfilled, i.e., not shipped), and make $a_{m+1} = d$, and solve the resulting $(m + 1) \times n$ balanced transportation problem.

6.11 Exercises

6.2 An oil company operates 3 refineries and sells its gas through 4 wholesale depots. Data is given below. Determine an optimum transportation policy.

	Cost (in money units) to ship/unit to depot				Available/week unit = 10^6 gal.
	1	2	3	4	
Refinery 1	4	7	9	10	8
2	6	4	3	6	10
3	9	6	4	8	6
Requirement (units/week)	5	3	8	4	

6.3 A dealer in peanuts stocks them at four depots, from where he supplies five different markets. Data relevent to the coming summer is given below.

	Shipping cost to mkt. (money units/unit)					Expected availability (units)
	1	2	3	4	5	
Depot 1	3	9	5	6	7	**10**
2	2	1	8	10	13	**50**
3	3	12	6	5	2	**25**
4	1	9	14	3	2	**30**
Requirement (units)	**20**	**30**	**40**	**15**	**25**	

Clearly, the total requirement exceeds the total supply. All the markets have the same priority standing. Determine how to distribute the available supply to meet as much of the total demand as possible at minimum transportation cost.

6.4 Show how to transform the following balanced transportation problem into an equivalent assignment problem. Obtain an optimum solution for it through the Hungarian method. c_{ij} is the shipping cost/unit from source i to market j, a_i is the number of units of supply available at source i, and b_j is the demand at market j in units.

Source i	c_{ij}				a_i
	1	2	3	4	
1	10	5	9	12	1
2	9	12	15	18	1
3	8	8	3	6	1
4	9	8	12	16	1
5	7	5	9	13	1
b_j	1	1	1	2	

6.5 Solve the following balanced transportation problem using $\{(1, 2), (1, 5), (2, 4), (2, 5), (3, 1), (3, 3), (4, 2), (4, 3)\}$ as the initial basic set of cells. c_{ij} is the shipping cost/unit from source i to market j.

Suppose the requirement at market 5, b_5, increases from its current value of 19. What is the best source, say source p, at which to create additional supplies to meet this extra demand? Explain the reasons for the choice of p clearly.

Source i	c_{ij}					a_i = supply (units)
	1	2	3	4	5	
1	12	6	12	8	8	20
2	15	9	14	8	10	21
3	12	9	13	10	9	8
4	11	7	11	9	12	13
b_j = requirement	6	14	11	12	19	

Increase both b_5 and a_p by δ, and obtain the new optimum solution as a function of δ, and find its optimality range.

For what range of values of c_{11} (cost of shipping from source 1 to market 1) will the original optimum solution remain optimal? Do the same for c_{31} (cost of shipping from source 3 to market 1). Also, find an optimum solution if c_{31} changes to 10.

6.6 Consider the balanced transportation problem with $m = 4$ sources, $n = 6$ markets, $a = (a_i) = (13, 31, 51, 21)$, $b = (b_j) = (17, 4, 16, 13, 54, 12)$; where a_i, b_j are the amounts available to be shipped out of source i, required at market j respectively. c_{ij} = cost of transporting from source i to market j/unit, and

$$c = (c_{ij}) = \begin{pmatrix} 10 & 2 & 9 & 1 & 11 & 12 \\ 12 & 9 & 3 & 11 & 4 & 15 \\ 3 & 7 & 10 & 9 & 6 & 6 \\ 12 & 9 & 11 & 3 & 5 & 18 \end{pmatrix}$$

Find an initial BFS to this problem using Vogel's method. Solve the problem beginning with that BFS.

6.7 An oil company imports crude from three foreign sources and refines it at five refineries. This question is concerned with minimizing the cost of transporting the crude from the sources to the refineries. Sources 1, 2, 3 can ship 20, 50, 20 units of crude respectively each week. Refineries 1 to 5 need 10, 24, 6, 20, 30 units of crude respectively every week. c_{ij} is the cost ($/unit) of shipping crude from source i to refinery j, and the matrix $c = (c_{ij})$ is given below.

$$c = \begin{pmatrix} 30 & 30 & 10 & 27 & 15 \\ 15 & 15 & 8 & 13 & 5 \\ 25 & 21 & 5 & 15 & 21 \end{pmatrix}$$

Solve this problem beginning with the basic set of cells $\{(1,3), (1,5), (2,1), (2,2), (2,5), (3,1), (3,4)\}$.

Does this problem have alternate optimum solutions? If so, find a formula that describes a general optimum solution.

A company called PSC headquartered in the same city as our company operates the shipping route from source 3 to refinery 1. Discuss how much our company would loose per unit of business given to PSC if our company wanted to patronize them.

Do cost ranging for cost coefficient c_{13}. From the present optimum solution, obtain an optimum solution for the problem if c_{13} becomes 14.

Consider the original problem again (i.e., $c_{13} = 10$). If the weekly supply at source 3, and the weekly demand at refinery 1 change to $20 + \delta$, $10 + \delta$ respectively; find an optimum solution as a function of δ and its optimality range.

Chapter 7

The Simplex Method for General LP

In this chapter we will discuss the primal simplex method for general LPs. It is a pivotal method that has proven to be quite efficient for solving LP models arising in real world applications, and is used quite heavily by most OR practitioners. The method is also suitable to solve small problems by hand computation. High quality and highly reliable software for this method is available from several commercial vendors as explained in Section 1.6.

Before applying the simplex method on an LP, all the constraints on which pivot operations are carried out must be transformed into equality constraints for reasons explained in Section 3.2.2. We will discuss the simplex method for solving LPs in standard form defined earlier, and show in the next section how every type of LP model can be transformed into this standard form.

7.1 How to Transform Any LP Model Into Standard Form?

As defined earlier, an LP is said to be in **standard form** if it is a minimization problem in which the constraints are a system of equations in nonnegative variables. Transforming any LP into standard form involves the following steps.

Step 1 Transform lower bounds on individual variables to zero If there is a lower bound condition on a variable, such as "$x_1 \geqq \ell_1$" where $\ell_1 \neq 0$, substitute $x_1 = y_1 + \ell_1$ wherever it appears in the problem. Here y_1 is a new variable required to be $\geqq 0$, it is the slack variable for this condition. Now x_1 is eliminated from the problem, the lower bound condition on it becomes the nonnegativity restriction on the new variable y_1.

Step 2 Transformation of a variable with an upper bound restriction but no lower bound restriction   If there is a variable x_j with an upper bound condition "$x_j \leqq k_j$", and no lower bound condition on its value, substitute $x_j = k_j - y_j$ wherever it appears in the problem. Here y_j is a new nonnegative variable replacing x_j; it is the slack variable for this condition. The upper bound condition on x_j is replaced by the nonnegativity restriction on y_j.

Step 3 Transformation of all inequalities (other than nonnegativity restrictions on individual variables) into equality constraints   Convert all inequality constraints other than the nonnegativity restrictions on variables (including any remaining upper bound conditions on individual variables) into equations by introducing an appropriate slack variable for each one. All these slack variables are nonnegative variables.

Step 4 Put objective in minimization form   If the objective function is required to be minimized, leave it as it is. If it is required to be maximized, replace it by the equivalent problem of minimizing its negative, subject to the same constraints.

Step 5 Eliminate any unrestricted variables   At this stage all the constraints are either equality constraints, or nonnegativity restrictions on individual variables. If all the variables in the problem at this stage are nonnegative variables, the problem is now in standard form; stop.

Otherwise there must be some unrestricted variables (i.e., those on whose value there is no explicit lower or upper bound stated at this stage) in the model. In this case, eliminate each of these unrestricted variables using an equality constraint in which it appears with a nonzero coefficient, one by one. As discussed in Section 3.4, this can be carried out efficiently by the following.

Suppose there are M equality constraints in N variables, $x_1, \ldots, x_N$, at this stage, of which some are unrestricted and the others are nonnegative. If the objective at this stage is to minimize $z = \sum_{j=1}^{N} c_j x_j + \alpha$, where α is a constant, augment the system of equality constraints by the additional equation $\sum_{j=1}^{N} c_j x_j - z = \alpha$ defining the objective function. Express all the equality constraints in a detached coefficient tableau form as given below.

Tableau 7.1: System of equality constraints and objective row

x_1	$\cdots$	x_N	$-z$	
a_{11}	$\cdots$	a_{1N}	0	b_1
$\vdots$		$\vdots$	$\vdots$	$\vdots$
a_{M1}	$\cdots$	a_{MN}	0	b_M
c_1	$\cdots$	c_N	1	α

Choose an unrestricted variable, x_N say, and select its column as the pivot column. Select a row with a nonzero coefficient in the pivot column as the pivot row, and perform the pivot step. The unrestricted variable x_N is now the dependent variable in the pivot row. As explained in Section 3.4, performing this pivot step is equivalent to eliminating the unrestricted variable x_N from the objective function and all the constraints other than that represented by the pivot row. In the tableau obtained after the pivot step, use the equation corresponding to the pivot row to express x_N as a function of the other variables. Store this expression, to obtain the value of x_N in the solution after the values of the other variables are obtained. Now delete the pivot row, and the column of the variable x_N from this tableau. Thus x_N is now removed from the optimization portion of the problem. Thus elimination of an unrestricted variable this way, reduces from the optimization portion of the problem, the number of variables by one and the number of equality constraints by one.

If there are some more unrestricted variables in the remaining tableau, repeat the same procedure on it.

When all the unrestricted variables are eliminated from the optimization portion of the problem this way, the remaining tableau represents an LP in standard form that is equivalent to the original one.

EXAMPLE 7.1

As an example, consider the following LP

$$
\begin{array}{rrrrrrrcr}
\text{Maximize } z' = & 2x_1 & +7x_2 & -8x_3 & +2x_4 & +4x_5 & -x_6 & & \\
\text{subject to} & x_1 & -x_2 & +x_3 & -x_4 & -x_5 & +x_6 & \geqq & 2 \\
& & x_2 & -2x_3 & +x_4 & +x_5 & -2x_6 & \leqq & 11 \\
& x_1 & +2x_2 & -x_3 & +x_4 & +x_5 & & = & 14 \\
& -x_1 & & +x_3 & +2x_4 & & +2x_6 & = & 5
\end{array}
$$

$$x_1 \geqq 2; \ x_2 \leqq 5; \ x_3, x_4 \geqq 0; \ x_5, x_6 \text{ unrestricted}$$

To transform this problem into standard form, substitute $x_1 = 2 + y_1$, $x_2 = 5 - y_2$, where $y_1, y_2 \geqq 0$, as in Steps 1, 2. After rearranging terms, this transforms the problem into

$$
\begin{array}{llllllll}
\text{Maximize} & 2y_1 & -7y_2 & -8x_3 & +2x_4 & +4x_5 & -x_6 & +39 \\
\text{subject to} & y_1 & +y_2 & +x_3 & -x_4 & -x_5 & +x_6 & \geqq 5 \\
& & -y_2 & -2x_3 & +x_4 & +x_5 & -2x_6 & \leqq 6 \\
& y_1 & -2y_2 & -x_3 & +x_4 & +x_5 & & = 2 \\
& -y_1 & & +x_3 & +2x_4 & & +2x_6 & = 7
\end{array}
$$

$$y_1, y_2, x_3, x_4 \geqq 0; \; x_5, x_6 \text{ unrestricted}$$

Introducing the nonnegative slack variables s_1, s_2, the first two constraints here become $y_1 + y_2 + x_3 - x_4 - x_5 + x_6 - s_1 = 5$, and $-y_2 - 2x_3 + x_4 + x_5 - 2x_6 + s_2 = 6$. And we replace the objective function by that of minimizing $z = -z' = -2y_1 + 7y_2 + 8x_3 - 2x_4 - 4x_5 + x_6 - 39$. In tabular form, here is the problem.

Tableau 7.2

				PC						
y_1	y_2	x_3	x_4	x_5	x_6	s_1	s_2	$-z$		
1	1	1	-1	$\boxed{-1}$	1	-1	0	0	5	PR
0	-1	-2	1	1	-2	0	1	0	6	
1	-2	-1	1	1	0	0	0	0	2	
-1	0	1	2	0	2	0	0	0	7	
-2	7	8	-2	-4	1	0	0	1	39	

$y_1, y_2, x_3, x_4, s_1, s_2 \geqq 0; \; x_5, x_6$ unrestricted; minimize z
PC = Pivot column, PR = Pivot row

We select the column of the unrestricted variable x_5 as the pivot column, and row 1 in which the pivot column has a nonzero coefficient as the pivot row. The pivot element is boxed. The pivot step leads to the following tableau.

					PC					
Dependent variable	y_1	y_2	x_3	x_4	x_5	x_6	s_1	s_2	$-z$	
x_5	-1	-1	-1	1	1	-1	1	0	0	-5
	1	0	-1	0	0	$\boxed{-1}$	-1	1	0	11	PR
	2	-1	0	0	0	1	-1	0	0	7
	-1	0	1	2	0	2	0	0	0	7
	-6	3	4	2	0	-3	4	0	1	19

PC = Pivot column, PR = Pivot row

In this tableau we select the column of the unrestricted variable x_6 as the pivot column, and row 2 as the pivot row. The pivot element is boxed. After this pivot step we get the next tableau.

Dependent variable	y_1	y_2	x_3	x_4	x_5	x_6	s_1	s_2	$-z$	
x_5	-2	-1	0	1	1	0	2	-1	0	-16
x_6	-1	0	1	0	0	1	1	-1	0	-11
	3	-1	-1	0	0	0	-2	1	0	18
	1	0	-1	2	0	0	-2	2	0	29
	-9	3	7	2	0	0	7	-3	1	-14

Now pivot steps have been performed in the columns of both the unrestricted variables. From the equation corresponding to row 1 in this tableau we have $x_5 = 2y_1 + y_2 - x_4 - 2s_1 + s_2 - 16$. And similarly from row 2 we have $x_6 = y_1 - x_3 - s_1 + s_2 - 11$. Now we delete these two rows, and the columns of x_5, x_6, and are left with the remaining problem in standard form.

Tableau 7.3

y_1	y_2	x_3	x_4	s_1	s_2	$-z$	
3	-1	-1	0	-2	1	0	18
1	0	-1	2	-2	2	0	29
-9	3	7	2	7	-3	1	-14

$y_1, y_2, x_3, x_4, s_1, s_2 \geqq 0$; minimize z

We get the values of $y_1, y_2, x_3, x_4, s_1, s_2$ in an optimum solution by solving the LP in standard form in this last tableau. By substituting their values in the expressions given above, we get the corresponding values of x_5, x_6, x_1, x_2 in this solution.

Exercise

7.1 Transform the following LP into standard form.

Maximize	$3x_1$	$-2x_2$	$+2x_3$	$-x_4$	$-2x_5$	$-2x_6$		
subject to	$-2x_1$	$+2x_2$	$+2x_3$	$-2x_4$	$+x_5$	$-2x_6$	$\leqq$	200
	$4x_1$		$+x_3$	$-x_4$	$+x_5$	$-3x_6$	$\geqq$	100
	$-x_1$	$-2x_2$	$+2x_3$	$+x_4$			$=$	25
	$2x_1$	$+2x_2$	$+x_3$	$+2x_4$		$+x_6$	$=$	80

$x_1 \geqq 20$; $x_2 \leqq 50$; $x_3, x_4 \geqq 0$; $x_3 \leqq 20$; x_5, x_6 unrestricted

7.2   The Primal Simplex Algorithm

Here we discuss the version of the simplex algorithm known as the **revised simplex algorithm maintaining explicit basis inverse** for solving an LP in standard form. This version was developed by George B. Dantzig and published in his 1953 paper. Let the problem to be solved be

$$
\begin{aligned}
\text{Minimize} \quad & z(x) = cx \quad +\alpha \\
\text{subject to} \quad & Ax \quad = b \\
& x \quad \geqq 0
\end{aligned}
\tag{7.1}
$$

where α is some constant, and without any loss of generality we assume that A is a matrix of order $m \times n$ and rank m. Here is the original tableau for this problem

Original tableau: Tableau 7.4

x	$-z$	
A	0	b
c	1	$-\alpha$

To apply this algorithm we need an initial primal feasible basic vector for (7.1). A method for handling the problem when a primal feasible basic vector is not known is discussed in the next section. Here we assume that a primal feasible basic vector, x_B say, is available to initiate the primal simplex algorithm on the problem. Compute the inverse tableau corresponding to x_B as discussed in Section 4.2.1, and let it be

Tableau 7.5   Inverse tableau wrt x_B

Basic var.	Inverse tableau		Basic values	Pivot col.	Ratios
			$\bar{b}_1$	$\bar{a}_{1s}$	
			$\vdots$	$\vdots$	$\vdots$
x_B	B^{-1}	0	$\bar{b}_i$	$\bar{a}_{is}$	$\bar{b}_i/\bar{a}_{is}$ *
			$\vdots$	$\vdots$	$\vdots$
			$\bar{b}_m$	$\bar{a}_{ms}$	
$-z$	$-\bar{\pi}$	1	$-\bar{z}$	$\bar{c}_s$	Min. $= \theta$

* Computed only if $\bar{a}_{is} > 0$

Let x_D be the vector of nonbasic variables. Then the primal and dual basic solutions corresponding to the basic vector x_B are

$$
\text{primal basic solution} \begin{pmatrix} x_B \\ x_D \end{pmatrix} = \begin{pmatrix} \bar{b} \\ 0 \end{pmatrix}
$$

$$
\begin{aligned}
\text{objective value} &= \bar{z} \\
\text{dual basic solution} &= \bar{\pi}
\end{aligned}
$$

We will first state all the steps to be carried out in the iteration of the algorithm when x_B is the primal feasible basic vector. This will be followed by an explanation of these steps and the rationale behind them using a numerical example.

THE PRIMAL SIMPLEX ALGORITHM

Iteration in the Primal Simplex Algorithm When x_B Is the Primal Feasible Basic Vector

Step 1 Compute nonbasic relative cost coefficients   For each nonbasic variable x_j compute its relative cost coefficient $\bar{c}_j = c_j - \bar{\pi} A_{.j} = (-\bar{\pi}, 1) \begin{pmatrix} A_{.j} \\ \dots \\ c_j \end{pmatrix}$, the dot product of the last row of the inverse tableau with the column of x_j in the original tableau.

Step 2 Check optimality   The optimality criterion is

$$\bar{c}_j \gtreqqless 0 \quad \text{for all nonbasic variables } x_j \tag{7.2}$$

If $\bar{c}_j \gtreqqless 0$ for every nonbasic variable x_j, the present dual basic solution is dual feasible. So, the present basic vector is both primal and dual feasible, and hence optimal by the results discussed in Section 4.2. Thus the present primal BFS is optimal to the LP (7.1), and the present dual basic solution is an optimum solution for the dual problem, terminate.

If the optimality criterion (7.2) is not satisfied, let $\mathbf{E} = \{j : \bar{c}_j < 0\}$. For each $j \in \mathbf{E}$, x_j is a nonbasic variable with a negative relative cost coefficient, and is said to be **eligible to enter the present basic vector**. Go to Step 3

Step 3 Select an entering variable   Select one of the nonbasic variables eligible to enter the present basic vector as the **entering variable**. This selection can be carried out arbitrarily, but a couple of rules which work well in computational experiments are the following.

First eligible variable encountered   While computing the relative cost coefficients of the nonbasic variables, $\bar{c}_j$, in Step 1, the moment a negative $\bar{c}_j$ turns up, select the corresponding variable x_j as the entering variable. You do not even have to compute the relative cost coefficients of the other nonbasic variables.

Dantzig's entering variable choice rule   Compute the relative cost coefficients, $\bar{c}_j$, of all nonbasic variables, and if the optimality criterion (7.2) is not satisfied, select the entering variable to be the nonbasic variable x_j which has the most negative $\bar{c}_j$. This rule is also called the **minimum $\bar{c}_j$ rule**, or the **most negative reduced cost column (edge) selection rule**. Go to Step 4.

Step 4 Compute the updated column of the entering variable and check for unboundedness Suppose x_s is the entering variable. Its column in the original tableau 7.1 is $\begin{pmatrix} A_{.s} \\ \cdots \\ c_s \end{pmatrix}$. Multiply it on the left by the present inverse tableau, this yields the updated column of x_s denoted by $\begin{pmatrix} \bar{A}_{.s} \\ \cdots \\ \bar{c}_s \end{pmatrix} =$ $(\bar{a}_{1s}, \ldots, \bar{a}_{ms}, \bar{c}_s)^T$. $\bar{c}_s < 0$, otherwise x_s would not be eligible to be selected as the entering variable. The unboundedness criterion is

$$\bar{a}_{is} \leqq 0, \quad \text{for all } i = 1 \text{ to } m \tag{7.3}$$

If (7.3) is satisfied, the objective function is unbounded below on the set of feasible solutions of (7.1), i.e., its minimum value is $-\infty$, and there is no finite optimum solution. Construct the feasible solution $x(\lambda) = (x_j(\lambda))$ depending on the nonnegative parameter λ, defined by

$$x_j(\lambda) = \begin{cases} 0 & \text{for all nonbasic } x_j, \ j \neq s \\ \lambda & \text{for entering variable } x_s \\ \bar{b}_i - \bar{a}_{is}\lambda & \text{if } x_j \text{ is the } i\text{th basic variable in } x_B \end{cases} \tag{7.4}$$

$x(\lambda)$ is a feasible solution for the problem for all $\lambda \geqq 0$, and the objective value at $x(\lambda)$, $z(x(\lambda)) = \bar{z} + \bar{c}_s\lambda$. Since $\bar{c}_s < 0$, $z(x(\lambda)) \to -\infty$ as $\lambda \to +\infty$. The half-line $\{x(\lambda) : \lambda \geqq 0\}$ is a half-line beginning at the present primal BFS. As we travel along this half-line by increasing λ from 0, the objective value continues to decrease indefinitely. This half-line is called an **extreme half-line** or **unbounded edge of the set of feasible solutions of (7.1) along which $z(x)$ diverges to $-\infty$** in this case. Terminate.

If (7.3) is not satisfied, go to Step 5.

Step 5 Minimum ratio test to identify pivot row for maintaining primal feasibility Make the updated column of the entering variable into the pivot column and enter it on the right hand side of the inverse tableau (already shown in Tableau 7.5). For each $i = 1$ to m, compute the ratio $\bar{b}_i/\bar{a}_{is}$ only if $\bar{a}_{is} > 0$, and enter these computed ratios in another column on the right (the ratio is 0 if $\bar{b}_i = 0$, otherwise it is > 0). Find the minimum of all the computed ratios. Suppose it is θ. Row i is said to be **eligible to be the pivot row**, if it ties for the minimum ratio, in this case the basic variable in it is said to be **eligible to be the dropping** or **leaving basic variable**. θ is known as the **minimum ratio** in this iteration. Select one of the eligible rows as the **pivot row** arbitrarily. The element in the pivot column and the pivot row is the **pivot element**. Here it will be > 0 by the way we selected the pivot row. The present basic variable in the pivot row is the **dropping basic variable** in this iteration. It will be replaced by the entering variable to yield the new basic vector. Go to Step 6.

Step 6 Pivot step   Perform the pivot step with the selected pivot element, carrying out the row operations on all the columns other than the ratios column. This transforms the pivot column into the column of the unit matrix of order $m + 1$ with a $+1$ entry in the pivot row. In the resulting tableau replace the basic variable in the pivot row by the entering variable x_s, this gives the new basic vector $x_{\hat{B}}$ say. In this tableau when you delete the pivot column just used, you are left with the inverse tableau of $x_{\hat{B}}$ as discussed in Section 3.3.8.

The objective value of the new BFS, $\hat{z} = \bar{c}_s \theta +$ (the previous objective value) $= \bar{z} + \bar{c}_s \theta$, and since $\bar{c}_s < 0$ and $\theta \geqq 0$, $\hat{z}$ will be $\leqq \bar{z}$.

If $\theta > 0$, this pivot step is called a **nondegenerate pivot step**. In this case $\hat{z} < \bar{z}$. So the pivot step produces a new and strictly better primal BFS.

If $\theta = 0$, this pivot step is called a **degenerate pivot step**. In this case $\hat{z} = \bar{z}$. In fact the new BFS will be the same as the previous one. The previous basic vector x_B and the new one $x_{\hat{B}}$ differ in exactly one zero valued basic variable. So, in this case there is no change in the objective value or the primal BFS, but we have a new basic vector.

With the new primal feasible basic vector and its inverse tableau, go to the next iteration.

EXAMPLE 7.2

We will now explain the various steps in the algorithm using a numerical example. Consider the following LP.

Original tableau: Tableau 7.6

x_1	x_2	x_3	x_4	x_5	x_6	x_7	$-z$	b
1	0	0	0	-1	1	1	0	2
0	1	0	0	1	-1	1	0	1
0	0	1	0	2	20	1	0	5
0	0	0	1	0	-1	1	0	0
0	0	1	1	-1	29	-8	1	0

$x_j \geqq 0$ for all j, minimize z

$x_B = (x_1, x_2, x_3, x_4)$ is a primal feasible basic vector. The augmented basis corresponding to it is the square matrix consisting of the column vectors of $x_1, x_2, x_3, x_4, -z$ in Tableau 7.6. Its inverse can be obtained by the formulae discussed in Section 3.3.9. This yields the first inverse tableau given below.

First inverse tableau

Basic var.	Inverse tableau					Basic values	PC x_5	Ratios
x_1	1	0	0	0	0	2	-1	
x_2	0	1	0	0	0	1	$\boxed{1}$	1/1 PR
x_3	0	0	1	0	0	5	2	5/2
x_4	0	0	0	1	0	0	0	
$-z$	0	0	-1	-1	1	-5	-3	Min. $= \theta = 1$

PC = pivot column, PR = pivot row

The primal BFS corresponding to x_B is $\bar{x} = (\bar{x}_1 \text{ to } \bar{x}_7)^T = (2, 1, 5, 0; 0, 0, 0)^T$, with objective value $\bar{z} = 5$. The dual basic solution corresponding to x_B is $\bar{\pi} = (0, 0, 1, 1)$ from the last row of the inverse tableau. The relative cost coefficients of the nonbasic variables, x_5, x_6, x_7, $(\bar{c}_5, \bar{c}_6, \bar{c}_7) = (-3, 10, -10)$, obtained by taking the dot product of the last row of the inverse tableau with the original columns of these variables. The optimality criterion, (7.2), is not satisfied, and x_5, x_7 are eligible to enter the basic vector x_B. Among these suppose x_5 is selected as the entering variable. Its updated column is its original column multiplied on the left by the inverse tableau, this is $(-1, 1, 2, 0, -3)^T$. Since there are positive entries in this column, the unboundedness criterion (7.3) is not satisfied. So, the updated column of x_5 is the pivot column, and we have to carry out the minimum ratio test to identify the pivot row.

We will now explain the rationale behind the minimum ratio test. Its purpose is to make sure that the next basic vector obtained after the pivot step will also be primal feasible, i.e., to maintain primal feasibility throughout the algorithm.

At this stage the simplex algorithm fixes all the nonbasic variables other than the entering variable at their present value of 0, gives the entering variable a tentative value, say λ which is a nonnegative parameter, and obtains the solution of the remaining system as a function of λ. Suppose this solution is denoted by $x(\lambda)$ and its objective value by $z(\lambda)$. In this example, the remaining system after fixing the nonbasic variables x_6, x_7 at 0, and rearranging variables is

x_1	x_2	x_3	x_4	$-z$	x_5	
1	0	0	0	0	-1	2
0	1	0	0	0	1	1
0	0	1	0	0	2	5
0	0	0	1	0	0	0
0	0	1	1	1	-1	0

$$x_6 = x_7 = = 0$$

Multiplying this system on the left by the present inverse tableau converts it into

x_1	x_2	x_3	x_4	$-z$	x_5	
1	0	0	0	0	-1	2
0	1	0	0	0	1	1
0	0	1	0	0	2	5
0	0	0	1	0	0	0
0	0	0	0	1	-3	-5

$$x_6 = x_7 = = 0$$

In this tableau, the column vector of x_5 is its updated column computed above. From this system we read out the solution $x(\lambda)$ discussed above to be

$$(x_1(\lambda), \ldots, x_7(\lambda)) = (2 + \lambda, 1 - \lambda, 5 - 2\lambda, 0, \lambda, 0, 0) \qquad (7.5)$$
$$z(\lambda) = 5 - 3\lambda$$

The coefficient of λ in $z(\lambda)$, -3, is $\bar{c}_5$, the relative cost coefficient of the entering variable x_5. Thus $\bar{c}_5$ is the rate of change in the objective value per unit change in the value of the nonbasic variable x_5 from its current value of 0 in the present BFS. This is the reason for selecting the entering variable among those with negative relative cost coefficients, because it helps to get better solutions with smaller values for z.

In the solution (7.5), $z(\lambda)$ decreases as λ increases from 0, so to get maximum possible reduction in this iteration, we should give λ the maximum possible value. As λ increases, the basic values, $x_1(\lambda), x_4(\lambda)$ (which are $2 + \lambda$, and 0) remain $\geqq 0$ always. These are the values of basic variables in rows in which the pivot column has entries $\leqq 0$. However the basic values $x_2(\lambda), x_3(\lambda)$ (which are $1 - \lambda, 5 - 2\lambda$) will become < 0 if λ becomes too large. For $x_2(\lambda) = 1 - \lambda$ to be $\geqq 0$ we need $\lambda \leqq 1/1 = 1$. This is the ratio in row 2 in which x_2 is the basic variable. This row has a positive entry in the pivot column. For $x_3(\lambda) = 5 - 2\lambda$ to be $\geqq 0$, we need $\lambda \leqq 5/2$, this is the ratio in row 3 in which x_3 is the basic variable, this row also has a positive entry of 2 in the pivot column. Thus each ratio computed in the ratio test is an upper bound for the value of λ to keep $x(\lambda)$ in (7.5) to be $\geqq 0$, i.e., feasible. Hence the maximum possible value that we can give to λ in this iteration is the smallest of these upper bounds for it, i.e., the minimum ratio, $\theta = \min\{1/1, 5/2\} = 1$ here.

Hence the minimum ratio test determines the maximum possible value for the entering variable that keeps the new solution $x(\lambda)$ primal feasible.

When $\lambda =$ the minimum ratio θ, the basic variable x_2 in whose row the minimum ratio is attained, will reach a value of 0. Any increase in λ beyond the minimum ratio θ will make $x_2(\lambda) < 0$, that is why x_2 is called a **blocking basic variable** in this step. When we fix $\lambda = \theta$, this blocking basic variable becomes 0, and can be dropped from the present basic vector to become a nonbasic variable at that time, and its place in the basic vector given to the entering variable. Since every basic vector for an LP always consists of the same number of basic variables, when the entering variable is brought into the present basic vector, one of the present basic variables must be dropped from it. Candidates eligible to be dropping basic variables are those whose value becomes 0 in $x(\theta)$.

By the results in Section 3.3.8, the inverse tableau corresponding to the new basic vector is obtained by performing a pivot step with the updated column of the entering variable as the pivot column, and the row of the dropping basic variable as the pivot row.

In the solution $x(\lambda)$ in (7.5), if we put $\lambda = 0$, we get the present BFS $\bar{x}$. If we put $\lambda = \theta$, the minimum ratio, we get the BFS associated with the next basic vector $x_{\hat{B}}$, with its objective value of $5 - 3\theta = 2$, since $\theta = 1$ here. The line segment joining these two solutions is called an **edge** of the set of feasible solutions of the problem. So, in this iteration, we can say that the simplex algorithm travels from one end of this edge, $x(0)$, to the other end, $x(\theta)$, attaining a strict decrease in the objective value in this process. See Figure 2.3. Geometrically, this move is like moving from the point $\tilde{x}$ to the point x^* in that figure, along the edge joining them.

Selecting the entering variable by Dantzig's rule minimizes its relative cost coefficients among the present nonbasics, and hence minimizes the change in objective value per unit change in the entering variable from its present value of 0.

The pivot column is entered on the first inverse tableau, and the ratios are computed. Row 2 is the pivot row, and the pivot element is boxed. Since the minimum ratio is $\theta = 1 > 0$, this pivot step is a nondegenerate pivot step. Performing the pivot step yields the second inverse tableau given below.

Second inverse tableau

Basic var.	Inverse tableau				Basic values	PC x_7	Ratios
x_1	1	1	0	0	0	3	2	3/2
x_5	0	1	0	0	0	1	1	1/1
x_3	0	-2	1	0	0	3	-1	
x_4	0	0	0	1	0	0	$\boxed{1}$	0/1 PR
$-z$	0	3	-1	-1	1	-2	-7	Min. $= \theta = 0$

PC = pivot column, PR = pivot row

The new BFS is $\hat{x} = (3, 0, 3, 0, 1, 0, 0)^T$ with an objective value of $\hat{z} = 2$. The relative cost coefficients of the nonbasic variables here are $(\bar{c}_2, \bar{c}_6, \bar{c}_7) = (3, 7, -7)$. So, x_7 is the entering variable in this iteration. Its updated column is $(2, 1, -1, 1, -7)^T$. It is the pivot column, and the minimum ratio is zero, and the pivot row is row 4. This pivot step is a degenerate pivot step. Performing this pivot step leads to the next inverse tableau.

Third inverse tableau

Basic var.	Inverse tableau					Basic values
x_1	1	1	0	-2	0	3
x_5	0	1	0	-1	0	1
x_3	0	-2	1	1	0	3
x_7	0	0	0	1	0	0
$-z$	0	3	-1	6	1	-2

Verify that the BFS and objective value are $\hat{x}, \hat{z}$, no change since the pivot step was a degenerate pivot step. The relative cost coefficients of the nonbasic variables wrt the new basic vector are $(\bar{c}_2, \bar{c}_4, \bar{c}_6) = (3, 7, 0)$. Since they are all $\geqq 0$, the optimality criterion is satisfied, and the present BFS $\hat{x} = (3, 0, 3, 0, 1, 0, 0)^T$ is an optimum solution with an optimum objective value of $\hat{z} = 2$. The dual solution $\hat{\pi} = (0, -3, 1, -6)$ from the third inverse tableau is an optimum dual solution. The primal basic vector here is primal degenerate, since the primal basic variable x_7 is 0 in the associated BFS. So, as discussed in Section 4.3, we cannot interpret the optimum dual solution ($\hat{\pi}$ here), as the marginal value vector in this problem.

EXAMPLE 7.3

We will now discuss another example illustrating the unboundedness termination. Consider the following LP.

Original tableau: Tableau 7.7

x_1	x_2	x_3	x_4	x_5	x_6	$-z$	b
0	0	1	1	-1	-5	0	7
1	0	0	-1	-1	-3	0	9
0	1	0	-1	-1	0	0	1
-1	-1	-1	10	6	4	1	0

$x_j \geqq 0$ for all j, minimize z

We will initiate the primal simplex algorithm on this problem with the primal feasible basic vector $x_B = (x_3, x_1, x_2)$. The inverse tableau corresponding to it is given below.

Inverse tableau

Basic var.	Inverse tableau				Basic values
x_3	1	0	0	0	7
x_1	0	1	0	0	9
x_2	0	0	1	0	1
$-z$	1	1	1	1	17

The primal BFS associated with x_B is $\bar{x} = (9, 1, 7, 0, 0, 0)^T$ with an objective value of $\bar{z} = -17$. The relative cost coefficients of the nonbasic variables are $(\bar{c}_4, \bar{c}_5, \bar{c}_6) = (9, 3, -4)$. So, we select x_6 as the entering variable. The updated column of x_6 is $(-5, -3, 0, -4)^T$. There is no positive entry in this column. Hence the unboundedness criterion (7.3) is satisfied, and hence the objective function $z(x)$ is unbounded below in this problem, and there is no finite optimum solution. Actually, the solution $x(\lambda)$ obtained by fixing the nonbasic variables x_4, x_5 at 0, giving the value λ to the entering variable x_6, and then finding the values of the basic variables as functions of λ, is the solution of $x_6 = \lambda$ and

x_3	x_1	x_2	$-z$	x_6	
1	0	0	0	-5	7
0	1	0	0	-3	9
0	0	1	0	0	1
0	0	0	1	-4	17

It is $x(\lambda) = (9+3\lambda, 1, 7+5\lambda, 0, 0, \lambda)^T$ with an objective value of $z(\lambda) = -17-4\lambda$. $x(0)$ is the present BFS $\bar{x}$. $\{x(\lambda) : \lambda \geq 0\}$ is a half-line beginning with the present BFS. As $\lambda \to \infty$, $z(\lambda) \to -\infty$ along this half-line. For example, if you want a feasible solution with an objective value of -4017, $z(\lambda) = -17 - 4\lambda$ is attained when $\lambda = 1000$, this corresponds to the point $x(1000) = (1009, 1, 5007, 0, 0, 1000)^T$ on this half-line. In the same manner, for any objective value however small, we can find a $\lambda > 0$ such that the point $x(\lambda)$ on this half-line is feasible to this problem and attains that objective value. So, we terminate with the unboundedness conclusion.

Practical Consequences of Satisfying the Unboundedness Criterion

Suppose the unboundedness criterion is satisfied while solving an LP model for a practical problem. Negative cost is profit, so this implies that we have found a way for making an unlimited profit! This infinite profit will be achieved if one implements the feasible solution at the infinite end of the half-line identified by the algorithm along which the cost function $z(x)$ diverges to $-\infty$. Clearly, some of the variables will have the value $+\infty$ in that solution. For a simple example of this consider a scheme popularly known in business circles as a daisy link. Here we are able to buy crude oil at a cost say of \$15/barrel from a middle east supplier, and sell it to a local company at \$25/barrel. To make an infinite profit from this deal, we have to buy and sell an infinite number of barrels at these rates. In practice, we are always limited by finite resources, and hence it is impossible to implement a solution in which some variables are $+\infty$. Thus, while infinite profit is a mathematical possibility, if the unboundedness criterion is satisfied while solving an LP model for a practical problem, it is probably an indication that some constraints on the decision variables have been forgotten in constructing the model; or there may be serious errors in the cost coefficients used in the model. So, one should review the model very carefully, and look for errors and omissions.

Features of the Simplex Algorithm

The primal simplex algorithm discussed in this section requires a primal feasible basic vector initially, and maintains primal feasibility of the basic vector throughout. Also, in every iteration, the primal and dual basic solutions satisfy all the complementary slackness conditions. So, of the three conditions required for optimality mentioned in Theorem 4.1 (primal and dual feasibility, and complementary slackness) it maintains two (primal feasibility and complementary slackness) and strives to attain the third one, dual feasibility.

7.3   The Primal Simplex Method

To solve an LP in standard form, the simplex algorithm discussed in Section 7.2 needs a primal feasible basic vector to start it off, and so it can be applied directly only if an initial primal feasible basic vector is available.

The simplex method is a method that can solve an LP whether a feasible basic vector is available for it or not. If a primal feasible basic vector is not available, it first tries to find one, ignoring the goal of minimizing the objective function temporarily. For this it constructs a new LP called the **Phase I problem** for which an artificial primal feasible basic vector is readily available. The simplex method then applies the simplex algorithm of Section 7.2 to solve the Phase I problem. The Phase I problem is so constructed that from its solution, we either have a proof that the original LP has no feasible solution, or we obtain a primal feasible basic vector for the original problem. If Phase I leads to a feasible basic vector for the original problem, starting with it, the simplex algorithm of Section 7.2 is applied to solve the original problem. This later part of the work in the simplex method is called **Phase II**.

So, if an initial feasible basic vector for the original problem is available, we go into Phase II directly and solve the original problem by the simplex algorithm of Section 7.2. If an initial primal feasible basic vector is not available, we have to use the simplex algorithm twice; first on a specially constructed Phase I problem (whose job is to find a primal feasible basic vector of the original problem, or determine that it has no feasible solution), and then on the Phase II problem if Phase I yields a primal feasible basic vector for the original problem.

Let the problem to be solved be the one in Tableau 7.4. We will first state the primal simplex method, and then follow up with explanation of the various steps using numerical examples.

THE PRIMAL SIMPLEX METHOD

Initialization

Step 1.1 Making the RHS constants vector nonnegative   Multiply each row in the original tableau in which the RHS constant b_i is negative, by -1. When this operation is completed, the RHS constants vector in the original tableau becomes $\geqq 0$.

Step 1.2 Selecting an initial basic vector by setting up the Phase I problem if necessary   For each $i = 1$ to m do the following. Search for $I_{.i}$, the ith column of the unit matrix of order m, among the column vectors of the coefficient matrix consisting of the constraint rows in the original tableau. If this column appears one or more times, select a variable associated with one of them (the one with the smallest cost coefficient if there are several) as the basic variable in the ith row.

In this process if basic variables have been selected in all the constraint rows, let x_B be the resulting basic vector. By its choice, the basis corresponding to x_B is the unit matrix, and since the RHS constants vector b is now $\geqq 0$, x_B is a primal feasible basic vector. Go over to Phase II with the basic vector x_B, i.e., construct the inverse tableau corresponding to it as in Sections 3.3.9, 4.2.1, and beginning with it solve the original problem by the simplex algorithm discussed in Section 7.2.

If basic variables have not been selected in some of the constraint rows in the above process, we will introduce artificial variables associated with the missing columns of the unit matrix to get an initial basis that is the unit matrix, and set up the Phase I problem. For this do the following for each i $= 1$ to m for which row i does not have a basic variable selected at this stage. Introduce a new nonnegative **artificial variable**, call it t_i, associated with the column $I_{.i}$ into the original tableau, and select this t_i as the basic variable in row i (this artificial variable t_i will remain the basic variable in row i as long as it remains in the problem).

We now have a full basic vector containing all these artificial variables, for which the basis is the unit matrix, and since the RHS constants vector $b \geqq 0$, it is a feasible basic vector for the Phase I problem. The variables in the original tableau, x_js, are now called **original problem variables** to distinguish them from the newly introduced artificial variables, t_is. During Phase I the original objective function is called the **Phase II objective function** and it is kept dormant, i.e., ignored. A new objective row called the **Phase I objective row** is opened as the $(m + 2)$th row in the original tableau, corresponding to the Phase I objective function which we denote by w. The various cost coefficients are defined as below.

Phase I cost coefficient of every $\begin{cases} \text{original problem variable } x_j \text{ is } 0 \\ \text{artificial variable } t_i \text{ is } 1 \end{cases}$

Phase II cost coefficient of every artificial variable t_i is 0

So, the Phase I objective function w = sum of all the artificial variables introduced, and since the artificial variables are all nonnegative, w is always $\geqq 0$. The original tableau for the Phase I problem is therefore of the form given below, where $t_{i_1}, \ldots, t_{i_r}$ are the artificial variables introduced.

Phase I original tableau: Tableau 7.8

x_1	$\ldots$	x_n	t_{i_1}	$\ldots$	t_{i_r}	$-z$	$-w$	
a_{11}	$\ldots$	a_{1n}	0	$\ldots$	0	0	0	b_1
			$\vdots$					
			1					
					$\vdots$			
					1			
$\vdots$		$\vdots$	$\vdots$		$\vdots$	$\vdots$	$\vdots$	$\vdots$
a_{m1}	$\ldots$	a_{mn}	0	$\ldots$	0	0	0	b_m
c_1	$\ldots$	c_n	0	$\ldots$	0	1	0	α
0	$\ldots$	0	1	$\ldots$	1	0	1	0

$x_j \geqq 0$ for all j; artificials $t_{i_1}, \ldots, t_{i_r} \geqq 0$; minimize w

Let x_{B_1} be the basic vector selected in the above process (it contains all the artificial variables), the basis corresponding to it among the constraint rows in Tableau 7.8 is the unit matrix of order m. The $(m+1)$th, $(m+2)$th rows in Tableau 7.8 are the Phase II and Phase I objective rows. During Phase I, $-z, -w$ will be treated as the basic variables in these rows; and the augmented basis corresponding to any basic vector x_B will be of order $(m+2) \times (m+2)$ consisting of the columns in Tableau 7.8 of the basic variables in x_B, and those of $-z$ and $-w$.

Any solution to the Phase I problem in which $w = 0$ must have all artificial variables $= 0$, and the x-part in it must therefore be feasible to the original problem. So, to find a feasible solution of the original problem, we need to look for a solution of the Phase I problem in which $w = 0$; this can be done by minimizing w in the Phase I problem. If the minimum value of w in the Phase I problem is > 0, then it is impossible to find a feasible solution for it which makes $w = 0$; this implies that the original problem has no feasible solution.

Step 1.3 Computing the inverse tableau corresponding to Phase I initial feasible basic vector Let x_{B_1} be the initial basic vector set up in Step 2 for the Phase I problem. Let c_{B_1}, d_{B_1} denote the Phase II, Phase I objective coefficient row vectors of the basic variables in x_{B_1}. Then the augmented basis corresponding to x_{B_1} is the matrix of order $(m+2) \times (m+2)$ given below.

$$
\begin{pmatrix}
I & \vdots & 0 & 0 \\
 & \cdots & & \\
c_{B_1} & \vdots & 1 & 0 \\
d_{B_1} & & 0 & 1
\end{pmatrix}
$$

Using the result in Section 3.3.9, here is the inverse tableau corresponding to this basic vector.

Basic var.	Inverse tableau			Basic values
x_{B_1}	I	0	0	b
$-z$	$-c_{B_1}$	1	0	$-z_1$
$-w$	$-d_{B_1}$	0	1	$-w_1$

During Phase I we solve the Phase I problem beginning with this inverse tableau using the simplex algorithm discussed in Section 7.2. The Phase II objective row plays no role during Phase I. The Phase I relative cost coefficient of any variable is the dot product of the last $((m+2)$th) row in the inverse tableau with the column vector of that variable in the Phase I original tableau, Tableau 7.8. These are used for determining Phase I termination, or for selecting entering variables during Phase I.

The artificial variables are introduced solely for providing a full basic vector to apply the simplex algorithm to move towards a feasible basic vector for the original problem. So, at some stage during Phase I, if an artificial variable becomes nonbasic after being replaced by an original problem variable, we delete it by erasing its column from the Phase I original tableau. So, an artificial variable exists in the problem only as long as it is a basic variable. Thus every nonbasic variable will always be an original problem variable, and the entering variables in every iteration, in fact every variable eligible to enter the basic vector in every iteration, will always be an original problem variable.

Since we become feasible to the original problem when the Phase I objective function w becomes 0, its value at any stage during Phase I provides a measure of how far away we are at that stage from feasibility to the original problem. That's why w is called an **infeasibility measure** for the current solution.

Since w is always $\geqq 0$, its value is bounded below by 0. In Phase I we minimize w and it can never be < 0. That is why during the process of applying the simplex algorithm of Section 7.2 on the Phase I problem, the unboundedness criterion can never be satisfied.

We will now describe the steps in an iteration in Phase I.

An Iteration During Phase I of the Simplex Method

Let x_B be the present feasible basic vector. Some of the basic variables in x_B may be artificial variables. Let the inverse tableau corresponding to this basic vector be the following.

Tableau 7.9   Inverse tableau wrt x_B

Basic var.	Inverse tableau			Basic values
x_B	B^{-1}	0	0	$\bar{b}$
$-z$	$-\bar{\pi}$	1	0	$-\bar{z}$
$-w$	$-\bar{\sigma}$	0	1	$-\bar{w}$

$\bar{\sigma}$ from the last row of the inverse tableau is the Phase I dual basic solution corresponding to x_B.

Step 2   Compute the Phase I relative cost coefficients   For each nonbasic variable x_j compute its Phase I relative cost coefficient, $\bar{d}_j = d_j - \bar{\sigma} A_{.j} =$ the dot product of the last $((m+2)$th$)$ row of the inverse tableau with the column of x_j in the Phase I original tableau, Tableau 7.8.

Step 3   Check Phase I termination criterion   The Phase I termination criterion is

$$\bar{d}_j \geqq 0 \quad \text{for all nonbasic variables } x_j \tag{7.6}$$

If (7.6) holds, $\bar{w}$, the present value of w, is its minimum. If $\bar{w} = 0$, the artificial variables, if any, are 0 in the present BFS; and hence the x-part of it is a feasible solution of the original problem. In this case go to Step 8 to move over to Phase II.

If (7.6) holds, but $\bar{w} > 0$, the original problem is infeasible. Terminate. See Section 7.4 for some suggestions on how the data can be modified to make the model feasible.

If (7.6) does not hold, let $\mathbf{E} = \{j : \bar{d}_j < 0\}$. For each $j \in \mathbf{E}$, x_j is a nonbasic variable with a negative Phase I relative cost coefficient, and is therefore eligible to enter the present basic vector. Go to Step 4.

Step 4 Select an entering variable   Select one of the variables x_j for a $j \in$ **E** as the entering variable. This can be done arbitrarily, or by one of the entering variable choice rules mentioned in Section 7.2 but using the Phase I relative cost coefficients. Go to Step 5.

Step 5 Compute the updated column of the entering variable   Suppose x_s is the nonbasic variable selected as the entering variable. Compute its updated column, $(\bar{a}_{1s}, \ldots, \bar{a}_{ms}, \bar{c}_s, \bar{d}_s)^T$ say, by multiplying its original column in the Phase I original tableau, Tableau 7.8, on the left by the present inverse tableau. Go to Step 6.

Step 6 Minimum ratio test to identify pivot row   Same as Step 5 in an iteration of the simplex algorithm discussed in Section 7.2.

Step 7 Pivot step   The pivot step is carried out exactly as in Step 6 in an iteration of the simplex algorithm discussed in Section 7.2 to obtain the new basic vector and the inverse tableau corresponding to it. If the dropping variable is an artificial variable, delete it from further consideration by erasing its column from the original tableau, Tableau 7.8. With the new inverse tableau, go to the next iteration.

Step 8 Moving over to Phase II   When we arrive at this step, the Phase I objective function w has become 0, this implies that if there are any artificial variables in the basic vector at this stage, their values in the present primal solution must be 0.

Suppose there are no artificial variables in the basic vector at this stage. Then, the present basic vector consists of original problem variables only, and is primal feasible to the original problem. Delete the $(m+2)$th row and column from the present Phase I inverse tableau, this leads to the inverse tableau of the present basic vector for the original problem. With this, apply the simplex algorithm of Section 7.2 to solve the original problem. This portion of the work is called Phase II of the simplex method.

Suppose there are some artificial variables in the basic vector at this stage. Their values in the primal solution are 0, and as long as this property holds, the x-part of the primal solution will be feasible to the original problem. Identify all nonbasic variables x_j whose present Phase I relative cost coefficient $\bar{d}_j > 0$, let **F** $= \{j : \bar{d}_j > 0\}$. These facts imply that for any $j \in$ **F**, if x_j assumes a positive value, then the infeasibility form w will become positive, making the solution infeasible to the original problem. Hence, in every feasible solution of the original problem, all the variables x_j for $j \in$ **F** must be $= 0$. Fix the values of these variables at 0, and delete them from the problem by erasing their columns from the original tableau, i.e., never select any of them as the entering variable. Now delete the $(m + 2)$th row and column from the present Phase I inverse tableau. This leads to the inverse tableau of the present basic vector for the Phase II problem, with it apply the simplex

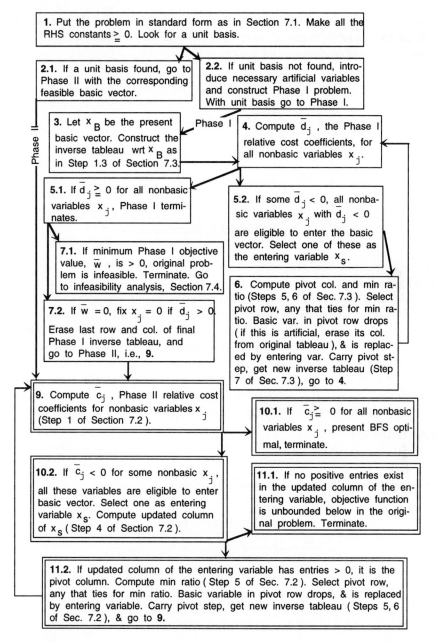

Figure 7.1   A flow chart for the primal simplex method. Phase II steps are outlined with double lines.

algorithm of Section 7.2 to solve the original problem. This work is called Phase II of the simplex method. During Phase II, the artificial basic variables that are in the basic vector may remain, but their values in the primal solution will always be 0 (if one of their values becomes > 0 during Phase II, it is an indication of computational errors).

We now present some numerical examples. As the first one, consider the following LP.

x_1	x_2	x_3	x_4	x_5	x_6	x_7	$-z$	b
1	0	1	-2	0	1	1	0	3
1	0	0	1	1	-2	0	0	4
-1	1	0	1	0	3	0	0	5
0	4	3	20	6	5	-2	1	0

$$x_j \geqq 0 \text{ for all } j, \text{ minimize } z$$

In this example, the RHS constants vector is already > 0, so we move to Step 2 of initialization, to select an initial basic vector. There are two variables, x_3, x_7 associated with the first unit column, and we select x_7 corresponding to the smallest cost coefficient among the two, as the first basic variable. x_5, x_2 are associated with the second and third unit vectors, we select them as the second and third basic variables. Thus we obtained a full primal feasible basic vector $x_B = (x_7, x_5, x_2)$, the inverse tableau corresponding to which can be computed using the formulae in Section 3.3.9. With it, we can apply the simplex algorithm of Section 7.2 to solve this problem.

EXAMPLE 7.4

Consider the following LP in standard form.

Tableau 7.10

x_1	x_2	x_3	x_4	x_5	$-z$	b
2	3	1	-1	0	0	10
1	2	-1	0	1	0	5
1	1	2	0	0	0	4
1	2	3	0	0	1	0

$$x_j \geqq 0 \text{ for all } j, \text{ minimize } z$$

The RHS constants vector is already > 0. We only have the second unit vector in the column of x_5. So, we select x_5 as the basic variable in row 2, and introduce artificial variables t_1, t_3 as basic variables in rows 1, 3 for the initial basic vector. The Phase I original tableau is

Tableau 7.11: Phase I original tableau

x_1	x_2	x_3	x_4	x_5	t_1	t_3	$-z$	$-w$	b
2	3	1	−1	0	1	0	0	0	10
1	2	−1	0	1	0	0	0	0	5
1	1	2	0	0	0	1	0	0	4
1	2	3	0	0	0	0	1	0	0
0	0	0	0	0	1	1	0	1	0

$x_j \geqq 0$ for all j; $t_1, t_3 \geqq 0$ artificials; minimize w

(t_1, x_5, t_3) is the initial feasible basic vector for the Phase I problem. The inverse tableau corresponding to it is given below. The value of w in the initial solution is 14; this is the infeasibility measure to the original problem of the initial solution. Subsequent inverse tableaus are listed one below the other, with all the information like $\bar{d}_j$s, the Phase I relative cost coefficients of x_1 to x_5, pivot element in a box, etc.

Basic var.	Inverse tableau					Basic values	PC x_1	Ratios	
t_1	1	0	0	0	0	10	2	10/2	
x_5	0	1	0	0	0	5	1	5/1	$\bar{d} = (-3, -4, -3, 1, 0)$
t_3	0	0	1	0	0	4	$\boxed{1}$	4/1	
$-z$	0	0	0	1	0	0	1		
$-w$	−1	0	−1	0	1	−14	−3		
							x_2		
t_1	1	0	−2	0	0	2	1	2/1	
x_5	0	1	−1	0	0	1	$\boxed{1}$	1/1	$\bar{d} = (0, -1, 3, 1, 0)$
x_1	0	0	1	0	0	4	1	4/1	
$-z$	0	0	−1	1	0	−4	1		
$-w$	−1	0	2	0	1	−2	−1		
t_1	1	−1	−1	0	0	1			
x_2	0	1	−1	0	0	1			$\bar{d} = (0, 0, 0, 1, 1)$
x_1	0	−1	2	0	0	3			
$-z$	0	−1	0	1	0	−5			
$-w$	−1	1	1	0	1	−1			

PC = pivot column

Phase I terminates when the basic vector (t_1, x_2, x_1) is reached. Since the value of w in the solution corresponding to this basic vector is $1 > 0$, the original problem has no feasible solution.

EXAMPLE 7.5

Consider the following LP in standard form

Tableau 7.12: Original tableau

x_1	x_2	x_3	x_4	x_5	x_6	$-z$	b
1	−1	0	0	2	0	0	0
−2	1	0	0	−2	0	0	0
1	0	1	0	1	−1	0	3
0	2	1	1	2	1	0	4
−40	−10	0	0	−7	−14	1	0

$x_j \geqq 0$ for all j, minimize z

Again the RHS constants vector is already > 0. Here we only have the fourth unit vector in the column of x_4. So, we select x_4 as the basic variable in row 4, and introduce artificial variables t_1, t_2, t_3 as as basic variables in rows 1, 2, 3 for the initial basic vector. The Phase I original tableau is

Phase I original tableau

x_1	x_2	x_3	x_4	x_5	x_6	t_1	t_2	t_3	$-z$	$-w$	b
1	−1	0	0	2	0	1	0	0	0	0	0
−2	1	0	0	−2	0	0	1	0	0	0	0
1	0	1	0	1	−1	0	0	1	0	0	3
0	2	1	1	2	1	0	0	0	0	0	4
−40	−10	0	0	−7	−14	0	0	0	1	0	0
0	0	0	0	0	0	1	1	1	0	1	0

$x_j \geqq 0$ for all j; $t_1, t_2, t_3 \geqq 0$ artificials; minimize w

(t_1, t_2, t_3, x_4) is the initial feasible basic vector for the Phase I problem. The various inverse tableaus obtained during Phase I are given below.

Basic var.	Inverse tableau						Basic values	PC x_3	Ratios	
t_1	1	0	0	0	0	0	0	0		
t_2	0	1	0	0	0	0	0	0		$\bar{d} = (0, 0, -1, 0,$
t_3	0	0	1	0	0	0	3	[1]	3/1	$-1, 1)$
x_4	0	0	0	1	0	0	4	1	4/1	
$-z$	0	0	0	0	1	0	0	0		
$-w$	−1	−1	−1	0	0	1	−3	−1		
t_1	1	0	0	0	0	0	0			
t_2	0	1	0	0	0	0	0			$\bar{d} = (1, 0, 0, 0,$
x_3	0	0	1	0	0	0	3			$0, 0)$
x_4	0	0	−1	1	0	0	1			
$-z$	0	0	0	0	1	0	0			
$-w$	−1	−1	0	0	0	1	0			

PC = pivot column

Phase I terminates and we have $w = 0$. The artificial variables t_1, t_2 are still in the basic vector, but their values in the final solution are 0. This final solution leads to the feasible solution $(x_1 \text{ to } x_6)^T = (0, 0, 3, 1, 0, 0)^T$ for the original problem.

Now we need to go to Phase II. We look for original problem variables with positive Phase I relative cost coefficient at Phase I termination. Only x_1 satisfies this property. So, $x_1 = 0$ in every feasible solution of the original problem. We fix $x_1 = 0$, and delete it from the problem. In Phase II we only consider variables x_2 to x_6 as candidates to enter the basic vector. From the Phase I terminal tableau we get the following inverse tableau to initiate Phase II.

Basic var.	Inverse tableau					Basic values	PC x_6	Ratios	
t_1	1	0	0	0	0	0	0		
t_2	0	1	0	0	0	0	0		$(\bar{c}_2 \text{ to } \bar{c}_6) = (-10,$
x_3	0	0	1	0	0	3	-1		$0, 0, -7, -14)$
x_4	0	0	-1	1	0	1	$\boxed{2}$	1/2	
$-z$	0	0	0	0	1	0	-14		
t_1	1	1	0	0	0	0			
t_2	0	1	0	0	0	0			$(\bar{c}_2 \text{ to } \bar{c}_6) = (4,$
x_3	0	0	1/2	1/2	0	7/2			$0, 7, 0, 0)$
x_6	0	0	$-1/2$	1/2	0	1/2			
$-z$	0	0	-7	7	1	7			

PC = pivot column

Since $(\bar{c}_2 \text{ to } \bar{c}_6) \geqq 0$ now Phase II terminates by satisfying the optimality criterion. The present BFS $\bar{x} = (\bar{x}_1 \text{ to } \bar{x}_6) = (0, 0, 7/2, 0, 0, 1/2)^T$ is an optimum solution to the problem with an optimum objective value of $\bar{z} = -7$.

Notice that even though the artificial variables t_1, t_2 were in the basic vector, their values in the solution remained 0 during Phase II.

How to Find a Feasible Solution to a System of Linear Constraints

In Section 3.4 we discussed the Gaussian elimination method for solving a system of linear equations.

Suppose we have to solve a system of linear constraints that consists of not just linear equations, but maybe some linear inequalities, sign restrictions or lower and/or upper bound constraints on variables. How does one solve such a general system? Notice that we are not required to optimize an objective function here, but just to find a feasible solution to the system if one exists. Using the techniques discussed in Section 7.1, this general system can be transformed into a system of linear equations in nonnegative variables. And the problem of finding a nonnegative solution to a system of linear equations can be solved by applying Phase I of the simplex method discussed above, to a Phase I formulation of the problem.

7.4   Infeasibility Analysis

An LP model for a real world problem may turn out to be infeasible, i.e., it may not have a feasible solution. There may be several reasons for this. For example, there may be constraints in the model specifying that demands on certain goods have to be met. And the resources needed to make these goods may not have been made available to make these goods in sufficient quantities to meet the demands. Then infeasibility will result.

When the model is infeasible, we should review it carefully and make modifications to the data to remove the causes for infeasibility (i.e., either increase resource availability, or reduce requirement levels for goods). What changes to make in the data to make the model feasible, is a practical issue on which there is likely to be heated debate. It is an issue that cannot be solved by mathematical techniques alone. Usually there are many different ways of changing the data to make the model feasible, this makes it even harder to settle this issue in a rational manner.

One simple way of making the model feasible that may be worth considering is the following. Suppose the model in Tableau 7.4 is the one being solved and suppose Phase I terminates with the infeasibility conclusion. For each $i = 1$ to m such that the artificial variable t_i is the basic variable in the ith row of the final Phase I inverse tableau, and is > 0 in the terminal Phase I solution, subtract its value from the original RHS constant b_i. Let $\hat{b}$ be the modified RHS constant obtained after these changes. Then if b in Tableau 7.4 is replaced by $\hat{b}$, it will no longer be infeasible.

As an example, consider the LP in Tableau 7.10 discussed in Example 7.4. The RHS constants vector in this problem is $b = (10, 5, 4)^T$. This model was shown to be infeasible in Example 7.4. In the final Phase I solution, only the artificial variable t_1, the basic variable in row 1 in the final Phase I inverse tableau, has a positive value of 1 in the final Phase I solution. When we subtract this value from b_1, the vector b changes to $\hat{b} = (9, 5, 4)^T$. So in Tableau 7.10 if b is replaced by $\hat{b}$ the model will have a feasible solution.

Suppose the modification of b to $\hat{b}$, as a way of making the model feasible, is acceptable. When this change from b to $\hat{b}$ is made, the final Phase I inverse tableau for the modified problem is obtained from the present one by changing the values of all the artificial basic variables, and the value of w, in the updated RHS constants column to zero; and leaving everything else unchanged. The algorithm can now move to Phase II to solve the modified problem.

As an example, when the b-vector in Tableau 7.10 for the LP discussed in Example 7.4 is modified to $(9, 5, 4, 0)^T$, the problem becomes feasible, as discussed above. The final Phase I inverse tableau for this modified problem is

Basic var.	Inverse tableau					Basic values
t_1	1	−1	−1	0	0	0
x_2	0	1	−1	0	0	1
x_1	0	−1	2	0	0	3
−z	0	−1	0	1	0	−5
−w	−1	1	1	0	1	0

and the primal simplex method now moves to Phase II to solve the modified original problem, beginning with this feasible basic vector.

7.5 Finding Alternate Optimum Solutions for LPs

In practical decision making, operating conditions usually change from the time a mathematical model for the problem is constructed, to the time an optimum solution is computed by solving the model. Because of these changes, or for some other reasons that may have come up since the model construction, one may then find that a computed optimum solution is not quite suitable. In this situation, one can revise the model and solve it again, but this takes additional time and computing expense. It is possible that the original model has alternate optimum solutions, and if so, it is always better to check whether one of them may be suitable to implement, before going through the expense of solving the revised model.

In linear programming there are efficient techniques to check whether an optimum solution computed by the algorithm is the unique optimum, and to generate other optimum solutions when there are alternate optima. We discuss these techniques here briefly.

Consider the LP (7.1) in standard form, and let x_B be an optimum basic vector associated with the BFS $\bar{x}$, obtained for it when it is solved by the simplex method.

If the relative cost coefficients wrt x_B of all the nonbasic variables are > 0, then $\bar{x}$ is the unique optimum solution of this LP.

On the other hand, if some nonbasic relative cost coefficients wrt x_B are 0, by bringing the corresponding nonbasic variables into this basic vector as in the simplex algorithm, we can get alternate optimum BFSs for this LP. Also, any convex combination of optimum solutions is also optimal.

As an example, consider the LP in Tableau 7.6 solved in Example 7.2. $x_{B_1} = (x_1, x_5, x_3, x_7)$ is the optimum basic vector obtained there for this problem, and the corresponding optimal primal BFS is $\bar{x} = (3, 0, 3, 0, 1, 0, 0)^T$. The third inverse tableau in Example 7.2 is the inverse tableau wrt x_{B_1}. The nonbasic relative cost coefficients wrt x_{B_1} are $(\bar{c}_2, \bar{c}_4, \bar{c}_6) = (3, 7, 0)$. Here $\bar{c}_6 = 0$, so, to get an alternate optimum solution we can bring this nonbasic variable into the basic vector x_{B_1}. The updated column of x_6 is $(2, 0, 21, -1, 0)^T$. We enter this on the right hand side of the inverse tableau of x_{B_1} as the pivot column, and perform the pivot step.

Basic var.	Inverse tableau					Basic values	PC x_6	Ratios
x_1	1	1	0	-2	0	3	2	3/2
x_5	0	1	0	-1	0	1	0	
x_3	0	-2	1	1	0	3	$\boxed{21}$	3/21
x_7	0	0	0	1	0	0	-1	
$-z$	0	3	-1	6	1	-2	0	
x_1	1	23/21	$-2/21$	$-44/21$	0	57/21		
x_5	0	1	0	-1	0	1		
x_6	0	$-2/21$	1/21	1/21	0	3/21		
x_7	0	$-2/21$	1/21	22/21	0	3/21		
$-z$	0	3	-1	6	1	-2		

PC = pivot column

The new basic vector $x_{B_2} = (x_1, x_5, x_6, x_7)$ is associated with the alternate optimum BFS, $\hat{x} = (57/21, 0, 0, 0, 1, 3/21, 3/21)^T$, for the problem. And any convex combination of $\bar{x}$ and $\hat{x}$, i.e., $\alpha\bar{x} + (1-\alpha)\hat{x} = (3\alpha + (57/21)(1-\alpha), 0, 3\alpha, 0, 1, (3 - 3\alpha)/21, (3 - 3\alpha)/21)^T$, for $0 \stackrel{\leq}{=} \alpha \stackrel{\leq}{=} 1$, is also an optimum solution for the problem.

Thus by starting with the inverse tableau of an optimum basic vector, and performing pivot steps with any nonbasic variable having zero relative cost coefficient as the entering variable, we can get alternate optimum BFSs, whenever this leads to a nondegenerate pivot step.

7.6 Marginal Analysis

As pointed out in Chapter 2, each constraint in an LP model for a system is the material balance equation or inequality associated with an item. From Section 4.3, we know that in general, there may be two quantities associated with an item; one called the **positive marginal value** (rate of change in the optimum objective value in the model, per unit **increase** in the RHS constant in the constraint corresponding to the item from its present level), and a **negative marginal value** (same rate as above, but per unit **decrease**). These values are called the marginal values of the item, or of the constraint corresponding to it, or of the RHS constant in that constraint. Marginal analysis is economic analysis of the various options available to the system based on these marginal values.

When the optimum dual solution for the LP is unique, for each item both the positive and negative marginal values are equal to the value of the dual variable associated with that item in the optimum dual solution; this common value is defined to be the **marginal value of that item, or of the constraint corresponding to it, or of the associated RHS constant in the model.** In this case we say that marginal values are well defined, and for each item its marginal value is the rate of change in the optimum objective value in the model, per unit change (positive or negative) in the RHS constant in the constraint corresponding to the item from its present level. We know that this case occurs if the LP model has a nondegenerate optimal primal BFS. In this case marginal analysis is simple and nice, and is based on the optimum dual solution as the marginal value vector. In Section 2.2 we presented an example (the fertilizer manufacturer's problem) of this case and discussed several planning applications of marginal analysis.

If the LP model does not have a nondegenerate optimal primal BFS, then it may have alternate optimum dual solutions, and in this case the positive and negative marginal values for an item may be different. In this case, to be correct, marginal analysis has to be based on these positive and negative marginal values. Efficient methods for computing both the positive and negative marginal values do exist and are discussed in [K. G. Murty, 1983 of Chapter 2]; they require extra computation on the dual problem. However, these methods are not known widely. None of the commercially available LP software packages have subroutines for them, and practitioners have never demanded that positive and negative marginal values be provided to them in this case.

In this book we only consider marginal analysis in the simpler case when the LP model has a nondegenerate optimal primal BFS, using the unique optimum dual solution as the vector of marginal values.

As an example, consider a company that needs three products for its internal use. There are four different processes that the company can use to make these products. When a process is run, it may produce one or more of these products as indicated in the following table.

Product	Output of product/unit time of process				Minimum daily requirement for product (in units)
	1	2	3	4	
P_1	1	2	0	1	17
P_2	2	5	1	2	36
P_3	1	1	0	3	8
Cost (in \$) of running process/unit time	28	67	12	35	

For $j = 1$ to 4, let x_j denote the units of time that process j is run daily. Let x_5, x_6, x_7 denote the slack variables corresponding to P_1, P_2, P_3 (these are the amounts of the product produced in excess of the minimum daily requirement). Then the model for meeting the requirements of the products at minimum cost is the following LP in standard form.

Tableau 7.13: Original tableau

Item	x_1	x_2	x_3	x_4	x_5	x_6	x_7	$-z$	b
P_1	1	2	0	1	-1	0	0	0	17
P_2	2	5	1	2	0	-1	0	0	36
P_3	1	1	0	3	0	0	-1	0	8
	28	67	12	35	0	0	0	1	0

$x_j \geq 0$ for all j; x_5, x_6, x_7 are P_1, P_2, P_3 slacks; minimize z

This problem has been solved by the simplex method, yielding the following optimum inverse tableau.

Tableau 7.14: Inverse tableau

Basic var.	Inverse tableau				Basic values
x_1	5	-2	0	0	13
x_2	-2	1	0	0	2
x_7	3	-1	-1	0	7
$-z$	-6	-11	0	1	-498

x_7 is P_3 slack

So, the optimum solution is to run processes 1, 2 for 13, 2 units of time daily. This solution attains the minimum cost of $498, and produces 17, 36, 15 units of P_1, P_2, P_3 respectively; meeting the minimum daily requirements of P_1, P_2 exactly, but leaving an excess of 7 units of P_3 after meeting its requirement. Since the optimal primal BFS is nondegenerate, the vector of marginal values of P_1, P_2, P_3 is the optimum dual solution $= (6, 11, 0)$.

So, the marginal value of P_3 is 0. This means that small changes in its daily requirement in the neighborhood of its present value of 7 units, does not change the cost. At the moment the requirement of P_3 is automatically covered while meeting the requirements of P_1, P_2, this actually produces an excess of 7 units of P_3 beyond its requirement.

P_2 has the highest marginal value of $11 among the three products. This means that small changes in its requirement from its present level of 36 units result in a change in the optimum cost at the rate of $11/unit. And if an outside supplier were to offer to supply P_2, it is worth considering if the rate is $\leq$ $11/unit. Since it has the highest marginal value, P_2 is a critical input for the company.

A similar interpretation can be made for P_1 and its marginal value of $6/unit.

Suppose the company's research lab has come up with a new process, process 5, which produces P_1, P_2 at the rate of 4, 9 units per unit time it is run, and does not produce any P_3. Let c_8 be the cost of running process 5 per unit time. For what values of c_8 is it desirable to run process 5? To answer this question, we evaluate the monetary benefit, in terms of the marginal values, of the output by running this process per unit time. Since it is 4, 9 units of P_1, P_2 respectively, and the marginal values of P_1, P_2 are 6, 11; this monetary benefit is $4 \times 6 + 9 \times 11 = $123/unit time.

Comparing this with the cost c_8 of running this process we conclude that process 5 is not worth running if, $c_8 > 123$, it breaks even with the present optimum solution if $c_8 = 123$, and can save cost if $c_8 < 123$.

Marginal analysis is this kind of cost-benefit analysis using the marginal values. It provides very valuable planning information.

Practitioners often use this kind of analysis using an optimum dual solution provided by the simplex method, even when the optimal primal solution is degenerate. As pointed out at the beginning of this section, this may lead to wrong conclusions in this case, so one should watch out.

7.6.1 How to Compute the Marginal Values in a General LP Model

The model for a real world problem may be a general LP model which may not be in standard form originally. Corresponding to each constraint in this model there will be a dual variable in its dual, which can be interpreted as the marginal value of the RHS constant in that constraint; it is the rate of change in the optimum objective value in the model per unit change in the RHS constant in that constraint.

To solve this model, we first put it in standard form as discussed in Section 7.1. In this process constraints are modified, some may even be eliminated if the original model contains some unrestricted variables.

To distinguish one from the other, let (P) refer to the original model with all the constraints as they were stated originally. Let (P') refer to the equivalent LP in standard form obtained by transforming (P) using the steps discussed in Section 7.1. Suppose (P') has a nondegenerate primal BFS. Then all the marginal values for it are well defined. Knowing the marginal values for (P'), how can we compute the marginal values for the original model (P)? We provide an answer to this question here. Let x_B denote an optimum basic vector for (P'). For (P') we assume that the relative cost coefficients wrt x_B of all nonbasic variables are computed. By definition, the relative cost coefficient of a basic variable is 0.

First suppose there are no unrestricted variables in the original model (P). Inequality constraints in (P) were transformed into equations by introducing appropriate slack variables in the process of getting (P'). Actually, the new variables y_1, y_2 introduced in Steps 1, 2 in Section 7.1 are the slack variables for the bound conditions on x_1, x_2.

The marginal value associated with any "$\geq$" ["$\leq$"] inequality constraint in (P) is the relative cost coefficient of the corresponding slack variable in (P') if the objective function in (P) is in minimization [maximization] form, or its negative if the objective function in (P) is in maximization [minimization] form.

The marginal value associated with any equality constraint in (P) is the same as the marginal value associated with this constraint in (P') if the objective function in (P) is in minimization form, or its negative if the objective function in (P) is in maximization form.

It is possible that there are still some equality constraints in (P) which were used to eliminate the unrestricted variables, marginal values associated with which are not determined by the above rules. If so, consider the system of equality constraints in Tableau 7.1, at the beginning of executing Step 5 of Section 7.1. When you augment the optimal basic vector x_B for (P'), with all the unrestricted variables that were eliminated in Step 5, you get a vector x_F say, which will be a basic vector for the system of equality constraints in Tableau 7.1. Treating the bottom row in Tableau 7.1 as the objective row, compute the dual basic solution associated with x_F for this system, as discussed in Section 4.2.1. The values of the dual variables associated with the eliminated constraints, in this dual solution, are the marginal values [negatives of marginal values] corresponding to these constraints in (P) if the objective function is in minimization [maximization] form.

7.7   Sensitivity Analysis

In applications of linear programming, the data such as the input-output coefficients, cost coefficients, and RHS constants, are normally estimated from practical considerations, and may have unspecified errors in them. Given an optimum basic vector for the LP model, the **optimality range** of a data element, is the interval within which that element can vary, *when all the other data remain fixed at their current values*, while keeping the present solution or basic vector optimal. In sensitivity analysis there is a technique called **ranging** that determines the optimality range of some of the data very efficiently. The width of the optimality range, and the position of the present value of the data element in this range, can be used to check the robustness of the present optimum solution or optimum basic vector, to possible errors in that data element.

Sensitivity analysis also has efficient techniques for finding a new optimum solution beginning with the current one, if simple changes occur in the model. Ranging and these other techniques in sensitivity analysis are all based on the optimality criterion (7.2). Here we discuss some sensitivity analysis techniques that proved to be useful in practice.

7.7.1   Introducing a New Nonnegative Variable

Let (7.1) be the original model for which we have an optimum basic vector x_B associated with the optimum primal BFS $\bar{x} = (\bar{x}_1, \ldots, \bar{x}_n)^T$, and optimum dual solution $\bar{\pi} = (\bar{\pi}_1, \ldots, \bar{\pi}_m)$.

After solving the original model (7.1), suppose a new activity has become available. Let x_{n+1} denote the decision variable corresponding to it, it is the level at which this new activity is to be carried out. Suppose the original column vector of x_{n+1} is $A_{.n+1}$ and its original cost coefficient is c_{n+1}. The following questions need to be answered at this stage. Is it worth performing the new activity at a positive level? If so, how can we find the new optimum solution beginning with the inverse tableau of x_B? If it is not worth performing the new activity at a positive level, how much should its cost coefficient change before it becomes worthwhile?

Enter the new variable x_{n+1}, with its column vector into the original tableau, Tableau 7.4. We will refer to the modified tableau as the augmented tableau. From the optimality criterion (7.2), and the fact that the basic vector x_B satisfied it before introducing x_{n+1} into the problem, we conclude that x_B remains optimal to the augmented problem if the relative cost coefficient of x_{n+1}, namely $\bar{c}_{n+1} = c_{n+1} - \pi A._{n+1}$ is $\geqq 0$. Thus if $c_{n+1} \geqq \pi A._{n+1}$, it is not worth performing the new activity. In this case $(\bar{x}_1, \ldots, \bar{x}_n, \bar{x}_{n+1} = 0)^T$ is an optimum solution of the augmented problem.

For the new activity to be worth performing, $\bar{c}_{n+1}$ has to be < 0, i.e., c_{n+1} has to decrease below $\pi A._{n+1}$.

If $\bar{c}_{n+1} < 0$, then x_B does not satisfy the optimality criterion for the augmented problem, and x_{n+1} is the variable eligible to enter it. To get an optimum solution of the augmented problem in this case, select x_{n+1} as the entering variable into x_B and continue the application of the simplex algorithm until it terminates again.

As an example, consider the LP model in Tableau 7.13 of the company trying to produce the required quantities of P_1, P_2, P_3 using four available processes at minimum cost, discussed in Section 7.6. The optimum basic vector for this problem is $x_B = (x_1, x_2, x_7)$ and its inverse tableau is given in Tableau 7.14. As discussed in Section 7.6, the new process 5 has become available. Let x_8 denote the level at which this process is operated. In Section 7.6, the column of x_8 in the model is given to be $(4, 9, 0, c_8)^T$, where c_8 is the cost coefficient. The relative cost coefficient of x_8 wrt x_B is

$$\bar{c}_8 = (-6, -11, 0, 1)(4, 9, 0, c_8)^T = c_8 - 123$$

If $\bar{c}_8 \geqq 0$, i.e., $c_8 \geqq 123$, it is not worth operating this process, and the present solution in Tableau 7.14 remains optimum with $x_8 = 0$.

If $\bar{c}_8 < 0$, i.e., $c_8 < 123$, then x_B does not satisfy the optimality criterion for the augmented problem. Suppose $c_8 = 122$. With this value for c_8 here is the original tableau for the augmented problem.

Original tableau

x_1	x_2	x_3	x_4	x_5	x_6	x_7	x_8	$-z$	b
1	2	0	1	−1	0	0	4	0	17
2	5	1	2	0	−1	0	9	0	36
1	1	0	3	0	0	−1	0	0	8
28	67	12	35	0	0	0	122	1	0

$x_j \geqq 0$ for all j; x_5, x_6, x_7 are P_1, P_2, P_3 slacks; minimize z

Here is the inverse tableau of $x_B = (x_1, x_2, x_7)$ with the updated column of x_8 selected as the pivot column.

Basic var.	Inverse tableau				Basic values	PC x_8	Ratios
x_1	5	-2	0	0	13	2	$13/2$
x_2	-2	1	0	0	2	$\boxed{1}$	$2/1$
x_7	3	-1	-1	0	7	3	$7/3$
$-z$	-6	-11	0	1	-498	-1	Min. $= \theta = 2$
x_1	9	-4	0	0	9		
x_8	-2	11	0	0	2		
x_7	9	-4	-1	0	1		
$-z$	-8	-10	0	1	-496		

PC = pivot column

In the new basic vector, the relative cost coefficients of the nonbasic variables are $(\bar{c}_2, \bar{c}_3, \bar{c}_4, \bar{c}_5, \bar{c}_6) = (1, 2, 7, 8, 10) > 0$. So, this basic vector is optimal to the augmented problem. The optimum BFS is $\hat{x} = (9, 0, 0, 0, 0, 0, 1, 2)^T$. It operates process 1 for 9 units of time, and the new process 5 (corresponding to x_8) for 2 units of time daily, and has a cost of $496.

7.7.2   Ranging the Cost Coefficient or an I/O Coefficient in a Nonbasic Vector

Let (7.1) be the original model for which we have an optimum basic vector x_B associated with the optimum primal BFS $\bar{x} = (\bar{x}_1, \ldots, \bar{x}_n)^T$, and optimum dual solution $\bar{\pi} = (\bar{\pi}_1, \ldots, \bar{\pi}_m)$.

Suppose x_j is a nonbasic variable now whose cost coefficient c_j is likely to change, while all the other data remain fixed at present levels. For what range of values of c_j does $\bar{x}$ remain an optimum solution to the problem?

To answer this question, notice that a change in c_j does not affect the primal or dual basic solutions associated with x_B, nor does it affect the primal feasibility of $\bar{x}$. However, for $\bar{\pi}$ to remain dual feasible, we need $\bar{c}_j = c_j - \bar{\pi}A_{.j} \stackrel{\geq}{=} 0$, i.e., $c_j \stackrel{\geq}{=} \bar{\pi}A_{.j}$. So, $\bar{x}$ remains an optimum solution to the problem as long as $c_j \stackrel{\geq}{=} \bar{\pi}A_{.j}$, this is the optimality range for c_j.

If the new value of c_j is $< \bar{\pi}A_{.j}$, then $\bar{c}_j < 0$, and the basic vector x_B is no longer dual feasible. In this case, x_j is eligible to enter x_B. To get the new optimum solution, correct the value of c_j in the original tableau, bring x_j into the basic vector x_B, and continue the application of the simplex algorithm until it terminates again.

As an example, consider the LP model in Tableau 7.13 of the company trying to produce the required quantities of P_1, P_2, P_3 using four available processes at minimum cost, discussed in Section 7.6. The optimum basic vector for this problem is $x_B = (x_1, x_2, x_7)$. Its inverse tableau is given in Tableau 7.14. Suppose the cost coefficient of x_4, the cost of running process 4 per unit time, is likely to change from its present value of $35, while all the other data remains fixed. Denote the new value of this cost coefficient by c_4. For what range of values of c_4 does the primal BFS in Tableau 7.14 remain optimal to the problem? The answer: As long as the relative cost coefficient of x_4, $\bar{c}_4 = (-6, -11, 0, 1)(1, 2, 3, c_4)^T = c_4 - 28$ is $\geqq 0$, i.e., as long as $c_4 \geqq 28$. This is the optimality range for c_4.

If the new value of c_4 is < 28, say $c_4 = 27$, the basic vector (x_1, x_2, x_7) in Tableau 7.14 is no longer dual feasible. x_4 is eligible to enter this basic vector. To get the new optimum solution, correct the original cost coefficient of x_4 to its new value of 27, bring x_4 into the basic vector (x_1, x_2, x_7) and continue the application of the simplex algorithm until it terminates again.

Now consider changes in an input-output coefficient in a nonbasic column. These coefficients may change due to changes in the corresponding process, or changes in technology. Again consider the LP (7.1), and an optimum basic vector x_B associated with the primal and dual optimum solutions, $\bar{x}, \bar{\pi} = (\bar{\pi}_1, \ldots, \bar{\pi}_m)$ for it. Let x_j be a nonbasic variable. Its column in (7.1) is $A._j = (a_{1j}, \ldots, a_{mj})^T$. Suppose the coefficient a_{rj} in this column is likely to change while all the other data remains fixed at present levels. Here also, the optimality range for a_{rj} is the set of all values of this coefficient that keeps the relative cost coefficient of x_j,

$$\bar{c}_j = c_j - \bar{\pi}A._j = c_j - (\sum_{\substack{i=1 \\ i \neq r}}^{m} \bar{\pi}a_{ij}) - \bar{\pi}_r a_{rj} \text{ nonnegative. In the expression for } \bar{c}_j$$

here, the only thing that varies is a_{rj}, all other entries are at their present values. So, if $\bar{\pi}_r = 0$, $\bar{c}_j$ remains at its present nonnegative value whatever a_{rj} is, so the optimality range for a_{rj} is the entire real line. If $\bar{\pi}_r > 0$, the optimality range for a_{rj} is $a_{rj} \leqq (c_j - (\sum_{\substack{i=1 \\ i \neq r}}^{m} \bar{\pi}a_{ij}))/\bar{\pi}_r$; and if $\bar{\pi}_r < 0$, this optimality range is

$$a_{rj} \geqq -(c_j - (\sum_{\substack{i=1 \\ i \neq r}}^{m} \bar{\pi}a_{ij}))/\bar{\pi}_r. \text{ If the new value for } a_{rj} \text{ is outside its optimality}$$

range, x_j becomes eligible to enter x_B. To get the new optimum solution, bring it in and continue the application of the simplex algorithm until it terminates again.

As an example, again consider the optimum inverse tableau in Tableau 7.14 for the LP in Tableau 7.13. Suppose by making changes in process 4, we can improve the output of P_3 from this process from its present value of 3 units/unit time. Will this change the optimum solution for the problem? The answer is no, since $\bar{\pi}_3 = 0$ in the dual optimum solution. The explanation for this is not hard to see. The present primal optimum solution which consists of running processes 1, 2 (corresponding to primal variables x_1, x_2) at 13, 2 units of time daily, already produces an excess of 7 units of P_3 over its requirement; and process 4 will not become economical even if the output of P_3 from it changes.

In the same example, suppose by making changes in process 4 we can improve the output of P_1 from it. Does this change the optimum solution for this problem, and if so how high should this output be before it does? Let a_{14} denote the new value of the output of P_1/unit time of process 4. With this value, the relative cost coefficient of x_4 is $\bar{c}_4 = (-6, -11, 0, 1)(a_{14}, 2, 3, 35)^T = 13 - 6a_{14}$. $\bar{c}_4 \geqq 0$ if

$a_{14} \leqq 13/6$, this is the optimality range for a_{14}. If $a_{14} > 13/6$, the basic vector (x_1, x_2, x_7) is no longer dual feasible to the problem, and x_4 becomes eligible to enter it. To get the new optimum solution, bring it in and continue the application of the simplex algorithm until it terminates again.

7.7.3   Ranging a Basic Cost Coefficient

Let (7.1) be the original model for which we have an optimum basic vector x_B associated with the optimum primal BFS $\bar{x} = (\bar{x}_1, \ldots, \bar{x}_n)^T$, and optimum dual solution $\bar{\pi} = (\bar{\pi}_1, \ldots, \bar{\pi}_m)$.

Suppose the cost coefficient of a basic variable in x_B, x_1 say, is likely to change, while all the other data remains fixed at present levels. Denote the cost coefficient of x_1 by c_1. For what range of values of c_1 does $\bar{x}$ remain optimal to the problem?

The simple technique used in Section 7.7.2 for ranging a nonbasic cost coefficient cannot be used here, because c_1 is now a basic cost coefficient, and any change in it results in a change of the dual basic solution associated with x_B.

Let c'_B denote the row vector of cost coefficients of basic variables in x_B, with that of x_1 recorded as c_1, and that of all others equal to their present values. Then the dual basic solution associated with x_B, as a function of c_1, is $\pi(c_1) = c'_B B^{-1}$, where B^{-1} is the inverse of the basis corresponding to x_B. This can be read from the inverse tableau for x_B.

For nonbasic variable x_j, its relative cost coefficient as a function of c_1 is $\bar{c}_j(c_1) = c_j - \pi(c_1)A_{.j}$. Express the condition that each of them should be $\geqq 0$. This identifies an interval for c_1 which is its optimality range.

Suppose the new value of c_1 is outside its optimality range. Correct the value of c_1 in the original tableau, and the dual solution in the last row of the inverse tableau of x_B. Now some nonbasic variables must be eligible to enter x_B. Select one of them to enter x_B and continue applying the simplex algorithm until it terminates again.

As an example, consider the optimum inverse tableau in Tableau 7.14 for the LP in Tableau 7.13. Suppose the cost coefficient of the basic variable x_1 is likely to change from its present value of 28. Denote its new value by c_1. The dual basic solution of the basic vector $x_B = (x_1, x_2, x_7)$ as a function of c_1 is

$$\pi(c_1) = (c_1, 67, 0) \begin{pmatrix} 5 & -2 & 0 \\ -2 & 1 & 0 \\ 3 & -1 & -1 \end{pmatrix} = (5c_1 - 134, 67 - 2c_1, 0)$$

So, the relative cost coefficient of a nonbasic variable x_j is the dot product of $(-\pi(c_1), 1)$ with the column of x_j in the original tableau. These are $(\bar{c}_3(c_1), \bar{c}_4(c_1), \bar{c}_5(c_1), \bar{c}_6(c_1)) = (2c_1 - 55, 35 - c_1, 5c_1 - 134, 67 - 2c_1)$. To keep all these relative cost coefficients $\geqq 0$, c_1 must satisfy $55/2 \leqq c_1 \leqq 67/2$, this is the optimality range for c_1.

If the new value of c_1 falls outside its optimality range, x_B is no longer dual feasible. Suppose this new value is 27. So, the new dual solution corresponding to x_B is $\pi(27) = (1, 13, 0)$. The original tableau for the modified problem, and the corrected inverse tableau wrt x_B are given below.

Original tableau

x_1	x_2	x_3	x_4	x_5	x_6	x_7	$-z$	b
1	2	0	1	−1	0	0	0	17
2	5	1	2	0	−1	0	0	36
1	1	0	3	0	0	−1	0	8
27	67	12	35	0	0	0	1	0

$$x_j \geqq 0 \text{ for all } j; \text{ minimize } z$$

Inverse tableau

Basic var.	Inverse tableau				Basic values
x_1	5	−2	0	0	13
x_2	−2	1	0	0	2
x_7	3	−1	−1	0	7
$-z$	−1	−13	0	1	−485

The relative cost coefficient of x_3 now $= \bar{c}_3(27) = -1$, so x_3 is eligible to enter this basic vector. Select x_3 as the entering variable, and continue applying the simplex algorithm until it terminates again.

7.7.4   Ranging the RHS Constants

Let (7.1) be the original model for which we have an optimum basic vector x_B associated with the optimum primal BFS $\bar{x}$, and optimum dual solution $\bar{\pi}$. Suppose one of the RHS constants, b_1, is changing, while all the other data remains fixed at present levels. When b_1 changes, the primal basic solution corresponding to x_B will change with it. So, in this case, the optimality range for b_1 is defined to be the set of values of b_1 for which the basic vector x_B remains optimal to the problem, even though the actual optimal primal BFS changes with b_1 in this interval.

A change in b_1 does not affect the dual feasibility of x_B, but it does affect its primal feasibility. So, x_B remains optimal as long as it is primal feasible. Let b' denote the new RHS constants vector with b_1 left as a parameter, but all the other entries being equal to their current values. Then, the basic vector x_B remains primal feasible if $B^{-1}b' \geqq 0$, where B^{-1} is the basis inverse corresponding to x_B read from the inverse tableau. Each entry in $B^{-1}b'$ is an affine function of b_1, express the condition that they must all be $\geqq 0$. This determines an interval for b_1, which is its optimality range. As long as b_1 is in this range, the optimum solutions are

$$\text{Primal optimum} \begin{cases} \text{all nonbasic variables} = 0 \\ \text{basic vector } x_B = B^{-1}b' \end{cases}$$

$$\text{Optimum dual solution} = \bar{\pi}$$

$$\text{Optimum objective value} = \bar{\pi}b'$$

Thus in this range, the optimum dual solution remains unchanged, but the basic variables in the primal optimum BFS, and the optimum objective value are linear functions of b_1.

As an example consider the optimum inverse tableau, Tableau 7.14, for the LP model in Tableau 7.13. Suppose the first RHS constant in this model, the minimum daily requirement for P_1, is likely to change from its current value of 17. Denote the new value by b_1. As a function of b_1, the primal basic solution wrt the basic vector (x_1, x_2, x_7) is

$$\text{Nonbasic variables } x_3, x_4, x_5, x_6 = 0$$

$$\text{Basic vector} \begin{pmatrix} x_1 \\ x_2 \\ x_7 \end{pmatrix} = B^{-1}b' = \begin{pmatrix} 5 & -2 & 0 \\ -2 & 1 & 0 \\ 3 & -1 & -1 \end{pmatrix} \begin{pmatrix} b_1 \\ 36 \\ 8 \end{pmatrix} = \begin{pmatrix} 5b_1 & -72 \\ -2b_1 & +36 \\ 3b_1 & -44 \end{pmatrix}$$

For this basic solution to be primal feasible, we need $5b_1 - 72 \geqq 0$ which implies $b_1 \geqq 72/5$; $36 - 2b_1 \geqq 0$ which implies $b_1 \leqq 18$; and $3b_1 - 44 \geqq 0$ which implies $b_1 \geqq 44/3$. The set of values of b_1 satisfying all these conditions is the interval $\max\{72/5, 44/3\} = 44/3 \leqq b_1 \leqq 18$. This is the optimality range for b_1; for b_1 in this range the optimum solution of the problem is the one given above, the optimum dual solution is $\bar{\pi} = (6, 11, 0)$ from the inverse tableau, and the optimum objective value is $\bar{\pi}b' = 6b_1 + 396$.

7.7.5   Features of Sensitivity Analysis Available in Commercial LP Software

Suppose an LP model is solved using a software package. When the model has an optimum solution, almost all commercially available LP software packages based on the simplex method will output: a list of the basic variables in the final optimum basic vector, together with their values in the primal optimum BFS; the corresponding dual optimum solution; the relative cost coefficients of nonbasic variables; and usually the optimality ranges of all the cost coefficients and RHS constants. Conspicuously absent are the positive and negative marginal values associated with the RHS constants when the primal optimum BFS is degenerate, mainly because practitioners have so far not asked for these quantities.

7.8 Exercises

7.2 Find a feasible solution to the following system of linear constraints using a method discussed in this chapter.

x_1	x_2	x_3	x_4	
1	0	1	-1	$= 3$
1	1	2	0	$= 10$
1	1	1	-2	$\geqq 14$

$$x_j \geqq 0 \text{ for all } j$$

If the system is infeasible, from the information in the final tableau show how the data in the original RHS constants vector can be modified to make the system feasible.

From your answer to the original question, show that every feasible solution of the system

$$
\begin{array}{llll}
x_1 & & +x_3 & -x_4 & = 3 \\
x_1 & +x_2 & +2x_3 & & = 10 \\
\end{array}
$$
$$x_1 \text{ to } x_4 \geqq 0$$

must satisfy $x_1 + x_2 + x_3 - 2x_4 < 14$.

7.3 Consider the diet problem with the following data involving two nutrients (vitamins A, K) with minimum daily requirements (MDR), and 5 different foods.

Nutrient	Nutrient units/unit food					MDR for
	1	2	3	4	5	nutrient
Vit. A	1	0	1	1	2	21
Vit. K	0	1	2	1	1	12
Cost (cents/unit)	20	20	31	11	12	

Formulate the problem of finding a minimum cost diet meeting the requirements, and find an optimum solution for it. Let $\bar{B}$ denote the optimum basis obtained.

For what range of values of c_2 (cost/unit of food 2) is the current diet optimal? Determine the optimum solution when c_2 is reduced to 8 cents/unit.

For what range of values of c_4 (cost/unit of food 4) is the current diet optimal?

For what range of values of the MDR of vitamin K (whose present value is 12) does the basis $\bar{B}$ remain optimal to the problem?

A local pharmacist is selling vitamin K pills at a cost of 12 cents/unit of vitamin K content. Is this price competitive with the available foods in meeting this vitamin requirement?

A delicious new food containing 3, 2 units of vitamins A, K respectively per unit has become available at a price of 28 cents/unit. How much is the urge to include at least 1 unit of this in the daily diet going to cost, over a minimum cost diet?

7.4 Consider the following LP

x_1	x_2	x_3	x_4	x_5	x_6	x_7	$-z$	b
1	0	0	0	-3	2	-2	0	3
0	1	0	-10	ξ	-1	-3	0	4
0	0	1	β	2	3	0	0	α
0	0	0	γ	δ	Δ	7	1	-10

$x_j \geqq 0$ for all j. z to be minimized.

The entries $\alpha, \beta, \gamma, \delta, \Delta, \xi$ in the tableau are real valued parameters. Let B_1 be the basis for this problem corresponding to the basic vector $x_{B_1} = (x_1, x_2, x_3)$. Write down the range of values of each of the parameters $\alpha, \beta, \gamma, \delta, \Delta, \xi$, which will make the conclusions in the following statements true.

(i) B_1 is not a primal feasible basis for this problem.

(ii) By itself, row 3 in this tableau leads to the conclusion that this problem has no feasible solution.

(iii) From this tableau it is possible to select an initial primal feasible basic vector and initiate Phase II of the simplex method to solve this problem.

(iv) B_1 is a primal feasible, but not dual feasible basis for this problem.

(v) B_1 is a primal feasible but non-optimal basis for this problem, but the updated column vector of x_4 indicates that this problem has no finite optimum solution.

(vi) B_1 is a primal feasible basis for this problem, x_5 is a candidate to enter the basic vector x_{B_1}; and when x_5 enters, x_2 is the unique dropping variable.

(vii) B_1 is a primal feasible basis for this problem, x_6 is the entering variable into the basic vector x_{B_1} by the minimum $-\bar{c}_j$ rule; but when x_6 enters this basic vector the objective value remains unchanged.

(viii) x_{B_1} is a primal feasible basic vector for this problem, and x_5 is a nonbasic variable eligible to enter it. When x_5 enters this basic vector x_3 is the dropping variable, and we get a new BFS with objective value of 5.

7.5 The following LP is a company's product mix model in which the RHS constants b_1, b_2, b_3 are the requirements for three products 1, 2, 3, the company has agreed to meet.

x_1	x_2	x_3	x_4	x_5	x_6	$-z$	b
5	2	1	10	13	11	0	$41 = b_1$
1	1	0	3	3	3	0	$12 = b_2$
4	2	1	9	11	10	0	$38 = b_3$
15	7	3	39	51	35	1	0

$x_j \geqq 0$ for all j. z to be minimized.

Compute the inverse tableau for this problem corresponding to the basic vector (x_1, x_2, x_3) and check whether it is optimal. If so, write down an optimal dual solution corresponding to this problem.

Are the marginal values associated with the three product requirements in this problem well defined? If so, what are they?

The company has a choice to decrease one of the b_is from its present value in the original tableau. Is it advantageous for the company to do so? If so which b_i should have higher priority for being decreased? Why?

Does this problem have alternate optimum solutions? Why?

Find the optimality range of b_1 for the basic vector (x_1, x_2, x_3) assuming that all the data other than the value of b_1 remains unchanged.

A new activity has cropped up. Denote its level by x_7. Its column in the original tableau, and the cost coefficient will be $A_{.7} = (7, 1, 6)^T$, $c_7 = 30$. Determine whether it is worthwhile for the company to perform this activity. If it is not, determine how much the value of c_7 should decrease from its present value of 30 before it becomes worthwhile for the company to perform this activity. If the new value of c_7 is this breakeven cost minus 1, find the new optimum solution from the present one.

7.6 Consider the following LP

x_1	x_2	x_3	x_4	x_5	x_6	$-z$	b
1	2	-1	1	2	2	0	3
0	1	1	2	1	2	0	6
-1	0	2	2	-2	1	0	5
-4	0	9	12	-3	7	1	0

$x_j \geqq 0$ for all j. z to be minimized.

Compute the inverse tableau for this problem corresponding to the basic vector (x_1, x_2, x_3) and verify that this basic vector is optimal.

Find out the optimality range of c_2 (the cost coefficient of x_2 with present value of 0) for the present basic vector. If the new value of c_2 is $1 +$ the upper bound of this optimality range, which variable becomes eligible to enter the present basic vector? Find the new optimum solution.

Referring to the original LP again (i.e., $c_2 = 0$ as in the original problem) find the optimality range of the RHS constant in the first constraint, b_1 (with present value 3), for the basic vector (x_1, x_2, x_3). What is the slope of the optimum objective value as a function of b_1 in this interval? Write down the optimum solution of the problem, and the optimum objective value, as functions of b_1, as it varies in this interval.

Consider the original LP again. A new activity has cropped up. Denote its level by x_7. Its column in the original tableau, and the cost coefficient will be $A_{.7} = (0, 2, 3)^T$, $c_7 = 18$. Determine whether it is worthwhile for the company to perform this activity. If it is not, determine how much the value of c_7 should decrease from its present value of 18 before it becomes worthwhile for the company to perform this activity. If the new value of c_7 is 12, find the new optimum solution from the present one.

7.7 A company manufactures products A to G using two types of machines P_1, P_2; and three raw materials R_1, R_2, R_3. Relevant data is given below.

Item	Item input (in units) to make one unit of							Max. available per day
	A	B	C	D	E	F	G	
R_1 (in units)	0.1	0.3	0.2	0.1	0.2	0.1	0.2	500
R_2	0.2	0.1	0.4	0.2	0.2	0.3	0.4	750
R_3	0.2	0.1	0.1	0.2	0.1	0.2	0.3	350
P_1 time (mc. hrs.)	0.02	0.03	0.01	0.04	0.01	0.02	0.04	60
P_2 time	0.04		0.02	0.02	0.06	0.03	0.05	80
Constraint on daily output	$\geqq 200$	$\leqq 800$			$\leqq 400$			
Profit (\$/unit)	10	12	8	15	18	10	19	

Formulate the product mix problem to maximize total daily profit as an LP, and solve it using one of the available software packages to obtain primal and dual optimum solutions. Also, answer each of the following questions about this original problem.

(i) Are the marginal values of the various items well defined in this problem? If so, what are they?

(ii) Is it worth increasing the supply of R_1 beyond the present 500 units/day? The current supplier for R_1 is unable to supply any more than the current amount. The procurement manager has identified a new supplier for R_1, but that supplier's price is $15/unit higher than the current suppliers'. Should additional supplies of R_1 be ordered from this new supplier?

(iii) The production manager has identified an arrangement by which 20 hours/day of either P_1- or P_2-time can be made available at a cost of $150/day. Is it worth accepting this arrangement?

(iv) The sales manager would like to know the relative contributions of the various products in the company's total profit. What are they?

(v) The sales manager believes that product C is priced too low for a good image. This manager claims that if the selling price of C were increased by $2/unit, the demand for it would be 600 units/day. What is the effect of this change? The production manager claims that the manufacturing process for G can be changed so that its need for P_1-time goes down by 50% without affecting quality, demand or selling price. What will be the effect of this change on the optimum product mix and total profit?

(vi) The production manager believes that by changing specifications, it should be possible to make product B with 33.3% less of R_1, this would have no effect on the saleability of this product. What will be the effect of this change on the optimum product mix and total profit?

(vii) The company's research division has formulated a new product, H, which they believe can yield a profit of $8-10/unit made. The input requirements to make one unit of this product will be

Item	R_1	R_2	R_3	P_1-time	P_2-time
Input	0.1	0.2	0.1	0.02	0.02

Is this product worth further consideration?

(viii) The sales manager feels that the selling price/unit of product F can be increased by $2 without affecting the demand for it. Would this lead to any changes in the optimum production plan? What is the effect of this change on the total profit?

(D. C. S. Shearn [1984])

7.8 Consider the following system of constraints

$$
\begin{aligned}
x_1 &+ x_2 - 2x_3 + x_4 &\leqq 1 \\
x_1 &+ 2x_2 - x_3 + 2x_4 &\leqq 3 \\
2x_1 &- x_2 + x_3 - x_4 &\leqq 4 \\
-2x_1 &+ x_2 + 5x_3 - x_4 &\leqq \alpha \\
& x_j \geqq 0, \quad j = 1 \text{ to } 4 &
\end{aligned}
$$

where α is a real valued parameter.

If $\alpha = -3$ show that the system has no feasible solution. Determine the range of values of α for which the system is feasible.

7.9 A company has two plants, plants 1, 2, manufacturing a product at the rates of 1500, 1800 units/week respectively. Due to an upward shift in demand the company has decided to open a new plant 3 to produce 1100 units/week. They have identified two sites 1, 2 where the new plant can be opened.

The company sells the product through distribution centers 1 to 4. The demand at distribution centers 1 to 4 is running currently at 1400, 1100, 900, 1000 units/week respectively. By opening plant 3, the company would like to meet the demand at each distribution center exactly.

Production requires skilled labor (SL), unskilled labor (UL), raw material 1 (R1), raw material 2 (R2), grade 1 coal (C1), and grade 2 coal (C2), as inputs. It takes 0.09, 0.40, 0.10, 0.04, 0.10, 0.08 units of SL, UL, R1, R2, C1, C2 respectively to manufacture one unit of product.

Skilled labor is more productive than unskilled labor. Plant operation permits substitution of unskilled labor by skilled labor at the rate of 1 unit of SL per 1.4 units of UL.

Similarly, grade 1 coal provides 20% greater energy output per unit than grade 2 coal. The plant's operation permits substitution of C2 by C1 at the rate of 1 unit of C1 per 1.2 units of C2.

The costs of the inputs and shipping costs at the two sites are given below.

Site	Cost ($/unit) of						Shipping cost ($/unit) to dist. center			
	SL	UL	R1	R2	C1	C2	1	2	3	4
1	15	13	3.5	2.3	2.7	1.8	0.3	0.5	0.4	0.2
2	17	14.5	2.8	2.1	3.1	2.4	0.4	0.6	0.1	0.3

Do a facility location analysis using linear programming to determine whether plant 3 should be built at site 1 or 2.

7.10 In a linear programming model in which the objective function is in minimization form and all the variables are nonnegative variables, the first constraint is

$$\sum_{j=1}^{N} a_{1j} x_j \overset{\geq}{=} b_1 \qquad (7.7)$$

In putting this LP in standard form, this constraint was transformed into

$$\sum_{j=1}^{N} a_{1j} x_j - x_n = b_1 \qquad (7.8)$$

where x_n is the slack variable corresponding to (7.7). The whole LP model in standard form is

$$
\begin{aligned}
\text{Minimize} \quad & z(x) = cx \\
\text{subject to} \quad & Ax \;=\; b \\
& x \;\overset{\geq}{=}\; 0
\end{aligned}
\qquad (7.9)
$$

where the coefficient matrix A is of order $m \times n$ and rank m. The first constraint in (7.9) is exactly the constraint (7.8).

Let $\bar{\pi} = (\bar{\pi}_1, \ldots, \bar{\pi}_m)$ be the dual basic solution associated with a given basic vector x_B for (7.9). Prove that $\bar{\pi}_1$ is exactly the relative cost coefficient of the variable x_n wrt the basic vector x_B in (7.9).

7.11 An LP in standard form has been solved by the revised simplex method, and the following optimum inverse tableau obtained

Basic variable					Updated RHS constants vector
x_1	3	-2	2	0	3
x_2	-1	1	0	0	0
x_3	4	0	-1	0	5
$-z$	-3	-9	-4	1	-100

Let $b = (b_1, b_2, b_3)^T$ denote the original RHS constants vector in the standard form of this LP.

What are the marginal values with respect to b_1, b_2, b_3 in this LP? Why?

Write down what all you know about rate of change in the optimum objective value when b_1 changes slightly from its current value, assuming that the problem remains feasible under this change.

7.12 In a blending problem, the decision variables are, for $j = 1$ to n

$$x_j = \text{units of component } j \text{ included in the mix}$$

The problem of finding a minimum cost mix lead to the following optimization problem

$$\text{Minimize}\quad z(x) = cx$$
$$\text{subject to}\quad Ax = b \tag{7.10}$$
$$\left(\sum_{j=1}^{n} d_j x_j\right) \Big/ \left(\sum_{j=1}^{n} x_j\right) = r$$
$$x_j \geqq 0 \text{ for all } j$$

where the coefficient matrix A is of order $m \times n$ and rank m. The single ratio constraint (7.10) specifies that if a positive quantity of the mixture is made, then a certain property of the mixture has to be equal to the specified positive quantity r. Here $d_j > 0$ is the value of this property per unit of component j.

To incorporate this problem into a linear programming model, the ratio constraint (7.10) has been linearized as discussed in Chapter 2, and the original problem (7.10) is transformed into the following LP.

$$\text{Minimize}\ z(x) = cx$$
$$\text{subject to}\ Ax = b \tag{7.11}$$
$$\sum_{j=1}^{n} d_j x_j - r \sum_{j=1}^{n} x_j = 0$$
$$x_j \geqq 0 \text{ for all } j$$

The LP (7.11) has been solved by the revised simplex method. It lead to the optimum solution $\bar{x} = (\bar{x}_1, \ldots, \bar{x}_n)^T$ which is nonzero, and the dual optimum solution $\bar{\pi} = (\bar{\pi}_1, \ldots \bar{\pi}_m, \bar{\pi}_{m+1})$. Here π_{m+1} is the dual variable associated with the last equality constraint in (7.11). Also, $\bar{x}$ is a nondegenerate BFS of (7.11).

It is required to determine the rate of change in the optimum objective value in (7.10), per unit change in the value of r from its present value. Determine this rate in terms of $\bar{x}, \bar{\pi}$ given above. Explain clearly and justify your answer carefully.

7.13 Consider the blending problem discussed in Exercise 7.13. Suppose the original blending problem with ratio constraints is

$$\text{Minimize}\quad z(x) = cx$$
$$\text{subject to}\quad Ax = b \tag{7.12}$$
$$r_1 \leqq \left(\sum_{j=1}^{n} d_j x_j\right) \Big/ \left(\sum_{j=1}^{n} x_j\right) \leqq r_2$$
$$x_j \geqq 0 \text{ for all } j$$

(7.12) is transformed into the linear program (7.13) given below.

$$\text{Minimize } z(x) = cx$$
$$\text{subject to } Ax = b \tag{7.13}$$
$$\sum_{j=1}^{n} d_j x_j - r_1 \sum_{j=1}^{n} x_j \geqq 0$$
$$\sum_{j=1}^{n} d_j x_j - r_2 \sum_{j=1}^{n} x_j \leqq 0$$
$$x_j \geqq 0 \text{ for all } j$$

Given the primal and dual optimum solutions $\bar{x}, \bar{\pi}$ for (7.13), if $\bar{x}$ is a nondegenerate BFS for (7.13), discuss how to obtain the marginal values in the original problem (7.12) wrt r_1, r_2. ([A. G. Munford, 1989]).

7.14 Solve the following linear program

$$
\begin{array}{rrrrl}
\text{minimize } z = & -10x_1 & -4x_2 & -3x_3 & \\
\text{subject to} & x_1 & & -x_3 & \leqq \; 0 \\
& x_1 & +x_2 & & \leqq \; 2 \\
& -x_1 & +x_2 & +3x_3 & \leqq \; 4 \\
& x_1, & x_2, & x_3 & \geqq \; 0
\end{array}
$$

7.15 Solve the following LP.

x_1	x_2	x_3	x_4	x_5	x_6	$-z$	b
1	2	-1	1	2	2	0	3
0	1	1	2	1	2	0	6
-1	0	2	2	-2	1	0	5
-4	0	9	12	-3	7	1	0

$$x_j \geqq 0 \text{ for all } j, \text{ minimize } z.$$

Do cost ranging for c_2, the cost coefficient of x_2.

Do ranging for the RHS constant in the first row, b_1.

A new nonnegative variable, x_7, associated with the original column vector $(0, 2, 3, 12)^T$ has to be incorporated into the model. Find the new optimum solution.

7.16 Solve the following LP.

$$\begin{array}{llllllll}
\text{Minimize} & -2x_1 & +2x_2 & +x_3 \\
\text{subject to} & & x_2 & +x_3 & -x_4 & +x_5 & +2x_6 & \leq 6 \\
& x_1 & & +x_3 & -x_4 & +x_5 & & = 5 \\
& -x_1 & +x_2 & -x_3 & +x_4 & & +x_6 & = 3 \\
& & & x_j \geq 0 \text{ for all } j
\end{array}$$

7.17 Solve the following LP.

$$\begin{array}{llllllll}
\text{Minimize} & -2x_1 & +2x_2 & +x_3 \\
\text{subject to} & & x_2 & +x_3 & -x_4 & +x_5 & +2x_6 & \leq 6 \\
& x_1 & & +x_3 & -x_4 & +x_5 & & = 5 \\
& -x_1 & +x_2 & -x_3 & +x_4 & & +x_6 & = -3 \\
& & & x_j \geq 0 \text{ for all } j
\end{array}$$

If possible, determine a feasible solution where the objective function has value $= -200$.

7.18 Solve the following LP.

$$\begin{array}{llllll}
\text{Minimize} & -x_1 & +x_2 \\
\text{subject to} & x_1 & +x_2 & -x_3 & +x_4 & \leq 4 \\
& & -x_2 & +x_3 & +x_4 & \leq 6 \\
& x_1 & & & +2x_4 & \geq 12 \\
& & & x_j \geq 0 \text{ for all } j
\end{array}$$

If the problem is infeasible, discuss how one of the RHS constants can be modified by the smallest amount to make the model feasible. Carry out this modification and obtain an optimum solution of the modified problem.

7.9   References

G. B. DANTZIG, 1953, "Computational Algorithm of the Revised Simplex Method", RM-1266, RAND Corporation, Santa Monica, CA.

A. G. MUNFORD, 1989, "Evaluating Marginal Costs Associated with Ratio and Other Constraints in Linear Programming", *Journal of the Operational Research Society*, 40, no. 10, 933-935.

D. C. S. SHEARN, March 1984, "Postoptimal Analysis in Linear Programming-The Right Example", *IIE Transactions*, 16, no. 1, 99-101.

Chapter 8

Algorithms for Multiobjective Models

As discussed in Chapter 1, many real world applications involve several objective functions simultaneously. For example, most manufacturing companies are intensely interested in attaining large values for several things, such as the company's net profit, its market share, and the public's recognition of the company as a progressive organization.

In all such applications, it is extremely rare to have one feasible solution which simultaneously optimizes all of the objective functions. Typically, optimizing one of the objective functions has the effect of moving another objective function away from its most desirable value. These are the usual **conflicts** among the objective functions in multiobjective models. Under such conflicts, a multiobjective problem is not really a well-posed problem unless information on how much value of one objective function can be sacrificed for unit gain in the value of another. Such **tradeoff information** is usually not available, but when it is available, it makes the problem easier to analyze.

A feasible solution x^1 to a multiobjective problem is said to be a **nondominated solution** or an **efficient point**, if there is no other feasible solution x^2 that satisfies: (i) x^2 has the same or better values as x^1 for every objective function in the model, and (ii) x^2 has a strictly better value than x^1 for at least one objective function in the model. In some multicriteria optimization problems there are usually two objective functions because there are two decision makers involved each with his or her own special interests, and in this case an efficient point is called a **pareto optimal point** by economists. We will use the names nondominated solution, efficient point, and pareto optimal point, synonymously.

For example, consider a problem in which two objective functions z_1, z_2 are required to be minimized simultaneously. If $\bar{x}$ is a feasible solution to the problem with values $\bar{z}_1, \bar{z}_2$ for the two objective functions, we represent $\bar{x}$ by the point

$(\bar{z}_1, \bar{z}_2)$ in the z_1, z_2-plane. In Figure 8.1 we mark the points in the z_1, z_2-plane corresponding to feasible solutions of the problem. They form the dotted region in the z_1, z_2-plane in Figure 8.1. A feasible solution corresponding to a point like $\hat{z} = (\hat{z}_1, \hat{z}_2)$ in the interior of the shaded region, is not a pareto optimum, since feasible solutions corresponding to points in the shaded region satisfying $z_1 \overset{\leq}{=} \hat{z}_1, z_2 \overset{\leq}{=} \hat{z}_2$ are strictly better wrt one or both the objective functions. So, for this problem, pareto optimum solutions are those corresponding to points on the thick boundary curve in Figure 8.1, and there are an infinite number of them.

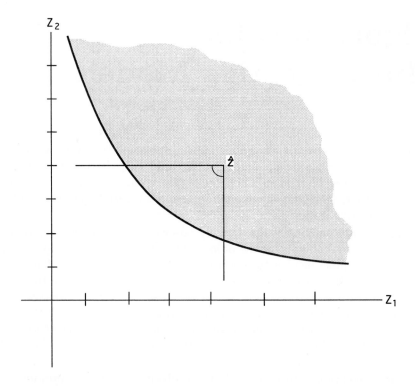

Figure 8.1 Dotted region consists of points in the objective plane corresponding to feasible solutions. $\hat{z}$ does not correspond to a pareto optimum point, since points in the cone region marked by the angle sign correspond to strictly superior feasible solutions on one or both objective functions. The thick boundary curve corresponds to the efficient frontier.

The set of all pareto optimum solutions in this problem is known as the **efficient frontier**. Feasible solutions in the efficient frontier correspond to points on the thick boundary curve in Figure 8.1. As points representing solutions trace out this efficient frontier, if there are gains in the value of one objective function, there will be losses in the value of the other.

The reader should not be fooled by the word **optimum** in the phrase **pareto optimum**. In a multiobjective model, a pareto optimum does not have the nice optimality properties that we have seen in single objective models. Remember that a pareto optimum point is just a feasible solution with the property that any move from it, if it leads to a gain in the value of one objective function, it also leads to a loss in the value of another objective function. For example, in a two objective model in which both objective functions z_1, z_2 are costs to be minimized, one pareto optimum $\bar{x}$ may have objective values $\bar{z} = (\bar{z}_1, \bar{z}_2) = (80, 65)$; and another pareto optimum $\hat{x}$ may have objective values of $\hat{z} = (\hat{z}_1, \hat{z}_2) = (100, 60)$. The move from $\bar{x}$ to $\hat{x}$ resulted in a decrease in the cost z_2 of 5 units, but an increase in the cost z_1 of 20 units. And among $\bar{x}, \hat{x}$, it is hard to determine which is better unless we have some idea of how much 1 unit decrease in the value of z_2 is worth in terms of units of z_1.

Obviously, a feasible solution which is not a pareto optimum is never desirable. But there are usually many many pareto optima, and unless the tradeoff information between objective functions mentioned above is available, there is no criterion available to select one of these pareto optima as the most desirable.

Mathematically, very efficient techniques are available for enumerating all the nondominated solutions in a linear multiobjective problem. But these techniques are of limited, if any, practical value, since there is no principle available for selecting the best among nondominated solutions.

The literature on multiobjective problems is large. But it is mostly of mathematical value and does not seem to offer any help for solving practical problems. We present two basic techniques that people seem to use most often for dealing with multiobjective problems in practice.

8.1 Combining the Various Objective Functions Linearly Into a Single Composite Objective Function

This strategy can be used if all the objective functions are defined using some common units (for example, in terms of money units, say dollars) and if explicit tradeoff information among the various objective functions is available. Express the objective functions, either all in the maximization form, or all in the minimization form. From the tradeoff information determine a positive weight for each objective function which measures its importance; i.e., the higher the weight, the more important the objective function is. Let $c_1(x), \ldots, c_k(x)$ be the various objective functions, and $w_1 > 0, \ldots, w_k > 0$ the weights associated with them. Without any loss of generality, we assume that the weights are scaled so that their sum is 1. We are discussing the general multiobjective problem here, so the objective functions

$c_1(x), \ldots, c_k(x)$ may be linear or nonlinear.

Let $\mathbf{X}$ denote the set of feasible solutions. $\mathbf{X}$ might be the set of all points $x \in \mathrm{I\!R}^n$ satisfying a specified system of constraints of the form

$$
\begin{aligned}
f_i(x) &\overset{\leq}{=} 0, \quad i = 1 \text{ to } m \\
h_p(x) &= 0, \quad p = 1 \text{ to } P
\end{aligned}
\tag{8.1}
$$

and may also include constraints that some variables only take values in a discrete set such as $\{0, 1\}$. In the general multiobjective problem, the constraint functions $f_i(x), h_p(x)$ may be nonlinear. If all the objective functions and constraint functions are linear, and all the variables are continuous variables, the problem is called a **multiobjective linear program**. If some of the functions are nonlinear, and all the variables are continuous variables, the problem is a **continuous multiobjective nonlinear program**. If some of the variables can only take values in a specified discrete set, the problem is a **discrete multiobjective program**.

In this approach we transform the original multiobjective problem into the following single objective problem

$$
\begin{aligned}
\text{Optimize} \quad & z(x) = \sum_{t=1}^{k} w_t c_t(x) \\
\text{subject to} \quad & x \in \mathbf{X}
\end{aligned}
\tag{8.2}
$$

In (8.2), $z(x)$ is to be maximized (minimized) if all the objective functions in the original multiobjective problem are in maximization (minimization) form.

There is only one objective function in (8.2), so it is a single objective optimization problem which can be solved by the techniques discussed in Chapters 5, 6, 7, 10, 11 or 14 depending on whether the original multiobjective problem is a linear, discrete, or nonlinear one. In (8.2), the objective function $z(x)$ may be viewed as a result of weighing the various objective functions in the original problem into a single function. In this approach, we accept an optimum solution of (8.2) as a compromise solution for the original multiobjective problem, determined by the vector of weights $w = (w_1, \ldots, w_k)^T$ selected.

It can be shown that any optimum solution of (8.2) is a pareto optimum point for the original multiobjective problem. But the solution obtained depends critically on the choice of the weights $w_1, \ldots, w_k$ used in combining the original objective functions $c_1(x), \ldots, c_k(x)$ into the composite objective function $z(x)$ in (8.2).

There may be several decision makers who have a stake in determining the optimum solution to be selected for implementation. They may not all agree on the choice of the weight vector to be used. It usually takes a lot of planning, discussion and negotiations, and many compromises, before a weight vector that everyone can agree upon is arrived at. For this negotiation process, it is often helpful to solve (8.2) with a variety of weight vectors and review the optimum solutions that come up, before selecting one of them for implementation.

EXAMPLE 8.1

As an example, we consider the problem of the fertilizer manufacturer to determine the best values for x_1, x_2, the tons of Hi-ph and Lo-ph fertilizer to manufacture daily, discussed in Section 2.1. There we considered only the objective of maximizing the net daily profit, $c_1(x) = \$(15x_1 + 10x_2)$.

Suppose the fertilizer manufacturer is also interested in maximizing the market share of the company, which is usually measured by the total tons of fertilizer the company sells, but in this case we will measure it by the daily sales revenue of the company in dollars in order to have all the objective functions in the same units. The selling prices of Hi- and Lo-ph fertilizers are \$300 and \$175/ton, respectively. So, the daily sales revenue associated with the feasible solution $x = (x_1, x_2)^T$ is $c_2(x) = 300x_1 + 175x_2$.

Suppose the fertilizer manufacturer is also interested in maximizing the public's perception of the company as being one on the forefront of technology. In fertilizers, Hi-ph fertilizer is considered to be a hi-tech product, and this third objective function can be measured by the Hi-ph tonnage sold by the company. But again, for the sake of measuring all the objective functions in common units, we will measure it by the sales revenue of the company from the sale of Hi-ph fertilizer, which is $c_3(x) = \$300x_1$.

So, the fertilizer manufacturer now has the three objective functions $c_1(x), c_2(x), c_3(x)$. We assume that the constraints on the variables are the same as those in (2.1). So, we now have a multiobjective linear program in which $c_1(x) = 15x_1 + 10x_2, c_2(x) = 300x_1 + 175x_2$, and $c_3(x) = 300x_1$ all of which are desired to be large, subject to the constraints in (2.1).

After extensive discussions with the board of directors and shareholder representatives, suppose the company has decided that the weights for the objective functions $c_1(x), c_2(x), c_3(x)$ should be .5, .25, .25, respectively. Then the resulting single composite objective function to optimize is $z(x) = .5(15x_1 + 10x_2) + .25(300x_1 + 175x_2) + .25(300x_1) = 157.5x_1 + 48.75x_2$. This leads to the single objective LP

$$
\begin{array}{llrcrcl}
\text{Maximize} & z(x) = 157.5x_1 & + & 48.75x_2 & & & \\
\text{Subject to} & 2x_1 & + & x_2 & \leq & 1500 & \\
& x_1 & + & x_2 & \leq & 1200 & \quad (8.3) \\
& x_1 & & & \leq & 500 & \\
& x_1 & \geq 0, & x_2 & \geq & 0 &
\end{array}
$$

for which the optimum solution is $\bar{x} = (\bar{x}_1, \bar{x}_2)^T = (500, 500)^T$. Under this approach, with the weight vector for the three objective functions selected as (.5, .25, .25), we are lead to the solution $\bar{x}$ as the best solution to implement.

8.2   The Goal Programming Approach

The goal programming approach is perhaps the most popular for handling multiobjective problems. It has the added conveniences that different objective functions can be measured in different units, and that it is not necessary to have all the objective functions in the same (either maximization or minimization) form.

As discussed in Section 8.1, let $c_1(x), \ldots, c_k(x)$ be the various objective functions to be optimized over the set of feasible solutions $\mathbf{X}$ defined by (8.1). In this approach, instead of trying to optimize each objective function, the decision maker is asked to specify a **goal** or **target value** that realistically is the most desirable value for that function. For $r = 1$ to k, let g_r be the specified goal for $c_r(x)$.

At any feasible solution $x \in \mathbf{X}$, for $r = 1$ to k, we express the deviation in the rth objective function from its goal, $c_r(x) - g_r$, as a difference of two nonnegative variables

$$c_r(x) - g_r = u_r^+ - u_r^- \tag{8.3}$$

where u_r^+, u_r^- are the **positive** and **negative parts of the deviation** $c_r(x) - g_r$ as explained in Section 2.7. That is

$$
u_r^+ = \left\{
\begin{array}{rl}
0 & \text{if } c_r(x) - g_r \leqq 0 \\
c_r(x) - g_r & \text{if } c_r(x) - g_r > 0
\end{array}
\right.
$$

$$
u_r^- = \left\{
\begin{array}{rl}
0 & \text{if } c_r(x) - g_r \geqq 0 \\
-(c_r(x) - g_r) & \text{if } c_r(x) - g_r < 0
\end{array}
\right.
\tag{8.4}
$$

The goal programming model for the original multiobjective problem will be a single objective problem in which we try to minimize a linear function of these deviation variables of the form $\sum_1^k (\alpha_r u_r^+ + \beta_r u_r^-)$, where $\alpha_r, \beta_r \geqq 0$ for all r, called a **penalty function**. We now explain how the coefficients α_r, β_r are to be selected.

If the objective function $c_r(x)$ is one which is desired to be maximized, then feasible solutions x which make $u_r^- = 0$ and $u_r^+ \geqq 0$ are desirable, while those which make $u_r^+ = 0$ and $u_r^- > 0$ become more and more undesirable as the value of u_r^- increases. In this case u_r^+ measures the (desirable) excess in this objective value over its specified target, and u_r^- measures the (undesirable) shortfall in its value from its target. To guarantee that the algorithm seeks solutions in which u_r^- is as small as possible, we associate a positive penalty coefficient β_r with u_r^-, and include a term of the form $\alpha_r u_r^+ + \beta_r u_r^-$ (where $\alpha_r = 0$, $\beta_r > 0$) in the penalty function that the goal programming model tries to minimize. $\beta_r > 0$ measures the loss or penalty per unit shortfall in the value of $c_r(x)$ from its specified goal of g_r. The value of β_r should reflect the importance attached by the decision maker for attaining the specified goal on this objective function (higher values of β_r represent greater importance). The coefficient α_r of u_r^+ is chosen to be 0 in this case because our desire is to see u_r^+ become positive as far as possible.

If the objective function $c_r(x)$ is one which is desired to be minimized, then positive values for u_r^- are highly desirable, while positive values for u_r^+ are undesirable. So, for these r we include a term of the form $\alpha_r u_r^+ + \beta_r u_r^-$, where $\alpha_r > 0$ and $\beta_r = 0$ in the penalty function that goal programming model minimizes. Higher values of α_r represent greater importance attached by the decision maker to the objective functions in this class.

There may be some objective functions $c_r(x)$ in the original multiobjective problem for which both positive and negative deviations are considered undesirable. For objective functions in this class we desire values which are as close to the specified targets as possible. For each such objective function we include a term $\alpha_r u_r^+ + \beta_r u_r^-$ with both α_r and β_r are > 0, in the penalty function that the goal programming model minimizes.

So, the goal programming model is the following single objective problem.

$$\text{Minimize} \quad \sum_{1}^{k} (\alpha_r u_r^+ + \beta_r u_r^-)$$

$$\begin{aligned}
\text{subject to} \quad c_r(x) - g_r &= u_r^+ - u_r^-, \quad r = 1 \text{ to } k \qquad (8.5) \\
u_r^+, \ u_r^- &\geqq 0, \quad r = 1 \text{ to } k \\
x &\in \mathbf{X}
\end{aligned}$$

Since all α_r and $\beta_r \geqq 0$, and from the manner in which the values for α_r, β_r are selected, solving (8.5) will try to meet the targets set for each objective function or deviate from them in the desired direction as far as possible. If the feasible set $\mathbf{X}$ is described through a system of linear constraints, and all the objective functions $c_r(x)$ are affine functions, (8.5) is a linear program which can be solved by the algorithms discussed earlier.

The optimum solution of (8.5) depends critically on the goals selected, and on the choice of the penalty coefficients $\alpha = (\alpha_1, \ldots, \alpha_k), \beta = (\beta_1, \ldots, \beta_k)$. Without any loss of generality we can assume that the vectors α, β are scaled so that $\sum_{r=1}^{k} (\alpha_r + \beta_r) = 1$. Then the larger an α_r or β_r, the more the importance the decision maker places on attaining the goal set for $c_r(x)$. Again, there may not be universal agreement among all the decision makers involved on the penalty coefficient vectors α, β to be used; it has to be determined by negotiations among them. Once α, β are determined, an optimum solution of (8.5) is the solution to implement. Solving (8.5) with a variety of penalty vectors α, β and reviewing the various optimum solutions obtained may make the choice in selecting one of them for implementation easier. One can also solve (8.5) with different sets of goal vectors for the various objective functions. This process can be repeated until at some stage, an optimum solution obtained for (8.5) seems to be a reasonable one for the original multiobjective problem. Exploring with the optimum solutions of (8.5) for different goal and penalty coefficient vectors in this manner, one can expect to get a practically satisfactory solution for the multiobjective problem.

It can be shown that this goal programming approach is equivalent to the positive linear combination approach when all the objective functions $c_r(x)$ are linear.

EXAMPLE 8.2

Consider the multiobjective problem of the fertilizer manufacturer discussed in Example 8.1. In this case we will measure the net daily profit by $c_1'(x) = \$(15x_1 + 10x_2)$ in dollars; the market share by $c_2'(x) = (x_1 + x_2)$tons, the total tonnage of fertilizer sold daily by the company; and the hi-tech image of the company by $c_3'(x) = x_1$tons, the tonnage of Hi-ph fertilizer sold by the company daily. All these objective functions are to be maximized. Suppose the company has decided to set a goal of $g_1 = \$13,000$ for daily net profit; $g_2 = 1150$ tons for total tonnage of fertilizer sold daily; and $g_3 = 400$ tons for Hi-ph tonnage sold daily. Also, suppose the penalty coefficients associated with shortfalls in these goals are required to be 0.5, 0.3, 0.2, respectively. With this data, the goal programming formulation of this problem, after transferring the deviation variables to the left hand side, is

$$
\begin{array}{llll}
\text{Minimize} & 0.5u_1^- & + 0.3u_2^- & + 0.2u_3^- \\
\end{array}
$$

subject to						
$15x_1$	$+ 10x_2$	$+ u_1^- - u_1^+$		$=$	13,000
x_1	$+ x_2$		$+ u_2^- - u_2^+$	$=$	1150
x_1				$+ u_3^- - u_3^+$	$=$	400
$2x_1$	$+ x_2$			$\leq$	1500
x_1	$+x_2$			$\leq$	1200
x_1				$\leq$	500
$x_1,$	$x_2,$	$u_1^-, u_1^+,$	$u_2^-, u_2^+,$	u_3^-, u_3^+	$\geq$	0

An optimum solution of this problem is $\hat{x} = (\hat{x}_1, \hat{x}_2)^T = (350, 800)^T$. $\hat{x}$ attains the goals set for net daily profit, and total fertilizer tonnage sold daily; but falls short of the goal on the Hi-ph tonnage sold daily by 50 tons. $\hat{x}$ is the solution for this multiobjective problem obtained by goal programming, with the goals and penalty coefficients given above.

8.3 Exercises

8.1 A marketing research firm has been commissioned to conduct a survey of households in a residential section of the city. There are three target populations: single persons, couples with no children, and couples with children. The contract calls for conducting at least 150, 200, 300 interviews in these populations, respectively.

Visits can be made either during the day or in the evening. Only a fraction of the visits are successful (i.e., find someone at home to be interviewed). Usually households consider the interviews a bother, and they generate negative reactions. Evening interviews generate more intense negative reactions than the day interviews. So the company allocates itself a penalty of 2 units for each interview conducted during the day, and 4 units for each interview conducted during the evening. They would like to see the total penalty minimized.

Other data is given below.

Population	Min. interviews needed	Fraction of successful visits	
		Day	Evening
Single	150	0.1	0.4
Couples, no children	200	0.2	0.6
Couples with children	300	0.4	0.8

By union regulations, the number of visits in the evening cannot exceed 80% of visits in the day. The company would like to minimize both the number of total visits, and the total penalty points for completing the contract. Formulate as a multiobjective problem, and develop a solution for it ignoring the integer requirements on the variables. Assume appropriate values for any other quantities whose values are not provided.

8.2 Agricultural land allocation In the state of West Bengal in India, the seasons are : summer- March to June, rainy season- July to October, winter- November

Crop	Yield (qtl/h)	Mkt. price (Rs/qtl)	Cash (Rs/h)	Inputs for cultivation Mc. time (mc. hrs./h)	Labor (man days/h)	Prod. target (qtl/year)
SB	45	190	6216	8	340	SB + RA
RA	27	225	3166	8	195	=10,500
SJ	22	336	2901	6	350	SJ + LJ
LJ	26	259	3266	6	375	= 3250
WP	283	113	5587	13	95	31350
WR	6	450	1292	6	47	170
WW	21	259	2601	6	156	1250
WM	9	516	1466	6	93	450

to February. In the district of Hoogly there, they grow: summer boro rice (SB) and early jute (SJ) in the summer season; late jute (LJ) over the summer and rainy seasons; rainy season amon rice (RA) in the rainy season; and potato (WP), ravi pulse (WR), wheat (WW), and mustard (WM) during the winter season. They have a total of 253,432 h (in units of hectares) of agricultural land available for cultivation. Whenever a crop is harvested, the crop of the next season can be grown on the same land. Thus it is possible to grow: (SB or SJ)-RA-(WP, or WR, or WW, or WM); or LJ-(WP, or WR, or WW, or WM) in succession over the year over the same piece of land. The yield of the various crops is measured in units called quintals (qtl). Data on the input requirements for cultivating the various crops, and the yield from them, is given above. Here the money unit is Rs (Indian Rupees).

They have a total of 154,215 man days of labor, 1100,000 mc. hrs. of machine time, and 3×10^6 Rs of cash available yearly to be used as inputs for the cultivation of this land. There is a target for the total market value of produce of 8×10^6 Rs/year.

The problem is to determine how much of the available agricultural land to allocate to the various crops in the various seasons. For each crop, the ideal situation is one in which the annual production of that crop is $\geqq$ its production target. However, for each crop, a goal less than the target level is given; this goal is the minimum level of annual production for this crop. When the annual production of a crop is greater than its goal but less than its target, a penalty is assessed for each unit the production is less than the target. These goals and penalty coefficients for the various crops are given below.

Crop	Goal (qtl)	Penalty coefficient
SJ + LJ	2275	5
SB + RA	8000	15
WP	22,000	5
WR	120	7
WM	315	7
WW	875	4
SB+RA+WW	9000	4

In the same way the usage of input resources can exceed their availabilities, but a penalty is assessed for each unit of usage over the availability. The penalty coefficients for the various resources are given below.

Resource	Labor	Mc. time	Input cash
Penalty coeff.	6	8	5

The above penalty scales are chosen according to the importance of achieving the production target of each crop, the requirements of different agricultural supporting resources, and considering the permissible marginal relaxations of different targets. Formulate and solve this problem using the goal programming approach. ([D. Ghosh, B. B. Pal, and M. Basu, 1993])

8.3 Machining of 390 Die Cast Aluminum Alloy   Important variables in machining are ν = cutting speed in sfm, f = feed rate in ipr, and d = depth of cut in in. Practical limits on these variables are

$$600 \leqq \nu \leqq 1200$$
$$0.002 \leqq f \leqq 0.0018 \tag{8.6}$$
$$0.050 \leqq d \leqq 0.100$$

Machining performance is measured in terms of part surface roughness (SR) measured in μ in., surface integrity (SI) measured by the percent of undamaged silicon in the alloy at 0.001 in. depth below the surface, tool life (TL) defined as the machining time in minutes to reach a fixed amount of uniform flank wear, and metal removal rate (MRR) expressed in in^3/min. Data on 8 tests conducted is given below.

Test no.	ν	f	d	SR	SI	TL	MRR
1	625	0.002	0.050	25	24	150.6	0.75
2	625	0.010	0.050	150	31	108.1	3.75
3	625	0.018	0.050	230	19	89.8	6.75
4	625	0.010	0.097	86	29	80.2	7.28
5	625	0.018	0.097	180	20	29.4	13.10
6	966	0.015	0.078	210	45	24.2	13.56
8	1200	0.010	0.050	95	30	30.5	7.20

Define $x_1 = \log_e \nu$, $x_2 = \log_e(1000f)$, $x_3 = \log_e(1000d)$. It has been observed that each of $z_1 = \log_e(\text{SR})$, $z_2 = \log_e(\text{SI})$, $z_3 = \log_e(\text{TL})$, $z_4 = \log_e(\text{MRR})$ can be approximated quite closely by affine functions of x_1, x_2, x_3. Using the data given above, find the coefficients in affine functions of x_1, x_2, x_3; say $a_{i0} + a_{i1}x_1 + a_{i2}x_2 + a_{i3}x_3$ that provides the best approximation for z_i, $i = 1$ to 3; use the goal programming approach.

The best fits for z_1, z_2, z_3 as affine functions of x_1, x_2, x_3 obtained by the least squares approach (see Chapter 14) are given below.

$$
\begin{aligned}
-z_1(x) &= -7.49 + 0.44x_1 - 1.16x_2 + 0.61x_3 \\
z_2(x) &= -4.13 + 0.92x_1 - 0.16x_2 + 0.43x_3 \\
z_3(x) &= 21.90 - 1.94x_1 - 0.30x_2 - 1.04x_3 \\
z_4(x) &= -11.33 + x_1 + x_2 + x_3
\end{aligned}
\tag{8.7}
$$

It is required to find an $x = (x_1, x_2, x_3)^T$ that maximizes each of $-z_1, z_2, z_3, z_4$ simultaneously subject to the constraints in (8.6), and the constraints SR $\leqq 75 \mu$

in, SI $\geqq$ 50%, TL $\geqq$ 30 min. Using the expressions in (8.7), these constraints lead to the system

$$
\begin{aligned}
6.40 &\leqq x_1 &&\leqq 7.09 \\
0.69 &\leqq x_2 &&\leqq 2.89 \\
3.91 &\leqq x_3 &&\leqq 4.61 \\
-0.44x_1 + 1.16x_2 - 0.61x_3 &&&\leqq -3.17 \\
-0.92x_1 + 9.16x_2 - 0.43x_3 &&&\leqq -8.04 \\
1.94x_1 + 0.30x_2 + 1.04x_3 &&&\leqq 18.50
\end{aligned}
\tag{8.8}
$$

Since z_1 to z_4 are in different units, it is not possible to measure all of them in the same units. Hence, to solve this multiobjective problem by the technique of combining them into a single composite function, we will use the technique of dimensionless weighting. The weights should reflect the relative importance of the objective functions, and should transform them into a common measure so that they can be added. For this we select aspiration levels for the various machining performance measures. These are: for SR 25.5 μ in, for SI 55%, for TL 38 mts., and for MRR 1.75 in^3/min. So, $(A_1, A_2, A_3, A_4) = (-\log_e(25.5), \log_e(55), \log_e(38), \log_e(1.75))$ is the vector of aspiration levels for $(-z_1, z_2, z_3, z_4)$. For $i = 1$ to 4, define $\ell_i = \min z_i(x), L_i = \max z_i(x)$ subject to (8.8). Then

$$
\begin{aligned}
y_1(x) &= (-z_1(x) + L_1)/(-\ell_1 + L_1) \\
y_i(x) &= (z_i(x) - \ell_i)/(L_i - \ell_i), \quad i = 2, 3, 4
\end{aligned}
$$

are the dimensionless transformations of the objective functions $-z_1(x), z_2(x), z_3(x), z_4(x)$.

The dimensionless weights associated with these objective functions can be taken to be $\alpha_1 = (A_1 + L_1)/(-\ell_1 + L_1), \alpha_i = (A_i - \ell_i)/(L_i - \ell_i)$, for i = 2,3,4.

Then $\sum_{i=1}^{4} \alpha_i y_i(x)$ is the composite function obtained by dimensionless weighting. An optimum solution x that maximizes $\sum_{i=1}^{4} \alpha_i y_i(x)$ subject to (8.8) is a compromise solution for this multiobjective problem. By obtaining such compromise solutions using different sets of aspiration levels for the various objective functions, the decision maker can make an informed choice through an iterative procedure.

Formulate the composite function and obtain the compromise solution using a software package to solve the resulting LP. Evaluate the solution in terms of the levels of the various objectives at it compared to their aspiration levels.

How would you modify the aspiration levels to get a better solution? Repeat until you reach a solution with objective values that compare favorably with original aspiration levels given. ([M. Ghiassi, R.E. DeVor, M.I. Dessouky, and B.A. Kijowski, 1984])

8.4 Nurse Scheduling This problem deals with scheduling nursing staff in a unit of a hospital. The unit employs 11 nurses in the day shift. The problem is to prepare their work schedule over the 14 days of a two-week period. The following goals have been set.

(a) Each nurse has a contracted number of days he/she is supposed to work during this period, as shown in the table given below. The actual number of days worked should be equal to this contracted number as far as possible.

(b) During each day of the period, it is desired to have 3 RNs, 2 LPNs, and 2 NAs. The actual number working should be equal to this desired number each day as far as possible.

(c) The number of days each nurse gets off during the four weekend days in this period should be at least two as far as possible.

Among the goals, the ones of highest priority are those in (a) for each of the 11 nurses. The next priority goes for goals (b) for each type of nurse, for each day of the period. Then goals (c) for each nurse come next in order of priority. Use the per unit penalty coefficients of 9, 7, 5 for violation of goal (a) for any nurse, goal (b) for any type of nurse any day, and goal (c) for any nurse, respectively. Defining the decision variables to be

$$x_{ij} = \begin{cases} 1 & \text{if nurse } i \text{ works on } j\text{th day} \\ 0 & \text{otherwise} \end{cases}$$

for $i = 1$ to 11, $j = 1$ to 14, give a goal programming formulation for the problem of scheduling these 11 nurses over the 14 days of the period as an integer program. ([A.A. Musa, and U. Saxena, 1984])

8.5 Wood pulp is produced by cooking hardwood and bamboo chips mixed in certain proportion in a **Kamyr digestor** with a cooking liquor consisting of chemicals. Three important quality characteristics of pulp are **k-number, burst factor,** and **breaking length**.

Overcooking results in excess disintegration of chips and low k-number, and consequently rejection of the material. Undercooking leads to high k-number, this will cause increased consumption of the bleaching agent when the pulp is processed into paper. Thus the k-number is the most important pulp characteristic that should be strictly maintained within set limits.

The burst factor and the breaking length of pulp are also important because they determine the strength of the finished paper into which pulp is processed.

The important raw material variables and process variables in the cooking process are

x_1 = Hardwood % (should be between 20 - 40).

x_2 = Upper cooking zone temperature in ^0C (should be between 140 - 175).

x_3 = Lower cooking zone temperature in ^0C (should be between 140 - 173).

x_4 = LP steam pressure in kg/cm^2 (should be between 2 to 4.4).

x_5 = HP steam pressure in kg/cm^2 (should be between 8 to 20.5).

x_6 = Active alkali as Na OH % (should be between 20 - 35).

x_7 = Sulphidity of white liquor in % (should be between 13 - 25).

x_8 = Alkali index number (should be between 12.5 - 18.7).

The output characteristics are

y_1 = k-number, which should be $\geq$ 16, and desired to be $\leq$ 18

y_2 = Burst factor, which is desired to be equal to 35 as far as possible

y_3 = Break length, which is desired to be equal to 5000 m as far as possible.

From data collected at the plant over a period of two weeks, the following relationship has been shown to exist between the output characteristics and the raw material variables and process variables.

y_1 = $22.84 + 0.06x_1 - 0.05x_2 + 0.004x_3 - 0.67x_4 + 0.24x_5 - 0.13x_6 + 0.19x_7 - 0.18x_8$

y_2 = $38.94 + 0.05x_1 - 0.02x_2 + 0.002x_3 + 1.67x_4 + 0.21x_5 + 0.06x_6 + 0.02x_7 - 0.69x_8$

y_3 = $3273.40 - 24.37x_1 + 9.997x_2 + 8.48x_3 - 268.68x_4 + 120.92x_5 + 67.27x_6 + 27.89x_7 - 138.46x_8.$

Formulate the problem of determining the best values for the controllable raw material and process variables x_1 to x_8 within their permissible limits, that yield the most desirable values for the output characteristics, as a goal programming problem. Take the penalty coefficients for unit deviation of y_1, y_2, y_3 from their desired ranges to be 10, 7, 4, respectively, in your formulation. Find an optimum solution by solving it using an LP software package and discuss the results. ([S. Sengupta, March 1981]).

8.6 A doll company makes two kinds of dolls; doll A, a high quality doll, and doll B, yielding profits of $2.5 and $1.5, respectively per doll sold. Each doll of type A requires twice as much labor as a doll of type B, and if all dolls made were type B, the company could make 500 per day. The supply of material is sufficient for only 400 dolls per day of both A and B combined.

The best customer has ordered 300 dolls of of type A for the coming season, an order which the company would like to meet as far as possible. Also, the company has a desire to earn a profit of $500 or more for the coming season as far as possible.

Ignoring the integer requirements on the number of dolls of the two types made in the coming season, formulate the problem of meeting both the goals as far as possible, using goal programming. Take the penalty for missing the profit goal per $ to be three times as much as the penalty for missing the best customer's order per doll in preparing your formulation. ([J. M. Martel, and B. Aouni, Dec. 1990]).

8.7 Nutrition and cost information on 6 foods is given below.

Nutrient	Nutrient content in food						MDR for adults
	Milk (pint)	Meat (lb.)	Eggs (dozen)	Bread (oz)	Salad (oz)	OJ (pint)	
Vit. A (IU)	720	107	7080	0	134	1000	5000
Energy (cal)	344	1460	1040	75	17.4	240	2500
Cholesterol (units)	10	20	120	0	0	0	
Protein (g)	18	151	78	2.5	0.2	4	63
Carbohydrates (g)	24	27	0	15	1.1	52	
Iron (mg)	0.2	10.1	13.2	0.75	0.15	1.2	12.5
cost ($/unit)	0.225	2.2	0.8	0.1	0.05	0.26	

It is required to find a diet that meets all the MDRs for nutrients for which the MDR are specified, but at the same time minimizes the cost, cholesterol intake, and the carbohydrate intake. Formulate this as a goal programming problem with the goals for the cost, cholesterol intake, and carbohydrate intake to be 2.4, 40, 150, respectively. Use the penalty coefficients of 5, 15, 10 for unit deviation in cost, cholesterol, carbohydrates from their goals, respectively. ([A. S. Masud, and C. L. Hwang, May 1981]).

8.8 State patrol manpower allocation Effective traffic enforcement consists of several components. One is vigilant traffic supervision which creates an impression of police omnipresence. Another is issuing as many warnings as possible.

Efficient allocation and assignment of highway patrol manpower, to achieve the departmental objectives, stands out as one of the most challenging responsibilities of the highway patrol administrator. It involves a variety of conflicting objectives such as reducing accidents, providing road coverage, reducing overtime, etc. The basic problem here is assigning a limited number of patrolmen in order to achieve maximum effectiveness.

Each patrolman in the Nebraska State Patrol works a 5-day, 50-hour week, and is assigned to specific road segments on a single patrolman per car basis. There are 22 men available for assignment to a region consisting of 6 road segments:

Segment   Description
S_1 US 77 from Lincoln to Wahoo (27 miles)
S_2 US 77 from Lincoln to Beatrice, and Nebraska 33 to Crete (52 miles)
S_3 Nebraska 2 from Lincoln to Syracuse (30 miles)
S_4 US 6 from Lincoln to Ashland (27 miles)
S_5 US 34 from Lincoln to Seward and Nebraska 79 from Lincoln to Hwy 92 (50 miles)
S_6 US 34 from Lincoln to Union Corner (33 miles)

There are seven different time periods during the day. These are

T_1 3 AM - 6 AM T_2 6 AM - 7 AM T_3 7 AM - 1 PM
T_4 1 PM - 4 PM T_5 4 PM - 5 PM T_6 5 PM - 2 AM
T_7 2 AM - 3 AM

The desired goals in decreasing order of priority are:

G_1 Assure that each patrolman is assigned to a 10 hour working shift.

G_2 Assign patrolmen to high traffic density segments and shifts.

G_3 Assign patrolmen to high fatality road segments.

G_4 Assign patrolmen to high accident segments and shifts.

G_5 Do not exceed normal manpower assignments of 15 patrolmen per 24-hour period.

Daytime troopers work a 10-hour shift : either 6 AM - 4 PM or 7 AM to 5 PM. Evening patrolmen work either 4 PM - 2 AM, or 5 PM - 3 AM. The overlapping shifts are designed to provide extra coverage during the evening rush hour of 4 - 5 PM. A patrolman handles a third shift for all the road segments in the region during 3 AM - 1 PM. The patrolmen report for duty at 3 AM, 6 AM, 7 AM, 4 PM, or 5 PM, and stay continuously for a 10 hour shift with coffee breaks and meals taken at sites along the patrol route. From this it is clear that any patrolman who works in a day would be on duty during time periods T_3 or T_6. Hence if we define

x_{ij} = Number of patrolmen assigned to road segment S_i during time period T_j; $i = 1$ to 6, $j = 1$ to 7

then $\sum_{i=1}^{6}(x_{i3} + x_{i6})$ = the number of patrolmen working on that day.

Since each patrolman works for 5 days in a 7 day week, each day the number of regular patrolmen working will be 22(5/7), approximately 15.

The following constraints are required to be satisfied.

$$x_{ij} \geqq 1 \quad , \quad i = 1 \text{ to } 6, \, j = 3 \text{ to } 6$$
$$x_{ij} \leqq 1 \quad , \quad i = 1 \text{ to } 6, \, j = 1, 2, 7$$
$$x_{ij} \leqq 2 \quad , \quad i = 1 \text{ to } 6, \, j = 3 \text{ to } 6$$

It is desired that the following constraints should hold as far as possible.

a) The total number of patrolmen working daily should be $\leqq 15$ as far as possible.

b) As far as possible the number of patrolmen working during time period 1 should be 1, during time period 2 should be 3, during time period 3 should be $\leqq 7$, during time period 6 should be $\leqq 6$, and during time period 7 should be 3.

c) The traffic density d_{ij} in million vehicle miles over the year per shift hour on road segment i during time period j is given below.

$j =$	1	2	3	d_{ij} 4	5	6	7
$i = 1$	0.284	0.814	1.485	1.943	2.163	0.979	0.218
2	0.477	1.366	2.492	3.261	3.631	1.644	0.366
3	0.371	1.062	1.936	2.534	2.822	1.278	0.284
4	0.271	0.788	1.437	1.881	2.095	0.948	0.211
5	0.339	0.970	1.769	2.315	2.578	1.167	0.260
6	0.316	0.495	1.649	2.158	2.403	1.088	0.242

$\sum \sum d_{ij} x_{ij}$ should be equal to a large positive number (denote it by M) as far as possible. This constraint forces patrolmen allocations into road segments with the highest traffic flow.

d) The accident rates a_{ij} per hour per million vehicular miles on road segment i during time period j are given below.

$j =$	1	2	3	a_{ij} 4	5	6	7
$i = 1$	2.11	0	0.40	1.18	2.77	1.84	9.17
2	0.63	0.73	0.60	1.01	0.28	1.28	0
3	4.31	0.94	0.26	1.03	0.71	1.41	7.04
4	1.09	2.54	1.04	1.59	2.39	1.90	0
5	0.88	1.03	1.24	0.95	1.55	3.00	15.38
6	0.95	1,11	1.40	0.46	0	1.10	0

$\sum \sum a_{ij}x_{ij}$ should also be equal to a large positive number as far as possible.

Formulate the problem of finding allocations of patrolmen to shifts to meet these constraints and desired goals, to the extent possible, using a goal programming approach. ([S. M. Lee, L. S. Franz, and A. J. Wynne, 1979]).

8.9 Goal programming is a very useful technique for decision making in the complex task of formulating a public-sector investment program, given the existence of multiple objectives. The mix of projects selected should typically ensure maximum contribution to the general developmental objectives of achieving sustained economic growth, income and wealth redistribution, increased self-reliance and foreign exchange generation. Quantitative targets need to be set to provide objective measures of the potential of each investment program towards the attainment of the nation's development objectives. The goal program then seeks to minimize the effect of the inevitable deviations from the desired target levels of attainment.

The following table provides information regarding 14 projects under consideration in Trinidad and Tobago on estimates of capital cost (CC) in \$mil., construction duration (CD) in months, present value of net benefits (NPV) in \$mil., employment during construction (E.C.) in man years, new permanent jobs created for operation on completion (J), power consumption (P) in mw, water consumption (W) in mil. gal., and natural gas consumption (NG) in mil. ft^3/day.

Project	Description	CC	CD	NPV	E.C.	J	P	W	NG
1	Tringen	50	3	60	15	120	5.5	0.8	45
2	Cement	150	24	100	658	260	4.5	0.2	10
3	Furfural	100	24	50	405	100	1.5	0.35	15
4	Polyester fiber	200	24	75	749	850	2.5	0.4	0.2
5	Iron & steel	600	36	200	3325	1070	80	1.6	28
6	Fertilizer	300	27	100	2360	282	6	0.8	78
7	Power plant	50	9	5	56	40	1	0.4	61
8	Caroniwater	700	42	50	6650	200	5	0	0
9	Marine facility	150	27	50	212	50	2	0	0
10	Ind. estate	100	9	10	135	40	1	0.1	0
11	Al. smelter	500	30	200	2080	1000	148	0.4	35
12	Petrochem.	1000	36	300	6575	2000	170	3.5	50
13	Transport link	20	6	-	-	-	-	0	0
14	Oil refinery	200	27	45	5365	250	15	0.5	15

Define the decision variables: x_j = level of acceptance of project j for implementation during plan period for $j = 1$ to 14; i.e., x_j between 0 and 1 is the fraction of project j to be implemented. The following are the various goals.

(i) Projects 1, 7, 10 are the unfinished projects at the start of the plan period. This goal is to complete each of these outstanding projects as far as possible.

(ii) The budget available for capital costs is 3500. It is desired to have the total capital costs to be $\overset{\leq}{=}$ this budget as far as possible.

(iii) This is the long term wealth goal, to have total NPV $\overset{\geq}{=}$ 1050 as far as possible.

(iv) Projects 8, 9, 13 are infrastructural projects. The goal is to complete each of them as far as possible.

(v) Total employment generated by construction is desired to be 20,000 man years or greater as far as possible.

(vi) Number of new jobs created for operation after selected projects are completed are desired to be $\overset{\geq}{=}$ 6000 as far as possible.

(vii) In order to utilize the natural resource base, it is desired to have NG exceed 350 as far as possible.

(viii) It is desired to invest in two or more major foreign exchange earning projects other than petrochemicals (these are projects 2, 4, 5, 6, 11 and 12) as far as possible.

(ix) There is a desire to keep the sum of the fractional investments in the two projects 5, 11 of the high-risk category $\overset{\leq}{=}$ 1 as far as possible.

(x) There is a desire to keep the power consumption for the selected projects to be $\overset{\leq}{=}$ 400 as far as possible.

(xi) There is a desire to keep the water consumption for the selected projects $\overset{\leq}{=}$ 9.0 as far as possible.

Formulate the goal programming for determining an optimal x vector in this problem, by taking the weights for deviations from goals (i) to (xi) to be 10, 10, 10, 8, 8, 8, 6, 6, 6, 4, 4. ([C. O. Benjamin, January 1985]).

8.10 This problem is concerned with determining the optimum mix of exports by the industrial sectors of a country, given a set of goals and constraints which are imposed on the exporting system. Data given pertains to South Korea for the period 1977-81. The industrial sectors identified are: 1) primary products, 2) processed foods, 3) textiles, 4) miscellaneous manufactured products, 5) non-metallic minerals, 6) chemicals, 7) steel and metal, 8) electronics, 9) ship building, 10) other machinery.

For industrial sector i we define

c_i = Capital-output ratio = (dollar value of capital engaged in exports of the industry)/(dollar value of industry's exports)

L_i = Labor-output ratio = (number of people engaged in the exports of the industry)/(amount of industry's exports)

p_i = Anticipated average export price index

I_i = (imported raw material used to produce exporting goods in industry)/(export value of industry)

s_i = Subsidy (in cents) per dollar value of export goods

u_i = Production capacity (in dollar value) of industrial sector i

ℓ_i = Dollar value of world demand for goods from industrial sector i of South Korea

Estimates of these quantities obtained by analysts are given below.

Sector	c_i	L_i	p_i	I_i	s_i	u_i	ℓ_i
$i=1$	0.91	0.74	188.1	0.15	19.9	1600	487
2	0.76	0.38	152.6	0.40	21.5	1170	283
3	1.77	0.60	179.5	0.40	20.2	9500	1817
4	1.08	0.45	198.7	0.35	23.8	5350	1069
5	2.31	0.26	181.3	0.39	36.0	1430	101
6	1.23	0.26	128.7	0.38	23.7	1970	188
7	1.26	0.35	146.1	0.40	26.9	2195	367
8	1.02	0.23	132.0	0.35	31.2	3105	409
9	1.38	0.40	121.9	0.36	26.0	1230	138
10	1.10	0.36	136.8	0.35	35.2	1685	222

Let x_i = decision variable representing the dollar value of exports from industrial sector i, for $i = 1$ to 10. The problem is to determine $x = (x_i)$ to achieve the following goals or targets.

(i)   As far as possible, the dollar value of exports is desired to be equal to \$15 billion.

(ii) Sectors 6, 7, 9, 10 are combined into a heavy industry group. As far as possible, the value of exports from this group is desired to be $\geqq 46\%$ of the total export value. Goals like this are set by government planners to promote industries that are skilled labor intensive, etc.

(iii) As far as possible, the weighted capital output ratio, $(\sum c_i x_i)/(\sum x_i)$, is desired to be $\geqq 0.25$.

(iv) It is desired to have the weighted labor-output ratio to be $\leqq 0.45$ (in units of 1000 persons to \$million) as far as possible.

(v)   It is desired to have the weighted export price index $\gtreqqless$ 161.8, as far as possible.

(vi)   As far as possible, the weighted ratio of the value of imported raw materials used to produce export goods to the value of exports, should be $\lesseqqgtr$ 0.37.

(vii)   Subsidies for exports should be kept as low as possible.

Take the penalties for deviations from desired ranges in goals (i) to (vii) to be 8, 7, 6, 6, 5, 4, 3 respectively, and formulate the problem of determining an optimal x using goal programming. (R. R. Levary, and T. S. Choi, Novenber 1983]).

8.4   References

C. O. BENJAMIN, January 1985, "A Linear Goal-Programming Model for Public Sector Project Selection", *Journal of the Operational Research Society*, 36, no. 1, 13-23.

D. GHOSH, B. B. PAL, and M. BASU, 1993, "Determination of Optimal Land Allocation in Agricultural Planning Through Goal Programming With Penalty Functions", *Opsearch*, 30, no. 1, 15-34.

M. GHIASSI, R. E. DeVOR, M. I. DESSOUKY, and B. A. KIJOWSKI, 1984, "An Application of Multiple Criteria Decision Making Principles for Planning Machining Operations", *IIE Transactions*, 16, no. 2, 106-114.

S. M. LEE, L. S. FRANZ, and A. J. WYNNE, 1979, "Optimizing State Patrol Manpower Allocation", *Journal of the Operational Research Society*, 30, 885-896.

R. R. LEVARY, and T. S. CHOI, November 1983, "A Linear Goal Programming Model for Planning Exports of Emerging Countries", *Journal of the Operational Research Society*, 34, no. 11, 1057-1067.

J.-M. MARTEL, and B. AOUNI, December 1990, "Incorporating the Decision-Maker's Preferences in the Goal Programming Model", *Journal of the Operational Research Society*, 41, no. 12, 1121-1132.

A. S. MASUD, and C. L. HWANG, May 1981, "Interactive Sequential Goal Programming", *Journal of the Operational Research Society*, 32, no. 5, 391-400.

A. A. MUSA, and U. SAXENA, 1984, "Scheduling Nurses Using Goal-Programming Techniques", *IIE Transactions*, 16, no. 3, 216-221.

S. SENGUPTA, March 1981, "Goal Programming Approach to a Type of Quality Control Problem", *Journal of the Operational Research Society*, 32, no. 3, 207-211.

Chapter 9

Modeling Integer and Combinatorial Programs

So far we considered continuous variable optimization problems. In this chapter we will discuss modeling **discrete** or **mixed discrete optimization problems** in which all or some of the decision variables are restricted to assume values within specified discrete sets, and **combinatorial optimization problems** in which an optimum combination out of a possible set of combinations has to be determined. Many of these problems can be modeled as LPs with additional integer restrictions on some or all of the variables. LP models with additional integer restrictions on decision variables are called **integer linear programming problems** or just **integer programs**. They can be classified into the following classes.

Pure (or, all) integer programs, if all the decision variables are restricted to assume only integer values.

0−1 pure integer programs, if they are pure integer programs, and in addition all decision variables are bounded variables with lower bound 0, and upper bound 1; i.e., in effect, if every decision variable is required to be either 0 or 1.

Mixed integer programs or MIPs, if there are some continuous decision variables and some integer decision variables.

0−1 mixed integer programs, if all the integer decision variables in an MIP are 0−1 variables.

Integer feasibility problems, if there is no objective function to be optimized, but the aim is to find an integer solution to a given system of linear constraints.

0−1 integer feasibility problems, if the aim is to find a 0−1 solution to a given system of linear constraints.

Many problems involve **yes - no decisions**, which can be considered as the $0-1$ values of an integer variable so constrained. Variables which are restricted to the values 0 or 1 are called **0−1 variables** or **binary variables** or **boolean variables**. And in many practical problems, activities and resources (like machines, ships, and operators) are indivisible, leading to integer decision variables in models involving them.

Many puzzles, riddles, and diversions in recreational mathematics and mathematical games are combinatorial problems that can be formulated as integer programs, or plain integer feasibility problems. We now provide a 0−1 integer feasibility formulation for a problem discussed in the superbly entertaining book [R. M. Smullyan, 1978].

EXAMPLE 9.1 An integer program from a play by Shakespeare

The setting of this problem from [R. M. Smullyan, 1978] is W. Shakespeare's play "The Merchant of Venice." In this play, a girl named Portia is the lead female character. She was a law graduate with an obsession for highly intelligent boys. Her life's ambition was to marry an extremely intelligent boy. For achieving this goal she devised a very clever scheme to choose her fiance. She purchased three caskets, one each of gold, silver, and lead, and hid a stunningly beautiful portrait of herself in one of them. The suitor was asked to identify the casket containing the portrait. If his choice is correct, he can claim Portia as his bride; otherwise he will be permanently banished to guarantee that he won't appear for the test again. To help the suitor choose intelligently, Portia put inscriptions on the caskets as in Figure 9.1. And she explained that at most one of the three inscriptions was true. She reasoned that only an intelligent boy could identify the casket with the portrait.

The portrait is in this casket	The portrait is not in this casket	The portrait is not in the gold casket
1 = Gold	2 = Silver	3 = Lead

Figure 9.1

Define, for $j = 1, 2, 3$,

$$x_j = \begin{cases} 1, & \text{if the } j\text{th casket contains the portrait} \\ 0, & \text{otherwise} \end{cases}$$

$$y_j = \begin{cases} 1, & \text{if the inscription on the } j\text{th casket is true} \\ 0, & \text{otherwise} \end{cases}$$

(9.1)

We will now formulate the problem of identifying the casket with the portrait as a $0-1$ integer feasibility problem. The model uses $0-1$ variables known as **combinatorial choice variables** corresponding to the various possibilities in this problem, defined above. These variables have to satisfy the following constraints

$$
\begin{array}{rcl}
x_1 \;+x_2 \;+x_3 & = & 1 \\
-x_1 \qquad\qquad +y_1 & = & 0 \\
x_2 \qquad\qquad +y_2 & = & 1 \\
x_1 \qquad\qquad\qquad +y_3 & = & 1 \\
y_1 \;+y_2 \;+y_3 & \leq & 1
\end{array}
\qquad (9.2)
$$

$$
x_j, y_j \;\; = \;\; 0 \;\; \text{or} \;\; 1 \quad \text{for all } j
$$

The first constraint in (9.2) must hold because only one casket contains the portrait. The second, third, fourth constraints must hold because of the inscriptions on caskets 1, 2, 3, and the definitions of the variables x_1, y_1; x_2, y_2; x_3, y_3. The fifth constraint must hold because at most one inscription was true.

In (9.2), the $0-1$ values for each variable denote the two distinct possibilities associated with that variable. Fractional values for any of the variables in (9.2) do not represent anything in the model, and hence do not make any sense for the problem. Also, given the definitions of the variables in (9.1), we cannot claim that a fractional value like 0.99 for one of these variables is closer to 1 than 0 for this variable. Due to this, we cannot take a fractional solution to the system consisting of the top 5 constraints in (9.2), and somehow try to round it to satisfy the integer requirements on the values of the variables also. The usual technique of rounding a fractional solution to a nearest integer point does not make any sense at all when dealing with integer models involving combinatorial choice variables like those in this problem.

Since there is no objective function to be optimized in (9.2), it is a $0-1$ integer feasibility problem. It is a formulation for Portia's casket problem involving $0-1$ integer variables.

We will now outline a very simple and direct method that can be used to solve any integer programming problem, provided there is no limit on the time taken to solve the problem.

The Total Enumeration Method for Solving Integer Programs

Since we are dealing with variables that take values from discrete sets, it is possible to enumerate all the possible solutions one by one and evaluate each separately, and identify the best solution encountered as the optimum solution for the problem. This is the fundamental, simple idea behind this method. The name of the method refers to the fact that the method examines every possible solution and selects the best feasible solution among them.

In a pure integer program there are only a finite number of solutions if all the variables have finite lower and upper bounds specified for them. A solution is obtained by giving each variable an integer value within its bounds. This solution is a feasible solution to the problem if it satisfies all the other equality and inequality constraints in the system; and if it is feasible, we evaluate the objective function at it. By examining each possible integer solution this way, and then comparing the objective values at the feasible solutions among them, we can find an optimum feasible solution if the problem has a feasible solution.

If there is no upper or lower bound on the value of one or more variables in a pure integer program, the number of solutions to examine in the above method could be infinite. But still they could be evaluated one by one as discussed above.

In MIPs there are some integer and some continuous decision variables. The total enumeration method uses the following enumeration scheme to solve an MIP. First, all the integer variables are given specific integer values within their bounds. The remaining problem, consisting of only the continuous decision variables, is a linear program, and can be solved by the simplex method of Chapter 7. If this LP is feasible, this yields the best possible feasible solution for the original problem with all the integer variables fixed at their current values. Repeat this process with each possible choice of integer values for the integer variables, and select the best among all the feasible solutions obtained as the optimum solution for the original problem.

As an example, we will illustrate how the total enumeration method solves Portia's casket problem discussed in Example 9.1. The only possible choices for the vector $x = (x_1, x_2, x_3)^T$ to satisfy the first and sixth constraints in (9.2) are $(1, 0, 0)^T$, $(0, 1, 0)^T$, and $(0, 0, 1)^T$. We try each of these choices and see whether we can generate a vector $y = (y_1, y_2, y_3)^T$ which together with this x satisfies the remaining constraints in (9.2).

If $x = (1, 0, 0)^T$, by the second and third constraints in (9.2), we get $y_1 = 1$, $y_2 = 1$, and the fifth constraint will be violated. So, $x = (1, 0, 0)^T$ cannot lead to a feasible solution to (9.2).

If $x = (0, 1, 0)^T$, from the second, third, and fourth constraints in (9.2), we get $y_1 = 0, y_2 = 0$, and $y_3 = 1$, and we verify that $x = (0, 1, 0)^T$, $y = (0, 0, 1)^T$ satisfies all the constraints in the problem, hence it is a feasible solution to (9.2).

In the same way we verify that $x = (0, 0, 1)^T$ does not lead to a feasible solution to (9.2).

Hence the unique solution of (9.2) is $x = (0, 1, 0)^T$, $y = (0, 0, 1)^T$; which by the definition of the variables implies that casket 2 (silver casket) must contain Portia's portrait.

Thus total enumeration involves checking every possibility. It is an extremely simple method, and it works well if the number of possibilities to be examined is small. In fact, this is the method that people use to solve problems of Category 1 discussed in Chapter 1. Unfortunately, in real world applications of integer programming and combinatorial optimization, the number of possibilities to check tends to be so huge, that even using the fastest and most sophisticated computers available today, the answer to the problem cannot be obtained by total enumeration within the lifetime of the decision maker, making it impractical.

As an example, consider the assignment problem discussed in Chapters 1 and 5, with applications in assigning machines to jobs, candidates to positions, etc. Since we desire only an integer feasible solution for it, it is a 0−1 integer program. The assignment problem of order n has $n!$ feasible solutions. Real world applications typically involve problems of order 100 to 10,000. To solve an assignment problem of order 100 by the total enumeration method involves comparing the cost of 100! assignments directly, a task which takes > 500 years by the fastest available computer. The efficient Hungarian method for the assignment problem discussed in Chapter 5 is not based on total enumeration; it is guaranteed to find an optimum assignment in a problem of order n with at most $O(n^3)$ effort. Commercial software based on the Hungarian method for the assignment problem typically solves problems of order 1000 in about a couple of minutes time, on workstations available today.

Thus for solving problems of Category 2 discussed in Chapter 1, total enumeration is not a practical approach. We need more efficient algorithms to handle these problems. An improved approach based on partial enumeration is presented in Chapter 10.

Conclusion of Portia's Story

As R. Smullyan reports in his 1978 book, an intelligent, nice, and handsome suitor showed up for Portia's test. He chose correctly and claimed Portia's hand in marriage, and they lived happily for a while. The sequel to this story is stated in the following exercise.

Exercises

9.1 This exercise is also adapted from the excellent 1978 book of R. Smullyan. As

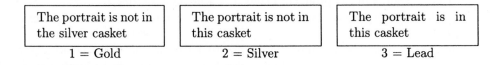

The portrait is not in the silver casket	The portrait is not in this casket	The portrait is in this casket
1 = Gold	2 = Silver	3 = Lead

Figure 9.2

discussed under the marriage problem in Chapter 1, familiarity breeds contempt, and after a brief blissful period of married life, Portia was haunted by the following thought: "My husband displayed intelligence by solving my casket problem correctly, but that problem was quite easy. I could have posed a much harder problem, and gotten a more intelligent husband." Because of her craving for intelligence, she could not continue her married life with this thought, and being a lawyer, she was able to secure a divorce easily. This time she wanted to find a more intelligent husband by the casket method again, and had the inscriptions put on the caskets as shown in Figure 9.2.

She explained to the suitors that at least one of the three inscriptions is true, and at least one of them is false.

Formulate the problem of identifying the casket containing the portrait in this situation as a $0-1$ integer feasibility problem, and solve it by total enumeration.

P.S. To complete the story, the first man who solved this casket problem turned out to be Portia's ex-husband. So, they got married again. He took her home, and being not only intelligent but also worldly-wise he was able to convince her that he is the right man for her, and they lived happily ever after.

9.2 Four persons, one of whom has committed a terrible crime, made the following statements when questioned by the police. Anita: "Kitty did it." Kitty: "Robin did it." Ved: "I didn't do it." Robin: "Kitty lied."

If only one of these four statements is true, formulate the problem of finding the guilty person as a $0-1$ feasibility problem, and find its solution by total enumeration.

Who is the guilty person if only one of the four statements is false? Formulate this as a $0-1$ feasibility problem and solve it.

In the following sections we show how a variety of combinatorial conditions arising in practical applications can be modeled through linear constraints involving $0-1$ integer variables. And we present several integer programming models that appear often in applications.

9.1 The Knapsack Problem

Knapsack problems (or **one dimensional knapsack problems** to be specific, see later on) are single constraint pure integer programs. The knapsack model refers to the following situation. Articles of n different types are available. Each article of type i has weight w_i kg. and value $\$v_i$. A knapsack that can hold a weight of at most w kg. has to be loaded with these articles so as to maximize the value of the articles included, subject to its weight capacity. Articles cannot be broken, only a nonnegative integer number of articles of each type can be loaded. For $j = 1$ to n define

$$x_j \quad = \quad \text{number of articles of type } j \text{ included in the knapsack}$$

In terms of these decision variables, the problem is

$$\text{Maximize} \quad z(x) = \sum_{j=1}^{n} v_j x_j$$

$$\text{subject to} \quad \sum_{j=1}^{n} w_j x_j \;\leqq\; w \tag{9.3}$$

$$x_j \geqq 0 \quad \text{and integer} \qquad \text{for all} \quad j$$

This is known as the **nonnegative integer knapsack problem**. It is characterized by a single inequality constraint of "$\leqq$" type, and all positive integral data elements.

If the last condition in (9.3) is replaced by "$x_j = 0$ or 1 for all j", the problem becomes the **0−1 knapsack problem**.

The knapsack problem is the simplest integer program, but it has many applications in capital budgeting, project selection, etc. It also appears as a subproblem in algorithms for cutting stock problems and other integer programming algorithms.

We now present an application for the knapsack model to a problem that arose at the University of Michigan Engineering Library in Ann Arbor. At that time the library was subscribing to 1200 serial journals, and the annual subscription budget was about $300,000. The unending battle to balance the serials budget, and the exorbitant price increases for subscriptions to scholarly journals, have made it essential for the library to consider a reduction in acquisitions. This lead to the problem of determining which subscriptions to renew and which to cancel, in order to bring the total serials subscription expenditure to within the specified budget. To protect the library's traditional strengths as a research facility, the librarian has set the goal of making these renewal/cancellation decisions in order to provide the greatest number of patrons the most convenient access to the serial literature they require within allotted budget. Anticipating this problem, the library staff have been gathering data on the use of journals for about four years. We constructed a sample problem to illustrate this application, using the data from 8 different journals. The value or the readership of a journal given below is the average number of unbound uses per year per title.

Journal j	Subscription $/year, w_j	Readership v_j
1	80	7840
2	95	6175
3	115	8510
4	165	15015
5	125	7375
6	78	1794
7	69	897
8	99	8316

Suppose the total budget available for subscriptions to these 8 journals is $670. Defining for $j = 1$ to 8

$$x_j = \begin{cases} 1, & \text{if subscription to journal } j \text{ is renewed} \\ 0, & \text{otherwise} \end{cases}$$

we get the following $0-1$ integer programming formulation for the library's problem of determining which journal subscriptions to renew, to maximize readership subject to the budget constraint.

Maximize $z(x)$ = $7840x_1 + 6175x_2 + 8510x_3 + 15015x_4 + 7375x_5 + 1794x_6 + 897x_7 + 8316x_8$

subject to $80x_1 + 95x_2 + 115x_3 + 165x_4 + 125x_5 + 78x_6 + 69x_7 + 99x_8 \leqq 670$

$x_j = 0$ or 1 for all j

Clearly, this problem is a $0-1$ knapsack problem.

The Multidimensional Knapsack Problem

Consider the knapsack problem involving n articles. Suppose we are given the value of each article, and both its weight as well as volume. And assume that the knapsack has a capacity on both the weight and the volume that it can hold. Then the problem of determining the optimum number of articles of each type to load into the knapsack, to maximize the total value loaded subject to both the weight and volume constraints, is a problem of the form (9.3) with two constraints instead of one. A problem of this form is in called a **multidimensional knapsack problem**. A general multidimensional knapsack problem is the following problem

$$\begin{aligned} \text{Maximize} \quad & z(x) = cx \\ \text{subject to} \quad & Ax \ \leqq \ b \\ & x \geqq 0 \qquad \text{and integer} \end{aligned} \qquad (9.4)$$

where A is an $m \times n$ matrix, and A, b, c are all > 0 and integer.

$0-1$ Multidimensional Knapsack Problem
with Additional Multiple Choice Constraints

Consider a multidimensional knapsack problem involving n articles, in which at most one copy of each article is available for packing into the knapsack. So, the decision variables in this problem are, for $j = 1$ to n

$$x_j = \begin{cases} 1, & \text{if } j\text{th article is packed into the knapsack} \\ 0, & \text{otherwise} \end{cases}$$

Let c_j be the value of article j, so $z(x) = \sum c_j x_j$ is the objective function to be maximized in this problem. Let $Ax \leqq b$ be the system of m multidimensional knapsack constraints in this problem.

In addition, suppose the n articles are partitioned into p disjoint subsets $\{1, \dots, n_1\}, \{n_1 + 1, \dots, n_1 + n_2\}, \dots, \{n_1 + \dots + n_{p-1} + 1, \dots, n_1 + \dots + n_p\}$ consisting of n_1, n_2, ..., n_p articles respectively, where $n_1 + \dots + n_p = n$, and it is specified that precisely one article must be selected from each of these subsets. These additional requirements impose the following constraints

$$
\begin{aligned}
x_1 + \dots + x_{n_1} &= 1 \\
x_{n_1+1} + \dots + x_{n_1+n_2} &= 1 \\
&\;\;\vdots \\
x_{n_1+\dots+n_{p-1}+1} + \dots + x_n &= 1
\end{aligned}
$$

A system of equality constraints of this type in $0-1$ variables is called a system of **multiple choice constraints**. Each constraint among these specifies that a single variable among a subset of variables has to be set equal to 1, while all the other variables in that subset are set equal to 0. The combined problem is the following

$$
\begin{aligned}
\text{Maximize} \qquad & z(x) = cx \\
\text{subject to} \qquad & Ax \leqq b
\end{aligned}
$$

$$
\begin{aligned}
x_1 + \dots + x_{n_1} &= 1 \\
x_{n_1+1} + \dots + x_{n_1+n_2} &= 1 \\
&\;\;\vdots \\
x_{n_1+\dots+n_{p-1}+1} + \dots + x_n &= 1 \\
x_j = 0 \text{ or } 1 \text{ for all } j
\end{aligned}
$$

This is the general $0-1$ multidimensional knapsack problem with additional multiple choice constraints.

Exercises

9.3 A Capital Budgeting Problem There is a total of $w_0 = \$35$ mil. available to invest. There are 7 independent investment possibilities, with the jth one costing w_j in mil.\$, and yielding an annual payoff of v_j in units of \$10,000, $j = 1$ to 7. The following table provides this data. Each investment possibility requires full participation, partial investments are not acceptable. The problem is to select a subset of these possibilities to invest, to maximize the total annual payoff (measured in units of \$10,000) subject to the constraint on available funds. Formulate as a knapsack problem.

Investment possibility j	cost w_j in $mil.	Annual payoff v_j in $10,000 units
1	3	12
2	4	12
3	3	9
4	3	15
5	15	90
6	13	26
7	16	112

9.4 A Multiperiod Capital Budgeting Problem   An investor who is expecting to receive sizable income annually over the next three years is investigating 8 independent projects to invest the spare income. Each project requires full participation, no partial participation is allowed. If selected, a project may require cash contributions yearly over the next 3 years, as in the following table. At the end of the 4th year, the investor expects to sell off all the selected projects at the expected prices given in the following tableau. The investor needs to determine the subset of projects to invest in, to maximize the total expected amount obtained by selling off the projects at the end of the 4th year, subject to the constraint on available funds in years 1, 2, 3. Formulate as a multidimensional knapsack problem.

Project	Investment needed in year in $10,000 units			Expected selling price in 4th year in $10,000 units
	1	2	3	
1	20	30	10	70
2	40	20	0	75
3	50	30	10	110
4	25	25	35	105
5	15	25	30	85
6	7	22	23	65
7	23	23	23	82
8	13	28	15	70
Funds available to invest	95	70	65	

9.5 Problem with Multiple Choice Constraints   Consider the investment problem discussed in Exercise 9.3. Projects 1, 2 there deal with fertilizer manufacturing; projects 3, 4 deal with tractor leasing; and projects 5, 6, 7, 8 are miscellaneous projects. The investor would like to invest in one fertilizer project, one tractor leasing project, and at least one miscellaneous project. Derive a formulation of the problem that includes these additional constraints.

9.2 Set Covering, Set Packing, and Set Partitioning Problems

Consider the following problem faced by the US Senate. They have various committees having responsibility for carrying out the senate's work, or pursuing various investigations. Membership in committees brings prestige and visibility to the senators, and they are quite vigorously contested. We present an example dealing with that of forming a senate committee to investigate a political problem. There are 10 senators, numbered 1 to 10, who are eligible to serve on this committee. They belong to the following groups.

Group	Eligible senators in this group
Southern senators	$\{1, 2, 3, 4, 5\}$
Northern senators	$\{6, 7, 8, 9, 10\}$
Liberals	$\{2, 3, 8, 9, 10\}$
Conservatives	$\{1, 5, 6, 7\}$
Democrats	$\{3, 4, 5, 6, 7, 9\}$
Republicans	$\{1, 2, 8, 10\}$

It is required to form the smallest size committee which contains at least one representative from each of the above groups. We will give a $0-1$ integer formulation for this problem. For $j = 1$ to 10, define

$$x_j = \begin{cases} 1, & \text{if senator } j \text{ is selected for the committee} \\ 0, & \text{otherwise} \end{cases}$$

Then the committee size is $\sum x_j$ which has to be minimized. From the definition of the decision variables, we see that the number of senators selected for the committee, from group 1 is $x_1 + x_2 + x_3 + x_4 + x_5$ and this is required to be $\geqq 1$. Continuing in the same way, we get the following integer programming formulation for this problem

$$\text{Minimize} \quad z(x) = \sum_{j=1}^{10} x_j$$

$$\begin{aligned}
\text{subject to} \quad x_1 + x_2 + x_3 + x_4 + x_5 &\geqq 1 \\
x_6 + x_7 + x_8 + x_9 + x_{10} &\geqq 1 \\
x_2 + x_3 + x_8 + x_9 + x_{10} &\geqq 1 \\
x_1 + x_5 + x_6 + x_7 &\geqq 1 \\
x_3 + x_4 + x_5 + x_6 + x_7 + x_9 &\geqq 1 \\
x_1 + x_2 + x_8 + x_{10} &\geqq 1 \\
x_j = 0 \text{ or } 1 \text{ for all } \quad j &
\end{aligned}$$

This is a pure $0-1$ integer program in which all the constraints are $\geqq$ inequalities, all the right hand side constants are 1, and the matrix of coefficients is a $0-1$ matrix. Each constraint corresponds to a group, and when a $0-1$ solution satisfies it, the associated committee has at least one member from this group. A $0-1$ integer program of this form is called a **set covering problem**. The general set covering problem is of the following form

$$
\begin{aligned}
\text{Minimize} \quad & z(x) = cx \\
\text{subject to} \quad & Ax \;\geqq\; e \\
& x_j = 0 \ \text{or} \ 1 \qquad \text{for all } j
\end{aligned}
\tag{9.5}
$$

where A is a $0-1$ matrix of order $m \times n$, and e is the vector of all 1s in $\mathbb{R}^m$. The set covering model has many applications. Here are some of them.

Delivery and Routing Problems These problems are also called **truck dispatching** or **truck scheduling problems**. A **warehouse** (or sometimes referred to as a **depot**) with a fleet of trucks has to make deliveries to m customers in a region. In Figure 9.3, the warehouse location is marked by "▢", and customer locations are marked by an "x". The problem is to make up routes for trucks which begin at the depot, visit customers to make deliveries, and return to the depot at the end. The input data consists of the cost (either distance or driving time) for traveling between every pair of locations among the depot and the customers, the quantity to be delivered to each customer in tons say, and the capacity of each truck in tons. A single truck route covering all the customers may not be feasible if the total quantity to be delivered to all the customers exceeds the truck capacity, or if the total distance or time of the route exceeds the distance or time that a truck driver can work per day by union regulations or company policy. So, the problem is to partition the set of all customers into subsets each of which can be handled by a truck in a feasible manner, and the actual route to be followed by each truck (i.e., the order in which the truck will visit the customers in its subset), so as to minimize the total cost incurred in making all the deliveries.

One approach for solving this problem generates a large number of feasible routes which are good (i.e., have good cost performance for the deliveries they make) one after the other using simple rules, and selects a subset of them to implement using a set covering model. In Figure 9.3 we show two routes, one with dashed lines, and the other with continuous lines. Let n denote the total number of routes generated, and c_j the cost of route j, $j = 1$ to n. Each customer may lie on several of the routes generated, in fact the larger n is, the better the final output. Let $\mathbf{F}_i$ denote the subset of routes among those generated, which contain the ith customer, $i = 1$ to m. Since each of the m customers has to be visited, at least one of the routes from the set $\mathbf{F}_i$ has to be implemented, for $i = 1$ to m. Define

$$
x_j = \begin{cases} 1, & \text{if the } j \text{ route is implemented by a truck} \\ 0, & \text{otherwise} \end{cases}
$$

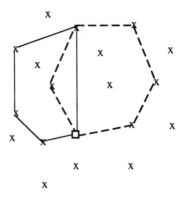

Figure 9.3 "☐" marks the location of the depot. Each customers location is marked by an "x". Two feasible truck routes are shown.

Then the problem of finding the best subset of routes to implement is

$$\text{Minimize} \quad \sum_{j=1}^{n} c_j x_j$$

$$\text{subject to} \quad \sum_{j \in \mathbf{F}_i} x_j \;\overset{\geq}{=}\; 1, \quad \text{for each } i = 1 \text{ to } m \qquad (9.6)$$

$$x_j = 0 \quad \text{or } 1 \qquad \text{for all } j = 1 \text{ to } n$$

(9.6) is a set covering problem. If $\bar{x}$ is an optimum solution of (9.6), the routes to implement are those in the set $\{j: \bar{x}_j = 1\}$. If only one route from this set contains a customer i on it, the truck following that route makes the delivery to this customer. If two or more routes from this set contain customer i on them, we select any one of the trucks following these routes to make the delivery to customer i, and the other trucks pass through customer i's location on their route without stopping.

Locating Fire Hydrants Given a network of traffic centers (nodes in a network), and street segments (edges in the network, each edge joining a pair of nodes), this problem is to find a subset of nodes for locating fire hydrants so that each street segment contains at least one fire hydrant. Suppose there are n nodes numbered 1 to n, and let c_j be the cost of locating a fire hydrant at node j. A subset of nodes satisfying the property that every edge in the network contains at least one node from the subset is called a **node cover** for the network. The constraint requires that the subset of nodes where fire hydrants are located should be a node cover. Thus the problem is to determine a minimum cost node cover. Define for $j = 1$ to n

$$x_j = \begin{cases} 1, & \text{if a fire hydrant is located at node } j \\ 0, & \text{otherwise} \end{cases}$$

Then the problem is to determine a $0-1$ vector x to minimize $\sum c_j x_j$ subject to the constraints that $x_i + x_j \geq 1$ for every edge $(i; j)$ in the network. For the network in Figure 9.4, the problem is the one following the figure.

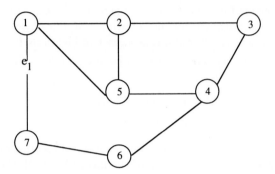

Figure 9.4   Street network for fire hydrant location problem. Nodes are traffic centers. Edges are street segments.

$$\begin{array}{lllllll}
\text{Minimize} & \multicolumn{6}{l}{\sum_{j=1}^{7} c_j x_j} \\
\text{subject to} & x_1 & & & & +x_7 & \geq 1 \\
& x_1 & +x_2 & & & & \geq 1 \\
& x_1 & & & +x_5 & & \geq 1 \\
& & x_2 & +x_3 & & & \geq 1 \\
& & x_2 & & +x_5 & & \geq 1 \\
& & & x_3 & +x_4 & & \geq 1 \\
& & & & x_4 & +x_5 & \geq 1 \\
& & & & x_4 & +x_6 & \geq 1 \\
& & & & & x_6 & +x_7 \geq 1
\end{array}$$

$$x_j = 0 \text{ or } 1 \text{ for all } j$$

Each constraint in this model corresponds to an edge in the network. For example, the first constraint requires that a fire hydrant should be located at at least one of the two nodes 1, 7 on the edge e_1 in the network in Figure 9.4. This is a set covering problem in which each constraint involves exactly two variables. Such problems are known as **node covering** or **vertex covering problems**.

Facility Location Problems   These problems have the following features. A region is partitioned into m neighborhoods, each of which requires the use of some facility (fire stations, snow removal equipment banks, etc.). There are n possible locations in the region for building these facilities. d_{ij} = the distance in miles between neighborhood i and location j, is given for all $i = 1$ to m, $j = 1$ to n. (A neighborhood could be a large area; the distance between a location and a neighborhood is usually defined to be the distance between the location and the population center of the neighborhood.) c_j = the cost of building a facility at location j, is given for $j = 1$ to n. There is a state restriction that every neighborhood must be within a distance of at most d miles from its nearest facility. The problem is to select a minimum cost subset of locations to build the facilities that meets the state's restrictions. For $i = 1$ to m define $\mathbf{F}_i = \{\text{location } j:\ d_{ij} \overset{\leq}{=} d\}$. Let

$$x_j = \begin{cases} 1, & \text{if a facility is built at location } j \\ 0, & \text{otherwise} \end{cases}$$

Then the problem of finding the optimum subset of locations to build the facilities is exactly the set covering problem of the form (9.6) with the definitions of x_j and $\mathbf{F}_i$ as stated here.

Airline Crew Scheduling Problem   This is a very important large scale

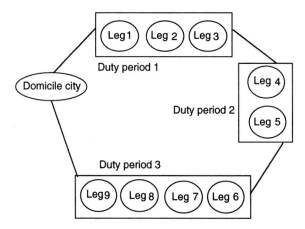

Figure 9.5 A pairing for a crew in airline operations.

application for the set covering model. The basic elements in this problem are **flight legs**. A flight leg is a flight between two cities, departing at one city at a specified time and landing next at the second city at a specified time, in an airline's timetable. A **duty period** for a crew is a continuous block of time during which the crew is on duty, consisting of a sequence of flight legs each one following the other in chronological order. A **pairing** for a crew is a sequence of duty periods that begins and ends at the same domicile. Federal aviation regulations, union rules, and company policies impose a complex set of restrictions in the formation of pairings. In particular, a duty period can contain no more than 7 flight legs, and cannot exceed 12 hours in duration; and a minimum rest period of 9.5 hours is required between consecutive duty periods in a pairing. And the crew can fly no more than 16 hours in any 48 hour interval. A pairing may include several days of work for a crew.

Even for a moderate sized airline the monthly crew scheduling problem may involve 500 flight legs. The problem of forming a minimum cost set of pairings which cover all the flight legs in a time table is a tantalizing combinatorial optimization problem.

One approach for handling this problem proceeds as follows. It generates a list consisting of a large number of good pairings using a pairing generator to enumerate candidate crew schedules, and computes the cost of each pairing. Suppose the list has n pairings. Each flight leg may appear in several pairings in the list. For $i = 1$ to m, let $\mathbf{F}_i$ = set of all pairings in the list that contain the ith flight leg. For $j = 1$ to n let c_j be the cost of the jth pairing in the list. Define for $j = 1$ to n

$$x_j = \begin{cases} 1, & \text{if the } j\text{th pairing in the list is implemented} \\ 0, & \text{otherwise} \end{cases}$$

Then the problem of selecting a minimum cost subset of pairings in the list to implement to cover all the flight legs is exactly the set covering problem of the form (9.6) with x_j and $\mathbf{F}_i$ as defined here.

The set covering model finds many other applications in such diverse areas as the design of switching circuits, assembly line balancing, etc.

The Set Packing Problem

A pure $0-1$ integer program of the following form is known as a **set packing problem**.

$$\begin{aligned} \text{Maximize} \quad & z(x) = cx \\ \text{subject to} \quad & Ax \overset{\leq}{=} e \\ & x_j = 0 \text{ or } 1 \qquad \text{for all } j \end{aligned} \qquad (9.7)$$

where A is a $0-1$ matrix of order $m \times n$, and e is the vector of all 1s in $\mathbb{R}^m$.

As an example of an application of the set packing problem, we now present a **meeting scheduling problem**. Large organizations such as big hospitals etc. are run by teams of administrators. In the course of the workweek these administrators attend several meetings where decisions are taken, and administrative and policy

problems are ironed out. This application is concerned with the timely scheduling of the necessary meetings. Suppose in a particular week there are n different
meetings to be scheduled. For the sake of simplicity assume that each meeting
lasts exactly one hour. Suppose we have T different time slots of one hour duration
each, available during the week to hold the meetings (for example, if meetings can
be held every morning Monday to Friday from 8 to 10 AM, we have $T = 10$ time
slots available). Suppose there are k administrators in all, and we are given the
following data: for $i = 1$ to k, $j = 1$ to n

$$a_{ij} = \begin{cases} 1, & \text{if the } i\text{th administrator has to attend the } j\text{th meeting} \\ 0, & \text{otherwise} \end{cases}$$

The $0-1$ matrix (a_{ij}) is given. If two meetings require the attendance of the
same administrator, they cannot both be scheduled in the same time slot, because
that will create a conflict for that administrator. On the other hand if there is no
common administrator that is required to attend two meetings, both of them can
be scheduled in the same time slot. The problem is to schedule as many of the n
meetings as possible in the T time slots available, subject to these conditions. For
$j = 1$ to n, $t = 1$ to T, define

$$x_{jt} = \begin{cases} 1, & \text{if meeting } j \text{ is scheduled for time slot } t \\ 0, & \text{otherwise} \end{cases}$$

Then the problem of scheduling as many of the meetings as possible in the available time slots without creating any conflicts for any administrator is the following
set packing model

$$\text{Maximize} \quad \sum_{j=1}^{n} \sum_{t=1}^{T} x_{jt}$$

$$\text{subject to} \quad \sum_{j=1}^{n} a_{ij} x_{jt} \;\leqq\; 1, \quad \text{for } i = 1 \text{ to } k,\, t = 1 \text{ to } T \qquad (9.8)$$

$$\sum_{t=1}^{T} x_{jt} \;\leqq\; 1, \quad \text{for } j = 1 \text{ to } n$$

$$x_{jt} \;=\; 0 \text{ or } 1 \text{ for all } j, t$$

The first set of constraints represents the fact that each administrator can attend
at most one meeting in any time slot. The second set of constraints assures that
each meeting is assigned at most one time slot.

The Set Partitioning Problem

A set partitioning problem is a $0-1$ pure integer program of the following form

$$\text{Minimize} \quad z(x) = cx$$
$$\text{subject to} \quad Ax \quad = \quad e \tag{9.9}$$
$$x_j = 0 \text{ or } 1 \qquad \text{for all } j$$

where A is a 0−1 matrix, and e is the column vector of all 1s of appropriate order. Notice the difference between the set covering problem (in which the constraints are of the form $Ax \stackrel{\geq}{=} e$), the set packing problem (in which the constraints are of the form $Ax \stackrel{\leq}{=} e$), and the set partitioning problem (in which the constraints are of the form $Ax = e$).

The set partitioning problem also finds many applications. One of them is the following. Consider a region consisting of many, say m, sales areas numbered 1 to m. These areas have to be arranged into groups to be called **sales districts** such that each district can be handled by one sales representative. The problem is to determine how to form the various sales areas into districts. One approach for handling this problem generates a list consisting of a large number of subsets of sales areas, each of which could form a good district (i.e., provides enough work for a sales representative and satisfies any other constraints that may be required). Let n be the number of such subsets generated, number them as subsets $1, \ldots, n$. For $i = 1$ to m, let $\mathbf{F}_i = \{j: \text{subset } j \text{ includes the } i\text{th sales area}\}$. For $j = 1$ to n let c_j denote the cost of forming the subset j in the list into a sales district. The approach now selects subsets from the list to form into sales districts using a set partitioning model. Define

$$x_j = \begin{cases} 1, & \text{if the } j\text{th subset in the list is formed into a sales district} \\ 0, & \text{otherwise} \end{cases}$$

Since each sales area must be in a district, our problem leads to the following set partitioning model.

$$\text{Minimize} \quad \sum_{j=1}^{n} c_j x_j$$
$$\text{subject to} \quad \sum_{j \in \mathbf{F}_i} x_j \quad = \quad 1, \quad \text{for each } i = 1 \text{ to } m \tag{9.10}$$
$$x_j = 0 \quad \text{or } 1 \qquad \text{for all } j$$

In a similar manner, the set partitioning model has applications in political districting, and in various other problems in which a set has to be partitioned at minimum cost subject to various conditions.

Exercises

9.6 A Facilities Location Problem   A residential region is divided into 8 zones. The best location for a fire station in each zone has been determined already. From these locations and the population centers in each zone, we have the following estimates for the average number of minutes of fire truck driving time to respond to an emergency in zone j from a possible fire station located in zone i. An estimate of more than 75 minutes indicates that it is not feasible to respond to an emergency within reasonable time using that route, so that cell is left blank. Because of traffic patterns etc., the estimate matrix is not symmetric.

| | Average driving time | | | | | | | |
to $j = 1$	2	3	4	5	6	7	8
from $i = 1$	10		25		40			30
2		8	60	35		60	20	
3	30		5	15	30	60	20	
4	25		30	15	30	60	25	
5	40		60	35	10		32	23
6		50	40	70		20		25
7	60	20		20	35		14	24
8	30		25		25	30	25	9

It is not necessary to have a fire station in each zone, but each zone must be within an average 25 minute driving time reach of a fire station. Formulate the problem of determining the zones in which fire stations should be located, so as to meet the constraint stated above with the smallest number of fire stations.

9.7 A Delivery Problem   A depot numbered 0 has to make deliveries to customers at locations numbered 1 to 8. A set of 9 good routes for delivery vehicles have been generated and given below. The first route 0-3-8-0 means that the vehicle starts at the depot 0, visits customer 3 first, from there goes to visit customer 8, and from there returns to the depot. The cost of the route given below is its expected driving time in hours. It is required to determine which of these 9 routes should be implemented so as to minimize the total driving time of all the vehicles used to make the deliveries. Formulate this as an integer program.

Route no.	Route	Cost
R_1	0-3-8-0	6
R_2	0-1-3-7-0	8
R_3	0-2-4-1-5-0	9
R_4	0-4-6-8-0	10
R_5	0-5-7-6-0	7
R_6	0-8-2-7-0	11
R_7	0-1-8-6-0	8
R_8	0-8-4-2-0	7
R_9	0-3-5-0	7

9.3 Plant Location Problems

Plant location problems are an important class of problems that can be modeled as MIPs. The simplest problems of this type have the following structure. There are n sites in a region that require a product. Over the planning horizon, the demand for the product in the area containing site i is estimated to be d_i units, $i = 1$ to n. The demand has to be met by manufacturing the product within the region. A decision has been taken to set up at most m plants for manufacturing the product. The set-up cost for building a plant at site i is \$$f_i$, and its production capacity will be at most k_i units over the planning horizon, $i = 1$ to n. \$$c_{ij}$ is the cost of transporting the product per unit from site i to site j. In practice, m will be much smaller than n, and the product will be shipped from where it is manufactured to all other sites in the region. The problem is to determine an optimal subset of sites for locating the plants, and a shipping plan over the entire horizon so as to meet the demands at minimum total cost which includes the cost of building the plants and transportation costs. To determine the subset of sites for locating the plants is a combinatorial optimization problem. Once the optimum solution of this combinatorial problem is known, determining the amounts to be transported along the various routes is a simple transportation problem. For $i, j = 1$ to n, define

$$y_i = \begin{cases} 1, & \text{if a plant is located at site } i \\ 0, & \text{otherwise} \end{cases}$$

$$x_{ij} \quad = \quad \text{units of product transported from site } i \text{ to } j \text{ over the planning horizon}$$

The MIP model for the problem is

$$\text{Minimize} \quad \sum_i f_i y_i \; + \; \sum_i \sum_j c_{ij} x_{ij}$$

$$\text{subject to} \quad \sum_j x_{ij} - k_i y_i \; \leqq \; 0 \quad \text{for all } i \tag{9.11}$$

$$\sum_i x_{ij} \; \geqq \; d_j \quad \text{for all } j$$

$$\sum_i y_i \; \leqq \; m$$

$$y_i = \; 0 \text{ or } 1, \quad x_{ij} \; \geqq \; 0 \quad \text{for all } i, j$$

Other plant location problems may have more complicated constraints in them. They can be formulated as integer programs using similar ideas.

The Uncapacitated Plant Location Problem

In some applications we also may have the freedom to select the production capacities of plants built. In this situation the production capacity constraints do not apply, and the problem is known as the **uncapacitated plant location problem**. Under this assumption, if a plant is built at site i there is no upper limit on how much can be shipped from this plant to any other sites. In this case the shipping cost is minimized if each site's demand d_j is completely satisfied from the plant at site i where i attains the minimum in $\min\{c_{rj} :$ over r such that a plant is built at site $r\}$. As an example, suppose plants are built at sites 1 and 2. To meet site 3's demand, if $c_{13} = 10$ and $c_{23} = 20$ since each plant can produce as much as necessary, we would not ship any product to site 3 from the plant at site 2, since it is cheaper to ship from the plant at site 1 instead. So, in this case there exists an optimum shipping plan, in which each site receives all its demand from only one plant. Using this fact we can simplify the formulation of the problem in this case. For $i, j = 1$ to n, define new variables

$$z_{ij} \quad = \quad \text{fraction of demand at site } j \text{ shipped from a plant at site } i$$

So, these variables satisfy $\sum_i z_{ij} = 1$ for each $j = 1$ to n. We can think of the variable z_{ij} to be equal to x_{ij}/d_j in terms of the variable x_{ij} defined earlier. Also, since the production level at each plant depends on which sites it is required to supply in this case, the cost of setting it up may depend on that level. So, we assume that the cost of setting up a plant of capacity α at site i is $f_i + s_i\alpha$, where f_i is a fixed cost, and s_i is the variable cost of setting up production capacity/unit at site i. With y_i defined as before, here is the formulation of the problem.

$$\text{Minimize} \quad \sum_{i=1}^{n} f_i y_i \;+\; \sum_{i=1}^{n} s_i \Big(\sum_{j=1}^{n} d_j z_{ij}\Big) + \sum_{i=1}^{n}\sum_{j=1}^{n} c_{ij} d_j z_{ij}$$

$$\text{subject to} \quad \sum_{i=1}^{n} z_{ij} \;=\; 1, \quad j = \; 1 \text{ to } n \tag{9.12}$$

$$\sum_{j=1}^{n} z_{ij} \;\leq\; n y_i, \quad i = \; 1 \text{ to } n$$

$$z_{ij} \geq 0, y_i \;=\; \quad 0 \text{ or } 1, \text{ for all } i, j$$

9.4   Batch Size Problems

In addition to the usual linear equality-inequality constraints and nonnegativity restrictions in a linear program, suppose there are constraints of the following form: variable x_j in the model can be either 0, or if it is positive it must be $\geq$ some specified positive lower bound ℓ_j. Constraints of this type arise when the model includes variables that represent the amounts of some raw materials used, and the suppliers for these raw materials will only supply in amounts $\geq$ specified lower bounds.

There are two conditions on the decision variable x_j here, $x_j = 0$, or $x_j \geq \ell_j$; and the constraint requires that one of these two conditions must hold. We define a $0-1$ variable y_j to indicate these two possibilities for x_j, as given below.

$$y_j = \begin{cases} 0, & \text{if } x_j = 0 \\ 1, & \text{if } x_j \geq \ell_j \end{cases} \tag{9.13}$$

Let α_j be a very large positive number or some practical upper bound for the value of x_j in an optimum solution of the problem. The constraint that x_j is either 0 or $\geq \ell_j$ is equivalent to

$$\begin{aligned} x_j - \ell_j y_j &\geq 0 \\ x_j - \alpha_j y_j &\leq 0 \\ y_j = 0 \quad &\text{or} \quad 1 \end{aligned} \tag{9.14}$$

(9.14) represents through linear constraints the definition of the binary variable y_j associated with the two conditions on x_j as defined in (9.13).

Constraints like this can be introduced into the model for each such batch size restricted variable in the model. This transforms the model into an integer program.

As an example, suppose we have the constraints

$$\text{Either } x_1 = 0, \text{ or } x_1 \geq 10; \quad \text{and either } x_2 = 0, \text{ or } x_2 \geq 25 \tag{9.15}$$

in a linear programming model. Suppose 1000 is an upper bound for both x_1, x_2 among feasible solutions of this model. Then defining the binary variables y_1, y_2 corresponding to the two possibilities on x_1, x_2 respectively as in (9.13), we augment the following constraints to the LP model to make sure that (9.15) will hold.

$$\begin{aligned} x_1 - 10y_1 &\geq 0 \\ x_1 - 1000y_1 &\leq 0 \\ x_2 - 25y_2 &\geq 0 \\ x_2 - 1000y_2 &\leq 0 \\ y_1, y_2 \quad \text{are both} \quad &0 \text{ or } 1 \end{aligned}$$

9.5 Other "Either, Or" Constraints

Let x be the column vector of decision variables in an LP, in which we have an additional constraint involving m conditions

$$g_1(x) \; \overset{\geq}{=} \; 0$$

$$\vdots \tag{9.16}$$

$$g_m(x) \; \overset{\geq}{=} \; 0$$

where each of these conditions is a linear inequality. The additional constraint does not require that all the conditions in (9.16) must hold, but only specifies that at least k of the m conditions in (9.16) must hold. To model this requirement using linear constraints we define binary variables $y_1, \ldots, y_m$ with the following definitions.

$$y_i = \left\{ \begin{array}{ll} 0, & \text{if the condition } g_i(x) \overset{\geq}{=} 0 \text{ holds} \\ 1, & \text{otherwise} \end{array} \right. \tag{9.17}$$

Let L_i be a large positive number such that $-L_i$ is a lower bound for $g_i(x)$ on the set of feasible solutions of the original LP model. Then the following system of constraints, augmented to the LP model, will guarantee that at least k of the conditions in (9.16) will hold.

$$g_1(x) + L_1 y_1 \; \overset{\geq}{=} \; 0$$

$$\vdots$$

$$g_m(x) + L_m y_m \; \overset{\geq}{=} \; 0 \tag{9.18}$$

$$y_1 + \ldots + y_m \; = \; m - k$$

$$y_i = \; 0 \text{ or } 1 \qquad \text{for all } i$$

In the same way, any restriction of the type that at least (or exactly, or at most) k conditions must hold in a given system of linear conditions, can be modeled using a system of linear constraints of the form (9.18) involving binary variables.

As an example, consider the system of linear constraints $0 \overset{\leq}{=} x_1 \overset{\leq}{=} 10$, $0 \overset{\leq}{=} x_2 \overset{\leq}{=} 10$, on the two variables x_1, x_2 in the two dimensional Cartesian plane. In addition, suppose we impose the constraint that "either $x_1 \overset{\leq}{=} 5$, or $x_2 \overset{\leq}{=} 5$" must hold. This constraint states that at least one of the following two conditions must hold.

$$g_1(x) = 5 - x_1 \; \overset{\geq}{=} \; 0 \tag{9.19}$$

$$g_2(x) = 5 - x_2 \; \overset{\geq}{=} \; 0$$

With this constraint, the set of feasible solutions of the combined system is the nonconvex dotted region in Figure 9.6

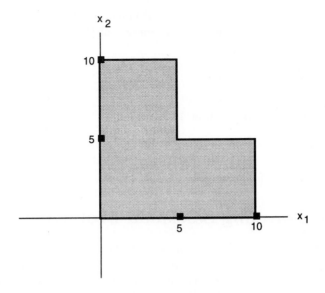

Figure 9.6

A lower bound for both $g_1(x), g_2(x)$ in (9.19) in the original cube is -15. So, the constraint that at least one of the two constraints in (9.19) must hold is equivalent to the following system

$$
\begin{array}{llll}
5 - x_1 & +15y_1 & & \geqq 0 \\
& 5 - x_2 & +15y_2 & \geqq 0 \\
& y_1 & +y_2 & =1 \\
y_1, y_2 = 0 \text{ or } 1
\end{array}
\qquad (9.20)
$$

When the constraints in (9.20) are augmented to the constraints $0 \leqq x_1 \leqq 10$, $0 \leqq x_2 \leqq 10$ of the original cube, we get a system that represents the dotted region in the x_1, x_2-plane in Figure 9.6, using the binary variables y_1, y_2. Using similar arguments, sets that are not necessarily convex, but can be represented as the union of a finite number of convex polyhedra, can be represented as the set of feasible solutions of systems of linear constraints involving some binary variables.

9.6 Discrete Valued Variables

Consider an LP model with the additional requirement that some variables can only lie in specified discrete sets. For example, in the problem of designing a water distribution system, one of the variables is the diameter of the pipe in inches; this variable must lie in the set $\{6, 8, 10, 12, 16, 20, 24, 30, 36\}$ because pipe is available only in these diameters.

In general, if x_1 is a decision variable that is restricted to assume values in the discrete set $\{\alpha_1, \ldots, \alpha_k\}$, this constraint is equivalent to

$$
\begin{aligned}
x_1 - (\alpha_1 y_1 + \ldots + \alpha_k y_k) &= 0 \\
y_1 + \ldots + y_k &= 1 \\
y_j &= 0 \text{ or } 1 \text{ for each } j
\end{aligned}
$$

Such constraints can be augmented to the other linear equality and inequality constraints in the model for each discrete valued variable. This transforms the problem into an integer program.

9.7   The Traveling Salesman Problem

A salesperson has to visit cities $2, \ldots, n$, and his/her trip begins at, and must end in, city 1. c_{ij} = the cost of traveling from city i to city j, is given for all $i \neq j = 1$ to n, and $c = (c_{ij})$ of order $n \times n$ is known as the **cost matrix** for the problem. Beginning in city 1, the trip must visit each of the cities $2, \ldots, n$ once and only once in some order, and must return to city 1 at the end. The cost matrix is the input data for the problem. The problem is to determine an optimal order for traveling the cities so that the total cost is minimized.

This is a classic combinatorial optimization problem that has been the object of very intense research over the last 40 years.

If the salesperson travels to the cities in the order i to $i + 1$, $i = 1$ to $n - 1$, and then from city n to city 1, this route can be represented by the order "1, 2, $\ldots, n; 1$". Such an order is known as a **tour** or a **hamiltonian**.

From the initial city 1 the salesperson can go to any of the other $n - 1$ cities. So, there are $n - 1$ different possibilities for selecting the first city to travel from the initial city 1. From that city the salesperson can travel to any of the remaining $n - 2$ cities, etc. Thus the total number of possible tours in a n city traveling salesman problem is $(n - 1)(n - 2) \ldots 1 = (n - 1)!$. This number grows explosively as n increases.

Let $\mathcal{N} = \{1, \ldots, n\}$ be the set of cities under consideration. Let $\mathcal{N}_1$ be a proper subset of $\mathcal{N}$. A tour covering the cities in $\mathcal{N}_1$ only, without touching any of the cities in $\mathcal{N} \backslash \mathcal{N}_1$ is known as a **subtour** covering or spanning the subset of cities $\mathcal{N}_1$. See Figure 9.7.

We will use the symbol τ to denote tours. Given a tour τ, define binary variables x_{ij} by

$$
x_{ij} = \begin{cases} 1, & \text{if the salesperson goes from city } i \text{ to city } j \text{ in tour } \tau \\ 0, & \text{otherwise} \end{cases}
$$

Then the $n \times n$ matrix $x = (x_{ij})$ is obviously an assignment; it is the assignment corresponding to the tour τ. Hence every tour corresponds to an assignment. For example, the assignment corresponding to the tour 1, 5, 2, 6, 3, 4; 1 in Figure 9.7 follows the figure.

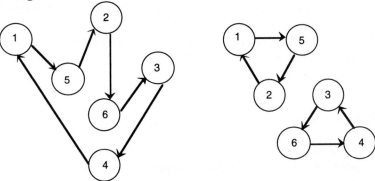

Figure 9.7 Each node represents a city. On the left is the tour 1, 5, 2, 6, 3, 4; 1. On the right we have two subtours 1, 5, 2; 1 and 3, 6, 4; 3.

to	$j=1$	2	3	4	5	6
from $i = 1$					1	
2						1
3				1		
4	1					
5		1				
6			1			

where the blank entries in the matrix are all zeros. In the notation of Chapter 5, this is the assignment $\{(1,5),\ (5,2),\ (2,6),\ (6,3),\ (3,4),\ (4,1)\}$. An assignment is called a **tour assignment** iff it corresponds to a tour.

As an example, the assignment $\{(1,5),\ (5,2),\ (2,1),\ (3,6),\ (6,4),\ (4,3)\}$ is not a tour assignment since it represents the two subtours on the right hand side of Figure 9.7.

So, the traveling salesman problem of order n with the cost matrix $c = (c_{ij})$ is

$$\text{Minimize } z_c(x) \quad = \quad \sum_{i=1}^{n}\sum_{j=1}^{n} c_{ij}x_{ij}$$

$$\text{Subject to } \sum_{j=1}^{n} x_{ij} \quad = \quad 1 \ \text{ for } i = 1 \text{ to } n \qquad (9.21)$$

$$\sum_{i=1}^{n} x_{ij} \quad = \quad 1 \ \text{ for } j = 1 \text{ to } n$$

$$x_{ij} \quad = \quad 0 \ \text{ or } 1 \ \text{ for all } i, j$$

$$\text{and } x = (x_{ij}) \qquad \text{is a tour assignment}$$

Since the salesperson is always going from a city to a different city, the variables x_{ii} will always be 0 in every tour assignment. To make sure that all the variables x_{ii} will be 0, we define the cost coefficients c_{ii} to be equal to a very large positive number for all $i = 1$ to n in (9.21)

There are $n!$ assignments of order n. Of these, only $(n-1)!$ are tour assignments. The last constraint that $x = (x_{ij})$ must be a tour assignment makes this a hard problem to solve.

9.8 Exercises

9.8 A Word Puzzle   Following is a list of 47 words each having three letters (these may not be words in the English language)

ADV	AFT	BET	BKS	CCW	CIR	DER	DIP	EAT	EGO
FAR	FIN	GHQ	GOO	HAT	HOI	HUG	ION	IVE	JCS
JOE	KEN	LKK	LIP	LYE	MOL	MTG	NES	NTH	OIL
OSF	PIP	PRF	QMG	QUE	ROE	RUG	STG	SIP	TUE
TVA	UTE	VIP	WHO	XIN	YES	ZIP			

Each letter corresponds to a unique numerical value, these letter values are: A $= 1$, B $= 2$, ..., Z $= 26$,   in the usual order.

It is required to select a subset of 8 words from the list given above to satisfy the following: let $s_t =$ sum of the letter values of the tth letter in the selected words, $t = 1, 2, 3$. Then s_1 must be less than both s_2 and s_3. The selection should maximize $s_1 + s_2 + s_3$ subject to these constraints.

(i)   Formulate this problem and solve it using an integer programming software package.

(ii)   If there are ties for the optimum solution of the above problem, find a solution that maximizes s_1 among those tied.

(iii) If there are still ties, it is required to find the solution that maximizes s_2 among those that tie for (ii). Give an argument to show that the solution found in (ii) meets this requirement.

([G. Weber, 1990]).

9.9 It is required to assign distinct values 1 through 9 to the letters A, E, F, H, O, P, R, S, T, to satisfy the conditions and achieve the objective mentioned below. Two groups of 6 words each are given below.

Group 1	Group 2
AREA	ERST
FORT	FOOT
HOPE	HEAT
SPAR	PAST
THAT	PROF
TREE	STOP

Let s_1, s_2 refer to the totals for groups 1, 2 using the letter values assigned.

(i) The letter value assignment should minimize $s_1 - s_2$. Formulate this problem and find an optimum solution for it. Use the decision variables

$$x_{ij} = \begin{cases} 1 & \text{if } i\text{th letter is given value } j \\ 0 & \text{otherwise} \end{cases}$$

(ii) It is required to find a letter-value assignment that maximizes s_1 subject to the constraint that $s_1 - s_2 = 0$. Formulate this problem and find an optimum solution for it using an integer programming software package.

([G. Weber, 1990]).

9.10 Ms. Skorean is famous for her parties in the Washington, D.C. area. In each four-year administration she throws k parties, sets m tables at each party and seats n people at each table. She makes a list of mn influential people at the start of the four year period and invites them to every party. Her tables are all distinct from each other (teak, oak, cherry etc.) and $k \leqq m$. It is rumored that her parties are famous because of her clever seating arrangements. She makes sure that each of her guests sits at each table at most once. And she believes very strongly that the atmosphere remains lively if each guest meets different people at his/her table at each party. So, every time any pair of guests find themselves at the same table after the first time, she awards herself an imaginary penalty of c units, where $c > 0$. Given k, m, n (all positive integers $\geqq 2$ and $k \leqq m$). Formulate the problem of determining the seating arrangement of her guests at the tables at the various parties, subject to the constraint that no guest sits at a table more than once and the overall penalty is as small as possible, as an integer program.

If $k = m$, what conditions should m, n satisfy to guarantee that a zero penalty seating arrangement exists?

When $k = m = n = 2$, show that there is no zero penalty seating arrangement using your formulation above. (Vishwas Bawle).

9.11 The foreign ministers of European countries (east and west) numbered 1 to 10 are planning a round table conference. In Figure 9.8 each minister is represented by a node, and a line joins two nodes if the corresponding ministers speak a common language. If there is no line joining two nodes, it means that the corresponding ministers cannot speak with each other without an interpreter.

An ideal seating arrangement around the table is one in which every pair of ministers occupying adjacent seats both know a common language. If an ideal one does not exist, an optimal arrangement should minimize the number of pairs of adjacent ministers who cannot speak with each other. Formulate the problem of finding an ideal or optimal seating arrangement as a traveling salesman problem, and solve it using the algorithm discussed in Chapter 10.

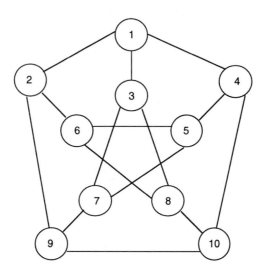

Figure 9.8

9.12 Subset-Sum Problem   We are given n positive integers $w_1, \ldots, w_n$; and another positive integer, w, called the **goal**. It is required to find a subset of $\{w_1, \ldots, w_n\}$ such that the sum of the elements in the subset is closest to the goal, without exceeding it. Formulate this as a special case of the knapsack problem. When the set of integers is $\{80, 66, 23, 17, 19, 9, 21, 32\}$, and the goal is 142, write this formulation, and see if you can solve it.

9.13 Bin Packing Problem n objects are given, with the ith object having the positive integer weight w_i kg., $i = 1$ to n. The objects need to be packed in bins, all of which are identical, and can hold any subset of objects as long as their total weight is $\overset{\le}{=} w$ kg., which is the positive integral weight capacity of each bin. Assume that $w \overset{\ge}{=} w_i$ for each $i = 1$ to n. Objects cannot be split, and each must be packed whole into a bin. Formulate the problem of determining the minimum number of bins required for this packing, subject to the bin's weight holding capacity, as an integer program. Write this formulation when there are 8 objects with weights 80,

66, 23, 17, 19, 9, 21, 32 kg, respectively, and the bins weight holding capacity is 93 kg. See if you can solve it.

9.14 A company can form 6 different teams using 10 experts it has available. In the following table if there is an entry of 1 in the first line in the row of team i and the column of expert j, then expert j has to be on team i if it is formed (in this case, the amount in $ that he/she is to be paid for being a member of this team, r_{ij}, is given just below this 1); he/she need not be included on this team otherwise. In addition to the team membership fee r_{ij}, if expert j is included in one or more teams that are formed, he/she has to be paid a *retainer fee* of c_j given in the last row of the table. If team i is formed, the company derives a gross profit of d_i given in the last column of the table. Also, because of other work commitments, company rules stipulate that no expert can work on more than two teams. It is required to determine which teams should be formed to maximize company's total net profit (gross profit from the teams formed, minus the retainer and team membership fees paid to the experts). Formulate this as an integer program.   (T. Ramesh).

					Expert						Gross
	1	2	3	4	5	6	7	8	9	10	profit d_i
Team 1	1			1	1			1			10,000
	200			200	300			200			
2		1	1				1			1	15,000
		200	400				200			300	
3	1					1		1	1		6,000
	300					200		250	150		
4		1	1		1					1	8,000
		200	400		250					200	
5	1					1	1		1		12,000
	200					200	150		150		
6		1		1				1		1	9,000
		200		200				150		200	
Retainer fee c_j	800	500	600	700	800	600	400	500	400	500	

9.15 There is an object whose weight is w kg. There are n types of stones, each stone of type i weighs exactly a_i kg., for $i = 1$ to n; and an unlimited number of copies of each are available. $w; a_1, \ldots, a_n$ are given positive integers. The object is placed in the right pan of a balance. It is required to place stones in the right and/or left pans of the balance so that it becomes perfectly balanced. It is required to do this using the smallest possible number of stones. Formulate this as an integer program. Write this formulation for the instance in which $n = 5, w = 3437$, and a_1 to a_5 are 1, 5, 15, 25, 57, 117, respectively.

9.16 Assembly line balancing An assembly line is being designed for manufacturing a discrete part. There are 7 operations numbered 1 to 7 to be performed on each part. Each operation can be started on a part any time after all its immediate predecessor operations given in the following table are completed, but not before. The table also gives the time it takes an operator to carry out each operation, in seconds. The cycle time of the assembly line will be 20 seconds (i.e., each operator will have up to 20 seconds to work on a part before it has to be put back on the line). An operator on the assembly line can be assigned to carry out any subset of operations, as long as the work can be completed within the cycle time and the assignments do not violate the precedence constraints among the operations. It is required to determine the assignment of operations to operators on this line, so as to minimize the number of operators needed. Formulate this as an integer program.

Operation	Immediate predecessors	Time in secs.
1		7
2		9
3	1,2	6
4	1	4
5	3	8
6	4	7
7	5,6	5

9.17 5000 acceptable units of a discrete product need to be manufactured in one day. There are 4 machines which can make this product, but the production rates, costs, and percentage defectives produced vary from machine to machine. Data is given below. Formulate a production plan to meet the demand at minimum cost.

Machine	Setup cost	Prod. cost/unit after mc. setup	Max. daily production	Expected defective %
1	400	4	2000	10
2	1000	6	4000	5
3	600	2	1000	15
4	300	5	3000	8

9.18 There are 5 locations where oil wells need to be drilled. There are two platforms from each of which any of these wells can be drilled. If a platform is to be used for drilling one or more of these wells, it needs to be prepared, the cost of which is given below. Once a platform is prepared, the drilling cost of each well depends on the drilling angle and other considerations. All the costs are in coded money units. Formulate the problem of determining which platform to use for drilling each well to complete all drilling at minimum total cost.

Platform	Preparation cost	Cost of drilling to location 1	2	3	4	5
1	15	10	8	30	15	19
2	20	14	25	25	15	16

9.19 A company has 5 projects under consideration for carrying out over the next 3 years. If selected, each project requires a certain level of investment in each year over the 3 year period, and will result in an expected yield annually after this 3 year construction period. In the data given below, expenditures and income are in units of $10,000. Formulate the problem of selecting projects to carry out, to maximize total return.

Project	Required expenditure in			Expected annual yield
	Year 1	Year 2	Year 3	after construction
1	10	5	15	3
2	5	5	11	2
3	15	20	25	10
4	20	10	5	7
5	10	8	6	5
Available funds	50	40	50	

9.20 There are 5 objects which can be loaded on a vessel. The weight w_i (in tons), volume v_i (in ft^3), and value r_i (in units of $1000) per unit of object i is given below for $i = 1$ to 5. Only 4 copies of object 1 and 5 copies of object 2 are available; but the other objects 3, 4, 5 are available in unlimited number of copies. For each object, the number of copies loaded has to be a nonnegative integer. The vessel can take a maximum cargo load of $w = 112$ tons, and has a maximum space of $v = 109$ ft^3. Formulate the problem of maximizing the value of cargo loaded subject to all the constraints.

Object i	w_i	v_i	r_i
1	5	1	4
2	8	8	7
3	3	6	6
4	2	5	5
5	7	4	4

9.21 The 800-telephone service of an airline operates round the clock. Data on

Period	Time of day	Min. operators needed
1	3 AM to 7 AM	26
2	7 AM to 11 AM	52
3	11 AM to 3 PM	86
4	3PM to 7 PM	120
5	7 PM to 11 PM	75
6	11 PM to 3 AM	35

the estimated number of operators needed during the various periods in a day, for attending to most of the calls in a satisfactory manner, is given above. Assume that each operator works for a consecutive period of 8 hours, but they can start work at the beginning of any of the 6 periods. Let x_t denote the number of operators starting work at the beginning of period t, $t = 1$ to 6. Formulate the problem of finding the optimum values for x_t, to meet the requirements in all the periods by employing the least number of operators.

9.22 Multiobjective Capital Budgeting Within a University   Within a university, different interest groups often have conflicting ideas concerning appropriate capital projects to undertake. During the past few years, inflation and state financial problems have focused increased attention on quantitative approaches to the problem of efficient allocation of scarce resources in universities. This exercise deals with decision making with respect to investment in fixed assets within a university under a capital rationing situation.

	Project	CC	AOS	Ly. sp.	\multicolumn{5}{ }{1000 QASDs in dept.}				
no.	Description	$1000s		10^3 ft^2	Bus.	Eng.	Che.	Phy.	Psyc.
1	New CDC CS	190	25		20	15	10	10	5
2	New IBM CS	230	30		30	23	15	15	8
3	Upgrade current CS	90	25		8	6	4	4	2
4	RB on NC	60	6				6	10	
5	RB in Eng. Bldg.	50	5			12			
6	RB in Bus. Bldg.	50	5		14				
7	Expand grad. Ly.	250	40	20	15	10	22	22	15
8	Expand Bus. Ly.	150	25	12	40				
9	Expand Eng. Ly.	200	25	10		25			
10	Expand NC Ly.	150	23	8			20	20	
11	Small expt. lab	50	10			2	10	6	
12	Large expt. lab	80	16			4	15	12	
13	Eng.Phy. Res. A	190	20				40	20	
14	Eng.Phy. Res. B	120	12				20	25	
15	Renovate Bus. Bldg.A	160	10		45				
16	Renovate Bus. Bldg. B	80	5		20				
17	Renovate Eng. Bldg.	80	5			20			
18	Behavioral lab plan A	120	10		20				18
19	Behavioral lab plan B	70	6		10				13
20	Improve Eng. lab	$\leq$30	*			0.17			
21	Improve Chem. lab	$\leq$25	*				0.15		
22	Improve Phy. lab	$\leq$35	*					0.20	
	Desired goal levels	$\leq$700	$\leq$180	$\geq$20	$\geq$50	$\geq$40	$\geq$15	$\geq$35	$\geq$30

* = 10% of CC, CC = Capital cost, AOS = Annual operating expenses, Ly. = Library
Sp. = space, CS = computer system, RB = Remote Batcher, Eng. = Engineering
Bus. = Business, Phy. = Physics, NC = North campus

The table given above presents data on 22 different projects being considered at a university. Projects numbered 1 to 19 cannot be carried out at a fractional level, they have to be either accepted and carried out fully as presented, or rejected altogether. Projects numbered 20, 21, 22 can be carried out at fractional level, i.e., the money spent on them can be anything between 0 and their capital cost; and the results achieved will be proportional to the actual money spent.

The improvement in each departmental or functional area as a result of carrying out a project is measured by QASDs (quality adjusted student days), the product of student days and a quality proxy (a subjective rating).

The following goals are set (money unit is $1000).

1. **Budget goal** The total expenditure for capital costs on all the accepted projects is desired to be at most 700 as far as possible.

2. **Operating expenses goal** The total annual operating expenses on the projects carried out among these 22 is desired to be at most 180 as far as possible.

3. **Library expansion goal** It is desired to increase the total library space in the four university libraries by at least 20 (in units of 1000 ft^2).

4. **Accreditation goal** Following a visit by a business school accreditation team several improvements in the school of business have been suggested. It is desired to accept at least two among the projects 1, 2, 3, 6, 8, 15, 16, 18, 19 that satisfy the accreditation team recommendations, as far as possible.

5. **Political-Social goal** Several state legislators, the business community, and alumni have exerted pressure to undertake at least one among the computer and library expansion projects 1, 2, 3, 4, 10, as far as possible.

6. **Improvement in departmental performance** It is desired to increase the performance measure in business, engineering, chemistry, physics, psychology departments by at least 50, 40, 15, 35, 30 QASDs (in units of 1000) respectively, as far as possible.

Constraints Constraints that must be satisfied are the following:

Five projects (5, 9, 13, 14, 17), all related to improvements in the engineering department have been designated as recipients of federal and private grants. The total capital cost spent on this group of projects must be at least 120 (in units of $1000).

Among the pair of projects (11, 12) at most one can be accepted. The same is true for project pairs (13, 14), (15, 16), and (18, 19). Also, among projects 1, 2, 3 at most one can be accepted.

The acceptance of either project 18 or 19 is contingent upon the prior acceptance of project 15.

Decision Variables For $j = 1$ to 19, define $x_j = 1$ if project j is accepted, 0 otherwise. For $j = 20, 21, 22$, let x_j = fraction of the capital cost of project j approved. Using these decision variables, formulate the problem of selecting the projects to accept as a 0−1 mixed integer goal programming model. Take the penalty for unit violation of goals 1 to 6 to be 10, 8, 7, 6, 4, 3, respectively. ([A. J. Keown; B. W. Taylor, III; and J. M. Pinkerton; 1981])

9.23 Design Problem in Food Industry   Consider the problem of improving the efficiency of a food processing plant discussed in Exercise 2.42. There we discussed the problem of finding weighing schedules corresponding to the shortest weighing time for any recipe, given the ingredients assigned to the various hoppers attached to each weighing machine.

That plant has 3 weighing machines each of which has 12 hoppers attached to it. It uses 17 different ingredients in all, and makes about 100 different recipes.

Clearly, the assignment of ingredients to hoppers has an effect on the weighing time for any recipe. Once the assignment of ingredients to the hoppers is made it is very costly to change it, so this assignment should be made so as to be optimal on a long term basis for the efficiency of the weighing system.

There are numerous recipes made, but about 20 recipes account for the bulk of the production at the plant. We are given the composition of each of these 20 major recipes and the number of batches of each to be made every week. It is required to find an assignment of the ingredients to the hoppers of the three weighing machines, so as to minimize the total time needed to make the required batches of the 20 major recipes each week. Formulate the problem of finding the optimal assignment of ingredients to hoppers, and the weighing schedules for each of the 20 major recipes, as an MIP. ([R. Benveniste, May 1986]).

9.24 The public works division in a region has the responsibility to subcontract work to private contractors. The work is of several types, and is carried out by teams, each of which is capable of doing all types of work.

The region is divided into 16 districts, and estimates of the amount of work to be done in each district are available.

There are 28 contractors of which the first 10 are experienced contractors. We have the following data.

i =   1 to 10 are indices representing experienced contractors; $i = 11$ to 28 are indices representing other contractors.

j =   1 to 16 are indices representing the various districts.

a_i =   number of teams contractor i can provide.

b_j =   number of teams required by district j.

e_j =   2 or 3, is the specified minimum number of contractors allotted to district j, this is to prevent overdependence on any one contractor.

c_{ij} =   expected yearly cost of a team from contractor i allotted to district j.

At least one experienced contractor must be appointed in each district, a precaution in case some difficult work arises. Enough teams must be allotted to meet the estimated demand in each district, and no contractor can be asked to provide more teams than it has available. Formulate the problem of determining the number of teams from each contractor to allot to each district, so as to satisfy the above constraints at minimum cost.

Also, develop a heuristic method to find a reasonably good solution to this problem. ([M. Cheshire, K. I. M. McKinnon, and H. P. Williams, August 1984]).

9.25 Apparel Sizing Problem   This exercise deals with the problem of a manufacturer of clothing or footware wishing to determine the design measurements of a given number of sizes for an item so as to maximize expected sales.

A *size* is an item having specified measurements along certain dimensions. For example, a size 10 in men's shoes is a pair having a certain length, width and height at the tip, ball, shank, breast, and heel, and will fit a person with measurements equal to those of the size perfectly.

A complete description of the human body requires measurements along a large number of dimensions. Even though some of these dimensions may be irrelevant for the fit of a particular item, and even though many persons may have identical measurements, a very large number of sizes would be required in order to fit everyone in a target population exactly. Since each size offered involves additional set-up and design costs, there is a strong incentive for attempting to fit a population with as few sizes as possible.

In practice, it is necessary to consider only one, two, or three *control dimensions*, i.e., dimensions judged important for the fit of an item.

Knowledge of the joint distribution of the target population according to the control dimensions is a prerequisite in any attempt to design a sizing system. For example, the Canadian standard for women's apparel is based on measurements of body dimensions taken on 10,000 women. Private manufacturers have even more extensive data bases on body measurements.

For each control dimension a *tolerance* is specified. For example, it may be decided that a tolerance of 1 inch is appropriate for waist girth, persons with a waist girth between 29 and 30 inches are expected to be satisfactorily fitted by a size having a waist girth of 30 inches, and so on. Consecutive intervals of this type are established for each control dimension. The consumer's purchase decision is of course a function of the quality of the fit. The closer a person's measurements are to those of an available item, the better the fit, and other things being equal, the greater the probability of purchase.

$$p_{ij} = \begin{cases} p \exp\left(\frac{x_1^i - x_1^j}{\sigma_1} + \frac{x_2^i - x_2^j}{\sigma_2}\right) & \text{if } x_1^i < x_1^j \text{ and } x_2^i < x_2^j \\ 0 & \text{otherwise} \end{cases}$$

Suppose there are two control dimensions x_1, x_2 (such as (bust girth, hip girth), or (length, width), etc.). Let p_{ij} be the probability of a person with measurements $x^i = (x_1^i, x_2^i)$ buying size $x^j = (x_1^j, x_2^j)$ if this is the only size offered. Clearly, if $x_1^i > x_1^j$ or $x_2^i > x_2^j$ this size x^j is too small for this person and p_{ij} will be zero. Also, even when $x_1^i \leqq x_1^j$ and $x_2^i \leqq x_2^j$, it is reasonable to assume that p_{ij} declines as $x_1^j - x_1^i$ or $x_2^j - x_2^i$ increases. A possible functional form for p_{ij} reflecting these observations is p_{ij} given above where p is a constant equal to the probability of purchase when the fit is perfect (i.e., $x_1^i = x_1^j$ and $x_2^i = x_2^j$), and σ_1, σ_2 are the standard deviations of x_1, x_2.

It is assumed that the joint distribution of x_1, x_2 in the target population is available in the form of frequencies for each pair of intervals of x_1 and x_2, as in the table given below. These intervals can be as narrow as the available information justifies. The cells in this table with positive frequencies correspond to possible size 'locations'. The midpoints, upper limits or lower limits of the intervals may be used as the design measurements of the size for each location as appropriate. We will use upper limits. Under this choice, the design measurements for size 1 in the following table are (length, width) = (50, 8), those for size 2 are (50, 10), and so on. Let m be the number of possible size locations, i.e., cells with positive frequency. For $t = 1$ to m, let r_t be the proportion of members of the target population in cell t.

If N is the size of the target population, and p_{st} is the probability that a person in cell s will purchase size t if that is the only size available, then the expected number of persons in cell s who will purchase size t if that is the only size offered is Nr_sp_{st}.

For example, for the data in the following table, it can be verified that σ_1, σ_2 standard deviations of length, width are 8.7, 1.6, respectively. Using the formula given above with $p = 1$, the probability that a person in cell 5 will buy size 2 is $\exp[((40 - 50)/8.7) + ((8 - 10)/1.6)] = 0.092$.

Let $c_{st} = r_s p_{st}$, for $s, t = 1$ to m; where p_{st} is computed as above with $p = 1$. So, c_{st} is the expected proportion of buyers in the target population belonging to cell s who will buy if size t is the only one offered. Let $\mathbf{S}$ denote the subset of possible size locations which are actually offered. Then the expected proportion of buyers in the target population will be $z = \sum_{s=1}^{m} \sum_{t \in \mathbf{S}} c_{st}$, and $\mathbf{S}$ should be selected so as to maximize this function.

The following table gives the frequency distribution for two control dimensions.

x_1 = length (cm)	x_2 = width (cm)				
	2 - 4	4 - 6	6 - 8	8 - 10	10 - 12
40 - 50			(1)	(2)	(3)
			4	6	3
30 - 40		(4)	(5)	(6)	(7)
		4	20	10	1
20 - 30		(8)	(9)	(10)	
		10	20	4	
10 - 20	(11)	(12)	(13)		
	2	3	6		

No. in brackets is size no. Other no. is cell frequency

There are 13 cells with positive frequencies. N = population size = sum of frequencies = 93. r_1 = the proportion in size number 1 = 4/93, etc. Using this data, and the formulae explained above, the following matrix (c_{st}) is computed.

				c_{st} for $t =$			
	1	2	3	4	5	6	7
$s = 1$	.0430	.0124	.0036	0	0	0	0
2	0	.0645	.0187	0	0	0	0
3	0	0	.0323	0	0	0	0
4	.0039	.0011	.0003	.0430	.0124	.0036	.0010
5	.0681	.0197	.0057	0	.2151	.0622	.0180
6	0	.0341	.0099	0	0	.1075	.0311
7	0	0	.0034	0	0	0	.0108
8	.0031	.0009	.0003	.0341	.0099	.0029	.0008
9	.0216	.0062	.0018	0	.0681	.0197	.0057
10	0	.0136	.0039	0	0	.0136	.0039
11	.0001	0	0	.0006	.0002	.0001	0
12	.0003	.0001	0	.0032	.0009	.0003	.0001
13	.0021	.0006	.0002	0	.0065	.0019	.0005

			c_{st} for $t =$			
	8	9	10	11	12	13
$s = 1$	0	0	0	0	0	0
2	0	0	0	0	0	0
3	0	0	0	0	0	0
4	0	0	0	0	0	0
5	0	0	0	0	0	0
6	0	0	0	0	0	0
7	0	0	0	0	0	0
8	.1075	.0311	.0090	0	0	0
9	0	.2151	.0622	0	0	0
10	0	0	.0430	0	0	0
11	.0020	.0005	.0002	.0215	.0062	.0018
12	.0102	.0030	.0009	0	.0323	.0092
13	0	.0204	.0059	0	0	.0645

It has been decided to manufacture at most 6 different sizes among the possible 13. Formulate the problem of selecting the subset of sizes to be manufactured to maximize z. ([P. Tryfos, October 1986]).

9.26 There are m sources for a material, with $b_i > 0$ as the capacity of source i, $i = 1$ to m. There are n users, with d_j as the demand for the material at user j. For each $i = 1$ to m, a fixed charge of $\$f_i$ is incurred if the ith source is assigned one or more users. c_{ij} is the variable cost of assigning user j to source i once the fixed charge for assigning any user to it is paid. It is required to find an assignment of each user to a source (a source may be assigned any number of users, subject to its capacity), so as to meet the demands of all the users at minimum total cost (= sum of fixed charges and the variable costs of assigning users to sources). Formulate this problem. ([A. W. Neebe and M. R. Rao, November 1983]).

9.27 There are m customers with d_i being the expected demand of customer i over a planning horizon which has to be met. There are n sites where plants can be set up to manufacture the product to meet this demand. Here is some data.

$$
\begin{aligned}
f_j &= \text{fixed cost of setting up a plant at site } j, \; j = 1 \text{ to } n. \\
k_j &= \text{capacity (over the planning horizon) of a plant located in site } j. \\
c_{ij} &= \text{cost of supplying all of the demand of customer } i \text{ from a plant located} \\
&\quad \text{site } j. \\
P_L, P_U &= \text{lower and upper bounds on the number of plants to be opened.}
\end{aligned}
$$

Also, there are subsets $N_1, \ldots, N_r$ of sites specified, and ℓ_t, u_t are specified lower and upper bounds on the number of plants to be opened among the sites in the set N_t for $t = 1$ to r. Defining the decision variables to be

$$
\begin{aligned}
x_{ij} &= \text{fraction of demand of customer } i \text{ supplied from a plant located at site } j. \\
y_j &= 1 \text{ if a plant is set up at site } j, \quad 0 \text{ otherwise.}
\end{aligned}
$$

formulate the problem of determining the best locations for the plants and the optimum supply pattern to meet all the customers demands at minimum total cost (the total cost is the sum of the supplying costs and the fixed costs of setting up the plants). ([R. Sridharan, July 1991]).

9.28 Equitable Distribution of Assets   There are n assets with the value of the ith asset being $\$a_i$ for $i = 1$ to n. $A = \sum_{i=1}^{n} a_i$ is the total value of all the assets. It is required to allot these assets to two beneficiaries in an equitable manner. Assets are indivisible, i.e., each asset has to be given completely to one beneficiary or the other, but cannot be split. Let A_1, A_2 be the total value of assets allotted to beneficiary 1, 2 respectively. It is required to distribute the assets in such a way that the difference between A_1 and A_2 is as small as possible. Formulate the problem of finding such a distribution as an integer program. Give this formulation for the numerical example with data $n = 10$, and $(a_i) = (14, 76, 46, 54, 22, 5, 68, 68, 94, 39)$. Do the same for the problem with data $n = 10$, and $(a_i) = (8, 12, 117, 148, 2, 85, 15, 92, 152, 130)$. Solve both these problems using an available integer programming package. ([H. M. Weingartner and B. Gavish, May 1993]

9.29 There are n assets with the value of the ith asset being $\$a_i$, $i = 1$ to n. These assets have to be distributed to two beneficiaries in such a way that the first beneficiary gets a fraction f of the total value of all the assets as closely as possible. The assets are indivisible, i.e., each asset has to be given to one beneficiary or the other and cannot be split. Formulate this as an integer program. Obtain an optimum solution for the problem with data $n = 10$, $f = 0.7$, and $(a_i) = (8, 12, 117, 148, 2, 85, 15, 92, 152, 130)$.)[H. M. Weingartner and B. Gavish, May 1993]).

9.30 There are 10 customers for a product and 9 potential locations where facilities for manufacturing it can be established. In the following table d_i = expected demand of customer i for the product (in units) over the lifetime of the facilities, k_j = expected production capacity of a facility if established in location j, c_{ij} = cost of transporting the product (per unit) to customer i from a facility established at location j, and f_j = cost of establishing a manufacturing facility at location j.

Cust. i	c_{ij} for $j =$									d_i
	1	2	3	4	5	6	7	8	9	
1	15	16	27	28	25	27	27	14	15	28
2	20	13	15	24	13	16	15	15	20	44
3	12	17	25	16	22	15	20	24	26	26
4	25	27	16	23	21	25	26	24	26	31
5	22	10	9	19	18	23	10	22	23	39
6	20	16	24	19	26	24	19	17	20	30
7	17	17	16	25	19	26	14	12	24	43
8	18	20	23	22	28	18	19	17	15	37
9	23	17	16	24	12	25	17	19	22	39
10	14	14	16	20	25	12	23	23	19	30
k_j	46	55	74	68	38	67	52	49	48	
f_j	727	547	674	501	605	482	382	442	606	

It is required to determine in which locations manufacturing facilities should be established, and the shipping pattern from the facilities to the customers, so as to meet the demands at minimum total cost which is = the cost of establishing the facilities + the cost of meeting the demand at the customers from the established facilities. Formulate this as an MIP. How does the formulation change if it is required to establish no more than 5 manufacturing facilities? ([K. Darby-Dowman and H. S. Lewis, November 1988]).

9.31 The Maximal Covering Location Problem There are m stations in a region with the ith station containing a population of a_i customers that need a service, $i = 1$ to m. There are n sites where facilities for providing this service can be located. If a facility is located at site j, it can serve at most k_j customers, $j = 1$ to n. The maximum number of facilities that can be opened is p. The travel distance (or travel time) from station i to site j is d_{ij}, $i = 1$ to m, $j = 1$ to n; and s is the maximum distance (or time) that a customer can travel for service.

(i) Formulate the problem of finding an optimal set of sites to locate the facilities, and allocations of stations to opened facilities, to maximize the total number of customers assigned to a facility within the coverage distance s.

(ii) Same as (i), but take the objective to be to minimize the total travel distance of all the customers who are assigned to a facility outside the covering distance of s from their station.

([H. Pirkul and D. A. Schilling, February 1991]).

9.32 Assigning Students to Field Study Projects   An MBA program requires all students to complete a group field study project during the summer between their first and second years in the program. The administrators of these projects have the conflicting objectives of attempting to satisfy student preferences while maintaining some control over the quality and composition of the resulting teams. In addition they want to balance the foreign/nonforeign student mix across team assignments, and the assignment process to be perceived as fair by the students. Student quality is evaluated by their academic performance during the first year in the program, called the student's class rank.

If T is the total number of available projects, students are asked to indicate their top choices (typically 3 to 5) by assigning a preference score of T to the top choice, $T-1$ to the next choice, and so on. Unranked projects are assigned a score of zero. Values for the following data elements are provided.

p_{ij} = Preference score of student i for project j (higher scores indicate greater preference)

N = Maximum number of students per team

S = Total number of students to be assigned

T = Total number of projects available (= number of teams to be formed)

F = Maximum number of foreign students allowed per team

f_i = 1 if student i is foreign,   0 otherwise

r_i = Class ranking of student i (lower value of rank indicates higher quality)

The overall quality of a team is measured by the sum of the class ranks of its members, with lower values indicating a better team. The overall quality of the weakest team is required to be $\overset{\leq}{=}$ a specified quantity Q.

Formulate the problem of assigning students to projects, giving each student his/her most preferred available project as far as possible, subject to all the constraints mentioned above. Use the decision variables

$$x_{ij} = \begin{cases} 1 & \text{if student } i \text{ assigned to project } j \\ 0 & \text{otherwise} \end{cases}$$

([G. R. Reeves and E. P. Hickman, Sept.-Oct. 1992]).

9.33 Generalized Assignment Problem   We have a set of n jobs, and n agents who can carry out these jobs. c_{ij} is the cost in \$ that agent i charges to perform job j. There is a resource needed to perform these jobs, and r_{ij} is the number of resource units that agent i needs to perform job j. b_i is the maximum number of resource units that are available to agent i.

The generalized assignment problem (GAP) is to find an assignment of jobs to agents, such that all jobs are performed, and the total resource requirements placed on any agent does not exceed its capacity, while minimizing the sum of the costs corresponding to the allocations. Formulate this problem.

9.34 There are three sites 1, 2, 3 for possible location of manufacturing facilities. There are two products P_1, P_2, and two types of manufacturing facilities (L = large with 4000 units/month production capacity of either product, and S = small with 2000 units/month production capacity of either product) that can be opened at any site.

The fixed monthly set-up cost of operating an L-facility (S-facility) at any site is \$4000 (\$2000). Once the fixed monthly set-up cost of operating a facility at a site is paid, there is an additional monthly set-up cost for equipping that facility to manufacture a product, these are given in the following table.

Site	Monthly set-up cost to make	
	P_1	P_2
1	\$1815	2255
2	1975	2015
3	2215	2575

Data on the unit transportation costs for shipping products is given below.

	Shipping cost/unit from site i to j					
	P_1			P_2		
	$j=1$	2	3	1	2	3
$i=1$	1.92	28.8	38.4	6.4	9.6	12.8
2	9.6	6.4	12.8	28.4	25.6	51.2
3	12.8	12.8	6.4	12.8	12.8	6.4

The monthly demand for each product at sites 1, 2, 3, of 500, 1000, 800 units respectively, has to be met. Formulate the problem of determining at which sites facilities have to be set up, of what types, and the shipping patterns, so as to meet the demand at minimum cost (= sum of monthly set-up costs and transportation costs).

9.35 A university library is considering 15 journals as candidates for weeding out of their collection to yield annual subscription savings of \$2000 to meet a proposed budget cut.

The citation counts CO_i in the subject, and CR_i in related area, of a journal J_i refer to the average number of times per year that articles appearing in J_i are referenced in the appropriate scientific literature. The faculty rating R_i of J_i is an average score between 1 and 5 given by the faculty of the university as an indication of the importance of journal J_i, in which the higher the score, the more important the journal is considered to be. The usage data u_i of J_i is the average number of times issues of journal J_i have been removed from the shelf for either borrowing or reading inside the library per quarter, obtained from data collected by the library. The availability rating a_i of journal J_i is an evaluation by the librarian on how easily available this journal is from other libraries; the smaller this rating the easier it is to obtain this journal from other sources. All this data is given below.

Journal	Subscription (\$/year)	CO_i	CR_i	R_i	u_i	a_i
J_1	300	200	110	5	70	2
J_2	220	50	120	5	70	1
J_3	400	400	200	2	90	2
J_4	700	60	80	4	90	2
J_5	350	70	100	3	60	2
J_6	260	160	210	5	90	3
J_7	250	351	152	2	105	2
J_8	360	130	111	1	85	2
J_9	250	85	95	6	70	2
J_{10}	210	70	65	5	40	3
J_{11}	260	215	98	3	65	2
J_{12}	320	45	35	4	45	1
J_{13}	200	66	43	5	50	2
J_{14}	520	130	120	4	70	3
J_{15}	200	312	110	1	90	4

The following goals, targets have been set.

(a) **Budget goal** Reduce the total subscription cost of journals among those subscribed from this list, to \$2000 or less as far as possible.

(b) **Citation count in subject** Keep the average of the citation count in subject, per canceled journal, to $\cong$ 800 as far as possible.

(c) **Citation count in related area** Keep the average of the citation count in related area, per canceled journal, to $\cong$ 500 as far as possible.

(d) **Faculty ratings** Keep the average of the faculty rating, per canceled journal, to $\overset{\leq}{=}$ 3 as far as possible.

(e) **Journal usage** Keep the total usage rate per quarter, of all the journals canceled from this list, at 500 or less as far as possible.

(f) **Availability from other sources** Keep the average of the availability rating, per canceled journal, to $\overset{\leq}{=}$ 2 as far as possible.

Take the penalty coefficient per unit violation of goals (a) to (f) to be 10, 8, 6, 5, 4, 2 respectively in that order. Using this information, formulate the problem of determining which journals to cancel with the goal programming method. ([M. J. Schniederjans and R. Santhanam, 1989]).

9.36 There are n jobs to be processed, and m parallel processors which can be used to process them. Each job takes exactly one unit of time to complete by any processor which can process it. Let

$$a_{ij} = \left\{ \begin{array}{ll} 1 & \text{if processor } i \text{ can handle job } j \\ 0 & \text{otherwise} \end{array} \right.$$

The $m \times n$ matrix $A = (a_{ij})$ is given. An allocation of jobs to processors is said to be *feasible* if every job is assigned to a processor that can execute it. Given a feasible allocation, the workload of a processor used in that allocation is the number of jobs assigned to it. A feasible allocation is called *balanced* if workloads of any pair of processors used in it differ by at most 1. Formulate the problem of finding a balanced allocation. ([Y. L. Chen and Y. H. Chin, 1989]).

9.37 One important job of a departmental chairperson is the semesterly course assignment of teaching staff. In the following table we have data on a department that uses both full-time and part-time faculty to staff their courses. In this example problem, after most of the full-time faculty were assigned their usual courses, there still remained a total of 21 sections open for assignment. There are 8 full-time and 4 part-time faculty available to staff these sections. The involved faculty were asked to rank their preferences in teaching these sections, these ranks ranged from 1 (most desired) to 4 (least desired). These rankings are given in the table below.

It is required to come up with an allocation of these courses to faculty that comes closest to satisfying most of the highest ranked faculty course preferences, considering the departmental need to offer these courses (all courses have one section, except 202 which has two sections), and faculty teaching load requirements. Formulate a goal programming model for this with the following goals.

(a) Each course section is to be offered as far as possible (if a section cannot be staffed, it can be canceled, but we want to minimize such cancellations).

(b) The allocation has to match the available (i.e., remaining) teaching load for each of these faculty members as far as possible.

| Section | Faculty course preference rankings | | | | | | | | | | | |
| | Full-time faculty | | | | | | | | Part-time faculty | | | |
	1	2	3	4	5	6	7	8	9	10	11	12
101		4	4	4	4	4	4		1			
105				4	2			3	1		2	1
120	3	3			1	4			1	1		
202-1		2	2	3	4		4	2		2	1	1
202-2		2	2	3	4		4	2		2	1	1
225		4			4			1				1
245	4	2			4							
275			2			1				2		
280			1							1		
290		1	3	4	2	4	4		2			
310		3	1		3	3	2		2	1		
320		3		2			3				2	
331				1		4						
340	1							2				
350				3		2	2					2
375		2			4	2	4	3			2	
580		3				4		1				
590			2			3						
650					3			3				
670		3			1	4		3				
696					3		1	4				
Available load	1	3	2	2	2	2	2	2	$\leqq 2$	$\leqq 2$	$\leqq 2$	$\leqq 2$

Use 5, 10 as the penalty coefficients for violating goals on (a), (b), in that order. ([M. J. Schniederjans and G. C. Kim, 1987]).

9.38 A map of a northern county of Greece, in which the inside lines represent the road network, follows. The point numbered 1 is the capital of the county, and the other numbered points are towns. The population of each town, and the distance between pairs of them (only when it is less than 20 km) are given in the following tables. d_{ij} is the distance from i to j, the distance matrix (d_{ij}) is symmetric, and the table provides the distance d_{ij} only for $i > j$ if that distance is $\leqq 20$ km.

Figure 9.9

(i)   It is required to set up facilities in some of these towns to serve the inhabitants
of this county. Each town is eligible as a location for a facility to be built,
and each facility constructed can serve any number of people who come for
service there. There is a constraint that each town must be within a 20 km
distance of the nearest facility to it. Formulate the problem of determining
where to locate the facilities so as to meet the above constraint using the
smallest number of facilities.

(ii) If the people in the community of town i are allotted to receive service from
a facility located in town j, the total distance that all of them together have
to travel yearly to receive service can be estimated by a function of the form
$cp_i d_{ij}$ where c is a constant and p_i is the population of town i. Suppose it is
required to locate the facilities so as to minimize the total yearly distance

i	j	d_{ij}	i	j	d_{ij}	i	j	d_{ij}	i	j	d_{ij}
5	1	17	17	7	18	25	11	18	30	23	16
	3	13		9	12		21	8		26	18
6	1	12	17	13	9		23	6	31	1	19
	4	16	18	2	5	26	4	14		6	15
7	1	13		8	10		6	19		9	11
	2	18		16	18		20	7		17	20
8	2	5	19	14	14		23	17		20	9
9	1	7		15	16		25	17		23	14
	3	18	20	1	16	27	2	9		30	9
	6	9		4	14		8	4	32	14	13
10	1	15		6	12		16	4	33	1	10
	3	10	21	3	13		18	14		9	17
	5	16		4	11	28	3	5		13	14
	6	13		11	10		5	8		17	5
	9	8		12	14		9	16		22	19
11	3	3	22	1	9		10	8		30	17
	5	16		3	16		11	8		31	15
	10	13		6	7		12	4	34	4	18
12	3	1		9	2		13	18		20	11
	5	12		10	6		21	18		26	4
	9	19		11	19		22	14		31	17
	10	11		12	17	29	2	20	35	1	18
	11	4		13	13		8	15		17	13
13	1	4		17	14		16	15		33	8
	5	14		20	19		27	11	36	1	15
	6	16	23	4	3	30	1	7		3	11
	7	17		6	19		4	19		5	2
	9	11		20	11		6	7		10	14
	10	11		21	14		9	4		11	14
15	14	12	24	7	18		10	12		12	10
16	2	13		14	8		13	11		13	12
	8	8		15	20		17	12		17	20
17	1	5	25	4	3		20	11		28	6
	6	17		6	19		22	14	37	24	13

traveled by all the people in the county to receive service from these facilities, subject to the constraint mentioned in (i). Formulate this problem.

Total population of community surrounding each town

i	p_i	i	p_i	i	p_i	i	p_i
1	33897	10	1665	19	2316	28	518
2	1316	11	1778	20	1118	29	235
3	1048	12	4442	21	833	30	567
4	1695	13	920	22	600	31	737
5	125	14	3123	23	660	32	1391
6	2476	15	1567	24	6420	33	3783
7	630	16	408	25	668	34	441
8	385	17	3588	26	1344	35	844
9	2608	18	365	27	577	36	2198
						37	1491

i is town i, p_i = population of i

([J. Darzentas and P. Bairaktaris, 1986]).

9.39 A geographical region is divided into n districts with one major city in each. There is demand for a product in this region. A company is planning to establish several plants for manufacturing this product in the region. The plan is to meet the entire demand from the production at these plants. The lifetime of a manufacturing facility is estimated to be 20 years, which is taken as the planning horizon. For i, j = 1 to n we have the following data.

a_i = Demand for product in district i, most of it in the major city in the district, over the planning horizon

d_{ij} = Distance between districts i and j (i.e., the distance between the major cities in these districts)

f_i = Cost of establishing a facility to manufacture the product in district i

$c > 0$ = Shipping cost per unit per unit distance

For i, j = 1 to n, define the decision variable x_{ij} to be 0 if a plant is not located in district j, or the fraction of the demand in district i met from the production at the plant located in district j otherwise.

Using these decision variables, formulate the problem of determining where to locate the plants, and an optimum shipping pattern, so as to meet all the demand at minimum cost (cost includes the expense of establishing the plants and the shipping costs over the planning horizon).

9.40 A construction company has 6 projects to be scheduled over the next 24 months. The duration of each project is known, and each needs both skilled and unskilled workmen over its duration and yields a known profit when it is completed.

The company has only 15 workmen during the first 12 months, which will go up to 20 in the 2nd year being augmented by 5 who will complete training by then. Unskilled workmen can be hired whenever the need arises, but the company wants to keep the number of unskilled workmen on their payroll to a maximum of 60. Once a project is started, work on it has to continue for its duration until it is finished. It is required to determine in which month of the planning horizon each project is to be started, so as to maximize the present value of expected profit (compute present value with interest rate of 8%). Formulate this problem. The following tables provide data on the number of skilled and unskilled workmen needed by each project in each month of its execution.

Project 1: 6 months, profit $100,000

Month	1	2	3	4	5	6
Skilled	6	8	8	8	8	6
Unskilled	30	30	30	20	20	20

Project 2: 9 months, profit $90,000

Month	1	2	3	4	5	6	7	8	9
Skilled	6	6	6	6	7	7	7	7	7
Unskilled	15	15	15	15	15	10	10	10	10

Project 3: 12 months, profit $120,000

Month	1	2	3	4	5	6	7	8	9	10	11	12
Skilled	2	2	2	3	3	3	4	4	4	5	5	5
Unskilled	40	40	40	40	50	50	50	30	30	30	30	30

Project 4: 10 months, profit $70,000

Month	1	2	3	4	5	6	7	8	9	10
Skilled	3	3	3	3	3	3	3	3	3	3
Unskilled	15	15	15	15	15	15	15	15	15	15

Project 5: 7 months, profit $90,000

Month	1	2	3	4	5	6	7
Skilled	8	8	8	8	6	6	6
Unskilled	4	4	4	8	8	8	8

Project 6: 5 months, profit $50,000

Month	1	2	3	4	5
Skilled	2	2	2	3	3
Unskilled	10	10	10	10	10

9.41 There are 5 projects being considered for approval. The following table presents the data on AR = expected annual return, FI = investment needed in first year, WC = working capital expenses, and SE = expected safety and accident expenses, on each project in some money units.

Project	AR	FI	WC	SE
1	49.3	150	105	1.09
2	39.5	120	83	1.64
3	52.6	90	92	0.95
4	35.7	20	47	0.37
5	38.2	80	54	0.44
Constraint on total	≥ 100	≤ 250	≤ 300	≤ 3.8

Formulate the problem of determining which projects to approve to maximize the expected annual return from the approved projects, subject to the constraints mentioned above.

9.42 There are 7 projects which are being considered for approval. Some projects can be approved only if other specified projects are also approved, as explained in the following table. Each project results in a profit or cost as indicated in the table. Formulate the problem of determining which projects to approve, so as to maximize the total net profit.

Project	Other projects that must be approved if this is	Profit or cost of this project
1	2	$10 m. profit
2		$8 m. cost
3	1, 5	$2 m. profit
4	2, 6	$ 4 m. profit
5		$5 m. cost
6		$ 3 m. profit
7	3	$ 2 m. profit

([H. P. Williams, Feb. 1982]).

9.43 There are 5 project proposals. If any of these proposals is accepted, the amount of investment money granted for its implementation must be 1, 2, 3, or 4 units (the money unit in this problem is $100,000). Rejecting a proposal is equivalent to accepting it with an investment grant of 0 units for its implementation. Data on the managing costs (MC), the expected annual returns (EAR), and the % interest expenses on investment (interest %) on each project are tabulated below.

Project	MC and EAR if amount granted is								Interest %
	1		2		3		4		
	MC	EAR	MC	EAR	MC	EAR	MC	EAR	
1	0.02	0.045	0.04	0.087	0.06	0.117	0.08	0.147	1
2	0.03	0.055	0.054	0.106	0.078	0.146	0.1	0.166	2
3	0.07	0.08	0.13	0.14	0.18	0.18	0.22	0.19	3
4	0.05	0.07	0.1	0.123	0.14	0.163	0.18	0.193	4
5	0.04	0.065	0.075	0.121	0.11	0.171	0.14	0.191	5

The total interest expenses cannot exceed 0.17, and the total managing costs have to be ≤ 0.2 in money units. Formulate the problem of determining which projects to accept, and how much investment money to grant to each of the accepted projects, so as to maximize the total expected annual return from the accepted projects, subject to these constraints, as an integer program with the smallest possible number of constraints. Find an optimum solution to the problem.

9.44 A public utility is divided into 6 regions. It has 10 power generation contracts to allocate to its regions as cheaply as possible. We have the following data.

c_{ij} = Cost per unit of meeting the requirement in contract j from power generated in region i

a_i = Capacity of region i to supply power

r_j = Total power requirement in contract j

ℓ_i = Minimum number of units of power that region i has to supply if it is selected to supply any power at all for these contracts

For $i = 1$ to 6, it is possible for region i not to supply any power towards these contracts, but if it is selected to supply some, the amount supplied by it has to be between ℓ_i and a_i. And for reliability reasons, no contract may be placed exclusively with only one region.

Formulate the problem of determining how much power to allocate from each region to each contract, so as to meet all the requirements at minimum cost subject to all the constraints. ([R. W. Ashford and R. C. Daniel, May 1992]).

9.45 There are n terminals to be connected to a central processor through m possible concentrator sites in a computer network. We have the following data.

k = Maximum number of terminals a concentrator can support

c_{ij} = Cost of connecting terminal i to concentrator j

f_j = Cost of placing a concentrator at site j, including the connection cost to the central site

Formulate the problem of determining at which sites concentrators are to be located, and the terminals to be connected to each concentrator, so as to make the connections at minimum total cost.

9.46 An investment firm is faced with a pool of 20 capital investment projects,

Project	Net cash flow in period									
	1	2	3	4	5	6	7	8	9	10
1	−50	−50	15	25	15	15	30	20	10	0
2	−80	−35	−35	40	70	70	70	0	0	0
3	−70	−60	−60	45	45	50	50	50	50	50
4	−35	−45	25	25	25	30	30	0	0	0
5	−50	−30	−30	0	0	0	40	20	20	20
6	−5	−45	−10	30	15	15	15	20	20	0
7	−100	−30	20	35	35	40	20	10	10	10
8	−50	10	−40	20	20	0	0	0	0	0
9	−20	−30	−40	−15	20	25	30	35	40	45
10	−55	0	5	35	15	10	0	0	0	0
11	−40	−10	25	10	10	15	15	10	5	5
12	−30	−50	−20	25	25	25	20	0	0	0
13	0	−20	−25	10	20	30	30	10	0	0
14	−60	−50	−20	−25	40	50	60	0	0	0
15	−10	−25	−20	20	25	35	20	0	0	0
16	−100	−20	15	20	20	30	0	0	0	0
17	−30	5	−10	15	15	15	15	15	15	15
18	−55	−10	10	15	20	20	20	0	0	0
19	−55	5	10	30	30	0	0	0	0	0
20	−75	−50	−30	20	30	40	40	20	20	20

and has as its constraints a limited capital budget of 600, 450, and 80 for periods 1, 2, 3, respectively; and target cash flow requirements of 240, 300, 320, 350, 130, 125, and 120 for periods 4 to 10, respectively. Formulate the problem of selecting the projects to invest in to maximize the present value of total return over the 10 year planning horizon with 8% as the interest rate for money per period, subject to these constraints.

9.47 A large bank has accounts information stored on a network of interconnected processors. There are m bank branches and n processors under this bank. Each branch has to be assigned to a processor. We are given the following data.

$$
\begin{aligned}
p_i &= \text{Processing demand from branch } i \\
v_i &= \text{Storage space demand from branch } i \\
t_i &= \text{Communication capacity needed by branch } i \\
P_j, V_j, T_j &= \text{Processing capacity, storage capacity, and communication capacity} \\
&\quad\ \text{of processor } j \\
k_j &= \text{Maximum number of branches that processor } j \text{ can handle} \\
c_{ij} &= \text{A measure of the communication cost of assigning branch } i \text{ to pro-} \\
&\quad\ \text{cessor } j
\end{aligned}
$$

Formulate the problem of finding a minimum cost assignment of branches to processors satisfying all the constraints.

9.48 Metal Ingot Production   A steel company has to fill orders for 4 types of ingots. For $i = 1$ to 4, r_i is the number of ingots of type i to be delivered, this and other data is given below.

Ingot type	1	2	3	4
Weight (tons)	7	11	15	23
r_i	53	84	117	243

They smelt the metal in vessels of fixed size of 100 tons, and then cast it into ingots. A vessel of liquid metal which is ready to be poured is called a *heat*, it is cast into as many full ingots as possible that can be made with it, and any leftover metal is poured out and has to be remelted - an expensive operation. That's why leftover metal at the end of pouring a heat is called *wastage* (measured in tons) and the company tries to minimize the total wastage generated.

The company prepares several combinations of ingots of various types that can be poured from a heat. For example, here are two combinations: Combination 1 - 4 ingots of type 4, and 1 ingot of type 1; Combination 2 - 5 ingots of type 3, 1 ingot of type 2, and 2 ingots of type 1. If Combination 1 is poured, the wastage is $100 - 4 \times 23 - 1 \times 7 = 1$ ton. If Combination 2 is poured, the wastage is $100 - 5 \times 15 - 1 \times 11 - 2 \times 7 = 0$.

Generate at least 10 different good combinations, i.e., those in which the wastage is reasonably small.

Construct a model to determine how many heats should be poured for each of the combinations generated by you so that the number of ingots of type i produced is $\geq r_i$ for all i, while minimizing 3(wastage) + (weight of ingots of all types left over after the order is filled). Find an optimum solution for this model using one of the available integer programming software packages.

Discuss how the company's Industrial Engineer should organize the weekly pouring schedule if there are many types of ingots to consider (as many as 50 different types), and the number of ingots of each type to be produced that week becomes known at the beginning of the week. ([R. W. Ashford and R. C. Daniel, May 1992]).

9.49 A firm is faced with a pool of 12 investment projects which are examined over a 5-year horizon, and have as constraints a limited capital budget in periods 1 and 2, as well as a target cash inflow for periods 3, 4, 5. The following table presents the data.

Project	Net cash flow in period				
	1	2	3	4	5
1	−45	−15	40	60	0
2	−90	25	35	70	0
3	−50	0	30	50	0
4	−60	−10	10	60	40
5	−100	−10	25	60	80
6	−40	−35	30	50	30
7	−60	5	10	30	50
8	−80	−15	35	40	60
9	−75	0	−10	50	75
10	−30	−20	−5	40	40
11	−35	−10	−5	30	40
12	−54	−20	−15	50	70
Available budget	500	80			
Target cash inflow			90	290	245

It is required to determine the projects to invest in, to maximize the total return while meeting all the constraints. Formulate this problem.

9.50 A Manpower Planning Problem Data on the number of workers needed by a company each month over a 10-month planning horizon is given below. No understaffing is allowed in any month. The company can recruit in any month to fulfill the staffing needs of that month or later months. There is a fixed cost s_t of recruiting in month t which is independent of the number of people recruited (this consists of the costs of advertising, traveling costs of interviewers, etc.). If a person is hired in a month t_1 for fulfilling the staffing requirements in month t_2 for some $t_2 > t_1$, then he/she has to be paid as surplus staff each month t for $t_1 \leq t \leq t_2$ at the overstaffing rate of c_t. r_t is the number of workers needed in month t. This data is given below.

Month t	1	2	3	4	5	6	7	8	9	10
r_t	79	34	52	61	25	89	56	29	48	34
s_t	728	705	698	714	708	739	695	704	689	712
c_t	15	12	16	14	16	12	15	17	12	13

The total cost consists of the fixed costs of recruiting in the months that recruiting is done, plus the surplus staff costs. Formulate the problem of finding the months to recruit, and the number to recruit in each of those months, so as to satisfy all requirements at minimum cost. ([P. P. Rao, Oct. 1990]).

9.51 A Problem in Transformer Design Transformer coil comprises a series of 'plates' of transformer steel packed together. The closer the 'packing' is to being circular, the better the design. Therefore, the two-dimensional version of the problem may be summarized as follows. Strips of steel of any of 14 widths between 10 and 140 cm (in 10 cm steps) have to be used to pack $\mathbf{S}$ = a semicircular section of a transformer of diameter 150 cm.

For any $i = 1$ to 14, let a_i be the maximum length in cm of a strip of width $10i$ that will fit in $\mathbf{S}$. It can be verified that a_1 to a_{14} are: 74.8, 74.3, 73.4, 72.2, 70.7, 68.7, 66.3, 63.4, 60.0, 55.9, 50.9, 45.0, 27.9, 26.9 cm respectively.

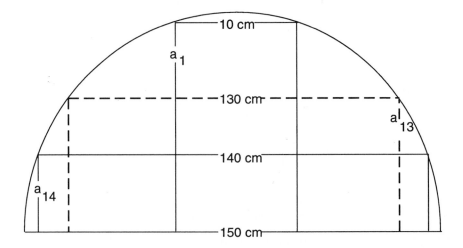

Figure 9.10.

If the maximum length strip of some width, 120 cm say, is packed, and the next smaller width strip packed has width 90 cm say, then the maximum area of $\mathbf{S}$ that is not covered by 120 cm width strip that the 90 cm strip can cover is $90(a_{12} - a_9)$. In this case $(a_{12} - a_9)$ is the maximum length that the strip of width 90 cm can cover that is not already covered by the 120 cm width strip.

It is required to pack strips of available widths in $\mathbf{S}$ to cover as much of its area as possible. However there is a restriction that at most 5 different widths can be used. Define for $i = 1$ to 14,

$$y_i = \begin{cases} 1 & \text{if the strip of width } 10i \text{ is used} \\ 0 & \text{otherwise} \end{cases}$$

x_i = length that strip of width $10i$ covers that is not already covered by a strip of larger width used

Using these decision variables, formulate the problem of finding at most 5 different widths among those available that cover as much area of **S** as possible as an MIP. ([B. Moores, 1986], [J. M. Wilson, Jan. 1988]).

9.52 There are n resources which can be allotted to m activities. For $i = 1$ to m, $j = 1$ to n, we have the following data.

a_j = Positive integer units of resource j available
α_{ij} = Units of output of activity i per unit of resource j allotted to this activity
r_i = Return per unit of output of activity i

Only integer quantities of any resource can be allotted to any activity. Formulate the problem of finding how many units of each resource to allot to each activity to maximize the total return.

9.53 A company uses an injection moulding machine to make 17 different plastic

i	a_i	s_i	h_i	I_i	Demand for component i in month $t =$					
					1	2	3	4	5	6
1	31.4	269	4.0	3000	1152	2235	360	0	5234	985
2	27.2	203	8.6	5000	538	650	8330	3770	1188	575
3	26.2	162	4.1	6000	0	0	5333	1667	7500	0
4	26.2	162	4.1	2000	0	0	6583	0	0	11667
5	26.3	169	4.6	10000	0	11000	0	0	0	3833
6	26.3	169	4.5	6000	0	0	5000	0	5000	0
7	12.7	190	1.4	15000	0	10500	8891	8002	11625	6039
8	12.7	190	1.4	5000	603	10543	9286	4925	6625	10608
9	11.5	207	1.7	40000	14040	38484	48627	17565	26881	14560
10	10.6	224	1.4	10000	951	27761	20064	12396	30397	4483
11	10.6	224	1.5	50000	60	42103	89748	15308	66872	27721
12	9.5	175	1.6	6000	0	6877	13615	648	14081	1314
13	9.4	175	1.5	15000	0	14369	8658	3047	13402	12302
14	27.3	212	8.5	4000	0	0	0	0	26270	6450
15	27.5	212	8.5	2000	0	0	0	0	0	6895
16	16.5	212	5.6	25000	0	7512	14862	2300	9550	7350
17	16.5	212	5.6	5000	0	0	1575	1280	1650	3045
		k_t			702	702	702	259	700	700

components. The demand for each component over the next 12 month planning horizon is given in the tables above and below. In the above table, a_i = time (in seconds) it takes on the machine to make one unit of component i, s_i = setup cost (in \$) to setup the machine for manufacturing component i (for simplicity, assume that setups can be carried out instantaneously without losing any time on the machine), h_i = cost (in cents) for storing component i per unit from one month to the next, I_i = stock of component i at the beginning of month 1, and k_t = machine time (in hours) available in month t.

The stock of component i at the end of month 12 is required to be $\geqq I_i$. The demand has to be met each month without backlogging. It is required to meet all these constraints at minimum cost (which consists of the cost for setups plus the inventory holding costs incurred during the horizon). Formulate as an MIP the problem of determining when and in what quantities to make each component to achieve these objectives. ([J. Maes and L. V. Wassenhove, Nov. 1988]).

i	Demand for component i in month $t =$					
	7	8	9	10	11	12
1	2220	1200	1152	2430	1728	1152
2	2213	3750	168	4493	3730	6075
3	5000	7333	0	0	5000	1886
4	0	5833	0	17500	3133	0
5	4000	0	0	8333	1100	4840
6	1667	5000	0	0	0	7180
7	5436	9784	10428	9128	9750	9834
8	11300	13187	12413	12452	12530	12375
9	57766	38847	30050	47056	47452	41147
10	28880	41054	17710	29550	45235	27072
11	81695	61088	69474	77415	85525	56687
12	19209	5510	6556	10910	7951	51333
13	18907	11654	4786	8616	26556	11810
14	12694	14560	0	0	29470	0
15	2000	0	4500	1750	0	7830
16	3357	552	2447	2447	7050	31587
17	5166	1150	2988	973	1190	2458
k_t	650	710	700	813	775	757

9.54 A company uses a very expensive machine to produce 11 different grades of plastic. The demand for these products over the next 10-month horizon is given below; it should be satisfied each month without backlogging. To change from the manufacture of one grade to another, the machine needs a set-up which costs 260 money units. Production is measured in units of 100 kg. I_i = the quantity of grade i in stock at the beginning of the first month, and h_i = the cost of storing grade i per unit per month, are given in the following table. The production of all grades put together can be at most 4600 units per month. At the end of the 10th month, the stock of grade i is required to be at least I_i for all i. Formulate as an MIP, the problem of determining when and in what quantities to manufacture each grade so as to minimize the sum of the set-up and inventory holding costs over the horizon, for satisfying all the constraints.

i	I_i	h_i	Demand for grade i in month									
			1	2	3	4	5	6	7	8	9	10
1	300	.25	58	117	110	110	75	75	198	198	198	198
2	150	.25	93	76	60	90	50	50	18	18	18	18
3	300	.25	124	179	109	160	106	145	300	300	300	300
4	100	.24	41	0	40	0	40	0	30	30	30	30
5	450	.23	116	141	161	161	161	161	169	169	169	169
6	1800	.23	578	935	1429	1429	1429	1429	2369	2369	2369	2369
7	1200	.17	313	398	1067	1067	1067	1067	728	728	728	728
8	100	.17	93	93	72	54	72	36	37	37	37	37
9	180	.17	0	103	55	55	55	55	84	84	84	84
10	60	.15	74	0	60	0	0	0	30	30	30	30
11	220	.14	0	0	66	66	66	102	102	102	102	110

([J. Maes and L. V. Wassenhove, Nov. 1988]).

9.55 Segregated Storage Problem   There are m different products to be stored, with $a_i > 0$ being the quantity of product i to be stored in some units. There are n different storage compartments, with $b_j > 0$ being the capacity in units of storage compartment j.

However, in each compartment, at most one product may be stored. A typical problem of this type is the 'silo problem' in which different varieties of grain are to be stored separately in the various compartments of the silo. Other examples are: different types of crude oil in storage tanks, customer orders on trucks with no mixing of different orders on any truck, etc.

We assume that there exists external storage space, available at premium cost, which is capable of storing any and all products. Call this external storage the $(n+1)$st compartment. For $i = 1$ to m, $j = 1$ to $n+1$, let c_{ij} denote the unit cost of storing product i in compartment j.

It is required to store the available quantities of the products in the compartments at minimum cost, subject to the storage capacities of compartments 1 to n, and the constraint that each of the compartments 1 to n can hold at most one product. Formulate this problem. Give this formulation for the numerical example with the following data in which the fourth compartment represents external storage ([A. W. Neebe, Sept. 1987]).

	c_{ij} for $j =$				
Compartment $j =$	1	2	3	4	a_i
Product $i = 1$	20	14	19	24	1
2	15	13	20	22	8
3	18	18	15	22	7
Capacity b_j	3	7	4	16	

9.56 The Army Training Mix Problem   The purpose of this exercise is to develop a mathematical model to assist the Army training developers in combining simulators, other training systems, and field training into least costly training strategies that can keep soldiers at an adequate level of skill; and to provide Army leadership with a basis for resource acquisition decisions.

The *training cycle* is a period of 6 months that is the planning horizon for training decisions. The *unit* is the group (we consider a platoon here, consisting of 16 soldiers) whose training activities over the training cycle are being considered. The purpose of training is to attain proficiency on CMTs (*critical mission tasks* which are those activities that are essential to the successful performance of the unit's assigned mission), which in this sample problem are coded as $P01$ to $P12$.

A *training mode* is an exercise in which the whole unit will participate, focusing on the accomplishment of a specific mission. During a training mode the various participants may be engaged in a variety of activities which train them for a variety of CMTs. Training modes are generated by carefully examining how the available training devices will be used in the training environment, and by considering logical groupings of related tasks with a view to determine how many tasks could be trained during a training session. Thus each mode provides training for a subset of CMTs (for example, in the sample problem presented later, mode D_1 provides training for tasks $P01$, $P02$, $P04$, $P06$, $P07$, $P11$, and $P12$ simultaneously, in each of its sessions). For the sample problem we consider training modes code named as BL, WG, LF, and D_1 to D_6 depending on the code name of the device used by them.

In order to measure the effectiveness of each mode for each CMT, a proficiency scale varying from 0 (no proficiency) to 1 (maximum proficiency) has been developed. On this scale, a *standard proficiency level* of 0.7 is specified for each CMT.

If no training were received on a task during a training cycle, the proficiency level of a well-trained soldier would decay to a base level called ZPM (*zero practice minimum*) for that task. The ZPM is somewhere between 0 and the standard level and depends on the complexity of the task. The ZPM values for all the CMTs were evaluated using responses of subject matter experts.

The goal of training is to make sure that every soldier in the participating unit retains by the end of the cycle a proficiency level $\geqq$ the standard level on every CMT.

For a task, mode pair (t, m) where m provides training for t, let $PR_{t,m}(r)$ denote the expected proficiency on task t retained by an average soldier by the end of the cycle if he/she repeats the training mode m a total of r times during the cycle.

The task, mode pair (t, m) where m trains for t is said to belong to *class 1* if $PR_{t,m}(r) \geqq$ the standard level for t for some number of repetitions $r \leqq 10$ of this mode m (10 is a practical upper bound on the number of times that a training mode can be repeated over the 6-month training cycle by a unit); or to *class 2* if $PR_{t,m}(10) <$ the standard level for t. An example of a class 2 task, mode pair is a training mode based on a simulator for a gunnery task. It does provide some training for this task, but cannot bring the proficiency to the standard level because a simulator cannot train for the important subtask of loading live ammunition into the gun in a gunnery task. Now define

u_{tm} = the smallest number of repetitions of m in a training cycle which will ensure reaching or exceeding the standard proficiency level for task t if (t, m) belongs to class 1; or = 8 if (t, m) belongs to class 2 (we noticed that $PR_{t,m}(r)$ changes very little beyond $r \overset{\geq}{=} 8$ when (t, m) belongs to class 2)

a_{tm} = gain in the proficiency retention by the end of the cycle, from each repetition of mode m in the cycle in the range 0 to u_{tm}

Thus we are assuming that in the range $0 \overset{\leq}{=} r \overset{\leq}{=} u_{tm}$, $PR_{t,m}(r) = a_{tm}r$. If (t, m) belongs to class 1 (class 2), $a_{tm}u_{tm} \overset{\geq}{=} 0.7$ $(a_{tm}u_{tm} < 0.7)$, the standard proficiency level.

t	m	ZPM	C	u_{tm}	a_{tm}	t	m	ZPM	C	u_{tm}	a_{tm}
P01	BL	0.06	1	4	0.16	P05	D_2	0.03	1	3	0.26
P01	WG	0.06	1	3	0.24	P05	D_3	0.03	2	8	0.07
P01	D_1	0.06	1	2	0.33	P06	D_4	0.43	1	2	0.17
P01	LF	0.06	1	2	0.33	P06	D_1	0.43	1	2	0.18
P01	D_2	0.06	1	2	0.33	P06	LF	0.43	1	2	0.19
P01	D_3	0.06	1	3	0.22	P07	D_2	0.43	1	2	0.19
P02	BL	0.36	1	4	0.09	P07	D_1	0.60	1	1	0.14
P02	WG	0.36	2	8	0.03	P07	LF	0.60	1	1	0.14
P02	D_4	0.36	1	2	0.21	P08	D_2	0.60	1	1	0.14
P02	D_5	0.36	1	5	0.07	P08	D_3	0.60	1	3	0.03
P02	D_1	0.36	1	2	0.22	P09	D_5	0.60	1	2	0.08
P02	LF	0.36	1	2	0.23	P09	LF	0.60	1	1	0.14
P02	D_2	0.36	1	2	0.23	P10	D_2	0.60	1	1	0.14
P02	D_3	0.36	2	8	0.03	P10	D_3	0.60	1	5	0.02
P03	BL	0.50	1	2	0.13	P11	D_4	0.24	1	2	0.28
P03	D_4	0.50	1	1	0.20	P11	D_5	0.24	1	5	0.10
P03	D_5	0.50	1	7	0.03	P11	D_1	0.24	1	2	0.28
P04	BL	0.60	1	2	0.07	P11	LF	0.24	1	2	0.28
P04	WG	0.60	1	1	0.11	P11	D_2	0.24	1	2	0.28
P04	D_1	0.60	1	1	0.14	P11	D_3	0.24	2	8	0.04
P04	LF	0.60	1	1	0.14	P12	D_4	0.24	1	2	0.27
P04	D_2	0.60	1	1	0.14	P12	D_5	0.24	1	4	0.12
P04	D_3	0.60	1	1	0.10	P12	D_1	0.24	1	2	0.28
P05	WG	0.03	1	5	0.14	P12	LF	0.24	1	2	0.28
P05	LF	0.03	1	3	0.26	P12	D_2	0.24	1	2	0.28

t = CMT, m = training mode, ZPM is for t, C = class of (t, m)

Proficiency for a task can be acquired by training with any of the available modes associated with it, and we make the additivity assumption that the proficiency retention for a task is the sum of the proficiency retention values from the selected modes which train this task.

For each training mode m we are given the operational cost per session associated with this mode (cost of ammunition, fuel etc.), the total number of sessions of this mode that is expected to be obtained over the lifetime of the device used (for this we assume a 15-year life for the device used, 240 working days per year, and 12 working hours per day), and the prorated equipment procurement cost per session of this mode. Define c_m = cost per session of mode m, this is the sum of the operational cost and the prorated equipment procurement cost. For the sample problem, the value of c_m for modes m = BL, WG, D_1, LF, D_2, D_3, D_4, D_5, D_6 is 0.4, 2.08, 20.54, 19.65, 17.78, 2.82, 17.30, 1.99, 6.89 respectively in units of $1000. The rest of the data for the sample problem relating to all the task, mode pairs (t, m) such that m trains for t, is given above. Define the decision variables to be

x_{tm} = the number of repetitions of mode m that count towards satisfying the proficiency constraint corresponding to CMT t, defined only for pairs (t, m) such that mode m trains for t

y_t = total number of repetitions of mode m per soldier during the cycle

We now explain the distinction between the decision variables x_{tm} and y_m. Consider a mode m_1, and two tasks t_1, t_2 both of which are trained by m_1. Suppose $u_{t_1 m_1} = 3$ and $u_{t_2 m_1} = 7$. Thus 3 repetitions of m_1 are enough to satisfy the proficiency retention constraint for task t_1 (hence we can impose an upper bound of $u_{t_1 m_1} = 3$ on $x_{t_1 m_1}$). But 7 repetitions of this mode are needed to satisfy the proficiency retention constraint for task t_2. If mode m_1 is in the optimum mix, we may have in the optimum solution $x_{t_1 m_1} = 3$, and $x_{t_2 m_1} = 7$. What this says is that even though this mode m_1 is repeated 7 (or more) times, the first 3 of these will satisfy the proficiency retention constraint on task t_1, while those beyond the first 3 are used to satisfy the constraint corresponding to task t_2 or other tasks.

Formulate the problem of determining how many times each of the 8 training modes have to be repeated during the training cycle, to satisfy the proficiency retention constraint on each of the 12 tasks at minimum cost. Solve this model using one of the available IP software packages.

Suppose it is required to train a whole battalion consisting of 12 platoons, up to standard. Also, we know that one copy of device BL, WG, LF, $D_1, D_2, D_3, D_4, D_5, D_6$ can deliver 360, 720, 720, 720, 240, 720, 240, 720, 1440 sessions respectively, of the corresponding modes during a 6-month cycle. Determine how many copies of each device are needed to implement the optimum solution obtained above for training the batallion.

Discuss the changes needed in the model to impose an additional constraint that exactly one of the gunnery modes LF, D_1, D_2 must occur exactly once during the 6-month training cycle when the unit goes to its gunnery density training. ([K. G. Murty, P. Djang, W. Butler, and R. R. Laferriere, 1993]).

9.57 A company is making plans to manufacture two products P_1, P_2. There are 6 operations O_1 to O_6, some are required by both products, others are required by only one product, as explained below. There are three types of machines, M_1, M_2, M_3 each of which can perform some of the operations on one or both of the products as explained below.

Product data

Product	Operations to be performed	Annual Prod. target (units)
P_1	O_1, O_2, O_6, O_5	48,000
P_2	O_2, O_3, O_5, O_4	38,000

Processing time (PT) data (minutes/unit)

Operation	P_1			P_2		
	\multicolumn{6}{c}{PT if carried out on}					
	M_1	M_2	M_3	M_1	M_2	M_3
O_1	5.2	6.2	7.3			
O_2	3.0	3.0		2.5	2.9	
O_3				4.0		5.2
O_4				2.0	2.0	2.4
O_5	2.0		2.2	5.9	7.1	8.1
O_6	7.0	8.2	9.0			

Blank indicates either that operation not needed for product,
or that mc. can't perform operation on product

Each machine will be available to work 1900 hours/year. Machine types M_1, M_2, M_3 cost \$96,000, \$82,000, \$70,000 respectively per copy. The company can buy a non-negative number of copies of each machine type. The problem is to determine how many copies of each machine type to buy, and how much of each product-operation combination to allot to each machine purchased, in order to carry out the yearly workload with minimum investment. Formulate as an MIP. (Y. Bozer).

9.58 MTBE (methyl tert butyle ether) is a popular additive in unleaded gasolines. The Saudi-European petrochemical company IBNZAHR produces 500,000 tons of MTBE per year. The production process is a hydrocarbon flow scheme involving three reactors as explained below.

A mixture of butane and very small quantities of CT (carbon tetrachloride) and H_2 (hydrogen) is fed into Reactor 1 known as the Isobutane reactor, which converts the input mixture into isobutane. The isobutane is fed into Reactor 2 which is a dehydrogenation reactor that converts isobutane into isobutylene. Only part (between 45-50%) of the input isobutane gets converted into isobutylene here, the remaining is recycled through this reactor again. The isobutylene is then mixed with methanol and isobutane, the mixture fed into Reator 3, the MTBE reactor, in which MTBE is produced by methanol-isobutylene synthesis process. Most of the input isobutane remains as it is at the end of processing in Reactor 3, and it is recycled again. In each reactor, besides the main products, some gas byproducts are also produced. These gas byproducts have potential uses, but at the moment the plant does not have the necessary equipment to process them, so they are being released into the atmosphere. These reactor releases are currently air pollutants.

Operating Parameters of Reactor 1

Parameters	Setup 1	Setup 2	Setup 3	Setup 4	Setup 5	Setup 6
Feed rates (tons/hr)						
Butane	67	68	69	70	71	73
CT	0.009	0.0092	0.009	0.009	0.009	0.009
H_2	0.065	0.067	0.069	0.07	0.075	0.066
Temp^{0C}	164	165	166	168	170	169
Pressure (kg/cm^2)	26	26	26	26	26	26
Isobutane output (tons/hr)	66.2	67.08	67.07	68.05	69.4	71.7
Byproduct gases (tons/hr)	0.4	0.42	0.43	0.45	0.6	0.5
Byproduct liquids (tons/hr)	0.4	0.5	0.6	0.7	1.0	0.8
Operating cost (SR/hr)	22,825	23149	23473	23797	24121	24769

SR = Saudi Riyals

Operating Parameters of Reactor 2

Parameters	Setup 1	Setup 2	Setup 3	Setup 4	Setup 5	Setup 6
Feed rate (tons/hr)						
Isobutane	82.4	82.7	86.3	85.6	84.9	81.5
Temp^{0C}	645	640	647	645	633	632
Pressure (kg/cm^2)	30	30	30	30	30	30
Isobutylene output (tons/hr)	42.34	40.23	42.75	40.55	39.62	33.89
Byproduct gases (tons/hr)	0.53	0.40	0.42	0.42	0.33	0.37
Remaining isobutane (tons/hr)	34.76	38.52	39.3	40.81	46.99	43.96
Operating cost (SR/hr)	33,747	33,866	35,284	35,006	34,725	33,399

Remaining isobutane is recycled through Reactor 2 again

Operating Parameters of Reactor 3

Parameters	Setup 1	Setup 2	Setup 3	Setup 4	Setup 5	Setup 6
Feed rates (tons/hr)						
Isobutylene	37	37.5	38	38.5	39	36
Methanol	21	21.2	21.5	22	23	24
Isobutane	34.8	38.5	39.3	40.8	48.6	44.0
Temp^{0C}	65	66	67	68	70	71
Pressure (kg/cm^2)	16	16	16	16	16	16
MTBE output (tons/hr)	57.8	58.4	59.1	60	61	62
Byproduct gases (tons/hr)	0.2	0.3	0.4	0.5	0.6	0.66
Byproduct liquids (tons/hr)	0.1	0.19	0.2	0.7	1.0	0.8
Remaining Isobutane (tons/hr)	34.8	38.9	39.3	40.8	48.6	41.0
Operating cost (SR/hr)	25,813	26,123	26,461	26,857	27,397	26,173

Remaining isobutane is recycled again

Reactor 1 also produces liquid byproducts. They also are useful, but are not being used currently. However, the liquid byproducts are disposed of safely, and hence do not contribute to pollution.

Each reactor can be operated at six different settings. The parameters which determine the reactor settings are the raw materials feed rate, the reactor pressure, and the reactor temperature. These parameters influence the production rates, and the cost of operation, as shown in the tables given above. The cost of operation in each reactor includes the cost of any fresh raw material inputs (i.e., butane, CT, and H_2 in Reactor 1; methanol in Reactor 3), and energy and labor costs.

MTBE sells at the rate of 1332 SR/ton. And the plant has a limit of 0.665 tons/hr of gas byproduct releases from each reactor to limit air pollution.

One setup has to be selected for each reactor. Once the setups are selected, they are fixed, and will not be changed. Formulate the problem of determining the setup at which each reactor has to be operated in order to maximize the plants net daily profit.

How does the formulation change if setups on reactors can be changed easily and quickly, and one is allowed to make any number of changes in setup on each reactor?([S. Duffuaa and N. H. Al-Saggaf, 1992] and [S. Duffuaa, Nov. 1991]).

9.59 An assembly line is being designed for the manufacture of a product. The manufacturing process consists of 9 operations. Some operations have to be finished before others can begin, i.e., there are precedence constraints among the operations. In Figure 9.11, an arc is drawn from node i to node j iff operation i has to be finished before operation j can begin. And the number on node i is the time (in time units) it takes to perform operation i.

The problem is to assign operations to workstations on the assembly line so that the precedence constraints are satisfied (i.e., if there is an arc (i, j) in Figure 9.11, then the workstation to which operation j is assigned has to be the same as, or one that comes later in the assembly line than, the one to which operation i is assigned).

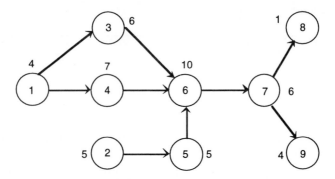

Figure 9.11

Given the assignment of operations to workstations, the workload of any workstation under this assignment is the sum of the time units for performing all the operations assigned to it. The cycle time of this assignment is the maximum workload among the workstations.

Single objective models for assembly line balancing have traditionally concentrated on the minimization of either the cycle time, or the number of workstations.

In this problem, if there is no constraint on cycle time, clearly no more than 5 workstations are needed. For $i = 1$ to 9, $j = 1$ to 5, define the decision variable

$$x_{ij} = \begin{cases} 1 & \text{if operation } i \text{ assigned to workstation } j \\ 0 & \text{otherwise} \end{cases}$$

Formulate a goal programming model to design an assignment of operations to workstations to achieve the following goals.

Goal 1 Limit number of workstations used to no more than 3 as far as possible.

Goal 2 Limit cycle time to no more than 15 as far as possible.

Goal 3 Limit the number of tasks per workstation to no more than 3 as far as possible.

Take the penalty coefficients per unit violation on goals 1, 2, 3 to be 10, 8, 3 respectively in your formulation. ([R. F. Deckro and S. Rangachari, 1990]).

9.60 Derive an integer programming formulation of the problem discussed in Exercise 12.14, for given t.

9.9 References

R. W. ASHFORD, and R. C. DANIEL, May 1992, "Some Lessons in Solving Practical Integer Programs", *Journal of the Operational Research Society*, 43, no. 5, 425-433.

Y. L. CHEN, and Y. H. CHIN, 1989, "Scheduling Unit-time Jobs On Processors With Different Capabilities", *Computers and Operations Research*, 16, no. 5, 409-417.

M. CHESHIRE, K. I. M. McKINNON, and H. P. WILLIAMS, Aug. 1984, "The Efficient Allocation of Private Contractors to Public Works", *Journal of the Operational Research Society*, 35, no. 8, 705-709.

K. DARBY-DOWMAN, and H. S. LEWIS, Nov. 1988, "Lagrangian Relaxation and the Single-source Capacitated Facility-location Problem", *Journal of the Operational Research Society*, 39, no. 11, 1035-1040.

J. DARZENTAS, and P. BAIRAKTARIS, 1986, "On the Set Partitioning Type Formulation for the Discrete Location Problem", *Computers and Operations Research*, 13, no. 6, 671-679.

S. DUFFUAA, Nov. 1991, "A Mathematical Optimization Model for Chemical Production at Saudi Arabian Fertilizer Company", *Applied Mathematical Modeling*, 15 (652-656).

S. DUFFUAA, and N. H. AL-SAGGAF, 1992, "An Integrated Optimization Model for Three Reactors Settings", Tech. report, SE, KFUPM, Dhahran 31261, Saudi Arabia.

A. J. KEOWN; B. W. TAYLOR,III; and J. M. PINKERTON; 1981, "Multiple Objective Capital Budgeting Within the University", *Computers and Operations Research*, 8, 59-70.

J. MAES, and L. V. WASSENHOVE, Nov. 1988, "Multi-item Single-level Capacitated Dynamic Lot-sizing Heuristics: A General Review", *Journal of the Operational Research Society*, 39, no. 11, 991-1004.

J. B. MAZZOLA, and A. W. NEEBE, 1993, "An Algorithm for the Bottleneck Generalized Assignment Problem", *Computers and Operations Research*, 20, no. 4, 355-362.

K. G. MURTY, P. DJANG, W. BUTLER, and R. R. LAFERRIERE, 1993, "The Army Training Mix Model", Tech. report, WSMR, NM.

A. W. NEEBE, Sept. 1987, "An Improved Multiplier Adjustment Procedure for the Segregated Storage Problem", *Journal of the Operational Research Society*, 38, no. 9, 815-825.

A. W. NEEBE, and M. R. RAO, Nov. 1983, "An Algorithm for the Fixed-charge Assigning Users to Sources Problem", *Journal of the Operational Research Society*, 34, no. 11, 1107-1113.

H. PIRKUL, and D. A. SCHILLING, Feb. 1991, "The Maximal Covering Location Problem With Capacities on Total Workload", *Management Science*, 37, no. 2, 233-248.

P. R. RAO, Oct. 1990, "A Dynamic Programming Approach to Determine Optimal Manpower Recruitment Policies", *Journal of the Operational Research Society*, 41, no. 10, 983-988.

G. R. REEVES, and E. P. HICKMAN, Sept.-Oct. 1992, "Assigning MBA Students to Field Study Project Teams: A Multicriteria Approach", *Interfaces*, 22, no. 5, 52-58.

M. J. SCHNIEDERJANS, and G. C. KIM, 1987, "A Goal Programming Model to Optimize Departmental Preference in Course Assignments", *Computers and Operations Research*, 14, no. 2, 87-96.

M. J. SCHNIEDERJANS, and R. SANTHANAM, 1989, "A 0-1 Goal Programming Approach for the Journal Selection and Cancellation Problem", *Computers and Operations Research*, 16, no. 6, 557-565.

R. M. SMULLYAN, 1978, *What Is the Name of this Book?*, Prentice Hall, Englewood Cliffs, NJ.

R. SRIDHARAN, July 1991, "A Lagrangian Heuristic for the Capacitated Plant Location Problem", *Journal of the Operational Research Society*, 42, no. 7, 579-585.

P. TRYFOS, Oct. 1986, "An Integer Programming Approach to the Apparel Sizing Problem", *Journal of the Operational Research Society*, 37, no. 10, 1001-1006.

G. WEBER, March-April 1990, "Puzzle Contests in MS/OR Education", *Interfaces*, 20, no. 2, 72-76.

H. M. WEINGARTNER, and B. GAVISH, May 1993, "How to Settle an Estate", *Management Science*, 39, no. 5, 588-601.

H. P. WILLIAMS, Feb. 1982, "Models With Network Duals", *Journal of the Operational Research Society*, 33, no. 2, 161-169.

Chapter 10

The Branch and Bound Approach

10.1 The Difference Between Linear and Integer Programming Models

The algorithms that we discussed in earlier chapters for linear programs, and some recently developed algorithms such as interior point methods not discussed in this book, are able to solve very large scale LP models arising in real world applications within reasonable times (i.e., within a few hours of time on modern supercomputers for truly large models). This has made linear programming a highly viable practical tool. If a problem can be modeled as an LP with all the data in it available, then we can expect to solve it and use the solution for decision making; given adequate resources such as computer facilities and a good software package, which are becoming very widely available everywhere these days.

Unfortunately, the situation is not that rosy for integer and combinatorial optimization models. The research effort devoted to these areas is substantial, and it has produced very fundamental and elegant theory, but has not delivered algorithms on which practitioners can place faith that exact optimum solutions for large scale models can be obtained within reasonable times.

Certain types of problems, like the knapsack problem, and the traveling salesman problem (TSP), seem easier to handle than others. Knapsack problems involving 10,000 or more $0-1$ variables and TSPs involving a few thousands of cities, have been solved very successfully in at most a few hours of computer time on modern parallel processing supercomputers by implementations of branch and bound methods discussed in this chapter custom-made to solve them using their special structure. But for many other types of problems discussed in Chapter 9, only moderate sized problems may be solvable to optimality within these times by existing techniques. Real world applications normally lead to large scale problems.

When faced with such problems, practitioners usually resort to heuristic methods which may obtain good solutions in general, but cannot guarantee that they will be optimal. We discuss some heuristic methods in Chapter 11.

The main theoretical differences between linear programs, and the discrete optimization problems discussed in Chapter 9 are summarized below.

Linear programs	There are theoretically proven necessary and sufficient optimality conditions which can be used to check efficiently whether a given feasible solution is an optimum solution or not (these are existence of a dual feasible solution that satisfies the complementary slackness optimality conditions together with the given primal feasible solution). These optimality conditions have been used to develop algebraic methods such as the simplex method and other methods for solving LPs.
Discrete and combinatorial optimization problems	For these problems discussed in Chapter 9, there are no known optimality conditions to check whether a given feasible solution is optimal, other than comparing this solution with every other feasible solution implicitly or explicitly. That is why discrete optimization problems are solved by enumerative methods that search for the optimum solution in the set of feasible solutions.

10.2 The Three Main Tools in the Branch and Bound Approach

The total enumeration method presented in Chapter 9 evaluates every feasible solution to the problem and selects the best. This method is fine for solving small problems for which the number of solutions is small. But for most real world applications, the total enumeration method is impractical as the number of solutions to evaluate is very large.

Branch and bound is an approach to search for an optimum feasible solution by doing only a **partial enumeration**. We will use the abbreviation "B&B" for "branch and bound". We will describe the main principles behind the B&B approach using a problem in which an objective function $z(x)$ is to be minimized (as before, a problem in which an objective function $z'(x)$ is to be maximized is handled through the equivalent problem of minimizing $-z'(x)$ subject to the same constraints). Let $\mathbf{K}_0$ denote the set of feasible solutions of the original problem, and z_0 the unknown optimum objective value in it. The main tools that the B&B approach uses to solve this problem are the following.

Branching or Partitioning   In the course of applying the B&B approach, $\mathbf{K}_0$ is partitioned into many simpler subsets. This is what one would do in practice if one is looking, say, for a needle in a haystack. The haystack is big and it is impossible to search all of it simultaneously. So, one divides it visually into approximately its right and left halves, and selects one of the halves to search for the needle first, while keeping the other half aside to be pursued later if necessary. Each subset in the partition of $\mathbf{K}_0$ will be the set of feasible solutions of a problem called a **candidate problem** abbreviated as "**CP**", which is the original problem augmented by additional constraints called **branching constraints** generated by the branching operation. This subset is actually stored by storing the CP, i.e., essentially storing the branching constraints in that CP.

In each stage, one promising subset in the partition is chosen and an effort made to find the best feasible solution from it. If the best feasible solution in that subset is found, or if it is discovered that the subset is empty (which happens if the corresponding CP is infeasible), we say that the associated CP is **fathomed**. If it is not fathomed, that subset may again be partitioned into two or more simpler subsets (this is the **branching operation**) and the same process repeated on them.

Bounding   The B&B approach computes and uses both upper and lower bounds for the optimum objective value.

The upper bound u, of which there is only one at any stage, is always an upper bound for the unknown z_0, the minimum objective value in the original problem. It is always the objective value at some known feasible solution. To find an upper bound one finds a feasible solution $\bar{x}$ (preferably one with an objective value close to the minimum) and takes $z(\bar{x})$ as the upper bound. When there are constraints, it may be difficult to find a feasible solution satisfying them. In that case we will not have an upper bound at the beginning of the algorithm, but the moment a feasible solution is produced in the algorithm we will begin to have an upper bound. At any stage, u, the current upper bound for z_0 is the least among the objective values of all the feasible solutions that turned up in the algorithm so far. The feasible solution whose objective value is the current upper bound u is called **the incumbent** at that stage. Thus the incumbent and the upper bound change whenever a better feasible solution appears during the algorithm.

In contrast to the upper bound of which there is only one at any stage, each candidate problem has its own separate lower bound for the minimum objective value among feasible solutions of that CP. For any CP, it is a number ℓ computed by a procedure called **the lower bounding strategy**, satisfying the property that every feasible solution for this CP has objective value $\geqq \ell$.

Pruning   Suppose we have an upper bound u for the unknown z_0 at some stage. Any CP associated with a lower bound $\ell \geqq u$ has the property that all its feasible solutions have objective value $\geqq u =$ objective value of the current incumbent, so none of them is better than the current incumbent. In this case the algorithm

prunes that CP, i.e., discards its set of feasible solutions from further consideration. Once an incumbent is obtained, if the lower bounding strategy used on each CP produces a **high quality lower bound** (i.e., a lower bound as high as possible, or close to the minimum objective value in this CP), then a lot of pruning may take place, thereby curtailing enumeration.

Thus the bounding step in the B&B approach contributes significantly to the efficiency of the search for an optimum solution of the original problem, particularly if the lower bounding strategy used produces high quality lower bounds without too much computational effort. The lower bounds are used in selecting promising CPs to pursue in the search for the optimum, and in pruning CPs whose set of feasible solutions cannot possibly contain a better solution than the current incumbent. Also, the lower bounding strategy applied on a CP may produce fortuitously the best feasible solution for it, thus fathoming it.

10.3   The Strategies Needed to Apply the Branch and Bound Approach

As before, we consider a problem in which an objective function $z(x)$ is to be minimized subject to a given system of constraints on the decision variables. We denote the set of feasible solutions of the original problem by $\mathbf{K}_0$, and the unknown minimum value of $z(x)$ in the original problem by z_0.

10.3.1   The Lower Bounding Strategy

z_0, the minimum value of $z(x)$, is obtained precisely if the original problem is solved, but it may be hard to solve. The purpose of applying the lower bounding strategy on the original problem is to compute a lower bound for z_0, i.e., a number ℓ satisfying $\ell \overset{\leq}{=} z(x)$ for all feasible solutions x of the original problem. It should be relatively easy to implement and computationally very efficient. Among several lower bounding strategies, the one which gives a bound closest to the minimum objective value without too much computational effort, is likely to make the B&B approach most efficient. Thus in designing a lower bounding strategy, we need to strike a balance between

the quality of the lower bound obtained (the larger the better)

the computational effort involved (the lesser the better)

There are several principles that can be used for constructing lower bounding strategies, but we will only discuss the most important one in this book. It is based on solving a **relaxed problem**.

In the lower bounding strategy based on relaxation, we identify the **hard** or **difficult constraints** in the problem. A subset of constraints is said to be hard if there is an efficient algorithm to solve the remaining problem after deleting these constraints. We select one such set, and relax the constraints in it. The remaining problem is called the **relaxed problem**. Since the relaxed problem has fewer (or less restrictive) constraints than the original, its set of feasible solutions contains $\mathbf{K}_0$ inside it. Hence the minimum objective value in the relaxed problem is a lower bound for the minimum objective value in the original problem.

The importance of computing good bounds for the minimum objective value in a minimization problem, particularly the lower bound, has been discussed in Chapter 5. In the branch and bound approach, computing a good lower bound is an essential component without which the approach degenerates into total enumeration and will be almost impractical.

Let $\bar{x}$ be the optimum solution of the relaxed problem. It is known as the initial **relaxed optimum**. Its objective value, $z(\bar{x})$, is the lower bound for z_0 obtained by the lower bounding strategy. If $\bar{x}$ satisfies the hard constraints that were relaxed, it is feasible to the original problem, and since $z(x) \overset{\geq}{=} z(\bar{x})$ for all feasible solutions x of the original problem, $\bar{x}$ is an optimum solution for the original problem. If this happens we say that the original problem is **fathomed**, and terminate. Otherwise, the algorithm now applies the branching strategy on the original problem.

10.3.2   The Branching Strategy

The branching strategy partitions the set of feasible solutions of the original problem into two or more subsets. Each subset in the partition is the set of feasible solutions of a problem obtained by imposing additional simple constraints called the **branching constraints** on the original problem. These problems are called **candidate problems**.

If there is a $0-1$ variable, x_1 say, in the problem, we can generate two candidate problems by adding the constraint "$x_1 = 0$" to the original problem for one of them, and "$x_1 = 1$" for the other. Clearly the sets of feasible solutions of the two CPs generated are disjoint, and since x_1 is required to be either 0 or 1 in every feasible solution of the original problem, the union of the sets of feasible solutions of the two CPs is $\mathbf{K}_0$. Thus this branching operation partitions $\mathbf{K}_0$ into the sets of feasible solutions of the two CPs generated.

Suppose there is a variable, x_2 say, in the problem which is a nonnegative integer variable, and the value of x_2 in the relaxed optimum is 6.4 say (in general assume it is $\bar{x}_2$). Then we can generate two candidate problems by adding the constraint "$x_2 \overset{\leq}{=} 6$" (in general "$x_2 \overset{\leq}{=} \lfloor \bar{x}_2 \rfloor$") to the original problem for one of them, and "$x_2 \overset{\geq}{=} 7$" (in general "$x_2 \overset{\geq}{=} \lfloor \bar{x}_2 \rfloor + 1$") for the other. Here again, clearly the set of feasible solutions of the two CPs is a partition of $\mathbf{K}_0$.

The variables x_1, x_2 used in the branching operations described above are known as the **branching variables** for those operations.

Now the lower bounding strategy is applied on each CP generated. The constraints relaxed in the CP for lower bounding will always be the hard constraints from the system of constraints in the original problem; the branching constraints in the CP are never relaxed for lower bounding because these constraints are usually simple constraints that can be handled easily by algorithms used to solve the relaxed problem.

$\hat{x}$, the computed optimum solution of the relaxed problem used for getting a lower bound for the minimum objective value in that CP, is known as the **relaxed optimum** for that CP. Then $z(\hat{x})$ is a lower bound for the minimum objective value in that CP. If $\hat{x}$ satisfies all the relaxed constraints (i.e., it is a feasible solution for the CP), then by the argument made earlier, $\hat{x}$ is in fact an optimum solution for this CP, and hence $z(\hat{x})$ is the true minimum objective value in this CP. If this happens we say that this CP is **fathomed**. In general, a CP is said to be fathomed whenever we find a feasible solution $\hat{x}$ for it with objective value equal to the computed lower bound for the minimum objective value in this CP, then $\hat{x}$ is an optimum solution for that CP. If this is the first CP to be fathomed, $\hat{x}$ becomes the **first incumbent**, and $z(\hat{x})$ the current upper bound for the unknown z_0, the minimum objective value in the original problem. If this is not the first CP to be fathomed, $\hat{x}$ replaces the present incumbent to become the new incumbent if $z(\hat{x}) <$ the present upper bound for z_0; and $z(\hat{x})$ becomes the new upper bound for z_0 (this operation is called **updating the incumbent**). Otherwise (i.e., if $z(\hat{x}) \geq$ the present upper bound for z_0), there is no change in the incumbent and the CP is **pruned**. Thus, the **current incumbent** at any stage is the best feasible solution obtained so far, and its objective value is the **current upper bound** for the unknown z_0.

If a candidate problem, P say, is not fathomed, we only have a lower bound for the minimum objective value in it. When this candidate problem P is pursued next, the branching strategy will be applied on it to generate two candidate subproblems such that the following properties hold.

1 Each candidate subproblem is obtained by imposing additional branching constraints on the candidate problem P.

2 The sets of feasible solutions of the candidate subproblems form a partition of the set of feasible solutions of the candidate problem P.

3 The lower bounds for the minimum objective values in the candidate subproblems are as high as possible.

The operation of generating the candidate subproblems is called **branching the candidate problem** P. The candidate problem P is known as the **parent problem** for the candidate subproblems generated; and the candidate subproblems are the **children** of P. The important thing to remember is that a candidate subproblem always has all the constraints in the original problem, and all the branching constraints of its parent, and the branching constraints added to the system by the branching operation which created it. Thus every candidate subproblem inherits all the constraints of its parent. Hence the lower bound for the minimum objective value in any candidate subproblem will be $\geqq$ the lower bound associated with its parent.

When there are several variables which can be selected as the branching variable, it should be selected among them so as to satisfy property 3 above as best as possible to increase the overall efficiency of the algorithm. The branching strategy must provide good selection criteria for branching variables. One way this is done is to compute an estimate for the difference between the lower bound for the most important among the candidate subproblems generated, and the lower bound for the parent, if a particular eligible variable is selected as the branching variable at that stage. An estimate like that is known as an **evaluation coefficient** for that variable. Then the branching variable can be selected to be the variable with the highest evaluation coefficient among the eligible variables.

10.3.3   The Search Strategy

Initially the original problem is the only candidate problem. Begin by applying the lower bounding strategy on it. Let x^0 be the relaxed optimum obtained, and L_0 the lower bound for z_0. If x^0 satisfies the hard constraints that were relaxed, it is an optimum solution for the original problem which is fathomed in this case, and the algorithm terminates. If x^0 violates some of the relaxed constraints, the situation at this stage can be represented as in Figure 10.1.

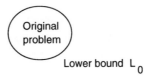

Figure 10.1

Now apply the branching strategy on the original problem, generating candidate problems CP 1 and CP 2. Apply the lower bounding strategy on these CPs and enter them as in Figure 10.2. This diagram is known as the **search tree** at this stage. The original problem has already been branched, and CP 1, CP 2 are its children. Nodes CP 1, CP 2 which have not yet been branched and hence have no children yet, are known as **terminal nodes** in the search tree at this stage.

At any stage of the algorithm, the **list** denotes the collection of all the unfathomed and unpruned CPs which are terminal nodes at that stage.

If either of CP 1, CP 2 is fathomed, the optimum solution in it becomes the incumbent and its objective value the upper bound u for the unknown z_0. There is no reason to pursue a fathomed CP, hence it is deleted from the list. If both CP 1, CP 2 are fathomed, the best of their optimum solutions is an optimum solution of the original problem, and the algorithm terminates.

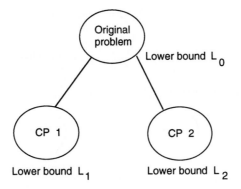

Figure 10.2

Suppose $L_1 \stackrel{\leq}{=} L_2$. If CP 1 is fathomed, then $L_1 =$ current upper bound $=$ objective value of the current incumbent which is the optimum solution for CP 1. In this case the lower bound L_2 for CP 2 is $\stackrel{\geq}{=}$ the objective value of the incumbent, and hence CP 2 is pruned; and the algorithm terminates again with the incumbent as the optimum solution for the original problem.

If CP 2 is fathomed but not CP 1 and $L_1 < L_2$, then we cannot prune CP 1 since it may contain a feasible solution better than the present incumbent. In this case CP 1 joins the list and it will be branched next.

Suppose both CP 1, CP 2 are unfathomed and $L_1 < L_2$. Now there is a possibility that the optimum solution for CP 1 has an objective value $< L_2$, and if so it will be optimal to the original problem. Thus at this stage both CP 1, CP 2 are in the list, but CP 1 is branched next, while CP 2 is left in the list to be pursued later if necessary. Any CP which is not branched yet, not fathomed and not pruned, is known as a **live node** in the search tree at this stage. It is a terminal node which is in the list.

When CP 1 is branched, suppose the candidate subproblems CP 11, CP 12 are generated. The search tree at this stage is shown in Figure 10.3. CP 1 is is no longer a terminal node, so it is deleted from the list.

Now the lower bounding strategy is applied on the new CPs, CP 11, CP 12. Any CP whose relaxed problem is infeasible cannot have any feasible solution, and so is pruned. If any of these CPs is fathomed, update the incumbent and the upper bound for z_0. Whenever a newly generated CP is fathomed, it is never added to the list, the optimum solution in it is used to update the incumbent. Any CP whose lower bound is $\stackrel{\geq}{=}$ the upper bound for z_0, is pruned and taken off the list. Therefore at any stage of the algorithm, the list consists of all the unpruned, unfathomed, and unbranched CPs at that stage. The following properties will hold.

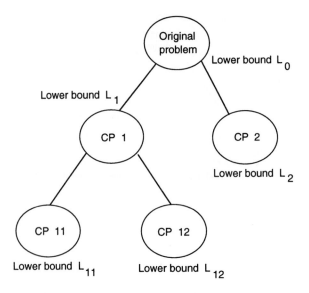

Figure 10.3

(i) The sets of feasible solutions of the CPs in the list are mutually disjoint.

(ii) If there is an incumbent at this stage, any feasible solution of the original problem that is strictly better than the current incumbent is a feasible solution of some CP in the list.

(iii) If there is no incumbent at this stage, the union of the sets of feasible solutions of the CPs in the list is the set of feasible solutions of the original problem.

In a general stage, identify a CP that is associated with the least lower bound among all CPs in the list at this stage. Let this be CP P. Delete P from the list and apply the branching strategy on it. Apply the lower bounding strategy on the candidate subproblems generated. If any of them turn out to be infeasible, prune them. If any of them is fathomed, update the incumbent. If there is a change in the incumbent, look through the list and prune. Add the unpruned and unfathomed among the newly generated candidate subproblems to the list. Then go to the next stage.

At the stage depicted in Figure 10.3, CP 2, CP 11, CP 12 are in the list. If $L_2 = \min\{L_2, L_{11}, L_{12}\}$, then CP 2 is branched next producing CP 21, CP 22 say with lower bounds L_{21}, L_{22} respectively, leading to the search tree in Figure 10.4. The search trees are drawn in this discussion to illustrate how the search is progressing. In practice, the algorithm can be operated with the list of CPs, and the incumbent when it is obtained, and updating these after each stage.

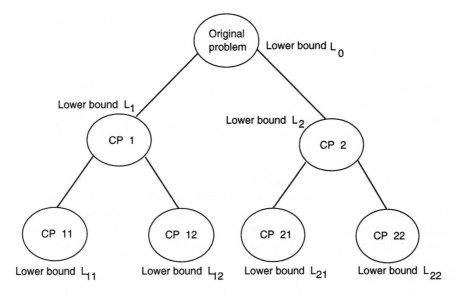

Figure 10.4

In general the search strategy specifies the sequence in which the generated CPs will be branched. We discussed the search strategy which always selects the next CP to be branched, to be the one associated with the least lower bound among all the CPs in the list at that stage. It is a **priority strategy** with the least lower bound as the priority criterion. Some people call it a **jump-track strategy** because it always jumps over the list looking for the node with the least lower bound to branch next. This search strategy with the least lower bound criterion seems to be an excellent strategy that helps to minimize the total number of nodes branched before the termination of the algorithm.

The algorithm terminates when the list of CPs becomes empty. At termination if there is an incumbent, it is an optimum feasible solution of the original problem. If there is no incumbent at termination, the original problem is infeasible.

Another search strategy that is popular in computer science applications is the **backtrack search strategy**. It keeps one of the CPs from the list for the purpose of the search and calls it the **current candidate problem**. The other CPs in the list constitute the **stack**.

If the current CP is fathomed, the incumbent is updated and the current CP is discarded. If the incumbent changes, the necessary pruning is carried out in the stack. Then a CP from the stack is selected as the new current CP, and the algorithm is continued.

If the current CP is not fathomed, the branching strategy is applied on it and the lower bounding strategy applied on the candidate subproblems generated. If both these candidate subproblems are fathomed, the incumbent is updated, pruning is carried out in the stack, and a new current CP is selected from the stack. If only one of the candidate subproblems generated is fathomed, the incumbent is updated, pruning is carried out in the stack, and if the other candidate subproblem is unpruned it is made the new current CP. If neither of the subproblems generated is fathomed, the more promising one among them (may be the one associated with the least lower bound among them) is made the new current CP, the other candidate subproblem is added to the stack and the algorithm is continued.

When the algorithm has to select a CP from the stack, the best selection criteria seems to be that of choosing the most recent CP added to the stack. This selection criterion is called LIFO (Last In First Out).

If this search strategy is employed, the algorithm terminates when it is necessary to select a CP from the stack, and the stack is found empty at that time. If there is an incumbent at that stage, it is an optimum solution of the original problem. If there is no incumbent at that stage, the original problem is infeasible.

In either search strategy, if a CP is sufficiently small that it is practical to search for an optimum solution for it by total enumeration, it is better to do it than to continue branching it further.

For the B&B approach to work well, the bounding strategy must provide a lower bound fairly close to the minimum objective value in the problem but with little computational effort. The branching strategy must generate candidate subproblems that have lower bounds as high as possible.

A well designed B&B algorithm makes it possible to do extensive and effective pruning throughout, and thus enables location of the optimum by examining only a small fraction of the overall set of feasible solutions. That is why B&B methods are known as **partial enumeration methods**.

Practical experience indicates that the search strategy based on the least lower bound allows for extensive pruning, and hence leads to more efficient algorithms.

In most well designed B&B algorithms, it often happens that an optimum feasible solution of the original problem is obtained as an incumbent at an early stage, but the method goes through a lot of computation afterwards to confirm its optimality. That is why a good heuristic to use on large scale problems is to terminate the algorithm when the limit on available computer time is reached, and take the current incumbent as a near optimum solution.

We will now formally state the basic step in the B&B approach to solve a problem. First, a lower bounding strategy, a branching strategy, and a search strategy have to be developed for the problem. If the problem size is large, good lower bounding and branching strategies are very critical to the overall efficiency of the algorithm; and almost always these strategies have to be tailormade for the problem to exploit its special nature, structure and geometry. Once these strategies are developed, the algorithm proceeds as follows.

THE BRANCH AND BOUND ALGORITHM

Initialization   Apply the lower bounding strategy on the original problem and compute a lower bound for the minimum objective value. If the original problem is fathomed, we have an optimum solution, terminate. If the relaxed problem used for lower bounding is infeasible, the original problem is infeasible too, terminate. If neither of these occur, put the original problem in the list and go to the general step.

General Step   If the list has no CPs in it; the original problem is infeasible if there is no incumbent at this stage; otherwise the current incumbent is an optimum solution for it. Terminate.

If the list is nonempty, use the search strategy to retrieve a CP from it for branching next. Apply the branching strategy on the selected CP, and apply the lower bounding strategy on each of the candidate subproblems generated at branching. Prune or discard any of them that turn out to be infeasible; and if any of them are fathomed, update the incumbent and the upper bound for the minimum. Any candidate subproblem, or CP in the list whose lower bound is $\geqq$ the present upper bound is now pruned. Add the unfathomed and unpruned candidate subproblems to the list, and go to the next step.

The application of the B&B approach will now be illustrated with several examples.

10.4   The Traveling Salesman Problem

The B&B approach was developed independently in the context of the traveling salesman problem (TSP) in [K. G. Murty, C. Karel, and J. D. C. Little, 1962], and in the context of integer programming in [A. H. Land and A. G. Doig, 1960]. Particularly, the important concept of bounding is from the former reference. Here we describe the B&B approach for the TSP from [K. G. Murty, C. Karel, and J. D. C. Little, 1962] with lower bounding based on the assignment problem relaxation. As a numerical example we will use the TSP involving 6 cities with the following cost matrix.

	Cost of traveling					
to $j =$	1	2	3	4	5	6
From city $i = 1$	$\times$	27	43	16	30	26
2	7	$\times$	16	1	30	25
3	20	13	$\times$	35	5	0
4	21	16	25	$\times$	18	18
5	12	46	27	48	$\times$	5
6	23	5	5	9	5	$\times$

$$(10.1)$$

Defining

$$x_{ij} = \begin{cases} 1, & \text{if the salesperson goes from city } i \text{ to city } j \text{ in tour } \tau \\ 0, & \text{otherwise} \end{cases}$$

the general TSP of order n with cost matrix $c = (c_{ij})$ is

$$\text{Minimize } z_c(x) \quad = \quad \sum_{i=1}^{n}\sum_{j=1}^{n} c_{ij}x_{ij}$$

$$\text{Subject to } \sum_{j=1}^{n} x_{ij} \quad = \quad 1 \text{ for } i = 1 \text{ to } n \qquad\qquad (10.2)$$

$$\sum_{i=1}^{n} x_{ij} \quad = \quad 1 \text{ for } j = 1 \text{ to } n$$

$$x_{ij} \quad = \quad 0 \text{ or } 1 \text{ for all } i,j$$

$$\text{and } x = (x_{ij}) \qquad \text{is a tour assignment} \qquad\qquad (10.3)$$

As mentioned in Chapter 9, it makes no sense in the TSP to say that the salesman goes from city i to i itself; i.e., all the variables x_{ii} must be $= 0$ in every tour assignment. This is achieved by making the cost coefficients of all the x_{ii} (i.e., the c_{ii}) very large positive numbers that will be denoted by "×" in the cost matrix.

In the actual traveling salesman context we would expect $c_{ij} = c_{ji}$ for all i,j (i.e., the cost matrix c to be symmetric), and to satisfy the **triangle inequality** (i.e., $c_{ik} \leqq c_{ij} + c_{jk}$ for all distinct i,j,k). A TSP in which the cost matrix satisfies these conditions (symmetry and the triangle inequality) is known as a **Euclidean traveling salesman problem**. But in many applications of the TSP in scheduling and other areas, the cost matrix may be symmetric but not satisfy the triangle inequality, or may not even be symmetric. A TSP with cost matrix c is said to be a **symmetric TSP** if c is symmetric; an **asymmetric TSP** otherwise.

The general problem is of course the asymmetric TSP. The B&B method that we discuss can be used to solve any TSP: Euclidean, symmetric, or asymmetric. We have taken the cost matrix in our example problem in (10.1) to be an asymmetric matrix to illustrate the fact that the approach will solve the general problem.

In the TSP (10.2), (10.3); the constraint (10.3) that the assignment must be a tour assignment is a hard constraint. If we relax it, the remaining problem (10.2) is an assignment problem for which we have the efficient Hungarian method discussed in Chapter 5 with a worst case computational complexity of $O(n^3)$. This leads to the following lower bounding and fathoming strategies for the TSP which we will use.

Lower bounding strategy	Relax the constraint that the assignment has to be a tour assignment, and solve the relaxed assignment problem. Minimum objective value in the relaxed assignment problem is a lower bound for the minimum tour cost.

Fathoming   If the relaxed optimum assignment is a tour, it is a mini-
strategy   mum cost tour and the problem is fathomed.

For our example problem with the cost matrix given in (10.1), the relaxed
optimum assignment is

$$x^0 = \{(1,4),(2,1),(3,5),(4,2),(5,6),(6,3)\} \qquad (10.4)$$

with an objective value of 54. There are two subtours in x^0, hence it is not a tour
assignment. The problem is not fathomed and 54 is a lower bound for the minimum
tour cost. The final reduced cost matrix for the relaxed assignment problem at the
termination of the Hungarian method on it is c_0 given below.

Final reduced cost matrix

c_0

$j =$	1	2	3	4	5	6
$i = 1$	×	7	23	0	10	11
2	0	×	11	0	25	25
3	13	8	×	34	0	0
4	3	0	9	×	2	7
5	0	36	17	42	×	0
6	16	0	0	8	0	×

$$(10.5)$$

Denoting by $z_c(x), z_{c_0}(x)$, the costs of a tour assignment x with c, c_0 as cost
matrices respectively, we have from the results discussed in Chapter 5

$$z_c(x) = 54 + z_{c_0}(x) \qquad (10.6)$$

Since the original problem is not fathomed, it has to be branched. We will
branch it using one of the $0-1$ variables x_{ij} as the branching variable.

Details of the Branching Strategy

Suppose the branching variable selected is x_{st}, and the CPs generated are CP
1, 2. CP 1 (CP 2) is the original problem with the branching constraint "$x_{st} = 0$"
("$x_{st} = 1$") as an additional constraint. In CP 1 the cell (s,t) is a forbidden cell as
defined in Chapter 5, i.e., the salesman cannot go from city s to city t in this CP.

Let n denote the number of cities. In CP 2, since x_{st} is constrained to be 1, the
salesman cannot travel from city s to any city other than city t. Since he has only
one choice open to him at city s, it can be verified that the number of feasible tours
for CP 2 is $(n-2)!$. And since the original problem has $(n-1)!$ tours in all, the
number of feasible tours for CP 1 is $(n-1)! - (n-2)! = (n-2)((n-1)!)$. Thus CP
1 has many more tours than CP 2; and hence among CP 1, CP 2, the branching

variable selection strategy should make sure that the lower bound associated with CP 1 becomes as large as possible.

Let c^* $[c_0^*]$ be the matrix obtained from c $[c_0]$ by changing c_{st} $[c_0(s,t)]$ into a very large positive number denoted by "$\times$". Then, if x is any tour assignment feasible to CP 1, we have from (10.6), $z_c(x) = z_{c^*}(x) = 54 + z_{c_0^*}(x)$. So, to get a lower bound for the minimum objective value in CP 1, we can solve the assignment problem with c_0^* as the cost matrix, and add 54 (this is the lower bound for the original problem, which is the parent problem of CP 1) to the minimum objective value in it. Let α_1 be the minimum objective value obtained, and c_1 the final reduced cost matrix, when the assignment problem with c_0^* as the cost matrix is solved by the Hungarian method. Then for any tour assignment feasible to CP 1, we have from (10.6) and the results discussed in Chapter 5

$$z_c(x) = z_{c^*}(x) = 54 + \alpha_1 + z_{c_1}(x) \tag{10.7}$$

Since $c_1 \geq 0$, $z_{c_1}(x) \geq 0$, and hence $54 + \alpha_1$ is a lower bound associated with CP 1. Hence α_1 is the amount by which the lower bound for CP 1 is higher than the lower bound of its parent. Therefore, the branching variable x_{st} should be selected in such a way that $\alpha_1 =$ cost of a minimum cost assignment with c_0^* as the cost matrix (c_0^* is obtained by changing $c_0(s,t)$ in c_0 to a very large positive number) is as large as possible. If (s,t) is not an allocated cell in the relaxed optimum assignment x^0, then all the allocated cells in x^0 have entries of 0 in c_0^*, and hence x^0 continues to be the relaxed optimum assignment associated with CP 1 and α_1 will be 0. So, in order to make α_1 as large as possible, the branching variable is always selected from those whose value is 1 in the relaxed optimum assignment x^0; we consider only these variables as being **eligible to be branching variables**.

Determining the quantity α_1 for each eligible branching variable requires the solution of an assignment problem. Thus selecting the best branching variable among those eligible by actually computing the value of α_1 for each requires too much computational effort. We will now develop a criterion which makes a reasonable selection without too much computation.

Suppose (s,t) is an allocated cell in the relaxed optimum assignment. c_0^* is obtained from c_0 by changing $c_0(s,t)$ into a very large positive number. So, $c_0^* \geq 0$ and every row other than row s, and every column other than column t, contains a 0 entry in c_0^* (in fact all the allocated cells in x^0 other than (s,t) have 0 entries in c_0^*). So, for solving the assignment problem with c_0^* as the cost matrix, the Hungarian method begins by subtracting $\min\{c_0(s,j) : j \neq t\}$ from every entry in row s, and then subtracting $\min\{c_0(i,t) : i \neq s\}$ from every entry in column t, thus obtaining the first reduced cost matrix. In this matrix, the partial assignment containing allocations in all allocated cells other than (s,t) in x^0 has allocations in admissible cells only. So, to get an optimum assignment in this problem, we need at most one allocation change step. And the first total reduction equal to $\beta = \min\{c_0(s,j) : j \neq t\} + \min\{c_0(i,t) : i \neq s\}$ is a lower bound on the cost of the minimum cost assignment in this problem. We define this β to be the **evaluation** of the eligible variable x_{st}. The quantity α_1, the amount by which the lower bound in CP 1 is higher than the lower bound in the original problem, is $\geqq$ this evaluation β of x_{st}. The evaluation is very easy to compute for each eligible cell, and we select the branching cell to be the eligible cell with the highest evaluation.

We will now describe how this branching strategy extends to a general CP obtained during the algorithm. The branching constraints in any CP will be of the form

$$x_{i_1,j_1} = x_{i_2,j_2} = \ldots = x_{i_r,j_r} = 1 \qquad (10.8)$$
$$x_{p_1,q_1} = x_{p_2,q_2} = \ldots = x_{p_u,q_u} = 0$$

The CP is the original problem with these branching constraints as additional constraints. These $r+u$ variables $x_{i_1,j_1}, \ldots, x_{i_r,j_r}, x_{p_1,q_1}, \ldots, x_{p_u,q_u}$ are called **fixed variables** in this CP because their values are fixed by the branching constraints in (10.8) defining this CP. In the same way, every CP obtained in this algorithm will fix a subset of variables at 1, and another subset of variables at 0. Any variable x_{pq}, the allocation of which completes a subtour with the allocations at $x_{i_1,j_1}, \ldots, x_{i_r,j_r}$ must be automatically fixed at 0 in the above constraints. This is done because our aim in the TSP is to find a minimum cost tour assignment, and a tour assignment cannot contain a subtour.

The cost of a minimum cost assignment satisfying the branching constraints in (10.8) is a lower bound w on the cost of every tour feasible for this CP. The branching constraints (10.8) have already fixed the allocations in rows $i_1, \ldots, i_r$ and columns $j_1, \ldots, j_r$ in this CP. So, the cost matrix for the relaxed assignment problem corresponding to this CP is a matrix of order $(n - r) \times (n - r)$ obtained by deleting the rows of cities $i_1, \ldots, i_r$ and the columns of cities $j_1, \ldots, j_r$ from the final reduced cost matrix associated with the parent problem of this CP; and then changing the entries in cells $(p_1, q_1), \ldots, (p_u, q_u)$ to a very large positive number which we denote by "$\times$". This cost matrix is called the **initial cost matrix**, D say, for this CP.

All variables associated with cells (i, j) not in rows $i_1, \ldots, i_r$ or columns $j_1, \ldots, j_r$; and not contained in the set of cells $\{(p_1, q_1), \ldots, (p_u, q_u)\}$ fixed at 0, are called **free variables** in this CP, because they are free to assume values of 0 or 1 in tour assignments feasible to this CP.

Solve the relaxed assignment problem of order $(n-r)$ with the initial cost matrix D for this CP as the cost matrix. The lower bound w for this CP is the sum of the optimum objective value in this relaxed assignment problem, and the lower bound of the parent node of this CP. When you combine the optimum assignment for this relaxed problem, with the fixed allocations in cells $(i_1, j_1), \ldots, (i_r, j_r)$ defined by the branching constraints (10.8) for this CP, we get the relaxed optimum assignment for this CP. If this relaxed optimum assignment is a tour, it is an optimum tour for this CP and this CP is fathomed. In this case, if this is the first CP to be fathomed, store the optimum tour in it as the first incumbent, and its objective value as the

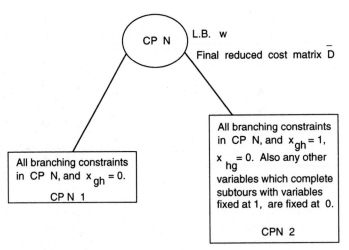

Initial cost matrix is E obtained by changing entry in cell (g, h) in $\bar{D}$ to a very large positive number.

Initial cost matrix is F obtained by striking off row of city g, and column of city h from $\bar{D}$; and changing entry in cell (h, g) and the cells of all other variables fixed at 0, to a very large positive number.

L.B. = w + cost of min. cost assignment with E as cost matrix. Let final reduced cost matrix for this problem be $\bar{E}$.

L.B. = w + cost of min. cost assignment with F as cost matrix. Let final reduced cost matrix for this problem be $\bar{F}$.

R.O.A. is obtained by combining variables fixed at 1 in this CP, with the optimum assignment obtained under the lower bounding strategy.

R.O.A. is obtained by combining variables fixed at 1 in this CP, with the optimum assignment obtained under the lower bounding strategy.

Final reduced cost matrix for CPN 1 is $\bar{E}$.

Final reduced cost matrix for CPN 2 is $\bar{F}$.

Fathomed if R.O.A. is a tour assignment.

Fathomed if R.O.A. is a tour assignment.

Figure 10.5   Details of branching a candidate problem CP N using x_{gh} as the branching variable. R.O.A. is "relaxed optimum assignment for the CP".

upper bound for the minimum objective value in the original problem. If this is not the first CP to be fathomed, use the optimum tour in it to update the incumbent and the upper bound for the minimum objective value in the original problem. If this CP is not fathomed, add it to the list. The final reduced cost matrix of order $(n - r) \times (n - r)$ for the relaxed assignment problem, $\bar{D}$ say, will be used at the time this CP is to be branched, so it may be useful to store this matrix with this CP.

When this CP is retrieved from the list for branching, the branching variable is selected using **evaluations**. Only variables which are not fixed in this CP, and which have allocations in the relaxed optimum assignment in this CP are eligible to be branching variables. For each of these variables (i, j), define its evaluation to be

Evaluation of (i, j) Sum of the minimum in row i and the minimum in column j of $\bar{D}$, the final reduced cost matrix for this CP, after changing $\bar{d}_{ij}$ to a very large positive number.

Among the free cells with an allocation in the relaxed optimum assignment for this CP, the one with the highest evaluation is selected as the branching cell (break ties arbitrarily). Suppose this is cell (g, h). When this CP, CP N say, is branched using x_{gh} as the branching variable, two candidate subproblems CPN 1, CPN 2 are generated as shown in Figure 10.5.

The selection procedure for the branching variable using evaluations given above tries to make the lower bound for CPN 1 as large as possible.

Termination occurs when the list of CPs becomes empty. The incumbent at that stage is an optimum tour.

Solution of the Example Problem

We now continue with the application of the algorithm on our example problem. The search tree at this stage is shown in Figure 10.6.

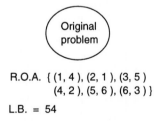

R.O.A. { (1, 4), (2, 1), (3, 5)
        (4, 2), (5, 6), (6, 3) }

L.B. = 54

Figure 10.6

From the final reduced cost matrix c_0 for the original problem given in (10.5), we have the following evaluations for cells corresponding to eligible branching variables.

Cell	(1, 4)	(2, 1)	(3, 5)	(4, 2)	(5, 6)	(6, 3)
Evaluation	7	0	0	2	0	9

So, we select x_{63}, the variable with the highest evaluation, as the branching variable. Branching generates CP 1, CP 2 as shown in Figure 10.7.

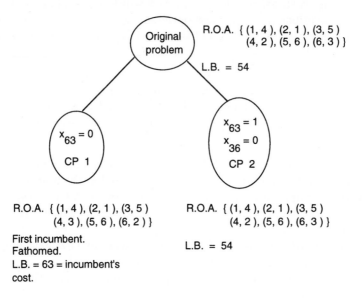

Figure 10.7

The initial cost matrix for CP 1 is obtained by changing the entry in cell (6, 3) in c_0 to "×" indicating a very large positive number. It is c_1 given below.

Initial cost matrix for CP 1

	$j =$	1	2	3	4	5	6
	$i = 1$	×	7	23	0	10	11
	2	0	×	11	0	25	25
c_1	3	13	8	×	34	0	0
	4	3	0	9	×	2	7
	5	0	36	17	42	×	0
	6	16	0	×	8	0	×

An optimum assignment with c_1 as the cost matrix, obtained by the Hungarian method is entered as the R.O.A. for CP 1 in Figure 10.7. Its objective value with c_1 as the cost matrix is 9, so its objective value wrt the original cost matrix c given in (10.1) is 9 + (lower bound of the parent of CP 1) = 9 + 54 = 63. This is a tour assignment, hence it is the first incumbent and CP 1 is fathomed. Its objective value 63 is the present upper bound for the minimum tour cost in the original problem.

In CP 2, x_{63} is fixed at 1, and since (3, 6) completes a subtour with (6, 3), we include "$x_{36} = 0$" as a branching constraint in this CP. The initial cost matrix for this CP is obtained by striking off the row of city 6 and the column of city 3 from c_0 and changing the entry in cell (3, 6) to "$\times$" indicating a very large positive number. It is c_2 given below.

Initial cost matrix for CP 2

	$j =$	1	2	4	5	6
	$i = 1$	$\times$	7	0	10	11
	2	0	$\times$	0	25	25
c_2	3	13	8	34	0	$\times$
	4	3	0	$\times$	2	7
	5	0	36	42	$\times$	0

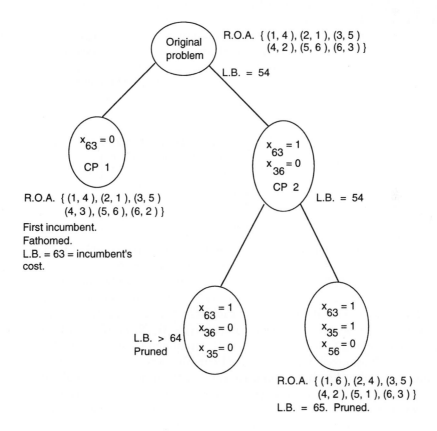

Figure 10.8

An optimum assignment with c_2 as the cost matrix is $\{(1, 4), (2, 1), (3, 5), (4, 2), (5, 6)\}$ with cost 0 wrt the cost matrix c_2. When combined with the variable fixed at 1 in this CP, this leads to the R.O.A. for this CP marked in Figure 10.7; this is the same as the R.O.A. for the original problem. The lower bound for CP 2 is therefore $0 + 54 = 54$. The final reduced cost matrix for CP 2 is c_2 itself.

Now CP 2 is the only CP in the list, and since it is not fathomed we have to branch it. Here are the evaluations for selecting the branching variable for it.

Variable	x_{14}	x_{21}	x_{35}	x_{42}	x_{56}
Evaluation	7	0	10	9	7

We select x_{35} with the highest evaluation as the branching variable for branching CP 2. This generates CP 21, CP 22 as shown in Figure 10.8. Since the evaluation of x_{35} was 10, the lower bound for CP 21 will be $\geqq 10 +$ (lower bound for the parent of CP 21) $= 10 + 54 = 64 > 63 =$ cost of the present incumbent. So, every feasible tour for CP 21 has a cost greater than that of the present incumbent, and therefore CP 21 is pruned.

In CP 22 the branching constraints require $x_{63} = 1$ and $x_{35} = 1$. Since both (5, 3) and (5, 6) complete subtours with (6, 3), (3, 5), we imposed the branching constraints $x_{53} = x_{56} = 0$ in this CP to avoid subtours. The initial cost matrix for this CP is obtained by striking off the row of city 3 and the column of city 5 from c_2 (the final reduced cost matrix of the parent node CP 2) and changing the entry in cell (5, 6) to "×" denoting a very large positive number. It is c_{22} given below.

Initial cost matrix for CP 22

	$j =$	1	2	4	6
	$i = 1$	×	7	0	11
$c_{22} =$	2	0	×	0	25
	4	3	0	×	7
	5	0	36	42	×

The optimum assignment with c_{22} as the cost matrix is $\{(1, 6), (2, 4), (4, 2), (5, 1)\}$ with a cost of 11. Combining this with the variables x_{63}, x_{35} fixed at 1 in this CP, the R.O.A. for this CP marked in Figure 10.8 is obtained; its cost wrt the original cost matrix c is $11 + 54 = 65$; this is the lower bound for CP 22. Since this lower bound is > 63, the cost of the present incumbent, this CP is pruned.

The list is now empty, so the incumbent $\{(1, 4), (2, 1), (3, 5), (4, 3), (5, 6), (6, 2)\}$ is an optimum tour for this TSP with a minimum tour cost of 63.

The B&B approach with the lower bounding strategy based on the relaxed assignment problem has been found to solve TSPs of orders up to about 80 arising in real world applications, in a few minutes on the workstations these days. Many new lower bounding strategies have been developed recently, some of them are computationally quite expensive and need large parallel processing supercomputers to implement, but do produce lower bounds much closer to the minimum objective value. Using them in the B&B approach, TSPs involving thousands of cities have been solved to optimality in a few hours of time on parallel processing supercomputers. With these new lower bounding strategies, the B&B approach has become a practical approach to solve large scale TSPs for those who have access to the necessary computing equipment and the software package.

10.5   The 0−1 Knapsack Problem

We consider the 0−1 knapsack problem in this section. As described in Chapter 9, in this problem there are n objects which can be loaded into a knapsack whose capacity by weight is w_0 weight units. For $j = 1$ to n, object j has weight w_j weight units, and value v_j money units. Only one copy of each object is available to be loaded into the knapsack. None of the objects can be broken; i.e., each object should be either loaded whole into the knapsack, or should be left out. The problem is to decide the subset of objects to be loaded into the knapsack so as to maximize the total value of the objects included, subject to the weight capacity of the knapsack. So, defining for $j = 1$ to n

$$x_j = \begin{cases} 1, & \text{if } j\text{th article is packed into the knapsack} \\ 0, & \text{otherwise} \end{cases} \tag{10.9}$$

the problem is

$$\text{Minimize} \quad z(x) = -\sum_{j=1}^{n} v_j x_j$$

$$\text{subject to} \quad \sum_{j=1}^{n} w_j x_j \overset{\le}{=} w_0 \tag{10.10}$$

$$0 \overset{\le}{=} x_j \overset{\le}{=} 1 \qquad \text{for all} \quad j$$

$$x_j \quad \text{integer} \qquad \text{for all } j \tag{10.11}$$

Here the objective function $z(x)$ is the negative total value of the objects loaded into the knapsack, it states the objective function in minimization form.

If there is an object j such that $w_j > w_0$, it cannot enter the knapsack because its weight exceeds the knapsack's weight capacity. For all such objects j, $x_j = 0$ in every feasible solution of (10.10), (10.11). Identify all such objects and fix all the corresponding variables at 0 and delete them from further consideration. To solve the problem, we need only find the values of the remaining variables x_j satisfying $w_j \overset{\le}{=} w_0$ in an optimum solution.

The remaining problem (10.10) is an LP; and so if we relax the integer requirements (10.11), we can solve the remaining problem by efficient LP methods. The lower bounding strategy based on relaxing the integer requirements on the variables is called the **LP relaxation strategy**. We will use it. Because of its special structure (only one constraint, and all the variables are subject to finite lower and upper bounds), the relaxed LP (10.10) can be solved very efficiently by the following special procedure. The objective value of the optimum solution of the relaxed LP is a lower bound for the minimum objective value in the original problem.

Special Procedure for Solving the LP Relaxation of the 0−1 Knapsack Problem

Suppose the knapsacks weight capacity is w_0 weight units; and there are n objects available for loading into it, with the jth object having weight w_j weight units and value v_j money units, for $j = 1$ to n. In the LP relaxation, variables are allowed to take fractional values. To find the optimum solution of the LP relaxation, first fix all variables x_j corresponding to j satisfying $w_j > w_0$ at 0 and remove them from further consideration. Then compute the **density** (value per unit weight, $d_j = v_j/w_j$ for object j) of each remaining object, and arrange the objects in decreasing order of this density from top to bottom. Begin making $x_j = 1$ from the top in this order until the weight capacity of the knapsack is reached, at that stage make the last variable equal to a fraction until the weight capacity is completely used up; and make all the remaining variables equal to 0.

EXAMPLE 10.1

Consider the journal subscription problem discussed in Section 9.1. The various journals are the objects in it, the subscription price of the journal plays the role of its weight, and the readership of the journal plays the role of its value. Here is the data for the problem from Section 9.1, with the objects arranged in decreasing order of density from top to bottom.

Object j (journal)	Weight w_j (annual subscription)	value v_j (annual readership)	Density v_j/w_j	Cumulative total weight
1	80	7840	98	80
8	99	8316	84	179
4	165	15015	74	344
3	115	8510	74	459
2	95	6175	65	554
5	125	7375	59	679
6	78	1794	23	757
7	69	897	13	826

From Section 9.1 the available budget for annual subscription to these journals,

$670 = w_0$ plays the role of the knapsack's capacity by weight in this example, and all objects have weight $< w_0$. In the last column of the table we provided the cumulative total weight of all the objects from the top and up to (including) that object. We begin loading objects into the knapsack (here it can be interpreted as renewing the subscriptions) from the top. By the time we come to object number 2, \$554 of the knapsack's capacity is used up, leaving $670 - 554 = 116$. The next journal, object 5, has a subscription price of \$125, the money left in the budget at this stage covers only $116/125$ of this journal's subscription. So, the optimum solution of the LP relaxation of this example problem is $\hat{x} = (\hat{x}_1, \hat{x}_8, \hat{x}_4, \hat{x}_3, \hat{x}_2, \hat{x}_5, \hat{x}_6, \hat{x}_7) = (1, 1, 1, 1, 1, 116/125, 0, 0)$. Or, arranging the variables in serial order of subscripts, it is $\hat{x} = (\hat{x}_1, \text{ to } \hat{x}_8) = (1, 1, 1, 1, 116/125, 0, 0, 1)$.

Fathoming Strategy

If the optimum solution, $\hat{x}$, of the LP relaxation is integral (i.e., every variable has a value of 0 or 1 in it), then that solution $\hat{x}$ is an optimum solution of the original 0−1 problem, and thus the original problem is fathomed.

The Branching Strategy

From the procedure described above, it is clear that if the optimum solution for the LP relaxation, $\hat{x}$, is not integral, there will be exactly one variable which has a fractional value in it. Suppose it is $\hat{x}_p$. A convenient branching strategy is to select x_p as the branching variable and generate two CPs, CP 1 (CP 2) by including the branching constraint "$x_p = 0$" ("$x_p = 1$") over those of the original problem. Since $\hat{x}_p$ is fractional, this branching strategy eliminates the present LP relaxed optimum $\hat{x}$ from further consideration as it is not feasible to either CP 1 or CP 2. We will use this branching strategy because it identifies the branching variable unambiguously, and its property of eliminating the current LP relaxed optimum from further consideration is quite nice.

The branching constraints in a general CP, CP N say, in this algorithm will be of the following form.

$$
\begin{aligned}
x_{j_1} = x_{j_2} = \ldots = x_{j_r} &= 0 \\
x_{p_1} = x_{p_2} = \ldots = x_{p_u} &= 1
\end{aligned}
\tag{10.12}
$$

CP N is the original problem with these branching constraints as additional constraints. The $r+u$ variables $x_{j_1}, \ldots, x_{j_r}, x_{p_1}, \ldots, x_{p_u}$ are called **fixed variables** in this CP N because their values are fixed in it by the branching constraints.

In the same way every CP obtained in this algorithm will fix a subset of variables at 0, and another subset of variables at 1. And the sum of the weights of the variables fixed at 1 in any CP will always be $\leqq$ the knapsack's weight capacity, as otherwise the CP will have no feasible solution.

In CP N, objects $p_1, \ldots, p_u$ are required to be included in the knapsack, and objects $j_1, \ldots, j_r$ are required to be excluded from it by the branching constraints. So, we only have $w_0^N = w_0 - (w_{p_1} + \ldots + w_{p_u})$ of the knapsack's weight capacity left to be considered in CP N; and objects in $\mathbf{\Gamma}_N = \{1, \ldots, n\} \backslash \{j_1, \ldots, j_r, p_1, \ldots, p_u\}$ available to load. Any object $j \in \mathbf{\Gamma}_N$ whose weight w_j is $> w_0^N =$ remaining knapsack capacity, cannot be included in the knapsack in this CP; hence the corresponding variable x_j must be fixed at 0 and removed from further consideration in this CP. We assume that these constraints are already included in (10.12).

All variables x_j for $j \in \mathbf{\Gamma}_N$ which are not fixed in CP N, are called **free variables** in this CP, since they are free to assume values of 0 or 1 in feasible solutions of this CP. So, the remaining problem in CP N is a smaller knapsack problem with choice restricted to objects in $\mathbf{\Gamma}_N$ and knapsack's weight capacity equal to w_0^N. It is the following problem.

$$\text{Minimize} \quad - \sum_{j \in \mathbf{\Gamma}_N} v_j x_j$$

$$\text{subject to} \quad \sum_{j \in \mathbf{\Gamma}_N} w_j x_j \;\leqq\; w_0^N \tag{10.13}$$

$$x_j \;=\; 0 \text{ or } 1 \text{ for all } j$$

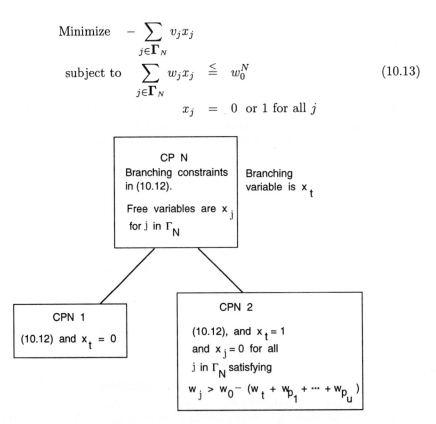

Figure 10.9 Candidate problems generated when CP N is branched using x_t as the branching variable.

So, to get a lower bound for the minimum objective value in CP N, we need to solve the LP relaxation of (10.13), for which the special procedure discussed earlier can be used. When the optimum solution for the LP relaxation of (10.13) is combined with the values of the fixed variables in the branching constraints in (10.12) in this CP N, we get the LP relaxed optimum, $\bar{x}$ say, for CP N; and its objective value is a lower bound for the minimum objective value in CP N. If $\bar{x}$ is integral, it is an optimum solution for CP N, and in this case CP N is fathomed. If this happens, we update the incumbent, prune CP N, and select a new CP from the list to branch next and continue the algorithm. If $\bar{x}$ is not integral, there will be a unique variable which has a fractional value in it, suppose it is x_t. When CP N is to be branched, we will choose x_t as the branching variable. This generates two CPs as shown in Figure 10.9.

Now the lower bounding strategy is applied on each of CPN 1, CPN 2, and the method is continued.

Fathoming a CP With Small Number of Free Variables by Enumeration

Consider CP N defined by the branching constraints (10.12). The number of free variables in it is $s = n - r - u$. The remaining problem in CP N, (10.13), is to decide which of the remaining free objects in Γ_N to load into the remaining part of the knapsack with residual capacity w_0^N. Since $|\Gamma_N| = s$, the optimum solution in this CP can be determined by evaluating each of the 2^s subsets of Γ_N to see which are feasible to (10.13), and selecting the best among those feasible. This becomes practical if s is small. Thus if s is small, we find the optimum solution of CP N by this enumeration instead of continuing to branch it. This is appropriately called **fathoming by enumeration**.

EXAMPLE 10.2

Object j	Weight w_j	Value v_j	Density $d_j = v_j/w_j$
1	3	21	7
2	4	24	6
3	3	12	4
4	21	168	8
5	15	135	9
6	13	26	2
7	16	192	12
8	20	200	10
9	40	800	

For a numerical example we consider a knapsack problem in which the knapsack's weight capacity is 35 weight units. There are 9 objects available for loading into the knapsack with data given above.

Defining the decision variables as in (10.9), here is the problem.

$$\text{Minimize}\ \ z(x) = -21x_1 - 24x_2 - 12x_3 - 168x_4 - 135x_5 -$$
$$26x_6 - 192x_7 - 200x_8 - 800x_9$$
$$\text{subject to}\ \ 3x_1 + 4x_2 + 3x_3 + 3x_4 + 15x_5 +$$
$$13x_6 + 16x_7 + 20x_8 + 40x_9 \ \ \leqq 35\ (10.14)$$
$$0 \leqq x_j \leqq 1\ \ \text{for all } j$$

$$x_j \text{ integer for all } j \qquad\qquad (10.15)$$

We fix $x_9 = 0$ because $w_9 = 40 > w_0 = 35$, and remove object 9 from further consideration. We need to find the values of the remaining variables x_1 to x_8 in an optimum solution with x_9 fixed at 0. The lower bounding strategy relaxes (10.15) and solves the LP relaxation (10.14) with x_9 fixed at 0. The densities of the objects are given in the last column in the above tableau. Using the procedure discussed above we find that the LP relaxed optimum is $x = (x_1 \text{ to } x_9) = (0, 0, 0, 0, 0, 0, 1,$ $19/20, 0)$ with an objective value of -382. Since x_8 is not integral in this solution, the original problem is not fathomed. A lower bound for the minimum objective value in the original problem is -382.

Now the original problem has to be branched. As discussed above, we use the variable x_8 with a fractional value in the LP relaxed optimum as the branching variable. CP 1, CP 2 with branching constraints "$x_8 = 0$", "$x_8 = 1$" respectively are generated. In CP 2 object 8 is already loaded into the knapsack, which leaves only 15 weight units of residual capacity in it. Hence object 7 with a weight of 16 cannot fit into the knapsack in CP 2. Thus in CP 2 "$x_7 = 0$" is an implied branching constraint (the constraint "$x_8 = 1$" implies "$x_7 = 0$" in this problem).

The entire search tree for the algorithm is shown in Figure 10.10. The branching constraints in each CP are recorded inside the node representing that CP. Besides each node the LP relaxed optimum for it is recorded by giving the values of the variables that are nonzero in this solution. The following abbreviations are used: LB = lower bound, BV = branching variable used for branching.

Here is an explanation of the various stages in the algorithm.

CP 2 is fathomed since the relaxed LP optimum for it is integral. This solution $x^1 = (0, 0, 0, 0, 1, 0, 0, 1, 0)^T$ is the first incumbent, and its objective value, $z^1 = -335$, is the present upper bound for the minimum objective value in the original problem.

Now CP 1 is the only CP in the list, so it is branched next. x_4, the fractional variable in its relaxed optimum, is used as the BV. This branching generates CP 3, CP 4. In CP 4 x_4 is fixed at 1, and x_8, x_9 are fixed at 0. So, this is a knapsack problem with residual capacity of $35 - 21 = 14$, and since w_5, w_7 are both > 14, we need to set $x_5 = x_7 = 0$ also as branching constraints in this CP. And since the lower bound for CP 4 , -233, is $>$ present upper bound of -335, it is pruned.

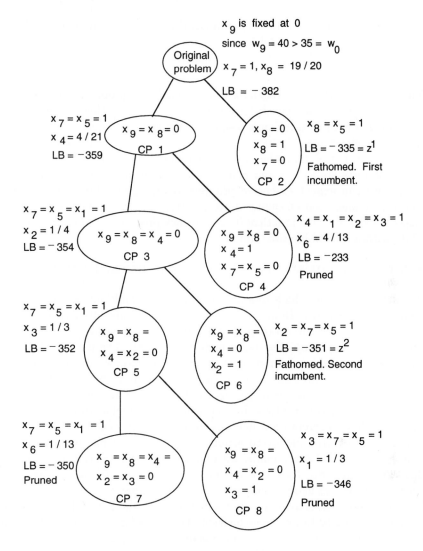

Figure 10.10

Now CP 3 is the only CP in the list, so it is branched next, resulting in CP 5, CP 6. CP 6 is fathomed, and the integral relaxed LP optimum in it, $x^2 = (0, 1, 0, 0, 1, 0, 1, 0, 0)^T$ replaces the present incumbent x^1 as the next incumbent since its objective value $z^2 = -351 < z^1$. z^2 is the new upper bound for the minimum objective value in the original problem.

CP 5, the only CP in the list now is branched next, resulting in CP 7, CP 8. Both these are pruned since their lower bounds are $> z^2$. The list is now empty, so the present incumbent $x^2 = (0, 1, 0, 0, 1, 0, 1, 0, 0)^T$ is an optimum solution for the original knapsack problem. This implies that an optimum choice to load into the knapsack is objects 2, 5, and 7.

This is the basic B&B approach for the $0-1$ knapsack problem. Recently, several simple mathematical tests have been developed to check whether a given CP in this algorithm has a feasible solution whose objective value is strictly better than that of the current incumbent. If one of these tests indicates that a CP cannot have a feasible solution better than the current incumbent, then the CP is pruned right away. These tests are simple and computationally inexpensive. By implementing such tests we can expect extensive pruning to take place during the algorithm, making the enumeration efficient. With a battery of such tests, modern software packages are able to solve practical $0-1$ knapsack problems involving thousands of variables within reasonable time limits.

The Greedy Heuristic for the $0-1$ Knapsack Problem

As mentioned above, high quality software is available for solving large scale $0-1$ knapsack problems. However, some practitioners are often reluctant to use such sophisticated techniques to solve their problems, preferring to obtain a near optimum solution by simple heuristic methods instead. The data in their models may not be very reliable, and may contain errors of unknown magnitudes. Or, the true data in the real problem may be subject to random fluctuations, and their model may have been constructed using numbers that represent the best educated guess about their expected values. In such situations, a global optimum solution for the model with the current data may not actually be an optimum solution for the real problem. Investing money to acquire a sophisticated but possibly expensive software package to solve the model with approximate data may not be worthwhile in these situations. So, they reason that it is better to obtain a near optimum solution for the model using a simple heuristic technique.

The most popular among the simple heuristic methods for the $0-1$ knapsack problem is the **greedy heuristic** which selects objects for inclusion in the knapsack using the density as the criterion to be greedy upon. It proceeds this way.

Consider the problem involving n objects with $w_j, v_j, d_j = v_j/w_j$ as the weight, value, density respectively of object j for $j = 1$ to n; and w_0 as the knapsacks's weight capacity. It first sets all x_j for j satisfying $w_j > w_0$ at value 0. Then it arranges the remaining objects in decreasing order of density from top to bottom. Starting from the top it begins to make $x_j = 1$ as it goes down until the weight capacity of the knapsack is reached. At some stage if the next object cannot be included in the knapsack because its weight exceeds the remaining capacity, it makes $x_j = 0$ for that object; then the process continues with the object below it. It terminates when either the knapsack's weight capacity is used up (in this case, x_j is made equal to 0 for all objects below the current one), or when all the objects have been examined in this way in decreasing order of density.

As an example consider the 0−1 knapsack problem involving a knapsack of weight capacity 40 weight units, and 7 objects with the following data.

Object j	Weight w_j	Value v_j	Density $d_j = v_j/w_j$
1	15	225	15
2	26	260	10
3	45	495	11
4	10	80	8
5	16	112	7
6	10	60	6
7	6	30	5

On this problem the greedy method selects the values of the variables in this order: $x_3 = 0$, $x_1 = 1$, $x_2 = 0$, $x_4 = 1$, $x_5 = 0$, $x_6 = 1$, $x_7 = 0$; leading to the solution $(x_1$ to $x_7) = (1, 0, 0, 1, 0, 1, 0)^T$.

The greedy method is not guaranteed to produce an optimum solution in general, but usually produces a solution close to the optimum. A mathematical upper bound for the difference between the value of the greedy solution and that of an optimum solution can be derived. For results on these, see [O. H. Ibarra and C. E. Kim, 1975].

0−1 Knapsack Problems with Flexible Data

In many applications we encounter 0−1 knapsack models in which slight changes in the value of $w_0 = $ the knapsack's weight capacity are entirely permissible. An example of this is the journal subscription problem discussed in Example 10.1. In this model, the knapsack's weight capacity is the budgeted amount of $670 for journal subscriptions. The financial VP will be delighted if the librarian wants to decrease this quantity by any amount; also he may not object to small increases in this quantity. If this quantity can be increased to $679 (a small increase of $9), then the solution $(x_1$ to $x_8) = (1, 1, 1, 1, 1, 0, 0, 1)^T$ becomes feasible (this solution is obtained by selecting journals in decreasing order of density, and uses up the budgeted quantity of $679 exactly) and is an optimum solution for the problem with this modification. Here, it makes sense to argue with the financial VP to agree to this slight modification.

In all such situations where the value of w_0 is flexible, one can look at two solutions to the original problem. One is $\hat{x}$, the solution of the original problem obtained by the greedy heuristic. The other is $\tilde{x}$ obtained by rounding up to 1 the value of the fractional variable in the LP relaxed optimum corresponding to the original problem. If one can increase w_0 to the capacity used by $\tilde{x}$, $\tilde{x}$ is an optimum solution for the modified problem. If w_0 can be decreased to the capacity used by $\hat{x}$, $\hat{x}$ is either optimal or near optimal to the modified problem. The decision makers can look at both $\hat{x}$ and $\tilde{x}$ and decide which solution is more desirable for the real problem, and make the appropriate change. In this situation, this may be the most appropriate way to handle this problem instead of trying to solve the model with the original value for w_0 to optimality using an expensive B&B package.

10.6   B&B Approach for the General MIP

We consider the following general MIP

$$
\begin{aligned}
\text{Minimize} \quad & z(x, y) = cx + dy \\
\text{subject to} \quad & Ax + Dy \;=\; b \qquad\qquad (10.16)\\
& x, y \;\geqq\; 0
\end{aligned}
$$

$$
y \quad \text{integer vector} \qquad\qquad (10.17)
$$

If there are no continuous variables x in the problem, it is a pure IP.

A lower bounding strategy for this problem is to solve the relaxed LP (10.16) obtained by relaxing the integer requirements (10.17), using LP techniques. If the relaxed LP (10.16) is infeasible, the MIP is clearly infeasible too, prune it and terminate. On the other hand if the relaxed LP has an optimum solution suppose it is (x^0, y^0) with an objective value of z^0. If y^0 satisfies the constraints (10.17) that were relaxed, (x^0, y^0) is an optimum solution of the MIP which is now fathomed, terminate. Otherwise, z^0 is a lower bound for the minimum objective value in the MIP.

A convenient branching strategy is to select one of the integer variables y_j whose value y_j^0, in the relaxed LP optimum is noninteger, as the branching variable. If y_j is a 0−1 variable, generate two CPs by imposing one additional constraint "$y_j = 0$" or "$y_j = 1$" on the original MIP. If y_j is a general nonnegative integer variable, generate two CPs by imposing one additional constraint "$y_j \leqq \lfloor y_j^0 \rfloor$" or "$y_j \geqq 1 + \lfloor y_j^0 \rfloor$" respectively on the original MIP.

If there are several j's such that y_j^0 is noninteger, the branching variable is selected from them so as to make the lower bounds for the CPs generated after branching as high as possible. The data in the optimum simplex tableau can be used to get estimates (called **penalties**) of the amount by which the lower bounds for the CPs are greater than the lower bound of their parent. See [G. L. Nemhauser and L. A. Wolsey, 1988] for a discussion of these penalties and their use in selecting the branching variable.

A consequence of selecting the branching variable among integer variables with fractional values in the relaxed LP optimum (x^0, y^0) is that this point (x^0, y^0) is eliminated from further consideration. This is a nice property.

The lower bounds for the newly generated CPs are computed by solving the relaxed LPs obtained by relaxing the integer requirements on the y's in them.

The relaxed LP corresponding to a CP contains just one additional constraint (the new branching constraint in it) over those in the relaxed LP for the parent problem. From the known relaxed LP optimum of the parent problem, a relaxed LP optimum for the CP can be obtained by using very efficient sensitivity analysis techniques (these techniques to handle the addition of a new constraint are not discussed in this book; see [K. G. Murty, 1983 of Chapter 2] for them).

A CP is fathomed when the relaxed LP optimum for it satisfies the integer requirements on the y's. The moment a CP is fathomed in the algorithm, we have an incumbent. Each time a new CP is fathomed, we update the incumbent. The current upper bound for the minimum objective value in the original MIP is always the objective value of the current incumbent. Any CP in the list whose lower bound is $\geqq$ the current upper bound is immediately pruned. Also, if the relaxed LP corresponding to a CP is infeasible, so is that CP, and hence that CP is pruned.

CPs for branching are selected from the list by the least lower bound criterion. The algorithm terminates when the list becomes empty. If there is no incumbent at termination, the original MIP is infeasible. Otherwise, the final incumbent is an optimum solution of the original MIP.

EXAMPLE 10.3

Consider the following MIP

y_1	y_2	x_1	x_2	x_3	x_4	$-z$	b
1	0	0	1	-2	1	0	$3/2$
0	1	0	2	1	-1	0	$5/2$
0	0	1	-1	1	1	0	4
0	0	0	3	4	5	1	-20

$y_1, y_2 \geqq 0$, and integer; x_1 to $x_4 \geqq 0$; z to be minimized

The optimum solution for the relaxed LP obtained by relaxing the integer requirements on y_1, y_2 is $(y^0, x^0) = (3/2, 5/2; 4, 0, 0, 0)$, with an objective value of $z^0 = 20$. Since this solution does not satisfy the integer requirements on y_1, y_2 the MIP is not fathomed. It has to be branched. Both y_1, y_2 have nonintegral values in the relaxed LP optimum solution. We selected y_2 as the branching variable (BV). Branching leads to CP 1, CP 2 shown in Figure 10.11 given below. The constraints inside a node in Figure 10.11 are the additional (branching) constraints in it over those of the original problem. By the side of each node in Figure 10.11 we give the relaxed LP optimum (RO) corresponding to that node

For example, the relaxed LP for CP 2 is the following, where s_1 is the slack variable $= y_2 - 3$ for the branching constraint $y_2 \geqq 3$ in it.

y_1	y_2	x_1	x_2	x_3	x_4	s_1	$-z$	b
1	0	0	1	−2	1	0	0	3/2
0	1	0	2	1	−1	0	0	5/2
0	0	1	−1	1	1	0	0	4
0	1	0	0	0	0	−1	0	3
0	0	0	3	4	5	1	−20	

$$y_1, y_2 \overset{\geq}{=} 0, s_1 \overset{\geq}{=} 0 \; ; \; x_1 \text{ to } x_4 \overset{\geq}{=} 0; \; z \text{ to be minimized}$$

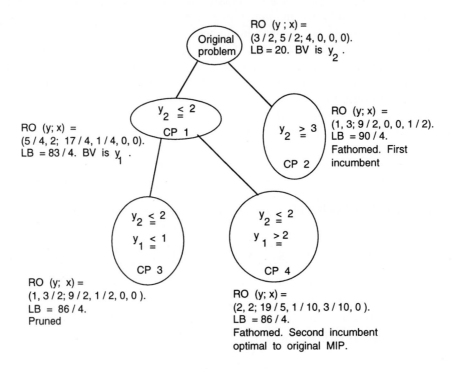

Figure 10.11

The optimum solution of the original MIP is the second incumbent $(y; x) = (2, 2; 19/5, 1/10, 3/10, 0)$, with an objective value of $86/4$.

10.7   B&B Approach for Pure 0−1 IPs

We consider the problem (10.18) where A is of order $m \times n$ and $x \in \mathbb{R}^n$. For some j if c_j is < 0, transform the problem by substituting $x_j = 1 - y_j$. In the transformed problem, the objective coefficient of y_j is > 0. After similar transformations as necessary, we get a problem of the same form as (10.18), but with $c \geqq 0$. In the rest of the section we assume that $c \geqq 0$.

$$
\begin{aligned}
\text{Minimize} \quad & z(x) = cx \\
\text{subject to} \quad & Ax \leqq b \\
& x_j = 0 \text{ or } 1 \text{ for all } j
\end{aligned}
\tag{10.18}
$$

In the B&B algorithm discussed below, CPs are obtained by selecting a subset of the variables x_j and fixing each of them at value 0 or 1. Any variable fixed at 0 (1) is called a **0-variable (1-variable)** in that CP. The 0-variables fixed at value 0, and the 1-variables fixed at value 1, constitute what is known as a **partial solution**. Each CP generated in the algorithm corresponds to a partial solution. Variables that are not fixed at 0 or 1 in a CP are called **free variables** in that CP. Given a partial solution, a **completion of it** is obtained by giving values of 0 or 1 to each of the free variables.

The B&B approaches discussed below for this pure 0−1 IP are called **implicit enumeration methods** in the literature. In general the name **implicit enumeration** is used for the class of B&B algorithms designed specifically for the pure 0−1 IP.

We will first discuss the analysis to be performed on a typical CP. Consider the CP in which $\mathbf{U}_0$, $\mathbf{U}_1$, $\mathbf{U}_f$ are the sets of subscripts of the 0-, 1-variables, and the free variables, respectively. $\mathbf{U}_f = \{1, \ldots, n\} \backslash (\mathbf{U}_0 \cup \mathbf{U}_1)$. Compute the vector $b' = (b'_i) = b - \sum_{j \in \mathbf{U}_1} A_{.j}$. The **fathoming criterion** is $b' \geqq 0$. If $b' \geqq 0$, the completion obtained by giving the value of 0 to all the free variables is optimal to the CP (because of our assumption that $c \geqq 0$), and the optimum objective value in it is $\sum_{j \in \mathbf{U}_1} c_j$.

If $b' \not\geqq 0$, several tests are available to check whether the CP is infeasible (i.e., has no feasible completion) and whether it has a feasible completion better than the current incumbent. Let $\bar{z}$ be the present upper bound for the minimum objective value in the original problem (i.e., the objective value of the current incumbent), or ∞ if there is no incumbent at this stage. For applying these tests on the CP, the system of constraints to be considered is

$$
\sum_{j \in \mathbf{U}_f} a_{ij} x_j \leqq b'_i = b_i - \sum_{j \in \mathbf{U}_1} a_{ij}, i = 1 \text{ to } m
\tag{10.19}
$$

$$
\sum_{j \in \mathbf{U}_f} a_{m+1,j} x_j \leqq b'_{m+1} = \bar{z} - \sum_{j \in \mathbf{U}_1} a_{m+1,j}
\tag{10.20}
$$

$$
x_j = 0 \text{ or } 1 \qquad \text{for all } j \in \mathbf{U}_f
\tag{10.21}
$$

where for notational convenience we denote c_j by $a_{m+1,j}$. The constraint (10.20) is omitted from this system if there is no incumbent at this stage.

Some of the tests examine each of the constraints in (10.19), (10.20) individually to check whether it can be satisfied in $0-1$ variables. For example, one of the tests is the following. In the ith constraint in (10.19), (10.20), if $\sum_{j \in \mathbf{U}_f}(\min\{a_{ij}, 0\}) > b_i'$, obviously it cannot be satisfied; and hence the system (10.19) to (10.21) is infeasible and the CP is pruned.

The tests may also determine that some of the free variables must have a specific value in $\{0, 1\}$ for (10-19)-(10.21) to be feasible. Here is an example of such a test. Suppose there is a $k \in \mathbf{U}_f$ and an i between 1 to $m + 1$ such that $\sum_{j \in \mathbf{U}_f}(\min\{a_{ij}, 0\}) + |a_{ik}| > b_i'$. Then obviously x_k must be 0 if $a_{ik} > 0$, or 1 if $a_{ik} < 0$; for (10.19)-(10.21) to be feasible. If such variables are identified by the tests, they are included in the sets of 0- or 1-variables in this CP, accordingly.

Let $\mu = (\mu_1, \ldots, \mu_{m+1})$ be a nonnegative vector. Any solution satisfying (10.19), (10.20) must obviously satisfy

$$\sum_{i=1}^{m+1} \mu_i \left(\sum_{j \in \mathbf{U}_f} a_{ij} x_j \right) \leqq \sum_{i=1}^{m+1} \mu_i b_i' \tag{10.22}$$

So, if (10.22) does not have a $0-1$ solution, (10.19)-(10.21) must be infeasible and the CP can be pruned. A constraint like this obtained by taking a nonnegative linear combination of constraints in the system (10.19), (10.20), is known as a **surrogate constraint**. For example from the system

$$x_1 - x_2 \leqq -1$$
$$-x_1 + 2x_2 \leqq -1$$

we get the surrogate constraint $x_2 \leqq -2$ by taking the multiplier vector $\mu = (1, 1)$. From this surrogate constraint we clearly see that the system has no $0 - 1$ solution, even though we cannot make this conclusion by considering any one of the two original constraints individually. In the same way, often a surrogate constraint enables us to make some conclusions about the system (10.19)-(10.21), which are not apparent from anyone of the constraints in the system considered individually. If a surrogate constraint has no $0-1$ solution, the system (10.19)-(10.21) is infeasible and the CP is pruned. If some of the free variables must have specific values of 0, 1 in every $0-1$ solution for the surrogate constraint, those free variables must have the same specific values in every $0-1$ solution for the CP that is better than the current incumbent, and hence they are included accordingly in the sets of 0-, 1-variables defining the CP. For a detailed discussion of useful tests, and methods for generating useful surrogate constraints, see [F. Glover, 1968] and [G. L. Nemhauser and L. A. Wolsey, 1988].

Let $\pi = (\pi_1, \ldots, \pi_m)$ be a nonnegative vector. Since every feasible solution of (10.19), (10.21) is a feasible solution for the problem (10.23), $\sum_{j \in \mathbf{U}_1} c_j +$ (minimum objective value in (10.23)) is a lower bound for the minimum objective value in

$$\text{Minimize} \quad \sum_{j \in \mathbf{U}_f} c_j x_j$$

$$\text{subject to} \quad \sum_{i=1}^{m} \pi_i \left(\sum_{j \in \mathbf{U}_f} a_{ij} x_j \right) \; \leqq \; \sum_{i=1}^{m} \pi_i b_i' \qquad (10.23)$$

$$x_j = 0 \text{ or } 1 \text{ for all} \quad j \; \in \; \mathbf{U}_f$$

the CP. (10.23) is a 0—1 IP with a single constraint, and hence can be solved by algorithms discussed for the knapsack problem. By applying a few steps of the knapsack algorithm on (10.23) if we can determine that the lower bound for the minimum objective value in the CP is > the cost of the present incumbent, then the CP can be pruned. It has been proved that the best π-vector to use for forming the problem (10.23) is the negative dual optimum solution associated with the relaxed LP for the CP. See [G. L. Nemhauser and L. A. Wolsey, 1988] for a proof of this result, and other ways of generating and using surrogate constraints effectively.

If all this work determines that (10.19)-(10.21) is infeasible, the CP is pruned. Otherwise let $\mathbf{U}_0', \mathbf{U}_1'$ be the sets of subscripts of the 0- and 1-variables respectively in the CP after augmenting the 0- and 1-variables determined by the tests to $\mathbf{U}_0, \mathbf{U}_1$ respectively. The set of free variables is $\mathbf{U}_f' = \{1, \ldots, n\} \backslash (\mathbf{U}_0' \cup \mathbf{U}_1')$. A lower bound for the minimum objective value in the CP is $\sum_{j \in \mathbf{U}_1'} c_j$.

THE ALGORITHM

We now state the algorithm completely. It uses the backtrack search strategy mentioned in Section 10.3.3 with the LIFO selection criterion. Initially the original problem is the current CP with both the subscript sets of 0- and 1-variables empty. The stack is empty initially.

In a general stage of the algorithm suppose the current CP is defined by the subscript sets $\mathbf{U}_0, \mathbf{U}_1$ for 0- and 1-variables respectively. Do the following.

1. If the current CP is fathomed update the incumbent, prune the stack and go to 2. If the current CP is pruned go to 2.

2. If the stack is empty at this stage, the incumbent is an optimum solution of the original problem, terminate. If the stack is empty and there is no incumbent at this stage, the original problem is infeasible, terminate. If the stack is nonempty, retrieve a CP from the stack using the LIFO selection criterion, and make it the new current CP, and go to 3.

3. If the current CP is not fathomed apply the tests on it. If the current CP is pruned by the tests, go to 2. If it is not pruned by the tests, let $\mathbf{U}_0', \mathbf{U}_1'$ be the subscript sets of 0- and 1-variables in the problem after augmenting

the new 0- and 1-variables identified by the tests, to $\mathbf{U}_0, \mathbf{U}_1$ respectively. $\mathbf{U}_f' = \{1, \dots, n\} \backslash (\mathbf{U}_0' \cup \mathbf{U}_1')$ is the subscript set of free variables in the problem.

If $\mathbf{U}_f' = \emptyset$, the current CP is fathomed; go to 1. If $\mathbf{U}_f' \neq \emptyset$, select an x_j with $j \in \mathbf{U}_f'$ as the branching variable. Branching generates two candidate subproblems. CP 1 has $\mathbf{U}_0', \mathbf{U}_1' \cup \{j\}$ as the subscript sets for 0- and 1-variables respectively. CP 2 has $\mathbf{U}_0' \cup \{j\}, \mathbf{U}_1'$ as the subscript sets for 0- and 1-variables respectively. Add CP 2 to the stack. Make CP 1 the new current CP and continue by applying this step 3. on it.

For efficient branching variable selection criteria in this algorithm see [G. L. Nemhauser and L. A. Wolsey, 1988].

10.8   Advantages and Limitations of the B&B Approach

In this chapter we discussed the general B&B approach and its application to solve a variety of problems. Various techniques for developing bounding, branching, and search strategies have also been illustrated in these applications. The examples provide insight into how B&B algorithms can be developed for solving integer programs and combinatorial optimization problems.

On some problems with nice mathematical structure (such as the TSP, the knapsack problem, and certain special types of pure $0-1$ IPs) great strides have been made recently in developing B&B algorithms for solving large scale instances of the problem by exploiting the special structure. The key behind all the success on these problems is the new and very deep branch of mathematics called **polyhedral combinatorics** whose aim is to identify families of linear constraints for characterizing the convex hull of the set of feasible solutions of these problems through a system of linear constraints. The known families of linear constraints are used to construct LP relaxations for these problems which produce lower bounds very close to the minimum objective value in the original problem. However, since the number of constraints in these relaxed LPs tends to be very large, their solution depends critically on recent breakthroughs in solving very large scale LPs, and the availability of expensive computer hardware (parallel processing supercomputers).

Outside of this class of problems with nice mathematical structure, the performance of B&B algorithms is uneven, particularly as the size of the instance (as measured by say, the number of $0-1$ variables in the model) becomes large. On such problems the B&B algorithm may require an enormous amount of computer time, as the number of nodes examined in the search tree grows exponentially with the size of the instance. However, it usually produces very good incumbents early in the search effort. Even though these early incumbents are not guaranteed to be optimal to the problem, they usually turn out to be very close to the optimum. This is what makes the B&B approach useful in applications.

10.9 Exercises

10.1 The new American President-elect has appointed a team of experts to visit

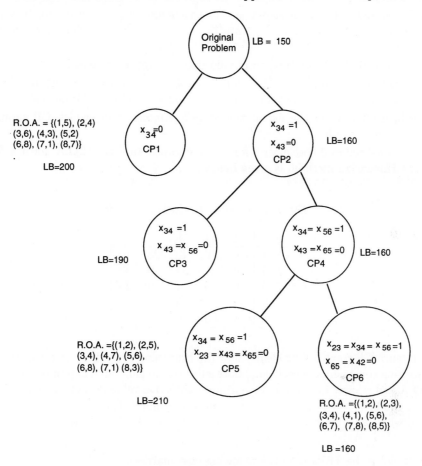

Figure 10.12: Search tree

7 major cities in the country (cities numbered 2 to 8) beginning in city 1 = Washington, D.C. to assess the problems in urban areas. The team has to begin their trip in city 1, visit each of the other cities 2 to 8 once and only once in some order, and return to city 1 at the end. c_{ij} = the cost for the team to travel from city i to city j, in some coded money units, is given for all i, j. The local cost for the team's stay in each city will be borne by the city itself. The problem is to determine the order of the cities for the team to visit, so as to minimize the total travel cost of the trip. This is a TSP involving cities 1 to 8, with $C = (c_{ij})$ as the cost matrix. The problem of finding an optimum tour is being solved by the B&B approach using the lower bounding strategy based on the relaxed assignment problem. The search tree

at some stage of this algorithm is given in Figure 10.12. Here R.O.A. is "Relaxed Optimum Assignment for the candidate problem," and LB is "Lower bound for the candidate problem = the cost of the R.O.A. wrt the original cost matrix."

The various CPs in the search tree at this stage are given in Figure 10.12. The constraints inside each node representing a CP are the branching constraints in that CP.

Are any of the CPs fathomed at this stage? If so, which ones?

Is there an incumbent at this stage? What is the current incumbent? Is it an optimum tour? Why? If not, give an upper bound for the minimum cost of a tour.

Which CPs are in the list at this stage? Among these, which CP should be branched next? Why?

The final reduced cost matrix for CP 6 obtained after the R.O.A. for it is found by the Hungarian method, is given below.

Final Reduced Cost Matrix for CP6

$j =$	1	2	5	7	8
$i = 1$	×	0	20	25	10
4	0	×	30	40	50
6	15	10	×	0	35
7	35	5	15	×	0
8	0	50	0	10	×

Branch CP 6 explaining clearly how the branching variable is selected. Mark the branching constraints in each candidate subproblem generated carefully. Obtain the R.O.A and a lower bound for the optimum objective value in each of them. Update the incumbent if necessary, and carry out any pruning. Is the problem solved now? If so, what is an optimal tour for the problem, and how much does it cost?

10.2 Solve the TSPs with the following cost matrices.

(a)

$j =$	1	2	3	4	5	6
$i = 1$	×	0	2	5	11	4
2	0	×	0	17	12	13
3	0	0	×	9	4	7
4	7	11	6	×	0	12
5	18	19	15	21	×	0
6	8	3	13	0	23	×

(b)

$j =$	1	2	3	4	5	6	7
$i = 1$	×	13	9	19	22	3	7
2	6	×	7	8	39	4	8
3	18	3	×	5	18	4	3
4	4	16	6	×	5	13	16
5	14	22	99	17	×	8	10
6	3	21	7	3	10	×	9
7	11	19	8	19	8	3	×

(c)

$j =$	1	2	3	4	5	6	7	8	9	10
$i = 1$	×	51	55	90	41	63	77	69	0	23
2	50	×	0	64	8	53	0	46	73	72
3	30	77	×	21	25	51	47	16	0	60
4	65	0	6	×	2	9	17	5	26	42
5	0	94	0	5	×	0	41	31	59	48
6	79	65	0	0	15	×	17	47	32	43
7	76	96	48	27	34	0	×	0	25	0
8	0	17	9	27	46	15	84	×	0	24
9	56	7	45	39	0	93	67	79	×	38
10	30	0	42	56	49	77	72	49	23	×

(d) Now consider the 10 city TSP in which city 1 is the starting and the terminal city, with the 2nd matrix given above as the cost matrix. However, in this problem, the salesman is required to visit cities 2, 3, 4 before going to 9. Find a minimum cost tour satisfying this property, by the B&B approach.

10.3 The Symmetric Assignment Problem Consider the problem of forming a set of 10 objects into 5 pairs. Each object has to be in exactly one pair, and each pair must contain 2 distinct objects. The cost of pairing objects i, j is d_{ij}, tabulated below for $j > i$.

$j =$	1	2	3	4	5	6	7	8	9	10
$i = 1$	×	4	6	110	116	126	118	120	116	114
2		×	8	118	114	124	106	102	118	106
3			×	112	116	112	110	122	116	124
4				×	2	4	218	216	226	230
5					×	6	212	212	234	232
6						×	338	306	316	308
7							×	4	2	16
8								×	6	8
9									×	10
10										×

Let $n = 10$, $c_{ii} = \infty$ for $i = 1$ to n, and let c_{ij} be real numbers satisfying $c_{ij} + c_{ji} = d_{ij}$ for all $j > i$. Define decision variables x_{ij} for $i, j = 1$ to n by

$$x_{ij} = x_{ji} = \begin{cases} 1, & \text{if objects } i, j \text{ are formed into a pair} \\ 0, & \text{otherwise} \end{cases} \qquad (10.24)$$

With the definition of decision variables as in (10.24) prove that the problem of finding a minimum cost pairing is equivalent to the following problem known as the **symmetric assignment problem**.

$$\text{Minimize } z(x) = \sum_{i=1}^{n} \sum_{j=1}^{n} c_{ij} x_{ij}$$

$$\text{Subject to } \sum_{j=1}^{n} x_{ij} = 1 \text{ for } i = 1 \text{ to } n \qquad (10.25)$$

$$\sum_{i=1}^{n} x_{ij} = 1 \text{ for } j = 1 \text{ to } n$$

$$x_{ij} = 0 \text{ or } 1 \text{ for all } i, j$$

$$\text{and } x_{ij} = x_{ji} \qquad \text{for all } i, j \qquad (10.26)$$

This problem is the standard assignment problem discussed in Chapter 5, with the additional constraints (10.26) known as the **symmetry conditions**. A feasible solution $x = (x_{ij})$ to it is called a **symmetric assignment**.

As an example, among the assignments x^1, x^2 of order 4 given below, x^1 is not a symmetric assignment because $x_{12} = 1$, $x_{21} = 0$ and hence $x_{12} \neq x_{21}$ in it, violating the symmetry condition. x^2 is a symmetric assignment because it satisfies all the symmetry conditions. In x^2, $x_{13} = x_{31} = 1$; and $x_{24} = x_{42} = 1$; so x^2 is the symmetric assignment representing the pairing of objects $\{1, 2, 3, 4\}$ into the two pairs $(1, 3)$ and $(2, 4)$.

$$x^1 = \begin{pmatrix} 0 & 1 & 0 & 0 \\ 0 & 0 & 1 & 0 \\ 0 & 0 & 0 & 1 \\ 1 & 0 & 0 & 0 \end{pmatrix}, \qquad x^2 = \begin{pmatrix} 0 & 1 & 0 & 0 \\ 1 & 0 & 0 & 0 \\ 0 & 0 & 0 & 1 \\ 0 & 0 & 1 & 0 \end{pmatrix}$$

Show that a lower bound for the minimum objective value in the symmetric assignment problem (10.25), (10.26) can be obtained by solving the the standard assignment problem (10.25) with the symmetry conditions (10.26) relaxed. Make an argument to show that the best lower bound is obtained by taking $c_{ij} = c_{ji} = d_{ij}/2$ for all $j > i$.

Develop a B&B approach for solving the symmetric assignment problem (10.25), (10.26). Develop an efficient branching strategy based on selecting a **branching pair of variables** using **evaluations** as in the TSP, but tailormade to this problem using its special nature. Solve the problem of finding a minimum cost way of forming 10 objects into 5 pairs mentioned above, using this algorithm.

Note   Here we asked the reader to develop an enumerative B&B algorithm to solve the symmetric assignment problem, mainly to give him/her experience in designing B&B algorithms for discrete optimization problems. However, it is not necessary to use enumerative algorithms to solve symmetric assignment problems, as there is a highly efficient nonenumerative primal-dual algorithm that is a generalization of the Hungarian method of Chapter 5 to solve it directly. See [K. G. Murty, 1992 of Chapter 5] for details of this algorithm known in the literature as a **blossom algorithm**.

10.4 There are 6 students in a projects course. It is required to form them into groups of at most 2 students each (so a single student can constitute a group by himself/herself). Here is the cost data. Find a minimum cost grouping using a B&B approach.

Cost of forming students i, j into a group, for $j \overset{\geq}{=} i$

$j =$	1	2	3	4	5	6
$i = 1$	16	10	8	58	198	70
2		10	6	72	50	32
3			15	26	198	24
4				15	14	18
5					13	6
6						10

10.5 Solve the 0−1 knapsack problem with 9 available objects with the following data, to maximize the value loaded into a knapsack of weight capacity 40 weight units.

Object	Weight	Value
1	19	380
2	15	225
3	20	320
4	8	96
5	5	70
6	7	126
7	3	30
8	2	22
9	4	68

10.6 Consider an undirected network consisting of nodes (represented by little circles with its number entered inside, in a figure of the network); and edges, each of which is a line joining a pair of distinct nodes. A **clique** in such a network is a subset of nodes **N** satisfying the property that every pair of nodes in **N** is joined by an edge in the network.

As an example, in the network in Figure 10.13 with 11 nodes; the subset of nodes {1, 10, 9, 2} is not a clique because nodes 1 and 9 in this subset are not joined by an edge in the network. But the subset of nodes {1, 2, 3, 5} is a clique because every pair of nodes in this subset is joined by an edge in the network.

The cost of including each node in a clique is given. Typically, these cost coefficients are negative. The cost of a clique is defined to be the sum of the cost coefficients of nodes in it. For example, the cost of the clique {1, 2, 3, 5} in the network in Figure 10.13 is $-2 - 5 - 7 - 1 = -15$.

Develop a B&B algorithm for finding a minimum cost clique. And find a minimum cost clique in the network in Figure 10.13 using your algorithm.

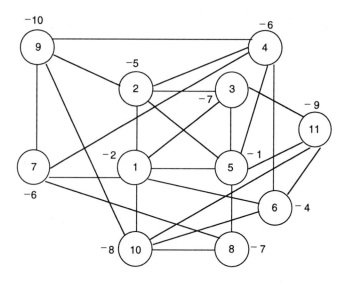

Figure 10.13   The negative number by the side of each node is its cost coefficient.

10.7 The Asymmetric Assignment Problem   Let $C = (c_{ij})$ be an $n \times n$ cost matrix for an assignment problem, with $c_{ii} = \infty$ for all i (i.e., all cells (i, i) are forbidden cells). Here we want a minimum cost assignment satisfying the additional constraints "if $x_{ij} = 1$, then $x_{ji} = 0$, for all $i \neq j$." These conditions are called **asymmetry constraints** since they force the feasible assignment to be asymmetric. Clearly these conditions are equivalent to "$x_{ij} + x_{ji} \overset{\leq}{=} 1$ for all $i \neq j$." Here is the problem.

$$\text{Minimize } z(x) \;=\; \sum_{i=1}^{n}\sum_{j=1}^{n} c_{ij}x_{ij}$$

$$\text{Subject to } \sum_{j=1}^{n} x_{ij} \;=\; 1 \;\; \text{for } i = 1 \text{ to } n \tag{10.27}$$

$$\sum_{i=1}^{n} x_{ij} \;=\; 1 \;\; \text{for } j = 1 \text{ to } n$$

$$x_{ij} \;=\; 0 \;\; \text{or } 1 \text{ for all } i, j$$

$$\text{and } x_{ij} + x_{ji} \stackrel{\leq}{=} 1 \qquad \text{for all } i \neq j \tag{10.28}$$

Because of (10.28), this problem is known as the **asymmetric assignment problem**. Develop a B&B algorithm to solve (10.27), (10.28) that takes advantage of its special structure. Solve the asymmetric assignment problem of order 6 with the following cost matrix, using your algorithm.

	$j =$	1	2	3	4	5	6
	$i = 1$	×	15	10	14	13	20
	2	15	×	16	18	18	8
$C = (c_{ij})$	3	10	16	×	28	25	24
	4	14	18	28	×	3	17
	5	13	18	25	3	×	13
	6	20	8	24	17	13	×

10.8 A 9 city cost minimizing TSP is being solved by the B&B approach discussed in Section 10.4. Some information on the search tree in Figure 10.14 in an intermediate stage of this algorithm is given below.

The final reduced cost matrix corresponding to CP 5 is

$j =$	1	2	3	4	5	6
$i = 1$	×	0	5	2	0	4
2	0	×	8	12	9	10
3	8	14	×	13	8	0
4	1	0	6	×	4	0
5	1	4	5	0	×	1
6	9	9	0	10	9	×

CP 1 to CP 4, and CP 6 are not fathomed so far.

Is CP 5 fathomed? Write down the candidate problems in the search tree which are in the list at this stage.

Branch CP 5 explaining all the work clearly. Are any of the new CPs generated fathomed? If so, write down the incumbent, and carry out any pruning that is possible. What is the new list? Is the incumbent optimal to the original problem? If not, explain which candidate problem should be selected for branching next if you use the a) least lower bound selection strategy, b) backtrack search strategy.

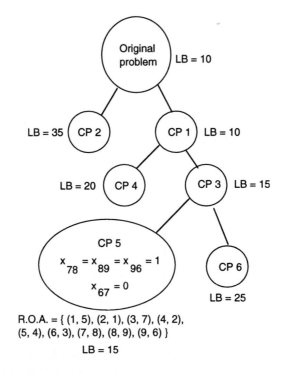

R.O.A. = { (1, 5), (2, 1), (3, 7), (4, 2),
(5, 4), (6, 3), (7, 8), (8, 9), (9, 6) }
LB = 15

Figure 10.14   The search tree. Complete information is only provided for CP 5.

10.9 A 7 city traveling salesman problem is being solved by the branch and bound algorithm using the lower bounding strategy based on the relaxed assignment problem. The definitions of the decision variables in the problem are: for $i, j = 1$ to 7

$$x_{ij} = \begin{cases} 1 & \text{if the salesman goes from city } i \text{ to city } j \\ 0 & \text{otherwise} \end{cases}$$

The search tree at some stage of the algorithm is given in Figure 10.15. Inside the node representing each CP, we list all the branching constraints which define that CP.

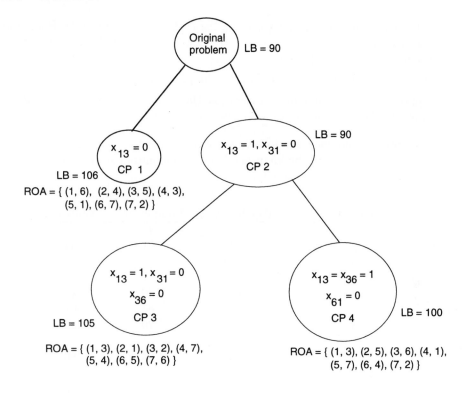

Figure 10.15

And the final reduced cost matrix at the end of solving the relaxed assignment problem corresponding to CP4 by the Hungarian method in C^4 given below.

$$C^4 =$$

$j =$	1	2	4	5	7
$i = 2$	8	x	3	0	1
4	0	10	x	4	2
5	2	3	5	x	0
6	x	8	0	4	7
7	3	0	6	2	x

1. Are any of the CPs fathomed so far? If so which one? Explain clearly. Is there an incumbent at this stage? What is an upper bound for the minimum cost of a tour in the original problem? Can any pruning be carried out?

2. What is the list at this stage?

   Which CP should we branch at this stage? Why? What is the best branching variable for branching it? Explain clearly showing your work carefully.

Carry out branching of that CP using the best branching variable. Record all the branching constraints in each of the new CPs generated carefully, explain them clearly.

If any of the newly generated CPs can be pruned, prune it.

Apply the lower bounding strategy on the unpruned new CPs. Continue applying the algorithm until an optimum solution is obtained.

10.10 Formulate and solve the following multiconstraint 0−1 knapsack problem. The total value included in the knapsack is to be maximized subject to the knapsack's weight and volume constraints.

Object	Weight (lbs.)	Volume (ft^3)	Value ($)
1	20	41	84
2	12	51	34
3	7	24	31
4	75	40	14
5	93	84	67
6	21	70	65
7	75	34	86
8	67	41	98
9	34	49	50
10	28	27	7
Knapsack's capacity	190	250	

([W. Shih, April 1979])

10.11 Solve the following MIPs by the B&B approach

$$\text{Maximize} \quad 2x_1 + x_2 + 3y_1 + 4y_2$$
$$\text{subject to} \quad x_1 + 3x_2 - y_1 + 2y_2 \leqq 16$$
$$-x_1 + 2x_2 + y_1 + y_2 \leqq 4$$
$$x_1, \quad x_2, \quad y_1, \quad y_2 \geqq 0$$
$$x_1, \quad x_2 \quad \text{are integer}$$

$$\text{Maximize} \quad 4y_1 + 5x_1 + x_2$$
$$\text{subject to} \quad 3y_1 + 2x_1 \leqq 10$$
$$y_1 + 4x_1 \leqq 11$$

$$3y_1 + 3x_1 + x_2 \lesseqgtr 13$$

$$y_1, \quad x_1, \quad x_2 \gtreqless 0$$

x_1, x_2 are integer

10.10 References

O. H. IBARRA and C. E. KIM, October 1975, "Fast Approximation Algorithms for the Knapsack and Sum of Subset Problems", *Journal of the ACM*, 22, no. 4 (463-468).

F. GLOVER, 1968, "Surrogate Constraints", *Operations Research*, 16, no. 4, (741-749).

A. H. LAND, and A. G. DOIG, 1960, "An Automatic Method for Solving Discrete Programming Problems", *Econometrika*, 28 (497-520).

K. G. MURTY, C. KAREL, and J. D. C. LITTLE, 1962, "The Traveling Salesman Problem: Solution by a Method of Ranking Assignments", Case Institute of Technology.

G. L. NEMHAUSER, and L. A. WOLSEY, 1988, *Integer and Combinatorial Optimization*, Wiley, NY.

W. SHIH, April 1979, "A Branch and Bound Method for the Multiconstraint Zero-one Knapsack Problem", *Journal of the Operational Research Society*, 30, no. 4, 369-378.

Chapter 11

Heuristic Methods for Combinatorial Optimization Problems

11.1 What Are Heuristic Methods?

The word **heuristic** comes from the Old Greek word **heuriskein** which means "discovering new methods for solving problems" or "the art of problem solving." In computer science and artificial intelligence, the term "heuristic" is applied usually to methods for intelligent search. In this sense "heuristic search" uses all the available information and knowledge to lead to a solution along the most promising path, omitting the least promising ones. Here its aim is to enable the search process to avoid examining dead ends, based on information contained in the data gathered already.

However, in operations research the term "heuristic" is often applied to methods (which may or may not involve search) that are based on intuitive and plausible arguments likely to lead to reasonable solutions but are not guaranteed to do so. They are methods for the problem under study, based on rules of thumb, common sense, or adaptations of exact methods for simpler models. They are methods used to find reasonable solutions to problems that are hard to solve exactly. In optimization in particular, a heuristic method refers to a practical and quick method based on strategies that are likely to (but not guaranteed to) lead to a solution that is approximately optimal or near optimal. So, while discussing these heuristic methods, the verb "solve" has the connotation of "finding a satisfactory approximation to the optimum." Thus heuristic methods can, but do not guarantee the finding of an optimum solution; although good heuristic methods in principle determine the best solution obtainable within the allowed time. Many heuristic methods do involve some type of search to look for a good approximate solution.

410

11.2   Why Use Heuristic Methods?

Heuristic methods are as old as decision making itself. Until the 1950s when computers became available and machine computation became possible, inelegant but effective heuristics were the only methods used to tackle large scale decision making.

By an **exact algorithm** for an optimization problem, we mean an algorithm that is guaranteed to find an optimum solution if one exists, within a reasonable time. In the 1960s and 70s exact algorithms based on sophisticated mathematical constructs were developed for certain types of optimization problems such as linear programs, convex quadratic programs, and nonlinear convex programming problems. The special distinguishing feature of all these problems is that optimality conditions providing efficient characterizations for optimum solutions for them are known. The exact algorithms for them are based on these optimality conditions. Because of this special feature, these problems are considered to be **nice problems** among optimization models. In the 1980s and recently, software packages implementing these sophisticated algorithms, and computer systems that can execute them, became very widely available. So, now-a-days there is no reason to resort to heuristic methods to solve instances of these problems, as they can be solved very efficiently by these exact algorithms.

The development of exact algorithms for these nice problems has been a significant research achievement. Unfortunately, this research did not lead to any reliable exact solution methods for optimization problems such as the discrete and integer programming problems and combinatorial optimization problems discussed in Chapters 9 and 10 that are not in this nice class. The B&B approach of Chapter 10 based on partial enumeration can solve instances of moderate sizes of these problems, but in general the time requirement of this approach grows exponentially with the size of the instance. And real world applications of combinatorial optimization usually lead to large scale models. We illustrate this with an application in the automobile industry.

EXAMPLE 11.1   A task allocation problem

This problem, posed by K. N. Rao, deals with determining a minimum cost design for an automobile's microcomputer architecture. In the modern automobile, many tasks such as integrated chassis and active suspension monitoring, etc. are performed by microcomputers linked by high speed and/or slow speed communication lines. The system's cost is the sum of the costs of the processors (microcomputers), and of the data links that provide inter-processor communication bandwidth.

Each task deals with the processing of data coming from sensors, actuators, signal processors, digital filters, etc., and has a throughput requirement in KOP (kilo operations per second). Several types of processors are available. For each, we are given its cost, maximum number of tasks it can handle, and its throughput capacity in terms of the KOP it can handle.

The tasks are inter-dependent. To complete a task we may need data from another. So, the typical communication pattern between tasks is that if two of them are assigned to different processors, they need communication link capacity (in bits/second) between them. Tasks executing in the same processor do not have communication overhead. Here is the notation for the data.

n = number of tasks to be performed (varies between 50 - 100 in applications)

a_i = throughput requirement (in KOP) of task i, $i = 1$ to n

T = maximum number of processors that may be needed

$\rho_t, \gamma_t, \beta_t$ = cost (\$), capacity in KOP, and upper bound on the number of tasks to be allotted, to processor t, $t = 1$ to T

c_{ij}, d_{ij} = low speed and high speed communication link capacity (in bits/second) needed for task pair i, j if they are assigned to different processors

L, H = unit cost of installing low speed, high speed communication bandwidth

To model this problem, we define the following decision variables for $i, j = 1$ to n, $t = 1$ to T.

$$\text{For } i \neq j, x_{ijt} = \begin{cases} 1, & \text{if both tasks } i \text{ and } j \text{ are assigned to processor } t \\ 0, & \text{otherwise} \end{cases}$$

$$x_{iit} = \begin{cases} 1, & \text{if task } i \text{ is assigned to processor } t \\ 0, & \text{otherwise} \end{cases}$$

$$y_t = \begin{cases} 1, & \text{if processor } t \text{ is used (i.e., it is allotted some tasks)} \\ 0, & \text{otherwise} \end{cases}$$

In terms of these decision variables the model for the minimum cost design is

$$\text{Minimize} \sum_{i=1}^{n-1} \sum_{j=i+1}^{n} (Lc_{ij} + Hd_{ij})(1 - \sum_{t=1}^{T} x_{ijt}) + \sum_{t=1}^{T} \rho_t y_t$$

$$\text{subject to} \quad \sum (x_{ijt} : \text{over } j \neq i) - (\beta_t - 1)x_{iit} \lesseqgtr 0, \text{for } i = 1 \text{ to } n, t = 1 \text{ to } T$$

$$\sum_{i=1}^{n} a_i x_{iit} \lesseqgtr \gamma_t y_t, \quad \text{for } t = 1 \text{ to } T$$

$$\sum_{i=1}^{n} x_{iit} \lesseqgtr \beta_t y_t, \quad \text{for } t = 1 \text{ to } T$$

$$\sum_{i=1}^{n} x_{iit} \gtreqless y_t, \quad \text{for } t = 1 \text{ to } T$$

$$\sum_{t=1}^{T} x_{iit} \;=\; 1, \quad \text{for } i = 1 \text{ to } n$$

$$x_{ijt}, y_t \quad \text{are all} \qquad 0 \text{ or } 1$$

The first constraint guarantees that the processor t to which task i is assigned, is not assigned more than $\beta_t - 1$ other tasks. The second constraint guarantees that the total KOP requirements of all the tasks assigned to processor t is $\leq$ its KOP capacity of γ_t. The third and fourth constraints together guarantee that processor t is either not used, or if it is used then it is assigned no more than β_t tasks. The fifth constraint guarantees that each task is assigned to a processor.

This is a 0−1 IP model with $T(n^2 + 1)$ integer variables. Even for $n = 50$, and $T = 10$, the number of 0−1 variables in the model is over 25,000, which is very large.

In the same manner, problems in the optimum design of many manufactured items, in telecommunication system design, and other areas, lead to large scale combinatorial optimization models.

Research carried out in computational complexity and NP-completeness theory since the 1970s has shown that many of the integer programming and combinatorial optimization problems discussed in Chapters 9 and 10 are hard intractable problems. It has provided evidence that there may be no effective exact algorithms to solve large scale versions of these problems, i.e., algorithms which can find optimum solutions to these problems within acceptable computer time. As a consequence, it has been recognized that the only practical alternative to attack large scale instances of these problems is through good heuristic methods. In fact, practitioners facing these problems have always had an interest in heuristics as a means of finding good approximate solutions. And experience indicates that there are many heuristic methods which are simple to implement relative to the complexity of the problem, and although they do not necessarily yield a solution close to the optimum always, they quite often do. Moreover, at the termination of a heuristic method, we can always improve performance by resorting to another heuristic search algorithm to resume the search for a better solution.

For some of the hard combinatorial optimization problems such as the TSP, a detailed study based on their mathematical structure has made it possible to construct special bounding schemes. B&B algorithms based on them have successfully solved several large scale instances of these problems within reasonable times. There is no guarantee that these special algorithms will give the same effective performance on all large scale instances of these problems, but their record so far is very impressive. However, many practitioners still seem to prefer to solve these problems approximately using much simpler heuristic methods. One reason for this is the fact that real world applications are often messy, and the data available for them is liable to contain unknown errors. Because of these errors in the data, an optimum solution of the model is at best a guide to a reasonable solution for the real problem, and an approximate solution obtained by a good but simple heuristic would serve the same purpose without the need for expensive computer hardware and software for a highly sophisticated algorithm.

For all these reasons, heuristic methods are the methods of choice for handling large scale combinatorial optimization problems.

11.3 General Principles in Designing Heuristic Methods

Heuristic methods are always problem-specific, but there are several widely applicable principles for designing them. A popular one is the **greedy principle** which leads to greedy methods, perhaps the most important methods among **single pass heuristics** that create a solution in a single sweep through the data. Each successive step in these methods is taken so as to minimize the immediate cost (or maximize the immediate gain). The three characteristic features of greedy methods are the following.

The incremental feature   They represent the problem in such a way that a solution can be viewed either as a subset of a set of elements, or as a sequencing of a set of elements in some order. The approach builds up the solution set, or the solution sequence, one element at a time starting from scratch, and terminates with the first complete solution.

The no-backtracking feature   Once an element is selected for inclusion in the solution set (or an element is included in the sequence in the current position) it is never taken back or replaced by some other element (or its position in the sequence is never altered again). That is, in a greedy algorithm, decisions made at some stage in the algorithm are never revised later on.

The greedy selection feature   Each additional element selected for inclusion in the solution set, or selected to fill the next position in the sequence, is the best among those available for selection at that stage by some criterion, in the sense that it contributes at that stage the least amount to the total cost, or the maximum amount to the total gain, when viewed through that criterion.

Several different criteria could be used to characterize the "best" when making the greedy selection, depending on the nature of the problem being solved. The success of the approach depends critically on the choice of this criterion.

Thus the greedy approach constructs the solution stepwise, in each step selecting the element to include in the solution to be the cheapest among those that are eligible for inclusion at that time. It is very naive. The selection at each stage is based on the situation at that time, without any features of look-ahead, etc. Hence greedy methods are also known as **myopic methods**.

Another approach for designing heuristic methods is based on starting with a complete solution to the problem, and trying to improve it by a local search in the neighborhood of that solution. The initial solution may be either a randomly generated solution, or one obtained by another method like the greedy method. Each subsequent step in the method takes the solution at the end of the previous step, and tries to improve it by either exchanging a small number of elements in the solution with those not in the solution, or some other technique of local search. The process continues until no improving solution can be found by such local search, at which point we have a local minimum. These methods are variously known as **interchange heuristic methods** or **local search heuristics** or **descent methods**. A local optimum is at least as good or better than all solutions in its neighborhood, but it may not be a global optimum, i.e., it may not be the best solution for the problem. One of the shortcomings of a descent method is the fact that it obtains a local minimum which in most cases may not be a global minimum. To overcome this limitation people normally apply the descent method many times, with different initial solutions, and take as the final output the best among all the local minima obtained. This restart approach is known as **the iterated descent method**.

The general design principle of local improvement through small changes in the feasible solution is also the principle behind the **simulated annealing technique**, which admits also steps that decrease solution quality based on a probabilistic scheme. After reaching a local optimum, simulated annealing moves randomly for a period, and then resumes a trajectory of descent again.

And then there are heuristic methods known as **genetic algorithms** which are probabilistic methods that start with an initial population of likely problem solutions, and then evolve towards better solution versions. In these methods new solutions are generated through the use of genetic operators patterned upon the reproductive processes in nature.

Sometimes several heuristic methods may be applied on a problem in a sequence. If the first heuristic starts from scratch to find an initial solution, the second may have the aim of improving it. And when this heuristic comes to its end, a third may succeed it. This could continue until all the heuristics in the list fail in a row to improve the current solution.

There are major differences between the techniques appropriate to different problems. As in the B&B approach, details of a heuristic algorithm depend on the structure of the problem being solved. In the following sections we discuss the essential ideas behind the popular heuristic methods, and illustrate their application on several problem types discussed in Chapter 9.

11.4 The Greedy Approach

We will now discuss greedy methods for various problems from Chapter 9.

11.4.1   A Greedy Method for the 0−1 Knapsack Problem

Consider the 0−1 knapsack problem in which there are n objects that could be loaded into a knapsack of weight capacity w_0 weight units. For $j = 1$ to n, v_j in money units is the value, and w_j in weight units is the weight, of object j. The problem is (10.10), (10.11) to determine the subset of objects to be loaded into the knapsack to maximize the value loaded subject to the knapsack's weight capacity.

For applying the greedy approach on this problem, the criterion to be greedy upon for selecting objects to include in the knapsack, could be either the value of the object or its density = value/weight. Once this criterion is decided, the objects are arranged in decreasing order of the criterion and loaded into the knapsack in this order. At some stage, if an object's weight is > remaining knapsack's weight capacity, we leave it out and continue the process with the next object in this order, until all the objects are examined. The set of objects loaded into the knapsack at the end of this process is the solution set determined by the greedy algorithm with the selected criterion.

As an example, consider the knapsack problem (10.14), (10.15) with knapsack's weight capacity of 35 weight units, and 9 different objects available for loading into it, discussed in Example 10.2. The solution set obtained by the greedy algorithm with object's value as the criterion to be greedy upon is {objects 8, 5} using up the knapsack's weight capacity completely, and attaining the value of 335 money units for the total value of objects loaded into the knapsack.

The solution set obtained by the greedy method with density as the criterion to be greedy upon is {objects 7, 5, 1} with a total value of 348 money units.

The optimum objective value in this problem, found by the B&B algorithm in Example 10.3 is 351 money units. So, neither of the solution sets obtained by the greedy algorithms above are optimal. However, the greedy algorithm with the density criterion yielded a much better solution than the one with the object's value criterion. In general, the greedy algorithm with object's density as the criterion to be greedy upon yields much better solutions than the one with the object's value as the criterion. Thus, for the 0−1 knapsack problem, the greedy algorithm is always implemented with object's density as the criterion to be greedy upon. And the greedy solution for this problem usually means the solution obtained by this version of the greedy algorithm. For this algorithm, the following result has been proved.

THEOREM 11.1 *Consider the 0−1 knapsack problem (10.14), (10.15), with w_0 = knapsack's weight capacity, n = number of available objects; and $w_j, v_j, d_j = v_j/w_j$, as the weight, value, and density of object j, for $j = 1$ to n. Eliminate all objects j with $w_j > w_0$ since they won't fit into the knapsack (i.e., fix $x_j = 0$ for all such j). So, assume $w_j \overset{\leq}{=} w_0$ for all $j = 1$ to n. Let $\hat{x} = (\hat{x}_j)$ be the solution obtained by the greedy algorithm with object's density as the criterion to be greedy upon (i.e., $\hat{x}_j = 1$ if object j is included in the knapsack by this algorithm, $\hat{x}_j = 0$ otherwise), and $\hat{v} = \sum_{j=1}^{n} v_j x_j$, $\hat{w} = \sum_{j=1}^{n} w_j x_j$.*

(i)   The greedy solution $\hat{x}$ is an optimum solution for the original problem
(10.14), (10.15), if the following conditions hold:

$\hat{w} = w_0$ (i.e., the greedy solution uses up the knapsack's weight capacity exactly),

and all the objects j left out of the greedy solution set (i.e., with $\hat{x}_j = 0$) have
density $d_j \leqq$ the density of every one of the objects in the greedy solution.

(ii)   Let v_* denote the unknown optimum objective value in (10.14), (10.15).
If the conditions in (i) are not satisfied $\hat{x}$ may not be optimal to (10.14),
(10.15), but $v_* - \hat{v} \leqq \max\{v_1, \ldots, v_n\}$.

For a proof of Theorem 11.1, see [G. L. Nemhauser and L. A. Wolsey, 1988
of Chapter 10]. It gives an upper bound for the difference between the optimum
objective value and the objective value of the greedy solution.

Exercise

11.1 Consider the 0−1 knapsack problem with $w_0 = 16 =$ knapsack's weight
capacity, and 4 objects with data given below, available to load into the knapsack.

Object j	Weight w_j	Value v_j
1	2	16
2	15	105
3	1	6
4	13	13

Find the optimum solution of this problem by total enumeration. Apply the
greedy heuristic with density as the criterion to be greedy upon and obtain the
greedy solution for the problem. Verify that the greedy solution uses up the knapsack's weight capacity exactly, but that it is not optimal because the second condition in (i) of Theorem 11.1 does not hold.

In practice the greedy heuristic with density as the criterion to be greedy upon,
usually yields solutions close to the optimum, and hence is very widely used for
tackling 0−1 knapsack problems.

11.4.2   A Greedy Heuristic for the Set Covering Problem

The set covering problem discussed in Section 9.2 is a pure 0−1 IP of the following
form:

$$\text{Minimize } z(x) = cx$$

$$\text{subject to } Ax \; \overset{\geq}{=} \; e \qquad\qquad (11.1)$$

$$x_j \;=\; 0 \text{ or } 1 \text{ for all } j$$

where $A = (a_{ij})$ is a 0–1 matrix of order $m \times n$ and e is the column vector of all 1s in $\mathbb{R}^m$. We will use the following problem as an example.

$$\text{Minimize } z(x) = 3x_1 + 2x_2 + 5x_3 + 6x_4 + 11x_5 + x_6$$

$$+ 12x_7 + 7x_8 + 8x_9 + 4x_{10} + 2x_{11} +$$

$$6x_{12} + 9x_{13} - 2x_{14} + 2x_{16}$$

$$\text{subject to } x_7 + x_9 + x_{10} + x_{13} \; \overset{\geq}{=} \; 1$$

$$x_2 + x_8 + x_9 + x_{13} \; \overset{\geq}{=} \; 1$$

$$x_3 + x_9 + x_{10} + x_{12} \; \overset{\geq}{=} \; 1$$

$$x_4 + x_5 + x_8 + x_9 \; \overset{\geq}{=} \; 1$$

$$x_3 + x_6 + x_8 + x_{11} \; \overset{\geq}{=} \; 1 \qquad (11.2)$$

$$x_3 + x_6 + x_7 + x_{10} \; \overset{\geq}{=} \; 1$$

$$x_2 + x_4 + x_5 + x_{12} \; \overset{\geq}{=} \; 1$$

$$x_4 + x_5 + x_6 + x_{13} \; \overset{\geq}{=} \; 1$$

$$x_1 + x_2 + x_4 + x_{11} \; \overset{\geq}{=} \; 1$$

$$x_1 + x_5 + x_7 + x_{12} \; \overset{\geq}{=} \; 1$$

$$x_{14} + x_{16} \; \overset{\geq}{=} \; 1$$

$$x_{15} + x_{16} \; \overset{\geq}{=} \; 1$$

$$x_j = 0 \text{ or } 1 \text{ for all } j$$

In (11.1) a variable x_j is said to **cover** the ith constraint if x_j appears with a $+1$ coefficient in this constraint. If x_j covers the ith constraint, any 0–1 vector x in which the variable $x_j = 1$ satisfies this constraint automatically. We will now discuss some results which help to fix the values of some of the variables at 1 or 0, and eliminate some constraints, and thereby reduce the problem into an equivalent smaller size problem.

RESULT 11.1 *If $c \overset{\leq}{=} 0$, an optimum solution for (11.1) is $x = e_n$, the vector in $\mathbb{R}^n$ with all entries equal to 1. Terminate.*

RESULT 11.2 *In (11.1) suppose $c \overset{\leq}{\neq} 0$. If j is such that $c_j \overset{\leq}{=} 0$ we can fix the corresponding variable x_j at 1 and eliminate all the constraints covered by this* *variable. If there are no more constraints left, fix all the remaining variables at 0, and this leads to an optimum solution to the problem in this case, terminate.*

RESULT 11.3 *If j is such that $c_j > 0$ and the variable x_j does not appear in any of the remaining constraints with a +1 coefficient, fix the variable x_j at 0.*

Apply Results 11.2, 11.3, as many times as possible and reduce the problem. At the end we are left with a reduced problem of the same form as (11.1), in which every variable has a positive coefficient in the objective function. The greedy method is applied on this reduced problem, and it consists of applying the following general step repeatedly. Here a **free variable** is one whose value is not fixed at 1 or 0 already.

GENERAL STEP In the remaining problem, for each free variable x_j, let d_j be the number of remaining constraints covered by x_j. c_j/d_j can be interpreted as the cost per constraint covered, associated with the free variable x_j at this stage. Find a free variable x_r which is associated with the smallest cost per constraint covered in the remaining problem. So, $c_r/d_r = \min\{c_j/d_j : j$ such that x_j is a free variable$\}$. Fix x_r at 1, and eliminate all the constraints covered by x_r. If there are no constraints left, fix all the remaining free variables at 0, and terminate with the vector obtained as the greedy solution vector. Otherwise apply Result 11.3 to the remaining problem and then go to the next step.

The solution vector at termination is the greedy solution for the set covering problem.

x_1	x_2	x_3	x_4	x_5	x_6	x_7	x_8	x_9	x_{10}	x_{11}	x_{12}	x_{13}	
					1		1	1			1		$\geqq 1$
	1						1	1			1		$\geqq 1$
		1					1	1		1			$\geqq 1$
			1	1			1	1					$\geqq 1$
		1			1		1			1			$\geqq 1$
		1			1	1			1				$\geqq 1$
	1		1	1							1		$\geqq 1$
			1	1	1						1		$\geqq 1$
1	1	1								1			$\geqq 1$
1				1		1					1		$\geqq 1$
3	2	5	6	11	1	12	7	8	4	2	6	9	$= z(x)$

$x_j = 0$ or 1 for all j. Minimize $z(x)$.

As an example, we will apply the greedy method on the problem (11.2). First, applying Result 11.2, we fix $x_{14} = x_{15} = 1$ since their coefficients in $z(x)$ are $\leqq 0$ and eliminate the last two constraints covered by these variables. Now applying

Result 11.3, we fix $x_{16} = 0$. The remaining problem is given above. All blank entries in the table are zero.

Letting d_j = number of remaining constraints covered by free variable x_j, we have the following information on the free variables at this stage.

Free var.	x_1	x_2	x_3	x_4	x_5	x_6	x_7	x_8	x_9	x_{10}	x_{11}	x_{12}	x_{13}
c_j	3	2	5	6	11	1	12	7	8	4	2	6	9
d_j	2	3	3	4	4	3	3	3	4	3	2	3	3
$\dfrac{c_j}{d_j}$	$\frac{3}{2}$	$\frac{2}{3}$	$\frac{5}{3}$	$\frac{6}{4}$	$\frac{11}{4}$	$\frac{1}{3}$	4	$\frac{7}{3}$	2	$\frac{4}{3}$	1	2	3

The free variable with the smallest c_j/d_j of $1/3$ at this stage is x_6. So we fix $x_6 = 1$, and eliminate constraints 5, 6, 8 in the above tableau covered by x_6. The remaining problem is given below.

x_1	x_2	x_3	x_4	x_5	x_7	x_8	x_9	x_{10}	x_{11}	x_{12}	x_{13}	
					1		1	1		1		≥ 1
	1					1	1			1		≥ 1
		1					1	1		1		≥ 1
			1	1		1	1					≥ 1
	1			1	1					1		≥ 1
1	1		1						1			≥ 1
1				1	1					1		≥ 1
3	2	5	6	11	12	7	8	4	2	6	9	$= z(x)$

$$x_j = 0 \text{ or } 1 \text{ for all } j. \text{ Minimize } z(x)$$

We have the following information on the free variables at this stage.

Free var.	x_1	x_2	x_3	x_4	x_5	x_7	x_8	x_9	x_{10}	x_{11}	x_{12}	x_{13}
c_j	3	2	5	6	11	12	7	8	4	2	6	9
d_j	2	3	1	3	3	2	2	4	2	1	3	2
$\dfrac{c_j}{d_j}$	$\frac{3}{2}$	$\frac{2}{3}$	5	2	$\frac{11}{3}$	6	$\frac{7}{2}$	2	2	2	2	$\frac{9}{2}$

The free variable with the smallest c_j/d_j of $2/3$ at this stage is x_2. We fix $x_2 = 1$, and eliminate constraints 2, 5, 6 in the above tableau covered by x_6. x_{11} with a cost coefficient of 2 does not appear in any of the remaining constraints, so we fix it at 0. The remaining problem is given below.

x_1	x_3	x_4	x_5	x_7	x_8	x_9	x_{10}	x_{12}	x_{13}	
				1		1	1		1	≥ 1
	1					1	1	1		≥ 1
		1	1		1	1				≥ 1
1			1	1				1		≥ 1
3	5	6	11	12	7	8	4	6	9	$= z(x)$

$$x_j = 0 \text{ or } 1 \text{ for all } j. \text{ Minimize } z(x)$$

We have the following information on the free variables at this stage.

Free var.	x_1	x_3	x_4	x_5	x_7	x_8	x_9	x_{10}	x_{12}	x_{13}
c_j	3	5	6	11	12	7	8	4	6	9
d_j	1	1	1	2	2	1	3	2	2	1
$\frac{c_j}{d_j}$	3	5	6	$\frac{11}{2}$	6	7	$\frac{8}{3}$	2	3	9

The free variable with the smallest c_j/d_j of 2 at this stage is x_{10}. We fix $x_{10} = 1$ and eliminate constraints 1, 2 in the above tableau. x_3, x_{13} with positive cost coefficients do not appear in any of the remaining constraints, so we fix them at 0. The remaining problem is given below.

x_1	x_4	x_5	x_7	x_8	x_9	x_{12}	
	1	1		1	1		≥ 1
1		1	1			1	≥ 1
3	6	11	12	7	8	6	$= z(x)$

$$x_j = 0 \text{ or } 1 \text{ for all } j. \text{ Minimize } z(x)$$

We have the following information on the free variables at this stage.

Free var.	x_1	x_4	x_5	x_7	x_8	x_9	x_{12}
c_j	3	6	11	12	7	8	6
d_j	1	1	2	1	1	1	1
$\frac{c_j}{d_j}$	3	6	$\frac{11}{2}$	12	7	8	6

The free variable x_1 has the smallest c_j/d_j of 3 at this stage. We fix $x_1 = 1$ and eliminate constraint 2 in the above tableau covered by it. x_7, x_{12} which do not appear in the remaining constraint are fixed at 0. The remaining problem is

$$\text{Minimize} \quad 6x_4 + 11x_5 + 7x_8 + 8x_9$$
$$\text{subject to} \quad x_4 + \ x_5 + \ x_8 + \ x_9 \ \geqq \ 1$$
$$x_j = 0 \text{ or } 1 \text{ for all } j$$

The remaining problem has only one constraint. The free variable x_4 has the smallest c_j/d_j of 6, so we fix it at 1 and the remaining free variables x_5, x_8, x_9 at 0. Collecting the values given to the variables at various stages, we see that the greedy solution obtained is $(x_1, \text{ to } x_{16}) = (1, 1, 0, 1, 0, 1, 0, 0, 0, 1, 0, 0, 0, 1, 1, 0)^T$, with an objective value of 14.

11.4.3   Greedy-Type Methods for the TSP

An n-city TSP with cost matrix $c = (c_{ij})$ is the problem of determining a minimum cost tour which is an order of visiting n cities each once and only once, beginning with a starting city and terminating at the initial city in the end. Hence it is a sequencing problem. The greedy methods for the TSP try to construct a near-optimal tour by building the sequence one element at a time using a greedy approach. Hence they are classified as **tour construction procedures or heuristics**. We describe some of the popular ones here.

1. **Nearest neighbor heuristic**   Starting with an initial city, this procedure builds a sequence one city at a time. The next city in the sequence is always the closest to the current city among the unincluded cities. In the end the last city is joined to the initial city.

   Typically, this process is repeated with each city selected as the initial one. The best among the n tours generated in this process is selected as the output of this algorithm.

   It can be proved [D. Rosenkratz, R. Sterns, and P. Lewis, 1977] that for the Euclidean TSP (i.e., the distance matrix is positive, symmetric, and satisfies the triangle inequality), the following result holds.

$$\frac{\text{Length of the nearest neighbor tour}}{\text{Length of an optimum tour}} \ \leqq \ \frac{1}{2}(1 + \log_2 n)$$

EXAMPLE 11.2

Consider a 6 city TSP with the following cost matrix:

$$c = (c_{ij}) =$$

to $j =$ from $i =$	1	2	3	4	5	6
1	×	14	23	25	36	42
2	14	×	17	23	30	36
3	23	17	×	29	35	28
4	25	23	29	×	17	11
5	36	30	35	17	×	6
6	42	36	28	11	6	×

Here are the tours obtained by the nearest neighbor heuristic in this problem.

Starting city	Nearest neighbor tour	cost
1	1, 2, 3, 6, 5, 4; 1	107
2	2, 1, 3, 6, 5, 4; 2	111
3	3, 2, 1, 4, 6, 5; 3	108
4	4, 6, 5, 2, 1, 3; 4	113
5	5, 6, 4, 2, 1, 3; 5	112
6	6, 5, 4, 2, 1, 3; 6	111

So, the output of this algorithm is the tour 1, 2, 3, 6, 5, 4; 1 with a cost of 107.

2. **The Clark and Wright savings heuristic**   Select an initial city, say city 1. Think of the initial city as a central depot, beginning at which all the cities have to be visited. For each ordered pair of cities not containing the initial city, (i, j) say, compute the savings s_{ij} of visiting the cities in the order 1, i, j, 1 as opposed to visiting each of them independently from 1 as in the orders 1, i, 1 and 1, j, 1. This savings s_{ij} is therefore equal to $(c_{1i} + c_{i1}) + (c_{1j} + c_{j1}) - (c_{1i} + c_{ij} + c_{j1})$. If the cost matrix $c = (c_{ij})$ is symmetric, we have $s_{ij} = s_{ji} = c_{i1} + c_{1j} - c_{ij}$.

Order these savings values in decreasing order from top to bottom. Starting at the top of the savings list and moving downwards, form ever larger subtours by inserting new cities, one at a time, adjacent to the initial city on either side of the subtour, as indicated by the pair corresponding to the present savings, whenever it is feasible to do so. Repeat until a tour is formed.

Typically this process is repeated with each city as the initial one, and the best of all the tours obtained is taken as the output.

As an example, consider the TSP of order 6 with the cost matrix given in Example 11.2. Since the cost matrix is symmetric, the savings $s_{ij} = s_{ji}$ for all i, j; so we need to compute them only for $j > i$. Suppose city 1 is selected as the initial city. $s_{23} = c_{12} + c_{13} - c_{23} = 14 + 23 - 17 = 20$. The savings coefficients computed this way are given below.

	s_{ij} for $j > i$				
$j =$	2	3	4	5	6
$i = 2$	×	20	16	20	20
3		×	19	24	37
4			×	44	56
5				×	72
6					×

The savings coefficients arranged in decreasing order, and the subtour grown are shown below (here we used the fact that the cost matrix is symmetric).

Savings coeff.	Its value	Present subtour
s_{56}	72	1, 5, 6, 1
s_{46}	56	1, 5, 6, 4, 1
s_{45}	44	"
s_{36}	37	"
s_{35}	24	1, 3, 5, 6, 4, 1
s_{23}	20	1, 2, 3, 5, 6, 4; 1

So, the tour 1, 2, 3, 5, 6, 4; 1 with a cost of 108 is obtained by this procedure beginning with city 1 as the initial tour. The same process can be repeated with other cities as initial cities. The best of all the tours generated is the output of the algorithm.

3. **Nearest insertion heuristic**   The insertion procedure grows a subtour until it becomes a tour. In each step it determines which node not already in the subtour should be added next, and where in the subtour it should be inserted.

The algorithm selects one city as the initial city, say city i. Then find $p \neq i$ such that $c_{ip} = \min\{c_{ij} : j \neq i\}$. The initial subtour is i, p, i.

Given a subtour, S say, find a city r not in S, and a city k in S such that $c_{kr} = \min\{c_{pq} : p \in S, q \notin S\}$. City r is known as the **closest or nearest city to S** among those not in it. It is selected as the city to be added to the subtour at this stage. Find an arc (i, j) in subtour which minimizes $c_{ir} + c_{rj} - c_{ij}$. Insert r between i and j on the subtour S.

Repeat until the subtour becomes a tour.

As an example, consider the TSP of order 6 with the cost matrix given in Example 11.2. Suppose city 1 is selected as the initial city. Min$\{c_{1j} : j \neq 1\}$ is c_{12}. So, the initial subtour is 1, 2, 1. The closest outside city to this subtour is city 3, and by symmetry inserting it on any arc of the subtour adds the same to the cost, so we take the next subtour to be 1, 3, 2, 1. The nearest outside city to this subtour is city 4. Min$\{c_{14} + c_{43} - c_{13}, c_{34} + c_{42} - c_{32}, c_{24} + c_{41} - c_{21}\}$ $= c_{14} + c_{43} - c_{13} = 31$. So, the new subtour is 1, 4, 3, 2, 1. Continuing this way we get the subtour 1, 4, 6, 3, 2, 1; and finally the tour 1, 4, 5, 6, 3, 2; 1 with a cost of 107. The procedure can be repeated with each city as the initial city, and the best of the tours obtained taken as the output of the algorithm.

It has been proved [D. Rosenkratz, R. Sterns, and P. Lewis, 1977] that on a Euclidean TSP, the tour obtained by this method has a cost no more than twice the cost of an optimum tour.

4. **Cheapest insertion heuristic**   This procedure also grows a subtour until it becomes a tour. It is initiated the same way as the nearest insertion heuristic. Given a subtour S say, it finds an arc (i, j) in S and a city r not in S such that $c_{ir} + c_{rj} - c_{ij}$ is minimal, and then inserts r between i and j. Repeat until a tour is obtained. This procedure also can be repeated with each city as the initial city, and the best of the tours obtained taken as the output. On Euclidean TSPs, this method has the same worst case bound as the nearest insertion heuristic.

5. **Nearest merger heuristic**   This procedure is initiated with n subtours, each consisting of a single city and no arcs. In each step it finds the least costly arc, (a, b) say, that goes between two subtours in the current list, and merges these two subtours into a single subtour.

If a, b are two single city subtours in the current list, their merger replaces them with the subtour a, b, a.

If one of a, b is in a single city subtour, say a, and the other in a multi-city subtour; insert a into the subtour containing b using the cheapest way of inserting it as discussed under the cheapest insertion heuristic.

If the subtours in the current list containing cities a and b each have two or more cities, find the arc (p_1, q_1) in the first subtour, and the arc (p_2, q_2) in the second subtour, such that $c_{p_1 p_2} + c_{q_2 q_1} - c_{p_1 q_1} - c_{p_2 q_2}$ is minimized. Then merge these subtours by deleting arcs $(p_1, q_1), (p_2, q_2)$ from them, and adding $(p_1, p_2), (q_2, q_1)$ to them.

We discussed a variety of greedy-type single pass heuristics for the TSP to give the reader some idea of how greedy methods can be developed for combinatorial optimization problems. All the methods discussed here for the TSP produce reasonably good tours with objective values usually close to the optimum objective value.

11.4.4   A Greedy-Type Method for the Single Depot Vehicle Routing Problem

This problem is concerned with delivering goods to customers at various locations in a region by a fleet of vehicles based at a depot. The index $i = 0$ denotes the depot, and $i = 1$ to n denote the customer locations. N denotes the number of vehicles available at the depot. The following data is given.

c_{ij} =   distance (or cost, or driving time for traveling) from i to j, for $i, j = 0$ to n.

k_v =   capacity of vehicle v in tons or some other units, $v = 1$ to N.

T_v =   maximum distance (or cost, or driving time) that vehicle v can operate, $v = 1$ to N.

d_i =   demand or amount of material (in tons or other units in which vehicle capacities are also measured) to be delivered to customer i, $i = 1$ to n; $d_0 = 0$.

All customer demands need to be delivered. The problem is to determine: (i) the subset of customers to be allotted to each vehicle that is used, and (ii) the route that each vehicle should follow (i.e., the order in which it should visit its allotted customers) so as to minimize the total distance (or cost or driving time) of all the vehicles used to make the deliveries. This is a prototype of a common problem faced by many warehouses, department stores, parcel carriers, and trucking firms and is therefore a very important problem. We will discuss a greedy-type method known as the **Clarke and Wright savings heuristic** [G. Clarke and J. Wright, 1964] that is very popular. It is an exchange procedure, which in each step exchanges the current set of routes for a better set. Initially, think of each customer being serviced by a separate vehicle from the depot. See the left part of Figure 11.1.

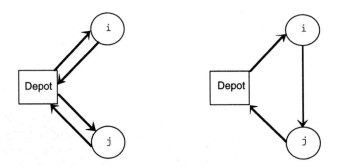

Figure 11.1 On the left, customers i and j are serviced by two vehicles from the depot. On the right, they are both serviced by the same vehicle.

If it is feasible to service customer j by the same vehicle which serviced customer i (i.e., if the vehicle capacity and maximum distance constraints are not violated by doing this) before returning to the depot (see the right part of Figure 11.1), the savings in distance will be $s_{ij} = c_{0i} + c_{i0} + c_{0j} + c_{j0} - (c_{0i} + c_{ij} + c_{j0}) = c_{i0} + c_{0j} - c_{ij}$. These savings coefficients s_{ij} are computed for all $i \neq j = 1$ to n, and ordered in decreasing order from top to bottom. Starting at the top of their savings list, form a route for vehicle 1 (which will be a subtour beginning and ending at the depot) by inserting new customers, one at a time, adjacent to the depot on either side of the depot as discussed in the second algorithm in Section 11.4.3, until either the vehicle capacity is used up or the maximum distance it can travel is reached. Now delete the customers allotted to vehicle 1 from the list.

Repeat the same process to form a route for vehicle 2 with the savings coefficients for pairs of remaining customers; and continue in the same way until all the customers are allotted to a vehicle.

The above process determines the subset of customers to be serviced by each vehicle used, and a tour to be followed by each vehicle to service its customers. One can now try to find a better tour for each vehicle to service its allotted customers using some of the algorithms discussed in Sections 11.4.3 and 11.5.

As an example consider the problem involving 12 customers and the following data. In this problem there is no limit on the distance that a truck can travel. For $i = 1$ to 12, d_i is the amount to be delivered to customer i in gallons.

Customer i	1	2	3	4	5	6
d_i	1200	1700	1500	1400	1700	1400
Customer i	7	8	9	10	11	12
d_i	1200	1900	1800	1600	1700	1100

Truck capacity (gallons)	Up to 4000	4000-5000	5000-6000
Number available	10	7	4

Symmetric distance matrix (miles) $= (c_{ij})$

to from	0	1	2	3	4	5	6	7	8	9	10	11	12
0	$\times$	9	14	21	23	22	25	32	36	38	42	50	50
1		$\times$	5	12	22	21	24	31	35	37	41	49	51
2			$\times$	7	17	16	23	26	30	36	36	44	46
3				$\times$	10	21	30	27	37	43	31	37	39
4					$\times$	19	28	25	35	41	29	31	29
5						$\times$	9	10	16	22	20	28	30
6							$\times$	7	11	13	17	25	27
7								$\times$	10	16	10	18	20
8									$\times$	6	6	14	16
9										$\times$	12	12	20
10											$\times$	8	10
11												$\times$	10
12													$\times$

Since the distance matrix is symmetric, the matrix of savings coefficients is also symmetric. For example $s_{2,1} = s_{1,2} = c_{1,0} + c_{0,2} - c_{1,2} = 9 + 14 - 5 = 18$. All the savings coefficients are computed in the same manner and are given below.

Symmetric savings matrix (miles) $= (s_{ij})$

$j =$	1	2	3	4	5	6	7	8	9	10	11	12
$i = 1$	×	18	18	10	10	10	10	10	10	10	10	8
2		×	28	20	20	16	20	20	16	20	20	18
3			×	34	22	16	26	20	16	32	34	32
4				×	26	20	30	24	20	36	42	44
5					×	38	44	42	38	44	44	42
6						×	50	50	50	50	50	48
7							×	58	54	64	64	62
8								×	68	72	72	70
9									×	68	76	68
10										×	84	82
11											×	90
12												×

The largest savings coefficient is $s_{11,12} = 90$. So, the initial subtour for vehicle 1 is 0, 11, 12; 0. The demand at customers 11, 12 put together is $1700 + 1100 = 2800$ gallons. The next biggest savings coefficient is $s_{10,11} = 84$. So, we insert customer 10 into the subtour for vehicle 1, leading to the new subtour 0, 10, 11, 12; 0 with a total demand of $1600 + 2800 = 4400$ gallons. Customers 10, 11, 12 are already assigned to vehicle 1. The next largest savings coefficient involving one of the remaining customers is $s_{9,11} = 76$, but adding customer 9 to vehicle 1 will make the total demand $= 1800 + 4400 = 6200$ gallons; but the depot has no vehicles of this capacity, so we drop customer 9 from consideration for vehicle 1. The next highest savings coefficients are $s_{8,10} = s_{8,11} = s_{8,12} = 72$, but again, customer 8 cannot be allocated to vehicle 1 because this allocation will make the total demand > the largest capacity of available vehicles.

The next highest savings coefficient is $s_{8,9} = 68$ and both customers 8 and 9 are presently unassigned. So, we select 0, 8, 9; 0 as initial subtour for vehicle 2. The combined demand of these two customers is $1900 + 1800 = 3700$ gallons.

The next highest savings coefficients are $s_{7,10} = s_{7,11} = s_{7,12} = 64$. So we insert customer 7 into the subtour for vehicle 1, leading to the new subtour 0, 7, 10, 11, 12; 0 with a total demand of 5600 gallons for vehicle 1. So, we make vehicle 1 to be one of the vehicles with capacity 5000-6000 gallons, and $\tau_1 = 0, 7, 10, 11, 12$; 0 as the subtour for it to follow. No more customers can be added to this vehicle because of the capacity constraint. Since this vehicle is now full, in the sequel we ignore all the savings coefficients involving one of the customers 7, 10, 11, or 12 assigned to this vehicle. And there are 3 vehicles of capacity 5000-6000 gallons still available.

The next highest savings coefficient involving an unassigned customer is $s_{6,8} =$ 50. So, we combine customer 6 in vehicle 2 leading to the new subtour for it of 0, 6, 8, 9; 0 with a total demand of 5100 gallons. No more customers can be assigned to vehicle 2 because of the capacity constraint. Thus we make vehicle 2 to be another vehicle with capacity 5000-6000 gallons, and $\tau_2 = $ 0, 6, 8, 9; 0 as the subtour for it to follow. And there are 2 more vehicles of capacity 5000-6000 gallons left.

The next highest savings coefficient involving unassigned customers is $s_{3,4} = 34$. So, we combine customers 3, 4 into the subtour 0, 3, 4; 0 which will be the initial subtour for vehicle 3 with a total demand of 2900 gallons. The next highest savings coefficient not involving a customer assigned to the already full vehicles 1, 2, is $s_{2,3} = 28$. So, we insert customer 2 into the subtour for vehicle 3, changing it into 0, 2, 3, 4; 0 with a total demand of 4600 gallons.

The next highest savings coefficients of $s_{45} = 26, s_{12} = 18$ cannot be used because adding any of customers 5 or 1 to vehicle 3 will exceed maximum available vehicle capacity. This leads to the next $s_{1,5} = 10$. Hence we combine customers 1, 5 into the subtour 0, 1, 5; 0 with a total demand of 2900 gallons for vehicle 4. Now all the customers are assigned. Here is a summary of the assignments.

Vehicle no.	Subtour to be followed	Total demand	Vehicle to be used
1	$\tau_1 = $ 0, 7, 10, 11, 12; 0	5600 gal.	5000-6000 gal. capacity
2	$\tau_2 = $ 0, 6, 8, 9; 0	5100 gal.	"
3	$\tau_3 = $ 0, 2, 3, 4; 0	4600 gal.	4000-5000 gal. capacity
4	$\tau_4 = $ 0, 1, 5; 0	2900 gal.	4000 gal. capacity

We should now try to find better tours for each vehicle to cover the customers assigned to it, using some of the other methods discussed in Sections 11.4.3 and the following sections.

11.4.5 General Comments on Greedy Heuristics

In this section we discussed a variety of ways of developing single pass heuristic methods for a variety of problems based on the greedy principle. One important point to notice is that heuristic methods are always tailormade for the problem being solved, taking its special structure into account. Practical experience indicates that for the problems discussed in this section, the heuristic methods discussed usually lead to satisfactory near-optimal solutions.

A point of caution. In Chapter 1 we presented a profit maximizing assignment problem for which the standard greedy procedure yielded the worst possible solution, the one minimizing the profit. It is perfectly reasonable to use greedy or other single pass heuristic methods if either theoretical worst case analysis, or extensive computational testing, has already established that the method leads to reasonable near optimal solutions. In the absence of encouraging theoretical results on worst case error bounds, or encouraging results from computational tests, one should be wary of relying solely on a greedy or any other single pass heuristic.

In this case it is always better to combine it with some heuristic search methods
discussed in the following sections.

Exercise

11.2 A bank account location problem   A business firm has clients in cities
$i = 1$ to m, and can maintain bank accounts in locations $j = 1$ to n. When the
payment for a client is mailed by a check, there is usually some time lag before the
check is cashed (time for the mail to reach back and forth), in that time the firm
continues to collect interest on that money. Depending on the volume of business
in city i, and the time it takes for mail to go between city i and location j, one
can estimate the float = expected benefit s_{ij} in the form of this interest if clients
in city i are paid by checks drawn out of a bank account in location j, $i = 1$ to m,
$j = 1$ to n. The following data is given.

$c_j=$ cost in money units for maintaining a bank account in
location $j = 1$ to n, per year

$s_{ij}=$ total float (= expected benefit in the form of interest on
money between the time a check for it is mailed, and the
time that check is cashed) per year, if payments due for
customers in city i are mailed in the form of checks drawn
out of a bank account in location j, $i = 1$ to m, $j = 1$ to
n.

$N=$ upper bound on the number of bank accounts that the firm
is willing to maintain in locations 1 to n.

(i) Formulate the problem of determining the subset of locations where bank
accounts should be maintained, and the bank accounts through which customers in
each city should be paid, so as to maximize (the total annual float earned − yearly
cost of maintaining the bank accounts), as a 0−1 pure IP.
(ii) Consider the numerical example in which $m = 7$, $n = 5$, $N = 3$, and

	$j =$	1	2	3	4	5
	$i = 1$	2	11	6	9	8
	2	7	1	8	2	10
	3	7	3	2	3	4
$(s_{ij}) =$	4	10	9	4	2	1
	5	3	8	5	6	2
	6	4	3	4	1	6
	7	6	5	1	8	4
	c_j	3	2	1	3	4

Suppose $\mathbf{J} \subset \{1, \ldots, n\}$, the subset of locations where bank accounts are to
be maintained, is given. Then clearly for each $i = 1$ to m, customers in city i

should be paid by checks drawn out of location r where r attains the maximum in $\max\{s_{ij} : j \in \mathbf{J}\}$, i.e., each customer should be paid from the bank account in location with the maximum float value in the row of the customer city, to maximize total float.

For illustration, if bank accounts are opened in locations $j = 1, 3$ only in the example given above: customers in cities $i = 1, 2, 5$ should be paid out of locations $j = 3$; customers in cities $i = 3, 4, 7$ should be paid out of location $j = 1$; and customers in city $i = 6$ can be paid out of locations $j = 1$ or 3 (the float values are equal). Also if a new bank account is opened in location 4, only customers in cities 1, 5, and 7 should be switched from their current account to this new account, because this will increase the float coming from them. Thus when bank accounts are already available in locations 1, 3, opening a new bank account in location 4 leads to a net extra profit of $(9 - 6) + (6 - 5) + (8 - 6) - 3 = 3$ money units (here the terms $9 - 6$, $6 - 5$, $8 - 6$ are the extra floats that will be obtained when customers in cities 1, 5, 7 are switched from their present account to this new account; and the last term 3 is the cost of maintaining an account in the new location). This net quantity 3 is called the **evaluation** of location 4 when bank accounts are already available in locations 1, 3. It measures the net extra profit that can be gained by opening a new account at location 4.

Using such evaluations as the criterion to be greedy upon, develop a greedy method for finding the subset of locations where bank accounts should be maintained in this problem. The method should open one new account at a time, until either N accounts are opened, or it turns out that opening a new account only decreases the net income.

Using this method, solve the numerical example with the data given above.

11.5 Interchange Heuristics

Interchange heuristics are local search methods that start with a solution and search for better solutions through local improvement, i.e., through small changes in the solution in each step. When the problem is represented as one of selecting an optimal subset from a set (or as one of arranging a set of objects in a sequence optimally), the method starts with an initial solution which may either be randomly generated or obtained by a single pass heuristic such as the greedy method; and attempts to improve it by exchanging a small number of elements in the solution with those outside it (or by changing the positions of a small number of objects in a sequence). If a better solution is found by such interchanges, the same process is repeated on it. This process is continued until a solution that cannot be improved by such interchanges is found. This final solution is a local minimum under such interchange operations. The procedure is usually repeated with several initial solutions, and the best of the local minima found is taken as the output of the algorithm. Heuristic methods based on this type of search are called **interchange** or **exchange heuristic methods**.

For example, consider the n-city TSP with cost matrix $c = (c_{ij})$. The interchange heuristic for the TSP begins with a tour τ, and searches for a better tour among all those that differ from τ in 2 or 3 or a small number of arcs. If such a tour is found, the method moves to that and continues in the same way.

In Figure 11.2 we display a 3-arc interchange. The nodes represent the cities with their numbers entered inside. The initial tour τ_1 is drawn in solid lines. The second tour τ_2 is obtained by exchanging the three thick arcs in τ_1 with the three dashed arcs.

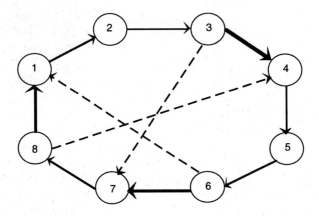

Figure 11.2 The three arc swap.

Deleting the thick arcs in the solid tour τ_1 in Figure 11.2 is equivalent to looking at the restricted problem with the arcs (1, 2), (2, 3); (4, 5), (5, 6); (7, 8) already fixed in the tour. To avoid subtours, this implies that arcs (3, 1), (6, 4), and (8, 7) are forbidden. Hence from the results in Chapter 10, we know that the three best outside arcs to replace the thick arcs in Figure 11.2 are the arcs in an optimum tour for the TSP of order 3 with the following cost matrix:

to	1	4	7
from 3	$\times$	c_{34}	c_{37}
6	c_{61}	$\times$	c_{67}
8	c_{81}	c_{84}	$\times$

which can be solved easily by inspection, since a TSP of order 3 has only 2 possible tours. If this produces a tour τ_2 with total cost less than that of τ_1, the choice of the set of three thick arcs to exchange from τ_1 has been successful, and the process is now repeated with the new tour τ_2. If the cost of τ_2 is $\geqq$ the cost of τ_1, the process is continued with τ_1 and a different subset of three arcs from it to exchange. If every subset of three arcs to exchange from τ_1 leads to a tour whose cost is $\overset{\geq}{=}$ that of τ_1, the three arc interchange heuristic terminates with τ_1 as a near optimum tour. To obtain a close approximation to an optimum tour, one should repeat this interchange procedure with a number of initial tours, and take the best of all the tours obtained as the final output.

We will now discuss an interchange heuristic method for a location problem.

11.5.1   An Interchange Heuristic for a Training Center Location Problem

A large company has offices in many cities around the country. Due to the continuing development of new technologies, they expect to have a steady demand in the future for the training of their employees. Hence they are embarking on a huge employee training and education program. They want to develop a few training centers, these will be located in a subset of cities where the company has offices. Once these centers are established, employees from various cities will be sent to these centers for training. We assume that each center will have the capacity to take an unlimited number of trainees.

All the employees needing training in a city will be assigned to the same training site (i.e., they will not be split between different training sites). Also, a trainee may have to make several trips back and forth before his training is complete. We are given the following data.

$$
\begin{aligned}
n &= \text{number of cities where offices are located} \\
s_i &= \text{expected number of employees at city } i \text{ needing training} \\
&\quad \text{annually, } i = 1 \text{ to } n \\
m_i &= \text{expected number of trips between city } i \text{ and training center} \\
&\quad \text{annually by trainees from city } i, \, i = 1 \text{ to } n \\
c_{ij} &= \text{cost per trip between cities } i \text{ and } j, \, i,j = 1 \text{ to } n \\
d_j &= \text{expected staying cost per trainee during training program,} \\
&\quad \text{if a training center is located in city } j, \, j = 1 \text{ to } n \\
r_{ij} &= s_i d_j + m_i c_{ij} = \text{total cost (travel + staying) incurred annu-} \\
&\quad \text{ally by trainees from city } i \text{ if they are assigned to a training} \\
&\quad \text{center located in city } j, \, i,j = 1 \text{ to } n \\
p &= \text{number of training centers to be opened}
\end{aligned}
$$

Suppose $V = \{j_1, \ldots, j_p\}$, the set of cities among 1 to n where training centers will be opened, is given. Then, to minimize the total cost, we should assign the trainees from city i to the cheapest training site in V, i.e., to $j_t \in V$ where j_t satisfies $r_{ij_t} = \min\{r_{ij_k} : k = 1 \text{ to } p\}$, for each $i = 1$ to n. Thus given the set of training sites $V = \{j_1, \ldots, j_p\}$, the minimum total annual cost (expected annual cost of travel + stay at assigned training centers during training for all the trainees) is $\sum_{i=1}^{n}(\min\{r_{ij_k} : k = 1 \text{ to } p\})$.

The problem is to find the set of training sites that minimizes the total annual cost. This problem is known as the p-**median problem**.

The interchange heuristic for this problem is initiated with a set of p sites for training centers, and applies the following general step repeatedly:

General Step   Let V be the present set of sites for training centers. For each $a \in V, b \notin V$ define Δ_{ab} as the change in the total cost if a in V is replaced by b. For each $a \in V$ define

$$T_a \quad = \quad \text{market set for city } a, \text{ i.e., the set of cities which send their trainees to } a, \text{ it is } \{i : r_{ia} = \min\{r_{ij} : j \in V\}\}.$$

For any $a \in V, b \notin V$, to compute Δ_{ab} it is necessary to find the "cheapest" new assignments for trainees from cities in T_a when b replaces a from V; and any other cities outside T_a which will also be switched from their present assignments to b.

Start computing Δ_{ab} for $a \in V, b \notin V$. In this process, if Δ_{gq} for $g \in V, q \notin V$ is the first negative quantity obtained, let $V' = \{q\} \cup (V \setminus \{g\})$. With V' as the new set of sites for training centers repeat this general step.

On the other hand, if $\Delta_{ab} \overset{\geq}{=} 0$ for all $a \in V, b \notin V$, accept the present set V as a near optimum set of training sites, and terminate.

EXAMPLE 11.3

As an example, consider the problem with the following data. $n = 8$, $p = 2$ [J. G. Klincewicz, 1980].

i	City	$s_i = $ no. trainees	$m_i = $ no. trips	$d_i = $ staying cost/trainee
1	Dallas	2	8	$1800
2	Denver	3	12	1590
3	G. Falls	6	24	1290
4	L.A.	8	32	2100
5	Omaha	5	20	1560
6	St. Louis	4	16	1650
7	S.F.	7	28	2130
8	Seattle	1	4	1680

$$c_{ij} = \text{travel cost/trip (symmetric)}$$

to j	1	2	3	4	5	6	7	8
from $i = 1$	0							
2	170	0						
3	266	163	0					
4	262	197	223	0				
5	158	141	202	273	0			
6	152	191	260	318	115	0		
7	301	216	204	113	292	343	0	
8	333	227	146	217	282	340	172	0

We compute $r_{ij} = s_i d_j + m_i c_{ij} = $ total cost of trainees from city i to be trained at a training center in city j, and give them below.

$$r_{ij}$$

$j =$	1	2	3	4	5	6	7	8
$i = 1$	3600	4510	4708	6296	4384	4516	6668	6024
2	7440	4770	5826	8664	6372	6242	8982	7764
3	17184	13452	7740	17952	14208	16140	17676	13584
4	22784	19024	17456	16800	21216	23376	20656	20384
5	12160	10770	10490	15960	7800	10550	16490	14040
6	9632	9506	9800	13488	8080	6600	14128	12160
7	21028	17178	14742	17864	19096	21154	14910	16576
8	3132	2498	1874	2968	2688	3010	2818	1680

Suppose the initial set of sites for training centers is $V_1 = \{6, 8\}$. With this set of training centers, since $r_{16} = 4516 < r_{18} = 6024$, trainees from city 1 will be assigned to the training center at city 6, i.e., city 1 is in the market set for training center at city 6. In the same way, we find that the market set for the center at city 6 is $T_6 = \{1, 2, 5, 6\}$; and the market set for the center at city 8 is $T_8 = \{3, 4, 7, 8\}$.

If 8 in V_1 is replaced by 1, we verify that cities 3 and 8 in T_8 will join the market set of 6 after the change, but 4 and 7 will join the market set of 1. Also, city 1 will move from the market set of 6 to that of 1. So, $\Delta_{8,1} = (3600 - 4516) + (16140 - 13584) + (22784 - 20384) + (21028 - 16576) + (3010 - 1680) = 9822 > 0$. So, replacing 8 by 1 in V_1 only increases the total cost.

Similarly we compute $\Delta_{8,2} = -4048 < 0$. Thus replacing 8 by 2 in V_1 reduces the total cost by \$4048. So, we make the exchange and have the new set of sites for training centers $V_2 = \{6, 2\}$.

The algorithm can be continued with the new set V_2 in the same way. It will terminate when a set of sites for training centers which cannot be improved by such interchanges is obtained.

To get even better solutions, the procedure should be repeated with different initial sets, and the best of all the solutions obtained is taken as the output.

Exercise

11.3 Formulate the problem of finding the best locations for training centers, and the assignment of cities to training centers, for the numerical example in Example 11.3 as a pure $0-1$ IP.

Practical experience indicates that the interchange heuristic discussed here for the training center location problem, and other p-median type location problems similar to it, gives excellent results. In a computational experiment, this heuristic obtained solutions verified to be optimal by the B&B approach in 26 out of the 27 cases tested, and within 1% of the optimum cost in the other case.

In general, a composite heuristic approach consisting of something like a greedy method to generate one or more good initial solutions, and an interchange method to search for better solutions by local improvement beginning with the initial solutions produced by the first method, leads to reasonable solutions for large scale combinatorial problems in applications.

11.6   Simulated Annealing

Simulated annealing (SA) is a type of local search heuristic involving some random elements in the way the algorithm proceeds.

For a problem in which the objective function is to be minimized, the simplest form of local search is a **descent method** that starts with an initial solution. The method should have a mechanism for generating a neighbor of the current solution. If the generated neighbor has a smaller objective value, it becomes the new current solution, otherwise the current solution is retained. The process is repeated until a solution is reached with no possibility of improvement in its neighborhood, such a point is a **local minimum**, and the descent method terminates. This is one of the disadvantages of simple local search methods. By requiring that the iterative steps move only downhill on the objective function surface, they may get stuck at a local minimum which may be far away from any global minimum. Simple local search methods try to avoid this difficulty by running the descent method several times, starting from different initial solutions, and finally taking the best of the local minima found.

On the other hand, SA avoids getting trapped at a local minimum by sometimes accepting a neighborhood move that increases the objective value, using a probabilistic acceptance criterion. These uphill moves make it possible to move away from local minima and explore the feasible region in its entirety. In the course of an SA algorithm, the probability of accepting such uphill moves slowly decreases to 0.

The motivation for the SA algorithm, and its name, come from an analogy with a highly successful Monte Carlo simulation model for the physical annealing process of finding low energy states of a solid. Physical annealing is the process of finding the ground state of a solid which corresponds to the minimum energy configuration, by initially melting the substance, and then lowering the temperature slowly, spending a long time at temperatures close to the freezing point. Metropolis et al. [1953] introduced the simple Monte Carlo simulation algorithm that modeled the physical annealing process very successfully. At each iteration of this algorithm, the system is given a small displacement, and the resulting change δ in the energy of the system is calculated. If $\delta < 0$, the resulting change is accepted, but if $\delta > 0$ the change is accepted with probability $\exp(-\delta/T)$ where T is a constant times the temperature, which we will refer to as the temperature. If a large number of iterations are carried out at each temperature, the model finds the thermal equilibrium that the system attains at that temperature. Simulating the transition to the equilibrium, and decreasing the temperature, one can find states of the system with smaller and smaller values of mean energy.

By first melting the model system at a high effective temperature, and then lowering the temperature in slow deliberate steps after waiting for equilibrium to be established at each temperature, one has in effect performed a simulated annealing procedure. Experimentally it is precisely such annealing that has the best chance of bringing a solid to a good approximation of its true ground state rather than freezing it into a metastable configuration that corresponds to a local but not global minimum energy level. The sequence of temperatures used, the number of rearrangements attempted to reach equilibrium at each temperature, and the criterion used for stopping, are collectively known as the **cooling or annealing schedule**.

In the analogy, the different feasible solutions of a combinatorial optimization problem correspond to the different states of the substance. The objective function to be minimized corresponds to the energy of the system. However, the concept of temperature in the physical system has no obvious equivalent in combinatorial optimization problems. In SA algorithms for optimization, this temperature is simply a control parameter in the same units as the cost function. The probability of accepting an uphill move which causes an increase $\delta > 0$ in the objective function, $\exp(-\delta/T)$, is called the **acceptance function**. This acceptance function implies that small increases in the objective function are more likely to be accepted than large increases, and that when T is high, most moves will be accepted; but as T approaches 0, most uphill moves will be rejected. So, in SA, the algorithm is started with a high value of T to avoid being permanently trapped at a local minimum. The algorithm drops the temperature parameter gradually, making a certain number of neighborhood moves at each temperature.

The simple local search method that accepts only rearrangements that lower the cost function, corresponds to extremely rapid quenching where the temperature is reduced quickly, so it should not be surprising that the resulting solutions are usually metastable. SA provides a generalization of iterative improvement in which controlled uphill moves are incorporated in the search for a better solution. This helps to attain some of the speed and reliability of descent algorithms while avoiding their propensity to stick at local minima.

Let $\mathbf{X}$ denote the set of feasible solutions of a combinatorial optimization problem, and $z(x)$ the objective function to be minimized over $\mathbf{X}$. $|\mathbf{X}|$ is exponentially large in terms of the natural measure of the size of the problem; for example, in the TSP of order n, $|\mathbf{X}| = (n-1)!$. To apply SA on this problem we need to define a **neighborhood** for each $x \in \mathbf{X}$. The essential feature of these neighborhoods is: from any point in $\mathbf{X}$ we should be able to reach any other point in $\mathbf{X}$ by a path consisting of moves from a point to an adjacent point. Also, usually neighborhoods are symmetric, i.e., y is in the neighborhood of x iff x is in the neighborhood of y. The efficiency of SA depends on the neighborhood structure that is used.

If the problem is posed as one of finding an optimum sequence of a set of elements, it is convenient to incorporate any constraints on the desired sequence, in the objective function using appropriate penalty function terms corresponding to them. Then $\mathbf{X}$ becomes the set of all permutations of the elements. The neighbors of a sequence could be considered as all those that can be obtained by interchanging the elements in two positions, or those obtained by reversing the order of the elements in a segment of the sequence, etc. By designing neighborhoods taking advantage of the problem structure, the efficiency of the SA algorithm can be improved substantially.

We also need an $x^0 \in \mathbf{X}$ to initiate the algorithm, the initial value T_0 of the temperature parameter T, the decreasing sequence $T_t, t = 0, 1, \ldots$ of values of temperature to be used, the number of iterations to be performed at each temperature (N_t at temperature T_t, $t = 0, 1, \ldots$), and a stopping criterion to terminate the algorithm. And we need a mechanism to select a solution y from the neighborhood of the current point x in each step of the algorithm. Once these choices are made, the algorithm proceeds as below.

GENERAL SA ALGORITHM

Initialization   Let x^0, T_0 be the initial solution and temperature, respectively.

General Step   When the temperature is T_t do the following. Set iteration counter n to 0. If x^i is the current solution, find a solution y in the neighborhood of x^i at random. If $z(y) \leqq z(x^i)$ make $x^{i+1} = y$. If $z(y) > z(x^i)$ make

$$
x^{i+1} = \begin{cases} y & \text{with probability } \exp\left(-\frac{z(y)-z(x^i))}{T_t}\right) \\ x^i & \text{with probability } 1 - \exp\left(-\frac{z(y)-z(x^i)}{T_t}\right) \end{cases}
$$

Increase the iteration count n by 1 and continue with x^{i+1} as the current solution. When $n = N_t$, change T to T_{t+1} and start the next step.

Continue until the stopping criterion is met.

Discussion

The cooling schedule may be developed by trial and error for a given problem, but a great variety of cooling schedules have now been suggested. $T_{t+1} = \alpha T_t$ where α is a number between 0.8 to 0.99 is sometimes used, with N_t being determined as a sufficient number of iterations subject to a constant upper bound. The cooling schedule $T_t = d/\log t$ where d is some positive constant, is also quite popular.

As an example we consider the TSP of order n with $c = (c_{ij})$ as the cost matrix. We will represent the tour $x = p_1, p_2, \ldots, p_n; p_1$ by the permutation $p_1, p_2, \ldots, p_n$, and its cost is $z(x) = \sum_{r=1}^{n-1} c_{p_r, p_{r+1}} + c_{p_n, p_1}$.

We take the neighborhood of a tour to be the set of all tours corresponding to permutations obtained by selecting a pair of positions in its permutation and reversing the segment between them. For example, consider $n = 7$, and the tour x^0 corresponding to the permutation 6, 3, 7, 2, 5, 4, 1. The tour x^1 corresponding to the permutation 6, 4, 5, 2, 7, 3, 1 is obtained by reversing the segment between positions 1 and 6 in the permutation for x^0; it is a neighbor of x^0.

We now describe the various steps in the SA algorithm for the TSP based on this definition of neighborhood. Here the symbol n denotes the number of cities, i.e., the order of the TSP. We take N_t, a target for the number of iterations to be performed at temperature T_t, to be n for all t. The actual value of N_t used may be more than n depending on the observed performance during the algorithm. We use the symbol i as an iteration counter, and also use it in defining the neighborhood of the current tour from which the next tour will be selected.

AN SA ALGORITHM FOR THE TSP

Step 1   Select the initial permutation $x^0 = p_1^0, \ldots, p_n^0$ and initial temperature T_0.

Step 2   Let $x = p_1, \ldots, p_n$ be the present permutation, $z(x)$ the cost of the corresponding tour, and T the current temperature.

Step 3   Set $i = 1$.

Step 4   Let $x = p_1, \ldots, p_n$ be the present permutation. Select an integer $j \neq i$ between 1 to n at random. Define $a = \min\{i, j\}, b = \max\{i, j\}$. Define y to be the permutation obtained by reversing the segment between a and b in the present permutation x, and $z(y)$ the cost of the tour corresponding to y.

If $z(y) \overset{\le}{=} z(x)$ accept y as the new current permutation. If $z(y) > z(x)$, let

$$\text{New current permutation} = \begin{cases} y & \text{with probability } \exp\left(-\frac{z(y)-z(x)}{T}\right) \\ x & \text{with probability } 1 - \exp\left(-\frac{z(y)-z(x)}{T}\right) \end{cases}$$

where T is the current temperature. Go to Step 5.

Step 5   If $i < n$, increase it by 1 and go back to Step 4. If $i = n$ and enough number of iterations have been performed at the current temperature, go to Step 6; otherwise, go to Step 3.

Step 6   If the temperature has reached the smallest value, terminate with the best tour obtained so far. Otherwise, change the temperature to the next value in the temperature sequence and go back to Step 2.

One has to repeat the iterations at each temperature until an equilibrium seems to have been reached. Then the temperature is decreased and the process repeated. Repeating this, solutions of improved cost will result, and one has to decide suitable stopping criteria. In the end, one can perform a deterministic local search beginning with the best solution obtained in the algorithm and continue as long as better solutions are found.

The attraction of SA is that it is general, yet simple to apply. Solving a problem with it requires a neighborhood structure to be specified, and a procedure for generating neighbors of solution points at random. Researchers are using SA extensively on various problems and obtaining good results.

11.7   Genetic Algorithms

Inspired by biological systems that adapt to the environment and evolve into highly successful organisms over many generations, J. Holland [1975] proposed heuristic search methods for hard combinatorial optimization problems based on operations called **mating, reproduction, cloning, crossover**, and **mutation**; these are patterned upon biological activities bearing the same names. Hence these methods are appropriately called **genetic algorithms** (GA). GAs are robust and effective iterative adaptive search algorithms with some of the creativity of human reasoning.

The first step to develop a GA for an optimization problem is to represent it so that every solution for it is in the form of a string of bits (integers or characters), all of them consisting of the same number of elements, n say. Each candidate solution represented as a string is known as an **organism** or a **chromosome**. So each chromosome is a bit string of length n. The variable in a position on the chromosome is called the **gene** at that position, and its value in a particular chromosome is called its **allele** in that chromosome. For example, if $n = 3$, a general chromosome is $x = (x_1, x_2, x_3)$ where $x_1, x_2,$ and x_3 are the genes on this chromosome in the three positions. In the chromosome (3, 8, 9), the second gene has allele 8.

We will discuss GAs as they apply to minimization problems. We assume that the objective function value at every chromosome is positive; this can be arranged by adding a suitable positive constant to the objective function value of every chromosome, if necessary.

To develop a GA for it, the problem has to be transformed into an unconstrained optimization problem so that every string of length n can be looked upon as a solution vector for the problem. For this purpose, a penalty function, consisting of nonnegative penalty terms corresponding to each constraint in the original problem, is constructed. Each penalty term is always 0 at every point satisfying it, and positive at every point violating it. So, the penalty function has value 0 at every feasible solution to the original problem, and a positive value at every infeasible point. Also, the value of the penalty function at an infeasible point increases rapidly as the point moves farther away from the feasible region. The construction of the penalty function is illustrated later with an example. The **fitness measure** is defined to be the objective function plus the penalty function. It is also called the **evaluation function**. Thus at every feasible solution to the original problem, the fitness measure is equal to the objective function value at that point. Hence associated with each chromosome is an objective function value and a fitness measure. From the way the fitness measure is defined, among two points the one with a smaller fitness measure is better than the other.

GAs start with an initial population of likely problem solutions, and evolve towards better solutions. The population changes over time, but always has the same number of members. New solutions are generated through operations resembling reproductive processes observed in nature. To evolve towards better solutions, it is necessary to reject the worst solutions and only allow the best ones to survive and reproduce. This incorporates nature's law of survival of the fittest which only allows organisms that adapt best to the environment to thrive. When applying GA to a minimization problem, the role of the environment is played by the fitness measure, the degree of adaptation of a solution point to the environment is interpreted as getting better as its fitness measure decreases. In successive generations, solutions improve until the best in the population is near-optimal.

We will now discuss the essential components for applying a GA on a minimization problem. After each component is discussed, we show how it applies on two problems; one is the TSP, and the other the task allocation problem modeled in Example 11.1.

Genetic representation of problem solutions   As mentioned above, the problem is transformed and represented in such a way that every solution can be represented by a string of bits. All strings corresponding to solution vectors of the problem contain the same number of bits, say n.

For some problems, developing this representation may be a nontrivial effort requiring careful thought, but for many others, a natural representation is usually available. For example, if the problem is one of finding an optimum sequence for n elements numbered 1 to n, every solution is a permutation of $\{1, \ldots, n\}$. In this case the permutation of $\{1, \ldots, n\}$ provides a string representation for solutions to the problem. Valid strings are those which are permutations of $\{1, \ldots, n\}$; i.e., strings in which each of the symbols $1, \ldots, n$ appears once and only once.

For the TSP involving cities 1 to n; a tour $i_1, i_2, \ldots, i_n; i_1$ can be represented by the permutation $i_1, \ldots, i_n$; i.e., the sequence of cities in the order in which they are visited. So, here again, valid strings are those in which each of the symbols $1, \ldots, n$ appears once and only once.

For the task allocation problem involving the allocation of n tasks to T processors discussed in Example 11.1, a solution can be represented as a string of n numbers $x_1, \ldots, x_n$ where for each $j = 1$ to n, x_j is the number of the processor to which task j is allotted. So, here valid strings are all sequences of the form $x_1, \ldots, x_n$ where each x_j is an integer between 1 to T. As an example, if the number of tasks $n = 6$, and the number of processors $T = 4$, the string 1, 1, 3, 2, 1, 3 is a valid chromosome. It represents allocating tasks 1, 2, 5 to processor 1; task 4 to processor 2; and tasks 3, 6 to processor 3; and not using processor 4 at all.

Developing evaluation function verifying the fitness of a solution   We assume that the objective function to be minimized has a positive value at every solution. GAs deal with a relaxed version of the problem in an unconstrained form, to allow the search to be carried out among all valid strings. The evaluation function or the fitness measure of any valid chromosome is the sum of its objective function value and that of the penalty function providing an infeasibility measure of the corresponding solution to the constraints in the original problem. GAs use the evaluation function value at a chromosome to verify its degree of fitness to the environment, and to lower the probability allotted to undesirable chromosomes to survive and to reproduce. This evolutionary aspect of the algorithm provides for the elimination of trial solutions that are relatively unsuccessful. Hence the choice of the evaluation function has a great influence on the overall performance of the algorithm.

For the TSP involving n cities and positive cost matrix (c_{ij}), when solutions are represented by permutations of $\{1, \ldots, n\}$, the evaluation function value of a chromosome can be taken to be the cost of the corresponding tour which is $\sum_{j=1}^{n-1} c_{x_j, x_{j+1}} + c_{x_n, x_1}$. There are no penalty terms needed here as every valid string corresponds to a feasible tour.

Now consider the task allocation problem involving the allocation of n tasks to T processors discussed in Example 11.1. As discussed above, we represent each solution by a string $x_1, \ldots, x_n$ where x_j is the number of the processor to which task j is allotted. A string $x_1, \ldots, x_n$ is valid if x_j is an integer between

1 to T for all j. A valid string $x = x_1, \ldots, x_n$ is infeasible to the problem if either (i) it allots a processor more tasks than it can handle, or (ii) if the sum of the KOP requirements of tasks assigned to a processor exceeds its throughput capacity. So, to represent this problem in an unconstrained fashion, we need two penalty terms, one for each of the above types of infeasibility. We can use **quadratic penalty functions** in which

Penalty for exceeding the capacity on no. of tasks allotted $= \delta_1(\text{excess no. of tasks})^2$

Penalty for exceeding throughput capacity $= \delta_2(\text{excess throughput over capacity})^2$

where δ_1, and δ_2 are appropriate positive penalty coefficients. Such penalty terms are commonly used, as they seem to produce good results. We get the total penalty function value at x by summing the above penalties for each processor for which infeasibility of type (i) or (ii) mentioned above, or both, occur in x.

The objective function value corresponding to this string x is the sum of the costs of the processors used plus the sum of the costs of data link capacity needed by various pairs of tasks allotted to different processors in it. And the evaluation function for x is the sum of the objective value and the penalty function value at x. As an illustration, we will now provide a numerical example to show how to compute the evaluation function in this problem.

Consider the instance with number of tasks $n = 6$, number of processors $T = 4$, and the following data:

Processor t	Cost ρ_t	Max. no. of tasks β_t	Throughput capacity γ_t
1	40	1	425 KOP
2	30	3	300
3	20	1	350
4	45	2	500

Task i	1	2	3	4	5	6
Throughput requirement a_i	150	150	250	250	150	80

	Cost of data link capacity for task pair i, j if allotted to different processors (symmetric)					
$j =$	1	2	3	4	5	6
$i = 1$	×	2	3	1	5	7
2		×	4	2	3	2
3			×	5	2	4
4				×	4	1
5					×	3
6						×

Consider the string $x = 1, 1, 3, 2, 1, 3$ discussed above. It allots tasks 1, 2, 5 to processor 1 with a total throughput requirement of $150 + 150 + 150 = 450$ KOP, exceeding the throughput capacity of 425 KOP of this processor. Also, the number of tasks allotted to this processor, 3, exceeds the capacity of 1 task that it can handle. So the penalty terms for processor 1 total to $\delta_1(3 - 1)^2 + \delta_2(450 - 425)^2 = 4\delta_1 + 625\delta_2$. Similarly, x allotted only task 4 with a throughput requirement of 150 KOP to processor 2, this is within the specified capacity of this processor, so there is no penalty from processor 2 for x. x has allotted tasks 3, 6 with a total throughput requirement of $250 + 80 = 330$ KOP to processor 3 which has a throughput capacity of 350 KOP, but it can handle only 1 task. So, the penalty from processor 3 for x is $\delta_1(2 - 1)^2 = \delta_1$. And processor 4 is not used. Hence, the overall penalty function value at x is $5\delta_1 + 625\delta_2$.

The objective function value at x is the sum of the costs of the processors used + the data link costs. Since tasks 1, 2, 5 are allotted to processor 1; task 4 to processor 2; and tasks 3, 6 to processor 3; data link costs are incurred for the pairs of tasks (1, 4), (1, 3), (1, 6), (2, 4), (2, 3), (2, 6), (5, 4), (5, 3), (5, 6), (4, 3), (4, 6) allotted to different processors. Thus, the objective function value at x is $(40 + 30 + 20) + (1 + 3 + 7 + 2 + 4 + 2 + 4 + 2 + 3 + 5 + 1) = 124$.

So, the overall evaluation function value at x is $124 + 5\delta_1 + 625\delta_2$. Given appropriate positive values to the penalty coefficients δ_1 and δ_2, this fitness measure can be computed.

Initial population   An initial population of solutions is created usually randomly. In some applications, the initial population is generated by using some other method.

The population size is maintained constant through successive generations. It is usually 40 to 250 or larger, depending on the size of the problem being solved.

Developing genetic operators, reproduction, cloning, crossover, and mutation A GA evaluates a population and generates a new one iteratively. Each successive population is called a **generation**.

Individuals in the population are selected for survival into the next generation, or for mating, according to certain probabilities. This probability is increased as the individuals fitness measure gets better. In our case smaller values of the evaluation function are more desirable, so we make the probability of selection of a chromosome to be inversely proportional to its evaluation function value. Through this artificial evolution, GAs seek to breed solutions that are highly fit (i.e., optimal or near-optimal).

A certain percentage (typically between 10% to 40%) of the chromosomes in the population are usually copied as they are into the next generation. There are two possible ways (called **reproduction** and **cloning** or **clonal propagation**) for selecting these individuals. We discuss them below.

> **Reproduction** This operation is probabilistic; it selects individuals from the current population according to probabilities inversely proportional to their evaluation function value as discussed above, and copies the selected individuals into the next generation. The process is repeated until the required number of individuals are selected.
>
> **Cloning** This operation is deterministic. It selects the required number of individuals who have the best values for the evaluation function in the current population, and copies them as they are into the next generation. It is an elitist type of strategy. The advantage of using cloning over reproduction is that the best solution is monotonically improving from one generation to another.

A majority of the remaining individuals in the next population are generated by **mating**, and a small percentage by **mutation**. We discuss mating first. Two parent chromosomes are selected probabilistically as described above, from the current population, to mate. The mating operation is called **crossover**. It creates children whose genetic material resembles the parents genes in some fashion. Many different crossover mechanisms have been developed. We describe some of them.

> **One-point crossover** This operation generates two children. Given parent chromosomes $x = x_1, \ldots, x_n$; $y = y_1, \ldots, y_n$ to mate; this operation selects a position called the **crossover point**, r, between 1 to n at random. The two children are obtained by exchanging the blocks of alleles between positions r to n among the two parents. Thus the children are $c_1 = x_1, \ldots, x_{r-1}, y_r, \ldots, y_n$ and $c_2 = y_1, \ldots, y_{r-1}, x_r, \ldots, x_n$.
>
> Now we have a choice between two possible strategies. Strategy 1 includes both the children in the next generation. Strategy 2 includes only

the child with the best evaluation function value in the next generation, and discards the other.

Two-point crossover   Given parent chromosomes $x = x_1, \ldots, x_n$; $y = y_1, \ldots, y_n$ to mate; this operation selects two positions $r < s$ between 1 to n at random, and swaps the blocks of alleles between positions r to s among the two parents, to get the two children. So, the two children are $c_1 = x_1, \ldots, x_{r-1}, y_r, \ldots, y_s, x_{s+1}, \ldots, x_n$, and $c_2 = y_1, \ldots, y_{r-1}, x_r, \ldots, x_s, y_{s+1}, \ldots, y_n$. Either both the children, or the best among them, get included in the next generation as discussed above.

Random crossover   Given parent chromosomes $x = x_1, \ldots, x_n$; and $y = y_1, \ldots, y_n$; this operation creates children $u = u_1, \ldots, u_n$; and $v = v_1, \ldots, v_n$ where for $j = 1$ to n

$$u_j = \begin{cases} x_j & \text{with probability } \alpha \\ y_j & \text{with probability } 1 - \alpha \end{cases}$$

$$v_j = \begin{cases} y_j & \text{with probability } \alpha \\ x_j & \text{with probability } 1 - \alpha \end{cases}$$

for some preselected $0 < \alpha < 1$. Values of α between 0.5 to 0.8 are often used. Either both the children, or the best among them, get included in the next generation as discussed above.

In problems in which the order of the alleles in the chromosome has no significance, the above crossover operations produce valid child strings for the problem. For the task allocation problem with the representation discussed above, all the above crossover operations produce valid child strings.

However, for the TSP with each tour represented by a permutation of the cities, each of the above crossover operators may produce invalid child strings. As an example consider the two strings $x = 4, 5, 2, 1, 3$ and $y = 1, 2, 4, 3, 5$ for a 5-city TSP. With position 3 as the crossover point, the one-point crossover operator generates the children $c_1 = 4, 5, 4, 3, 5$ and $c_2 = 1, 2, 2, 1, 3$ both of which are invalid strings for this problem since neither of them is a permutation of $\{1, 2, 3, 4, 5\}$. So, for the TSP and for other problems in which solutions are represented by permutations of $\{1, \ldots, n\}$, the following custom designed crossover operator called **partially matched crossover operator** or **PMX** can be used.

PMX for permutation strings   Let $x = x_1, \ldots, x_n$; and $y = y_1, \ldots, y_n$; be two parent permutations. Select two crossover positions $r < s$ randomly as in the two-point crossover operator. To get child 1, do the following for each $t = r$ to s in this order: if $x_t \neq y_t$ swap x_t and y_t in the permutation x. To get child 2, carry out exactly the same work on the permutation y instead of on x. It can be verified that both the children produced are permutations and hence valid strings for the problem.

As an example, consider $n = 6$, and the parents

$$
\begin{array}{ccccccc}
x & = & 4, & 5, & 6, & 2, & 1, & 3 \\
y & = & 1, & 2, & 6, & 4, & 3, & 5
\end{array}
$$

Suppose the crossover positions are 2 and 5 marked by bars above. Then child 1 is obtained by swapping 5 and 2, 2 and 4, and then 1 and 3, in x. As we carry these operations in this order x changes to 4, 2, 6, 5, 1, 3; then to 2, 4, 6, 5, 1, 3 and finally to $p = 2$, 4, 6, 5, 3, 1. Carrying out the same operations on the permutation y we are lead to the second child $q = 3$, 5, 6, 2, 1, 4. So, p, q are the children produced when this crossover operator is carried out with the parental pair x, y.

The **crossover ratio** (typically between 0.6 to 0.9) is the proportion of the next generation produced by crossover. The operation of mating pairs of randomly selected pairs of parents from the present population is continued until enough children to make up the next generation are produced.

Mutation Mutation makes random alterations, such as changing one or more randomly chosen genes, or swapping positions of two randomly selected bits, etc., on a randomly selected chromosome. The probability of mutation is usually set to be quite low (e.g., 0.001). A small percentage of the next generation is produced by applying the mutation operator on randomly selected chromosomes from the current population.

The processes of crossover and mutation are collectively referred to as **recombination operations**.

When all these operations are completed we have the new population which constitutes the next generation, and the whole process is repeated with it.

Stopping criterion The process of producing successive generations is usually continued until there is no improvement in the best solution for several generations, or until a predetermined number of generations have been simulated.

Usually one applies a local search heuristic beginning with the best solution in the final population, to make any possible final improvement. The solution obtained at the end of this process is the output of the algorithm.

Discussion

When a GA works well, the population quality gradually improves over the generations. After many generations, the best individual in the population is likely to be close to a global optimum of the underlying optimization problem.

As an example, we solved an instance of the task allocation problem discussed in Example 11.1 involving $n = 20$ tasks and $T = 7$ processors by the GA [A. Ben Hadj-Alouane, J. C. Bean, and K. G. Murty, 1993]. The representation discussed above for the problem was used. We maintained the population size at 50, with the initial population consisting of randomly generated solutions. In each generation, 10% of the population was obtained by cloning the best solutions in the previous population; 85% was obtained by mating using random crossover; and 5% was obtained by mutation. All the chromosomes in the initial population corresponded to infeasible solutions with infeasibility due to exceeding the throughput capacity on some processors, and due to allotting more than the number of tasks they can handle on some others.

The positive values given to the penalty coefficients had an effect on the performance of the algorithm. Starting small, their values were increased until infeasible solutions which are at the top of the population due to small penalty became highly penalized and are replaced with feasible solutions reasonably rapidly. When the penalty coefficients are large, and solutions at the top of the population are feasible, it turned out to be advantageous to decrease their values. Best results were obtained by adjusting the values of the penalty coefficients adaptively in this manner.

After 10 generations, the population had chromosomes corresponding to feasible solutions for the problem. After 110 generations the best chromosome in the population gave a solution to the problem which was considered to be very satisfactory. This solution was obtained in a few minutes of cpu time on an IBM RS/6000-320H workstation. The $0-1$ IP formulation of this problem given in Example 11.1, has about 2800 integer variables. We tried to solve this $0-1$ IP using the OSL software package based on B&B, on the same workstation. This program did not terminate even after running for 3 days continuously, when it was stopped. The best incumbent at that time was not better than the solution that GA found for this problem in a few minutes of cpu time.

In summary, the essential feature of GAs is that they search using a whole population of solutions rather than a single solution as other methods do. There are three essential requirements to apply a GA on a problem. First, the problem must be represented in such a way that every solution can be represented by a string of constant length. Second, a fitness measure to evaluate potential solutions needs to be developed. This measure is usually the sum of the objective function in the problem and of penalty terms corresponding to the violation of any of the constraints in the problem. Third, a suitable crossover operator has to be developed. The success of GA depends critically on these items, so they have to be developed very carefully.

The crossover operator can be designed in many different ways. In some problems, standard crossover operations may produce children strings which are invalid, as was shown for the case of the TSP. In such problems the crossover operation should be specialized and customized.

Without an appropriate representation and an effective crossover operator, genetic search may be slow and unrewarding. But with the appropriate representation and suitable genetic operators, it can produce high quality solutions very fast.

11.8 The Importance of Heuristics

A consummate skill in modeling problems is a great help to anyone aspiring to be a practitioner of optimization methodology. Knowledge of exact algorithms for well solved problems such as linear programs and convex programming problems, and an understanding of how these algorithms work is of course very important. But in the increasingly complex world of modern technology, skill in designing good heuristic methods for problems for which no effective exact algorithms are known, is an essential component in a successful optimization analyst's toolbox. The development of heuristic methods is being driven by the ever increasing needs for them in many fields.

11.9 Exercises

11.4 Insertion plan for radial components on a PCB A PCB is a rectangular board with holes in which electronic components are to be inserted. There could be 20 - 70 different types of components. A PCB may need to have 1 to 5 components of some of the types inserted on it. Each component goes into a hole. The holes are in predetermined positions on the board as shown in Figure 11.3. The insertion of the components is carried out by a radial insertion machine. The machine consists of an X-Y table on which the board is placed. On top of the table there is an insertion gun. During the operation the X-Y table moves to position each hole right under the gun. The insertion gun itself never moves. There is a linear strip through the inserion gun containing a row of stations. The number of stations on the strip is always equal to the number of types of components to be inserted. We need to assign each component type to a station on the strip. Then each station will be stocked with an unlimited supply of components of the type assigned to it.

The work of inserting a component in a hole is carried out through the following steps. The X-Y table moves from its present position so as to bring the hole right under the insertion gun. While this is going one, the linear strip in the insertion gun moves so that the station containing the type of component to be inserted into the hole comes into position in the insertion gun. When both these moves are complete, the component is inserted by the gun right into the hole. The machine then moves on to the next hole in the sequence.

The X-Y table moves both in the X and Y directions simultaneously until the correct position is reached at a constant speed of d inches per time unit in each direction. Hence after inserting a component at the hole whose Cartesian coordinates are (x_1, y_1) on the board, the time that it takes for the X-Y table to come to the hole at the point (x_2, y_2) is: $(1/d)(\max\{|x_1 - x_2|, |y_1 - y_2|\})$ time units. As an example, if holes 1,2 have their centers at points (5, 7) and (12, 3) on the board, and the speed $d = 3$ inches/time unit, then bringing hole 2 in position after filling hole 1 will take the X-Y table $(1/3)(\max\{|5 - 12|, |7 - 3|\}) = 7/3$ time units.

The linear strip in the insertion gun can move left or right to bring the necessary station into the insert position, but its speed depends on the length of the move. To move a length of r stations, it takes t_r time units, $r = 1,\ldots$ where t_r/r is monotonically decreasing with r and all t_rs are given. Assume that the stations are numbered 1,2,... from left to right on the strip. Thus if station k_1 is currently in the insert position, it takes the strip $t_{|k_1 - k_2|}$ time units to move so as to bring station k_2 into the insert position.

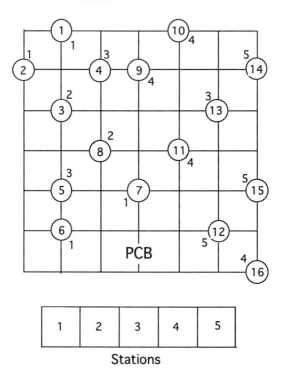

Figure 11.3 The holes on the PCB are marked with their numbers inside them. The type of component that goes in each hole is entered by its side. The stations on the linear strip in the insertion gun are shown with their number inside them.

Suppose the hole with its center at (x_1, y_1) has just been filled with the component stocked at station k_1; and the next hole to be filled has its center at (x_2, y_2), and the component that goes into this hole is stocked in station k_2 on the strip. Then the time lapse before the next hole filling is completed is $\max\{(1/d)(\max\{|x_1 -$

$x_2|, |y_1 - y_2|\}), t_{|k_1-k_2|}\}$ time units, since the X-Y table and the strip of stations move simultaneously and independently.

Consider a PCB with n holes numbered $1, 2, ...n$, and p types of electronic components numbered $1, ..., p$. We are given (x_i, y_i) = the coordinates of the center of the ith hole, and c_i = the type of component that goes into this hole for $i = 1, 2, ...n$. It is required to determine the insertion sequence of the holes, and the assignment of the component types to stations on the linear strip in the insertion gun, so as to minimize the total time it takes to complete all n insertions. Develop an efficient heuristic algorithm for this problem.

As a numerical example, consider the problem given in Figure 11.3, where n = 16, $\{(x_i, y_i) : i = 1$ to $16\} = \{(1,6), (0,5), (1,4), (2,5), (1,2), (1,1), (3,2), (2,3), (3,5), (4,6), (4,3), (5,1), (5,4), (6,5), (6,2), (6,0)\}$, $p= 5$, $d = 2$, $(t_1, t_2, t_3, t_4) = (3/4, 5/4, 7/4, 2)$, and $\{c_i :$ component type c_i goes into hole i for $i = 1$ to $16\}$ = $\{1,1,2,3,3,1,1,2,4,4,4,5,3,5,5,4\}$. Apply your algorithm on this numerical example and obtain the best insertion sequence, and the station assignment for each type of component. (S. Y. Chang)

11.5 In a textile firm there is a special loom for weaving extrawide fabrics of a special type. On the first day of a month the firm has 7 jobs or orders which can be processed on this loom. For $i = 1$ to 7, p_i, d_i, r_i are respectively the processing time in days, due date (day number), and profit from, job i. This data is given below.

i	1	2	3	4	5	6	7
p_i	9	10	12	5	11	8	13
d_i	4	13	15	8	20	30	30
r_i	90	130	85	35	77	68	100

If job i is accepted, the material has to be delivered on the due date d_i for that job (d_i is the day number counting from the first day of the month). Jobs are independent, and the loom can process only one job at a time. Formulate the problem of selecting the jobs to accept to maximize the total profit subject to the constraint that all the accepted jobs should be completed by their respective due dates. Develop a heuristic method for obtaining a good solution to this problem.

11.6 Consider a company producing a single product to meet known demand over a finite number of time periods. The cost function for producing x units of the product in a period may be written

$$f(x) = \begin{cases} 0 & \text{if } x = 0 \\ px + g & \text{if } x > 0 \end{cases}$$

where g is a fixed cost (or setup cost) that is incurred for producing a positive quantity of the product, and p is the variable cost for producing each unit of product once the setup cost is incurred.

Suppose the planning horizon consists of n time periods. For $i = 1$ to n we are given the following data: d_i = demand for the product in period i (in units) that must be met, k_i = production capacity in period i (in units), g_i = fixed (or setup) cost to be incurred to make a positive quantity of the product in period i, p_i = variable cost per unit of making product in period i after the fixed charge is incurred, c_i = holding or storage cost per unit for storing product from period i to period $i + 1$.

All demand has to be met exactly in each period. Product made in any period can be used to meet the demand in that period, or stored to fulfill the demand in later periods.

Develop a heuristic method to obtain good production-storage plans of minimal cost. Apply your method on the numerical problem in which $n = 4$, $k_i = 100$ for all i, $(d_1, d_2, d_3, d_4) = (50, 40, 30, 50)$; and for all i, $g_i = \$100$, $p_i = \$10$, $c_i = \$1$. ([T. E. Ramsay Jr., and R. R. Rardin, Jan. 1983]).

11.7 A large percentage of world seaborne trade in high value general cargo goods

Westbound values top half, Eastbound values bottom half							
	Max. cargo (TEU/week)				Revenue (\$100 units/TEU)		
	HFX	NYC	BLT	POR	HFX	NYC	BLT	POR
HAV	100	200	50	50	13	10	12	11
BRH	60	100	150	100	12	9	12	12
GOT	60	200	60	60	12	12	12	12
LIV	150	300	80	60	9	11	11	10
ROT	80	300	200	200	11	8	11	10
HAV	80	150	40	40	8	12	8	7
BRH	60	50	50	100	9	12	9	8
GOT	60	300	70	70	10	10	10	10
LIV	80	80	120	80	9	11	10	10
ROT	60	100	180	270	8	11	8	7

now moves in containers called TEU, because high port labor costs make capital intensive container operations much more economic than conventional methods. Purpose built, cellular container ships are used for this purpose. Consider a shipping company operating in the North Atlantic with container ships of capacity 1000 TEUs each. The ports that this company operates in Europe are HAV (Le Havre), BRH (Bremerhaven), GOT (Gothenburg); and in North America are HFX (Halifax), NYC (New York), BLT (Baltimore), and POR (Portsmouth). Assume that the travel time between any pair of ports in Europe is 10 hours, and between any pair of ports in North America is 8 hours; and that the travel time between the coasts is 150 hours. Also assume that the ships spend 24 hours at each port of call plus 6 hours of pilotage in and out of the port. The tables above and below give the cargo market data.

Critical time (hrs.), Westbound values left, Eastbound values right

	HFX	NYC	BLT	POR	HFX	NYC	BLT	POR
HAV	142	170	230	225	300	170	300	300
BRH	208	200	250	250	300	220	260	240
GOT	160	250	300	300	260	190	300	300
LIV	185	180	300	300	300	300	300	300
ROY	200	166	208	203	300	196	200	180

Assume that the demand for cargo on a ship's route drops by approximately 10% of the figure quoted above for each 24 hours that the transit time to destination exceeds the critical time given. Develop a heuristic method that builds good 3 week roundtrip ship routes for this company maximizing the revenue per roundtrip. The method can begin with 2 port routes and successively add one port at a time until the limit on roundtrip duration is reached. How many ships can the company operate profitably? Develop routes for all these ships using this heuristic method. ([T. B. Boffey, E, D, Edmond, A. I. Hinxman, and C. J. Pursglove, May 1979]).

11.8 A trucking company has a depot at location 1 from where they have to deliver a material to customers at locations 2, 3, 4, 5. Following table contains the data.

From	Driving time (mts.) to location						Units to deliver
	1	2	3	4	5	6	
Location 1		30	20	10	10	20	
2			10	20	40	50	100
3				10	30	40	10
4					20	30	20
5						10	100
6							170

Each truck can carry at most 200 units, and has a driving limit of 100 minutes. Develop an effective heuristic to find good routes for trucks in such a problem, and apply it on this numerical example. ([I. M. Cheshire, A. M. Malleson, and P. F. Naccache, Jan. 1982]).

11.9 One-dimensional Cutting Stock Problem   Material such as lumber, pipe, or cable is supplied in *master pieces* of a standard length C. Demands occur for pieces of the material of arbitrary lengths not exceeding C. The problem is to use minimum number of standard length master pieces to accommodate a given list of required pieces. Develop a heuristic method for producing a good solution for this problem. Apply your heuristic on the numerical problem in which $C = 100$, and one piece of length each 84, 63, 14, 33, 71, 94, 54, 39, 56, 41, 50 are required.

11.10 Single Machine Tardiness Sequencing   There are n jobs to be processed by a single machine. All the jobs are available for processing at time point 0. For $i = 1$ to n, p_i, d_i are the positive processing time and due date of job i, and w_i is a given positive weight. The machine processes only one job at a time without interruption.

Given the order or sequence in which the jobs are to be processed on the machine, the earliest completion time c_i and tardiness $t_i = \max\{c_i - d_i, 0\}$ of job i can be computed for all i. In the *total weighted tardiness problem*, the aim is to find a processing order for the jobs that minimizes $\sum_{i=1}^{n} w_i t_i$. When all the job weights are equal, minimizing $\sum_{i=1}^{n} t_i$ is called the *total tardiness problem*. Develop effective heuristics for solving both these problems. Apply your algorithm on the numerical problem with $n = 12$ and the following data.

Job i	p_i	d_i	w_i
1	33	35	2
2	17	110	1
3	6	43	3
4	89	119	1
5	5	23	3
6	13	36	4
7	21	74	1
8	15	69	2
9	63	210	3
10	34	184	4
11	12	39	1
12	9	51	2

([C. N. Potts and L. N. Van Wassenhove, Dec. 1991]).

11.11 Develop a heuristic method for obtaining a good solution to the multidimensional $0-1$ knapsack problem. Apply your method on the following problem.

$$
\begin{array}{llllll}
\text{Maximize} & 4x_1 & +3x_2 & x_3 & +6x_4 & +5x_5 \\
\text{subject to} & x_1 & +3x_2 & +4x_3 & +3x_4 & +2x_5 & \leq & 8 \\
& 8x_1 & +x_2 & +9x_3 & & +x_5 & \leq & 10 \\
& \multicolumn{5}{c}{x_j = 0 \text{ or } 1 \text{ for all } j}
\end{array}
$$

([A. Volgenant and J. A. Zoon, Oct. 1990]).

11.12 Nodes in the following network indicate the location of machines that require some service. There is a single service unit with home location at node 1, that is required to visit all the machines and provide the required service. Numbers on the nodes indicate the expected service time for the machine at the node, once the service unit gets to that node. The numbers on the edges of the network are the edge lengths in terms of the time in time units that it takes the service unit to traverse that edge. To travel from any node in the network to any other, the service unit always takes the shortest path in the network between these nodes. It is required to find the order in which the machines are to be serviced, so as to minimize the total waiting time of all the machines before their service is finished. Solve this problem by an effective heuristic method.

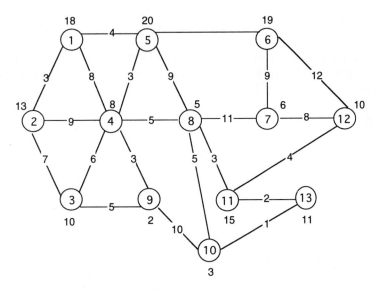

Figure 11.4

11.13 An adult training center provides education, training, and employment for handicapped adult trainees. There are 37 trainees for whom the center has to provide transportation back and forth each day. The center has four vehicles with seating capacities of 14, 13, 13, 15 that can be used for this task.

0 denotes the location of the center, and $i = 1$ to 37 denotes the location of the place where the ith trainee is to be picked up each day. The coordinates of these points on the two dimensional Cartesian plane are: 0 (69.5, 25), 1 (17.1, 36), 2 (1, 32.5), 3 (14.1, 13.6), 4 (22.5, 26.5), 5 (30, 17), 6 (41, 17), 7 (39, 8), 8 (60, 46), 9 (50, 45.5), 10 (34, 37), 11 (29, 42), 12 (25, 53), 13 (7, 51), 14 (9.5, 42), 15 (63, 56), 16 (60.5, 64), 17 (70.5, 50), 18 (71, 39.7), 19 (77, 18.2), 20 (84, 30), 21 (22, 59), 22 (19, 65.5), 23 (13, 70.3), 24 (19.5, 80), 25 (14, 82), 26 (57, 71), 27 (78, 89.7), 28 (90, 90), 29 (80.5, 77.7), 30 (74, 68), 31 (74, 31), 32 (81, 17), 33 (69.5, 8), 34 (82, 21.5), 35 (97, 18.5), 36 (89.5, 4), 37 (99.7, 59).

Determine routes for the vehicles so as to carry out this task while keeping the total distance traveled by all the vehicles as small as possible. ([C. Okonjo-Adigwe,

July 1989]).

11.14 Node Coloring Problem   The nodes of a network are to be colored. The same color can be used to color any number of nodes, but if there is an edge joining any pair of nodes, those two nodes must have different colors. It is required to find a node coloring satisfying this constraint that uses the smallest number of colors.

Let n be the number of nodes and p an upper bound on the number of colors needed (for example, $p = n$ will do). Define

$$x_{ij} = \left\{ \begin{array}{ll} 1 & \text{if node } i \text{ is assigned color } j \\ 0 & \text{otherwise} \end{array} \right.$$

Formulate the node coloring problem as an IP in terms of the decision variables x_{ij}. Develop a good heuristic algorithm for this node coloring problem. Apply the heuristic on the network in Figure 11.5.

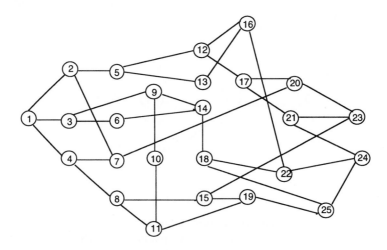

Figure 11.5

11.15 The Linear Placement Problem   This exercise is concerned with locating n facilities at n sites along a one dimensional line where adjacent sites are a unit distance apart. For $i \neq j$ between 1 to n, t_{ij} is the total traffic between facilities i and j, all these t_{ij} are given.

For $i = 1$ to n, if p_i is the number of the facility located at site i, then the distance between the facilities at sites i and j is $|i - j|$ and the cost incurred between them is $|i - j|t_{p_i p_j}$. Hence the total cost of the placement $(p_1, \ldots, p_n)$ is $\sum_{i=1}^{n} \sum_{j=i+1}^{n} (j - i)t_{p_i p_j}$. The problem is to find a placement of facilities to sites that minimizes this total cost

This problem has many applications. An example is the assignment of flights to gates in a horseshoe-shaped airport terminal. The traffic between two flights F_1

and F_2 would be defined as the number of passengers scheduled to fly F_2 following F_1 plus the number scheduled to fly F_1 following F_2. An optimum placement would minimize overall passenger inconvenience.

Develop a good heuristic method for this problem. Apply your heuristic method on the numerical problem in which the traffic data (t_{ij}) is given below.

$j =$	1	2	3	4	5	6
$i = 1$	0	1	5	5	7	8
2		0	3	4	1	5
3			0	7	8	1
4				0	6	4
5					0	10

11.16 A TSP With Side Constraints We are given n cities, in which 1 is the hometown. For $i \neq j = 1$ to n, v_i is the positive valuation for city i, d_i is the positive entrance fees for visiting city i, and c_{ij} is the positive airline fare to go from city i to city j. The problem is to find a roundtrip (either a tour or a subtour covering a subset of cities) starting and ending at the hometown that maximizes the sum of valuations of the cities visited, while satisfying a budget constraint that the total cost (total of airline fares plus the entrance fees for the cities visited) has to be $\leqq$ a specified budgeted amount b. Among subtours or tours having identical total valuation, the one with the least total cost is considered superior. Develop either an exact or a good heuristic algorithm for solving this problem. Apply your algorithm on the numerical problem with data $n = 11$, $b = 3000$, and the rest of the data given in the following table.

$j =$	1	2	3	4	5	c_{ij} 6	7	8	9	10	11
$i = 1$	0	320	220	250	330	220	600	310	150	420	550
2	270	0	290	410	460	230	780	310	360	580	620
3	190	250	0	230	260	100	640	130	240	380	450
4	220	490	270	0	200	330	400	250	150	250	430
5	390	550	300	230	0	370	290	240	300	190	340
6	190	200	120	280	310	0	600	170	270	440	490
7	500	320	270	340	250	510	0	290	430	140	270
8	260	370	140	290	210	200	340	0	290	340	370
9	170	430	280	170	350	310	520	340	0	410	630
10	490	690	450	300	220	520	150	410	350	0	360
11	660	750	530	520	280	590	320	440	530	310	0
d_j	0	100	100	100	100	100	100	100	100	100	100
v_j	6	10	16	8	6	10	20	6	6	6	6

([M. Padberg and G. Rinaldi, Nov. 89]).

11.17 A chemicals company manufactures 15 different products using a chemical reactor. This problem deals with planning the manufacture of these products over a 20-week planning horizon which is divided into 10 periods of two weeks each. On a three shift basis, 336 production hours are available on the reactor in each period. Quantities of products are measured in batches, the batch size being 60 tons for each product. The demand and relevant production data is given in the following tables.

Product	Demand in batches, in period									
	1	2	3	4	5	6	7	8	9	10
1	6.4	6.4	6.4	8.0	6.4	6.4	6.4	6.4	8.5	6.4
2	2.2	2.4	2.6	4.3	2.7	2.8	3.0	3.1	3.9	3.3
3	3.6	3.6	4.2	5.1	4.9	4.3	4.7	4.3	6.0	4.2
4	6.8	6.8	6.8	7.9	7.3	6.5	6.5	6.5	7.3	6.4
5	3.6	3.6	2.7	7.0	5.5	6.4	6.5	5.5	8.2	6.4
6	3.6	3.6	3.6	5.5	4.2	2.4	2.4	2.4	3.0	2.4
7	2.7	2.4	2.6	4.2	3.0	3.2	3.4	3.6	4.5	4.2
8	4.2	2.4	2.5	2.9	2.6	2.8	3.0	2.6	3.7	3.2
9	5.5	6.4	5.5	4.8	5.5	5.5	6.4	5.5	6.2	5.5
10	7.6	7.6	7.6	8.0	7.6	8.2	8.3	8.4	9.3	8.9
11	4.6	4.6	4.6	4.8	4.6	4.6	4.6	4.6	6.0	4.6
12	3.8	3.6	3.3	5.3	3.6	3.6	3.9	4.0	5.1	4.1
13	9.1	6.5	6.5	8.4	6.5	6.5	6.5	6.5	7.0	6.5
14	2.7	2.9	2.9	3.9	3.5	3.6	3.8	4.2	5.4	4.7
15	2.2	2.4	2.6	3.0	3.0	3.2	3.4	3.5	4.8	3.5

i	Switch-over time (in hours) from product i to product														
	1	2	3	4	5	6	7	8	9	10	11	12	13	14	15
1	0.5	2.0	6.6	8.0	7.6	3.1	3.7	8.0	6.3	8.0	6.6	5.9	5.6	6.9	2.0
2	2.4	0.5	3.6	8.0	8.0	4.3	2.5	3.2	3.8	3.3	5.6	2.2	3.3	5.1	2.0
3	6.1	5.1	0.5	3.8	8.0	2.0	2.3	2.0	6.3	7.8	5.6	7.1	2.0	3.4	8.0
4	5.0	8.0	2.0	0.5	6.7	3.1	8.0	5.4	5.1	6.1	2.0	8.0	8.0	2.0	3.3
5	7.4	2.1	4.8	8.0	0.5	2.8	7.9	6.8	2.7	8.0	5.3	5.8	4.6	6.7	2.8
6	6.0	2.5	8.0	2.0	4.1	0.5	2.0	2.8	8.0	5.4	5.4	5.8	5.8	6.2	8.0
7	2.1	2.0	2.6	4.8	2.0	8.0	0.5	3.2	3.8	8.0	4.4	8.0	6.7	7.5	5.9
8	6.6	3.1	8.0	4.5	6.7	2.7	5.1	0.5	7.9	8.0	2.8	2.4	3.8	2.0	7.4
9	4.6	8.0	6.5	5.6	5.2	3.6	6.1	7.8	0.5	2.7	8.0	4.8	5.7	4.4	4.9
10	2.0	2.6	8.0	5.8	8.0	8.0	2.5	5.4	8.0	0.5	8.0	7.8	2.1	8.0	8.0
11	2.0	4.6	4.9	5.5	4.5	4.9	2.0	2.1	4.8	5.8	0.5	7.0	8.0	6.0	2.0
12	5.2	3.0	5.2	7.0	8.0	8.0	2.0	6.9	8.0	7.5	4.4	0.5	8.0	3.1	5.9
13	3.5	8.0	5.0	8.0	4.8	4.4	4.4	7.6	8.0	2.5	6.2	2.4	0.5	3.3	5.2
14	2.0	5.3	3.9	8.0	5.2	4.6	6.4	5.1	2.0	6.2	2.2	2.1	8.0	0.5	2.5
15	7.6	8.0	8.0	8.0	8.0	3.0	4.8	4.2	3.5	2.0	4.5	2.0	2.0	2.0	0.5

Relevant production data

i	I_i	k_i	p_i	i	I_i	k_i	p_i	i	I_i	k_i	p_i
1	0	600	4	6	50	600	2.4	11	0	600	2
2	0	500	6	7	0	600	3.4	12	0	500	4
3	0	500	4	8	0	600	6	13	0	600	2
4	450	500	4	9	350	500	4	14	0	500	3.6
5	0	500	4	10	0	600	4	15	100	600	3.2

I_i = beginning inventory (tons), k_i = tank capacity (tons),
p_i = production time (hrs./batch), of product i

Inventory holding costs are \$1,000 per batch per period for each product. Opportunity cost for lost production on the reactor during time spent in switching over from one product to another is estimated at \$20,000/hour.

The operational planning problem in this plant is to establish a production plan that determines which products are to be manufactured in each period, the lot size for each, and the sequence in which these products are manufactured in each period. Develop a heuristic method that determines a good operational plan to minimize the total cost (inventory holding cost plus the opportunity costs due to setups between production runs) while meeting the demands for all the products. ([W. J. Selen and R. M. J. Heuts, Mar. 1990]).

11.18 Develop good heuristic approaches for handling the multi-item production scheduling problems discussed in Exercises 9.53, 9.54. Apply your methods on the numerical problems given in those exercises and review the solutions obtained.

11.10 References

A. BEN HADJ-ALOUANE, J. C. BEAN, and K. G. MURTY, 1993, "A Hybrid Genetic/Optimization Algorithm for a Task Allocation Problem", IOE Dept., University of Michigan, Ann Arbor, MI.

T. B. BOFFEY, E. D. EDMOND, A. I. HINXMAN, and C. J. PURSGLOVE, May 1979, "Two Approaches to Scheduling Container Ships With an Application to the North Atlantic Route", *Journal of the Operational Research Society*, 30, no. 5(413-425).

I. M. CHESHIRE, A. M. MALLESON, and P. F. NACCACHE, Jan. 1982, "A Dual Heuristic for Vehicle Scheduling", *Journal of the Operational Research Society*, 33, no. 1(51-61).

G. CLARKE, and J. WRIGHT, 1964, "Scheduling of Vehicles from a Central Depot to a Number of Delivery Points", *Operations Research*, 12(568-581).

G. CORNUEJOLS, M. FISHER, and G. NEMHAUSER, 1977, "Location of Bank Accounts to Optimize Float: An Analytic Study of Exact and Approximate Algorithms", *Management Science*, 23(789-810).

L. DAVIS, 1991, *Handbook of Genetic Algorithms*, Van Nostrand Reinhold, NY.

R. W. EGLESE, 1990, "Simulated Annealing: A Tool for Operational Research", *European Journal of Operational Research*, 46(271-281).

F. GLOVER, E. TAILLARD, and D. DE WERRA, 1993, "A User's Guide to Tabu Search", in

Tabu Search, Glover, Laguna, Taillard, and de Werra (eds.), *Annals of Operations Research*, Vol. 41.

D. GOLDBERG, 1989, *"Genetic Algorithms in Search, Optimization and Machine Learning"*, Addison-Wesley, Reading, MA.

J. HOLLAND, 1975, *"Adaptation in Natural and Artificial Systems"*, The University of Michigan Press, Ann Arbor, MI.

S. KIRKPATRICK, C.D. GELATT Jr., and M. P. VECCHI, 1983, "Optimization by Simulated Annealing", *Science*, 220(671-680).

J. G. KLINCEWICZ, 1980, "Locating Training Facilities to Minimize Travel Costs", Bell Labs. Technical Report, Holmdel, NJ.

J. R. KOZA, 1992, *"Genetic Programming: On the Programming of Computers by Means of Natural Selection"*, The MIT Press, Cambridge, MA.

C. OKONJA-ADIGWE, July 1989, "The Adult Training Center Problem: A Case Study", *Journal of the Operational Research Society*, 40, no. 7, (637-642).

M. PADBERG, and G. RINALDI, Nov. 1989, "A Branch-and-Cut Approach to a Traveling Salesman Problem With Side Constraints", *Management Science*, 35, 11 (1393-1412).

C. N. POTTS, and L. N. VAN WASSENHOVE, Dec. 1991, "Single Machine Tardiness Sequencing Heuristics", *IIE Transactions*, 23, no. 4, 346-354.

T. E. RAMSAY Jr., and R. R. RARDIN, Jan. 1983, "Heuristics for Multistage Production Planning Problems", *Journal of the Operational Research Society*, 34, no. 1 (61-70).

C. R. REEVES (ed.), 1993, *Modern Heuristic Techniques for Combinatorial Problems*, Blackwell Scientific Publications, Oxford, UK.

D. ROSENKRANTZ, R. STERNS, and P. LEWIS, 1977, "An Analysis of Several Heuristics for the Traveling Salesman Problem", *SIAM J. on Computing*, 6(563-581).

W. J. SELEN, and R. M. J. HEUTS, March 1990, "Operational Production Planning in a Chemical Manufacturing Environment", *European Journal of Operational Research*, 45, no. 1, (38-46).

A. VOLGENANT, and J. A. ZOON, Oct. 1990, "An Improved Heuristic for Multidimensional $0-1$ Knapsack Problems", *Journal of the Operational Research Society*, 41, no. 10, 963-970.

Chapter 12

Dynamic Programming

12.1   Sequential Decision Processes

So far, we have discussed methods for solving **single stage** or **static models**; i.e., we find a solution at one time for the model and we are done. But in many applications we need to make a **sequence of decisions** one after the other. These applications deal with a process or system that is observed at the beginning of a period to be in a particular **state**. That point of time may be a decision point where, one out of a possible finite set of **decisions** or **actions** is to be taken to move the system towards some goal. Two things happen, both depend on the present state of the system, and the decision taken:

(i)   an immediate cost is incurred (or reward earned)
(ii)   the action moves the system to another state in the next period.

And the same process is repeated over a finite number, n say, of periods. Thus, a sequence of decisions are taken at discrete points of time. The aim is to optimize an objective function that is additive over time, to get the system to a desired final state. The objective may be to minimize the sum of the costs incurred at the various decision points, or to maximize the sum of the rewards earned if the problem is posed that way. The important feature in such a sequential decision process is that the various decisions cannot be treated in isolation, since one must balance a desirable low cost at the time of a decision with the possibility of higher costs in later decisions.

Here we have a multistage problem involving a finite number, n, of stages. The system may be in several possible states. As time passes, the state of the system changes depending on the sequence of decisions taken and the initial state at the beginning. Because of these changing states of the system, the approach for optimizing the performance of such a system is called **dynamic programming** (DP).

A selection that specifies the action to take at each decision point is called a **policy**. The aim of DP is to determine an optimal policy that minimizes the total costs in all the stages (or maximizes the total reward if the problem is posed that way). DP solves such problems recursively in the number of stages n. At each decision point it selects an action that minimizes the sum of the current cost and the best future cost. We will now illustrate these basic concepts with some examples.

EXAMPLE 12.1

Consider a driver in his car, starting at his office in the evening, to get home as quickly as possible. The problem of finding an optimal route for this driver through the street network of the city, is known as a **shortest route problem** or **shortest chain problem**. The street network is represented by a directed network in which nodes correspond to major traffic centers or street intersections, and directed arcs joining pairs of nodes correspond to street segments joining the corresponding traffic centers; the orientation of the arc being specified by the segment's orientation if it is a one way street segment, or otherwise the direction in which our driver would normally travel that segment on his way home from work. For a picture of such a network, see Figure 12.7 in Section 12.3 later on.

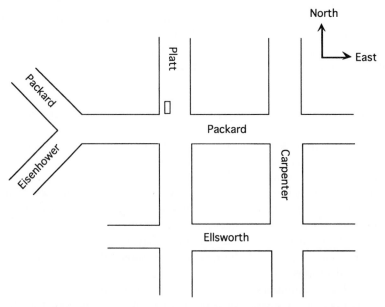

Figure 12.1 The car (system) at state "PL-PA on PL-S".

In this problem, the system is the car with the driver sitting behind the steering wheel. The states of the system are the various street intersections or nodes. We show some of the streets in Figure 12.1, and suppose at some stage the driver has just arrived at the Platt-Packard intersection on Platt South (called PL-PA on PL-S in Figure 12.2). So, the present state of the system is PL-PA on PL-S. There are 3 possible actions the driver can take now, they are: (a) to continue driving straight on Platt, (b) turn left onto Packard East, or (c) turn right onto Packard West. The result of each of these actions is to cause a transition of the system to the state which is the next intersection on the street along which the car continues to travel by that action; and the immediate cost incurred as a result of this action is the driving time in minutes it takes the car to reach that intersection. See Figure 12.2. The objective is to minimize the total driving time before reaching the "home" state.

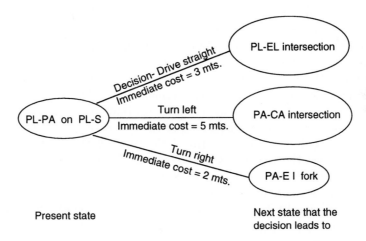

Figure 12.2 Choice between 3 possible decisions at present state. The outcome of each is a transition to the next state shown on the right. Immediate cost of taking the decision is the driving time incurred before reaching the next state.

The information needed to apply DP to solve a sequential decision problem such as the shortest route problem discussed in Example 12.1 is: the set of all possible states of the system (assumed to be a finite set); the set of all decisions that can be taken in each state (one of these decisions has to be taken when the system reaches this state); and the immediate cost incurred and the next state that the system will reach under each of these decisions. With this information we have total knowledge of the dynamics of the system. Since state transitions occur at discrete points of time, such a system is called a **discrete-time dynamic system**, and we assume that the cost function is additive over time. With this information, the problem of finding an optimum policy (one that specifies the optimum decision to be taken in each possible state of the system) can be solved by the DP approach. We will discuss this approach in the next section, but first we present some more examples to illustrate the basic concepts.

EXAMPLE 12.2 Solving nonnegative integer knapsack problem by DP

Consider the nonnegative integer knapsack problem discussed in Chapter 9. In this problem there are a set of n objects available to be loaded into a knapsack with a weight capacity of w_0, a positive integer. The aim is to determine how many copies of each object to load into the knapsack, to maximize the total value of all the objects loaded subject to the knapsack's weight capacity constraint. We assume that the weights of all the objects are positive integers.

This problem can be posed in a sequential decision format by considering the loading process as a sequential process loading one object at a time. In this format, the state of the system at any point of time in the loading process can be represented by the knapsack's remaining weight capacity. So, there are $w_0 + 1$ possible states of the system. At any stage, an object is considered available for loading into the knapsack iff its weight is $\leqq$ the knapsack's remaining weight capacity (i.e., the state of the system) at that stage. And the decisions that can be taken at that stage are to load one of the available objects into the knapsack. There is an immediate reward from that decision in the form of the value of the object loaded. This decision will reduce the knapsack's remaining weight capacity by the weight of the object loaded, and the next state of the system is determined from this.

As a numerical example, consider a point of time at which the knapsack's remaining weight capacity is 20 kg., and there are $n = 6$ objects according to the following data.

Data for the nonnegative integer knapsack problem

Object	Weight kg.	Value $
1	14	700
2	8	900
3	5	500
4	4	600
5	22	2700
6	25	3500

Knapsack's remaining weight capacity 20 kg.

Objects 5, 6 have weight > the knapsack's remaining weight capacity at this time, so they are not available for loading at this time; the other objects 1 to 4 are available now. Thus there are 4 possible decisions that can be taken in the present state, they correspond to loading one of objects 1 to 4. The results of these decisions are depicted in Figure 12.3

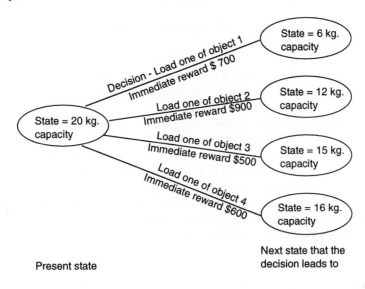

Figure 12.3 State transitions in a nonnegative integer knapsack problem.

In this problem the aim is to maximize the total value loaded, which is the sum of the rewards obtained over the entire process before it terminates. The DP approach for solving the nonnegative integer knapsack problem using this format is discussed later on.

EXAMPLE 12.3 Solving the 0−1 knapsack problem by DP

Here we consider the 0−1 knapsack problem discussed in Chapters 9, 10, 11. As in Example 12.2, there are n objects available to be loaded into a knapsack of weight capacity w_0, a positive integer; *but in this problem, only one copy of each object is available.* The aim is to determine the subset of objects to be loaded, so as to maximize the total value of the objects loaded subject to the knapsack's weight capacity.

In Example 12.2, any nonnegative integer number of copies of any of the objects could be loaded into the knapsack subject to its weight capacity, and we were able to represent the state of the system by the knapsack's remaining weight capacity. Here we can include only one copy of any object in the knapsack (that is why this is the 0−1 knapsack problem), and in this problem, the knapsack's remaining weight capacity does not include enough information to fully represent the state of the system and to decide what possible decisions can be taken in a state. For example, let the weight of object 1 be 14 kg. and at some stage, let the knapsack's remaining weight capacity be 20 kg. Because this is a 0−1 problem, at this stage object 1 is available for loading into the knapsack only if it is not already loaded into the knapsack. Thus in this problem, at any stage, an object is considered available for loading into the knapsack iff:

(i)   its weight is $\leqq$ knapsack's remaining weight capacity at this stage, and
(ii) the copy of the object is not already loaded into the knapsack.

In this format, the state of the system can be represented by the knapsack's remaining weight capacity and the subset of objects still available for loading into the knapsack at this stage by the above definition. And the decisions that can be taken in this state are to load one of the available objects into the knapsack. And the system moves forward.

As a numerical example, consider a point of time at which the knapsack's remaining weight capacity is 20 kg., and there are 6 objects not yet loaded into the knapsack, according to the following data.

Data on objects not yet loaded		
Object	Weight kg.	Value $
1	14	700
2	8	900
3	5	500
4	4	600
5	22	2700
6	25	3500

Knapsack's remaining weight capacity 20 kg.

Objects 5, 6 have weight greater than the knapsack's remaining weight capacity at this time, so they are not available for loading at this time; the other unincluded objects 1 to 4 are available now. Thus there are 4 possible decisions that can be taken in the present state. They correspond to loading one of objects 1 to 4. The results of these decisions are depicted in Figure 12.4

In this problem also, the objective is to maximize the total value loaded into the knapsack. Using the definition of states given here (characterized by the knapsack's remaining weight capacity and the subset of objects available for loading), the $0 - 1$ knapsack problem can be solved by the DP approach. This is discussed in Section 12.6.

The reader should pay careful attention to the difference in the definition of states in Example 12.2 and this example. To represent a nonnegative integer knapsack problem (any number of copies of any object could go into the knapsack subject only to its weight capacity) with n objects and knapsack's weight capacity w_0, in a sequential decision format, we needed $w_0 + 1$ states. The $0 - 1$ knapsack problem (only one copy of any object is available) with the same data may need $n(w_0 + 1)$ states to be represented in a sequential decision format, because here we need to

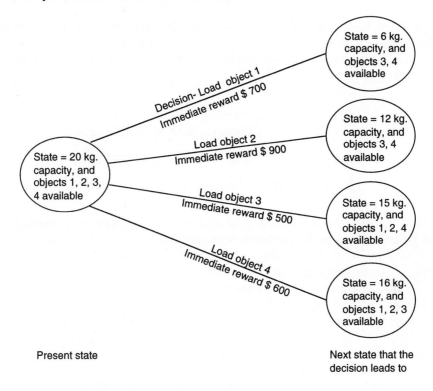

Figure 12.4 State transitions in a $0 - 1$ knapsack problem.

carry the subset of objects not yet loaded into the knapsack in the definition of the state.

Thus in posing a problem for solution by DP, one should formulate the definition of states very carefully taking the structure of the problem into account. The definition of states should always carry enough information so that the set of all possible decisions in any state can be determined unambiguously to continue the process till the end.

So far we assumed that the result of an action taken in a state is an immediate reward which is known with certainty, and transition to a known state. The branch of DP dealing with models in which there is no uncertainty, and we have perfect information about the effect of every possible action in every state of the system, is called **deterministic dynamic programming**.

In some applications the effects of actions may not be known with certainty. As an example, suppose the unemployment in the country is running around 7.5%, and the President is considering investing some federal money in public works programs to stimulate employment. Assume that the President has two possible options, to invest either \$100 billion or \$200 billion, over the next two years. The effect of either of these actions on the unemployment percentage cannot be predicted with certainty, but government economists have come up with the following estimates of the results from these investments.

Option	Estimated probability of unemployment % decreasing to		
	7.0	6.7	6.4
Invest \$100 bil.	0.70	0.20	0.10
Invest \$200 bil.	0.60	0.25	0.15

Here the state of the system is measured by the unemployment percentage. For each possible action we do not know with certainty to which state the system will move as a result of that action, but we have its probability distribution. Each of these actions may contribute some amount to the already high national debt; these contributions may not be known with certainty, but we can estimate their probability distributions. The President's goal may be to bring the unemployment percentage to a desirable level over the next 5 years, while minimizing the total contribution of the actions taken in this regard to the national debt.

In this situation, the total contribution incurred to the national debt to bring the unemployment percent to a desirable level is a random variable not completely under our control, and we can only hope to minimize its expected value.

The branch of DP which deals with models based on such probabilistic data to minimize total expected cost, is called **stochastic dynamic programming**.

In this chapter we treat deterministic DP, but the interested reader should consult the references at the end of this chapter for details of stochastic DP.

12.2   State Space, Stages, Recursive Equations

The set of all possible states of the system is called the **state space**. The definition of each state should contain all necessary information so that the set of decisions that can be taken when the system is in that state can be easily identified. Associated with each state s is the **decision set** $D(s)$ of decisions that can be taken at s.

States play a key role in DP. Transitions always occur from one state to another. In our deterministic DP models we assume that the definition of states is so formulated that the immediate cost or reward, and the next state that the system moves to after a decision, depend only on the current state and the decision, and not on the path of past states through which the system arrived at the current state. This property is known as the **Markovian property**.

Since states are the points where decisions are made; the sequential decision process evolves from one state to the next. The sequence of states visited by the system before the process ends, forms a path known as a **realization**; it depends on the initial state and the policy adopted (i.e., the decisions made at the various states along the path). This path can be represented as in Figure 12.5. Nodes in it represent states, and arcs correspond to decisions. There is an immediate cost incurred for each arc; the objective value of this realization is the sum of the costs incurred over all the arcs on the path.

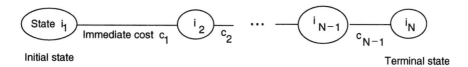

Figure 12.5 A realization of a sequential decision process. The total cost of this realization is $c_1 + c_2 + \ldots + c_{N-1}$.

To formulate an optimization problem for solution by DP requires the identification of the state space. This usually takes a lot of ingenuity. We will illustrate it with many examples to give the reader some experience.

Every DP model has states, and it can be solved using them. But in some DP models there are also the so-called **stages**. These are sequential decision models in which the states form groups called **stages** that appear in some order. In these models transitions always occur from a state in some stage, to a state in the next stage. The process always begins in some state in stage 1, then it moves to stage 2, and then to stage 3, etc. Such models usually arise in situations where decisions are taken on a periodic basis, say once every time period at the beginning of the period.

Thus, if a sequential decision process has a natural organization into stages, the state space $\mathbf{S}$ can be partitioned as $\mathbf{S}_1 \cup \mathbf{S}_2 \cup \ldots \cup \mathbf{S}_N$, and the system always moves from $\mathbf{S}_1$ to $\mathbf{S}_2$, from $\mathbf{S}_2$ to $\mathbf{S}_3$, etc. and finally from $\mathbf{S}_{N-1}$ to $\mathbf{S}_N$. All the states in $\mathbf{S}_N$ (stage N) are terminal states, i.e., the process terminates when a state in $\mathbf{S}_N$ is reached. In this case it is convenient to represent states so that the number of the stage to which they belong is apparent, since these stage numbers can be used to simplify the DP algorithm.

We will consider sequential decision processes that have definite ends, those at which decision making begins and ends; i.e., models with finite planning horizons. If the model is a staged model with N stages, it begins in stage 1 and terminates after $N - 1$ transitions by reaching stage N. If the model is not a staged model, it terminates whenever a desired terminal state is reached, or after some specified number of transitions take place.

Consider a general model in which the state space is the set $\mathbf{S}$. To apply DP we need to know, for each state $s \in \mathbf{S}$, the **decision set** $D(s)$ that can be taken at s, and the immediate cost (or reward) and the next state of the system that comes up as a consequence of selecting each of these decisions. Since a solution must specify the decision to be selected from $D(s)$ for each $s \in \mathbf{S}$, it is called a **policy**. A policy completely specifies the sequence of decisions to be taken after each transition in every possible realization.

Optimum Value Function

For each state $s \in \mathbf{S}$ define

$$f(s) \;=\; \begin{array}{l} \text{minimum total cost that is incurred (or maximum reward} \\ \text{that is obtained) by pursuing an optimum policy beginning} \\ \text{with } s \text{ as the initial state.} \end{array}$$

This function $f(s)$ defined over the state space $\mathbf{S}$ is called the **optimum value function** or OVF.

Suppose the problem specifies desired terminal states, i.e., the process terminates whenever the system reaches one of the states in this terminal set. No more decisions will be taken when a terminal state is reached, and the future cost (or reward) is 0. Hence, if the system is initiated in one of these terminal states, the optimum cost (or reward) is 0, i.e.,

$$f(s) \;=\; 0 \quad \text{if } s \text{ is a terminal state} \tag{12.1}$$

(12.1) are called the **boundary conditions** that the OVF satisfies.

Principle of Optimality

The DP technique rests on a very simple principle called the **principle of optimality**. We give several equivalent versions of it.

Principle of Optimality - Version 1 An optimum policy has the property that if s is a state encountered in an optimum realization obtained by pursuing an optimum policy beginning with an initial state s_0; then the portion of this realization from s till the end constitutes an optimum realization if the process is initiated with the system in s.

Principle of Optimality - Version 2 Given the current state at some point of time, the optimal decisions at each of the states encountered in the future do not depend on past states or past decisions made at them.

Principle of Optimality - Version 3 An optimum policy has the property that whatever the initial state and the initial decision are, the remaining decisions must constitute an optimum policy with regards to the state resulting from the first transition.

In other words, given the current state on an optimum realization at some time, an optimum strategy for the remaining time is independent of the policy adopted in the past. So, knowledge of the current state of the system conveys all the information about its previous behavior necessary for determining the optimum sequence of decisions henceforth. This is the consequence of the Markovian property mentioned above.

To explain the principle of optimality in terms of the shortest route problem, suppose we found the shortest route, call it $\mathcal{P}$, from Detroit to Seattle, and it passes through Chicago. Then the principle of optimality states that the Chicago to Seattle portion of this route, call it $\mathcal{P}_1$, is a shortest route from Chicago to Seattle. For, if $\mathcal{P}_1$ is not a shortest route from Chicago to Seattle, let $\mathcal{P}_2$ be a shorter route from Chicago to Seattle. Then by following the route $\mathcal{P}$ from Detroit until we reach Chicago, and then following the route $\mathcal{P}_2$ from Chicago to Seattle, we will get a route from Detroit to Seattle which is shorter than $\mathcal{P}$, contradicting that $\mathcal{P}$ is a shortest route from Detroit to Seattle.

The principle of optimality is a direct and simple consequence of the Markovian property and assumption that the objective function is the sum of the immediate costs incurred at each state along the optimal path (the additivity of the objective function).

The Functional Equation for the OVF

Consider the formulation in which the total cost is to be minimized. Let s_0 be the current state of the system at some time. Suppose there are k possible decisions available at this state, of which one must be chosen at this time. Suppose the immediate cost incurred is c_t and the system transits to state s_t if decision t is chosen at this time, for $t = 1$ to k. As defined earlier, for $t = 0, 1, \ldots, k$

$$f(s_t) \quad = \quad \text{minimum total cost incurred by pursuing an optimum policy beginning with } s_t \text{ as the initial state.}$$

For $t = 1$ to k, if we select decision t now, but follow an optimum policy from the next state onwards, the total cost from this point of time will be $c_t + f(s_t)$. The reason for this is the following: c_t is the immediate cost incurred as a result of the decision now, and this decision moves the system to state s_t. And $f(s_t)$ is the cost incurred by beginning with state s_t and following an optimum policy into the future. By the additivity hypothesis, the total cost from now till termination is the sum of these two costs, which is $c_t + f(s_t)$.

Hence, an optimum decision in the current state s_0 is the t between 1 to k which minimizes $c_t + f(s_t)$. Thus we have the equation

$$f(s_0) \quad = \quad \min\{c_t + f(s_t) : \quad t = 1 \text{ to } k\} \tag{12.2}$$

and an optimum decision at the current state s_0 is the decision t which attains the minimum in (12.2). Clearly (12.2) is a direct consequence of the additivity hypothesis through the principle of optimality. (12.2) is intimately related to version 3 of the principle of optimality because the sum $c_t + f(s_t)$ in it is the cost of the path that selects decision t now, and thereafter uses decisions dictated by an optimal policy.

(12.2) is known as a **functional equation** because it gives an expression for the value of the OVF at state s_0 in terms of the values of the same function at other states $s_1, \ldots, s_k$ that can be reached from s_0 by a single decision. It is also known as the **optimality equation** in the literature.

If the values of $f(s_1), \ldots, f(s_k)$ are all known, (12.2) can be used to determine the value of $f(s_0)$ and the optimum decision at s_0. This is called **recursive fixing** since it fixes the value of $f(s_0)$ from the known values of $f(s_1), \ldots, f(s_k)$.

By (12.1), $f(s) = 0$ for every terminal state s. Starting from the known values of $f(s)$ at terminal states s (obtained from the boundary conditions), we can compute the values of the OVF at all the states, using (12.2), by moving backward one state at a time. This method of evaluating the values of OVF at all the states is called the **recursive technique** or **backwards recursion** (because it starts at the terminal states and moves backward one state at a time), or **recursive fixing** (because it consists of evaluating the functional equations for the various states in a predetermined sequence). Some writers refer to the recursive technique itself as dynamic programming. DP finds an optimum policy by recursion.

If the problem is stated as one of maximizing the total reward, the OVF is defined as the total reward, and we get a functional equation similar to (12.2) with "maximum" replacing the "minimum."

If the states are grouped into stages in the problem, the boundary conditions state that $f(s) = 0$ for all states s in the terminal stage, stage N, say. In such staged problems, the recursive approach begins in stage N and moves backward stage by stage, each time finding the OVF value and the optimum decision for each state in that stage.

To solve a problem by DP, the functional equations have to be developed for it individually. It takes ingenuity and insight to recognize whether a problem can be solved by DP and how to solve it actually.

The final output from DP would be a list of values of the OVF and an optimum decision for each possible state of the system, an optimum policy. One should remember that all the states may not materialize in a particular realization, but an optimum policy provides complete information on what to do if any state in the state space were to materialize.

12.3   To Find Shortest Routes in a Staged Acyclic Network

A **simple circuit** in a directed network is a cycle beginning and ending at a node, with all the arcs in it having a compatible orientation making it possible to go around the circuit without violating the orientations on any arc. In Figure 12.6, we have a simple circuit on the left; and a simple cycle which is not a circuit on the right, because we cannot go around it without violating the orientation on at least one arc. A directed network is said to be **acyclic** if it does not have any simple circuits.

Figure 12.6 On the left we have a simple circuit. On the right we have a simple cycle that is not a simple circuit, because arc orientations are not compatible.

Here we consider the problem of finding a shortest route from an origin node to a destination node in a directed staged acyclic network. The length (or the driving time) of each arc is given and entered on the arc. We will illustrate the application of DP to solve this problem by backwards recursion on the network in Figure 12.7. The network is clearly acyclic, and it is staged. It is drawn in such a way that all the nodes in a stage are aligned on the same vertical line. As we move from the origin node, node 1 in stage 1, towards the destination node, node 14 in stage 6, we move from a node in a stage, to a node in the next stage.

Nodes in the network correspond to the states of the system. So in this problem there are 14 states in all, which are grouped into 6 stages. At each node the decisions correspond to which of the arcs incident out of it to travel next. The immediate cost of a decision is the length of the arc traveled, and this decision moves the system to the node at the other end of that arc. For example, when at node 3 there are three decisions to choose from, they are: travel along arc $(3, 6)$ (immediate cost 10, transit to node 6 next), or travel along arc $(3, 7)$ (immediate cost 4, transit to node 7 next), or travel along arc $(3, 8)$ (immediate cost 5, transit to node 8 next).

Now we define the OVF. For each $i = 1$ to 14, it is

$$f(i) \quad = \quad \text{length of the shortest route from node } i \text{ to the destination node 14.}$$

Since the destination node 14 represents the terminal state, the boundary condition in this problem is $f(14) = 0$. Moving backward one stage at a time, we determine the OVF and optimum decisions at the various nodes as shown below.

Stage 5
$f(12)$ =   min{14 + $f(14)$} = min{14 + 0} = 14. Opt. decision, travel along arc (12, 14).
$f(13)$ =   min{13 + $f(14)$} = min{13 + 0} = 13. Opt. decision, travel along arc (13, 14).

Stage 4
$f(9)$ =   min{19 + $f(12)$, 8 + $f(13)$} = min{19+14, 8+13} = 21. Opt. decision, travel along arc (9, 13).
$f(10)$ =   min{16+$f(12)$, 14+$f(13)$} = min{16+14, 14+13} = 27. Opt. decision, travel along arc (10, 13).
$f(11)$ =   min{12+$f(13)$} = min{12+13} = 25. Opt. decision, travel along arc (11, 13).

Stage 3
$f(5)$ =   min{12+$f(9)$} = min{12+21} = 33. Opt. decision, travel along arc (5, 9).
$f(6)$ =   min{6+$f(9)$, 4+$f(10)$} = min{6+21, 4+27} = 27. Opt. decision, travel along arc (6, 9).
$f(7)$ =   min{3+$f(10)$, 9+$f(11)$} = min{3+27, 9+25} = 30. Opt. decision, travel along arc (7, 10).
$f(8)$ =   min{7+$f(11)$} = min{7+25} = 32. Opt. decision, travel along arc (8, 11).

Stage 2
$f(2)$ =   min{6+$f(5)$, 4+$f(6)$} = min{6+33, 4+27} = 31.Opt. decision, travel along arc (2, 6).
$f(3)$ =   min{10+$f(6)$, 4+$f(7)$, 5+$f(8)$} = min{10+27, 4+30, 5+32} = 34. Opt. decision, travel along arc (3, 7).
$f(4)$ =   min{7+$f(7)$, 11+$f(8)$} = min{7+30, 11+32} = 37. Opt. decision, travel along arc (4, 7).

Stage 1
$f(1)$ =   min{8+$f(2)$, 3+$f(3)$, 9+$f(4)$} = min{8+31, 3+34, 9+37} = 37. Opt. decision, travel along arc (1, 3).

The optimum decisions, and the OVF values at the various nodes are shown on Figure 12.7. The shortest route from the origin node 1 to destination node 14 is marked with thick lines there; its length is 37. By following the optimum decisions determined at the nodes, we can also obtain the shortest route from any node in the network, to the destination node 14.

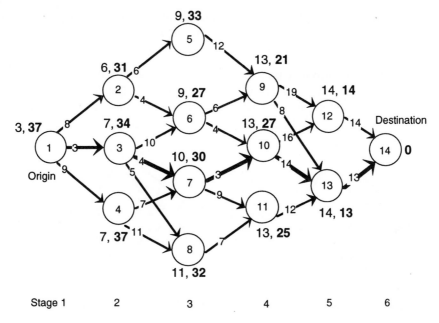

Figure 12.7 The staged acyclic network. Node numbers are entered inside them. Arc lengths are marked on them. By the side of each node we marked the next node to go to (to reach destination by a shortest route from that node), and in bold the length of the shortest route from that node to destination. The shortest route from the origin to the destination is marked in thick lines.

12.4 Shortest Routes in an Acyclic Network That is Not Staged

Let G be an acyclic network with $\mathcal{N}$ as the set of nodes, that is not staged. Let $|\mathcal{N}| = n$. Here we consider the problem of finding a shortest route in G from an origin node to a destination node. We show how this problem can be solved by DP treating the nodes in the network as states. Even though there are no stages, backwards recursion solves the functional equations beginning with the destination node, and moving backward one node at a time.

In this case it is possible to number the nodes in the network in such a way that for all arcs (i, j) in the network $i < j$. This is the special property of acyclic networks, see [K. G. Murty, 1992, of Chapter 5] for a proof. A numbering of nodes of the network satisfying this property is called **acyclic numbering of the nodes**. Such a numbering can be found by using the following procedure.

    1 Look for nodes which have no arcs incident into them in the remaining part of the network. If there are no nodes satisfying this property, the network is not

acyclic, terminate. Otherwise, number all these nodes serially in some order beginning with 1 if this is the first step, or beginning with the next unused integer if some nodes are numbered already. Go to 2.

2 If all the nodes are now numbered, we have the desired node numbering, terminate. Otherwise, consider all the newly numbered nodes and arcs incident at them as deleted, and go back to 1 with the remaining part of the network.

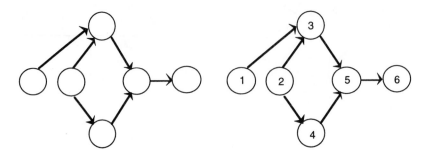

Figure 12.8 Acyclic numbering of nodes in an acyclic network.

As an example, consider the network on the left in Figure 12.8. In this network, the leftmost pair of nodes have no arcs incident into them. So, they are numbered 1, 2 first. Continuing this way, nodes get numbered by the above procedure from left to right, leading to the acyclic numbering of the nodes on the right in Figure 12.8.

As another example, consider the network in Figure 12.9. The one node in this network with no arc incident into it is numbered as node 1. The remaining network after node 1 and the thick arcs incident at it are deleted, does not have a node with no arcs incident into it. So, we terminate with the conclusion that the network in Figure 12.9 is not acyclic.

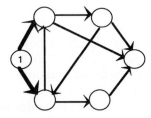

Figure 12.9 A network that is not acyclic.

The acyclic numbering of nodes is very convenient for solving the functional equations by recursion.

Assume that nodes in our network G are numbered serially using an acyclic numbering, i.e., for each arc (i, j) in the network $i < j$. Let nodes 1, n be the origin, destination respectively. For each arc (i, j) in the network G let c_{ij} be its length. As before, define the OVF

$$f(i) \quad = \quad \text{length of the shortest route from node } i \text{ to the destination node } n.$$

The boundary condition is $f(n) = 0$. Beginning with this, backwards recursion computes the values of the OVF in the order $f(n-1), f(n-2), \ldots, f(1)$, using the functional equation

$$f(i) = \min\{c_{ij} + f(j) : \quad j = i+1 \text{ to } n \text{ such that } (i, j) \text{ is an arc in } G\} \quad (12.3)$$

in the order $i = n - 1$ to 1. The j that attains the minimum on the right in (12.3) defines the next node to go to from node i.

As an example, consider the network in Figure 12.10, with arc lengths entered on the arcs, and nodes with an acyclic numbering. Clearly, the states (nodes) in this network do not form into stages, with arcs always going from one stage to the next. Here is how backwards recursion proceeds on this network. The boundary condition is $f(9) = 0$ since 9 is the destination node.

$f(8) \quad = \quad \min\{12 + f(9)\} = \min\{12 + 0\} = 12$. Opt. decision, travel along arc (8, 9).

$f(7) \quad = \quad \min\{3 + f(9)\} = \min\{3 + 0\} = 3$. Opt. decision, travel along arc (7, 9).

$f(6) \quad = \quad \min\{2 + f(7), \ 8 + f(9), \ 9 + f(8)\} = \min\{2+3, \ 8+0, \ 9+12\} = 5$. Opt. decision, travel along arc (6, 7).

$f(5) \quad = \quad \min\{7+f(6), \ 10+f(8)\} = \min\{7+5, \ 10+12\} = 12$. Opt. decision, travel along arc (5, 6).

$f(4) \quad = \quad \min\{6+f(7), \ 3+f(6)\} = \min\{6+3, \ 3+5\} = 8$. Opt. decision, travel along arc (4, 6).

$f(3) \quad = \quad \min\{11+f(4), \ 2+f(5)\} = \min\{11+8, \ 2+12\} = 14$. Opt. decision, travel along arc (3, 5).

$f(2) \quad = \quad \min\{4+f(4), \ 5+f(6), \ 7+f(5)\} = \min\{4+8, \ 5+5, \ 7+12\} = 10$. Opt. decision, travel along arc (2, 6).

$f(1) \quad = \quad \min\{3+f(2), \ 5+f(3)\} = \min\{3+10, \ 5+14\} = 13$. Opt. decision, travel along arc (1, 2).

The optimum decisions and the OVF values at the various nodes are shown on Figure 12.10. The OVF of node 1 is 13; it is the length of the shortest route from node 1 to node 9 in this network. By following the optimum decisions determined at the various nodes, we can also obtain the shortest route from any node in the network to the destination node 9.

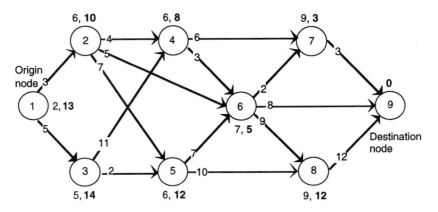

Figure 12.10 The length of each arc is entered on it. By the side of each node we marked the next node to go to on the shortest route from that node to the destination, and the length of that shortest route in bold face.

DP can also be applied to find shortest routes in directed networks that are not acyclic. We refer the reader to [K. G. Murty, 1992 of Chapter 5] for a discussion of DP based shortest routes algorithms in non-acyclic directed networks.

12.5   Solving the Nonnegative Integer Knapsack Problem By DP

Consider the nonnegative integer knapsack problem in which there are n objects available to load into the knapsack, with w_i, v_i being the weight and value of the ith object, for $i = 1$ to n. Let w_0 be the knapsack's weight capacity. All $w_0, w_1, \ldots, w_n$ are assumed to be positive integers. The problem is to determine the number of copies of each object to load into the knapsack to maximize the total value of all objects loaded, subject to the knapsack's weight capacity.

As discussed in Example 12.2, to solve this problem by DP we consider the loading process as a sequential process loading one object at a time, and represent the state of the system at any point of time in this process by the knapsack's remaining weight capacity. We define the OVF in state w to be

$$f(w) \quad = \quad \text{maximum possible value that can be loaded into the knapsack if its weight capacity is } w$$

When the knapsack's weight capacity is w, only objects i satisfying $w_i \overset{\leq}{=} w$, are available for loading into it. So, the functional equation satisfied by the OVF in this problem is

$$f(w) = \max\{v_i + f(w - w_i): \quad i = 1 \text{ to } n \text{ such that } w_i \overset{\leq}{=} w\} \qquad (12.4)$$

The operation in (12.4) is "max" instead of the usual "min" because our aim here is to maximize the total reward.

Clearly $f(0) = 0$, this is the boundary condition satisfied by the OVF in this problem. Beginning with this, we evaluate $f(w)$ for $w = 1, 2, \ldots, w_0$ in this order recursively using (12.4). The i attaining the maximum in (12.4) is the number of the object to be loaded into the knapsack when in state w, in an optimal policy.

As an example consider the problem with $n = 6$, and the following data.

Data for a nonnegative integer knapsack problem		
Object i	Weight w_i	Value v_i
1	3	12
2	4	12
3	3	9
4	3	15
5	7	42
6	9	18
Knapsack's weight capacity, $w_0 = 12$		

Since all objects have weights $\overset{\geq}{=} 3$, we have $f(0) = f(1) = f(2) = 0$ in this problem, these are the boundary conditions here.

So when the state of the system is 0, 1, or 2 (i.e., the remaining weight capacity of the knapsack is 0, 1, or 2) we just terminate, since no more objects can be loaded into the knapsack. When the state of the system is 3, objects 1, 3, 4 become available to be loaded into the knapsack, leading to the following equation for $f(3)$. Continuing in this way we evaluate $f(w)$ for higher values of w until $w_0 = 12$. As you can see, to evaluate an $f(W)$ say, the functional equation for $f(W)$ uses the values of $f(w)$ for $w < W$. That's why the procedure computes values of $f(w)$ in order of increasing w beginning with the known values of $f(0), f(1), f(2)$ given by the boundary conditions. This is the recursive feature of the DP algorithm.

$f(0) = f(1) = f(2) = 0$. Opt. decision - terminate.

$f(3) \quad = \quad \text{Max}\{12+f(0), \quad 9+f(0), \quad 15+f(0)\} \quad = \quad \max\{12+0, \quad 9+0, \quad 15+0\} = 15$. Opt. decision - load one of object 4 and continue as in state 0.

$f(4) \quad = \quad \text{Max}\{12+f(1), 12+f(0), 9+f(1), 15+f(1)\} = \max\{12+0, 12+0, 9+0, 15+0\} = 15$. Opt. decision - load one of object 4 and continue as in state 1.

$f(5)$ = Max$\{12+f(2), 12+f(1), 9+f(2), 15+f(2)\}$ = max$\{12+0,$ $12+0, 9+0, 15+0\}$ = 15. Opt. decision - load one of object 4 and continue as in state 2.

$f(6)$ = Max$\{12+f(3), 12+f(2), 9+f(3), 15+f(3)\}$ = max$\{12+15,$ $12+0, 9+15, 15+15\}$ = 30. Opt. decision - load one of object 4 and continue as in state 3.

$f(7)$ = Max$\{12+f(4), 12+f(3), 9+f(4), 15+f(4), 42+f(0)\}$ = max$\{12+15, 12+15, 9+15, 15+15, 42+0\}$ = 42. Opt. decision - load one of object 5 and continue as in state 0.

$f(8)$ = Max$\{12+f(5), 12+f(4), 9+f(5), 15+f(5), 42+f(1)\}$ = max$\{12+15, 12+15, 9+15, 15+15, 42+0\}$ = 42. Opt. decision - load one of object 5 and continue as in state 1.

$f(9)$ = Max$\{12+f(6), 12+f(5), 9+f(6), 15+f(6), 42+f(2),$ $18+f(0)\}$ = max$\{12+30, 12+15, 9+30, 15+30, 42+0,$ $18+0\}$ = 45. Opt. decision - load one of object 4 and continue as in state 6.

$f(10)$ = Max$\{12+f(7), 12+f(6), 9+f(7), 15+f(7), 42+f(3),$ $18+f(1)\}$ = max$\{12+42, 12+30, 9+42, 15+42, 42+15,$ $18+0\}$ = 57. Opt. decision - load one of object 4 and continue as in state 7.

$f(11)$ = Max$\{12+f(8), 12+f(7), 9+f(8), 15+f(8), 42+f(4),$ $18+f(2)\}$ = max$\{12+42, 12+42, 9+42, 15+42, 42+15,$ $18+0\}$ = 57. Opt. decision - load one of object 4 and continue as in state 8.

$f(12)$ = Max$\{12+f(9), 12+f(8), 9+f(9), 15+f(9), 42+f(5),$ $18+f(3)\}$ = max$\{12+45, 12+42, 9+45, 15+45, 42+15,$ $18+15\}$ = 60. Opt. decision - load one of object 4 and continue as in state 9.

By following the optimum decisions beginning with state 12, we see that an optimum strategy to maximize the value loaded when the weight capacity of the knapsack is 12, is to load four copies of object 4 into it, giving a total value of 60 for the objects loaded.

Since the value of $f(w)$ has to be computed for all $0 \leqq w \leqq w_0$ in this algorithm, it is not efficient for solving this problem, in comparison to B&B methods discussed in Chapter 10. We discussed this method mainly to illustrate an application of DP.

12.6   Solving the 0−1 Knapsack Problem by DP

Consider the 0−1 knapsack problem involving n objects. Let w_0 be the capacity of the knapsack by weight, and let w_i, v_i be the weight and value of the ith object, $i = 1$ to n. Here, only one copy of each object is available. The problem is to determine the subset of objects to be loaded into the knapsack to maximize the value loaded subject to its weight capacity. We assume that $w_0, w_1, \ldots, w_n$ are all positive integers.

To solve this problem by DP we consider the loading process as a sequential process loading one object at a time. However, as pointed out in Example 12.3, since there is only one copy of each object available, the definition of the state of the system at any point of time in this process must contain information on the remaining weight capacity of the knapsack at that time, and the set of objects not yet loaded. For this it is convenient to represent the process as a staged process with n stages. For each $k = 1$ to n, states in stage k will be denoted by the ordered pair $(\mathbf{k}, w)$ where $0 \leqq w \leqq w_0$ represents the knapsack's remaining weight capacity at that stage. In any state $(\mathbf{k}, w)$ in stage k, there are at most two possible decisions that can be taken, they are: (i) to decide not to include object k in the knapsack (in fact this is the only decision available if $w_k > w$) with an immediate reward of 0 and transition to state $(\mathbf{k + 1}, w)$ in stage $k + 1$; or (ii) to load object k into the knapsack (this decision is only available if $w \geqq w_k$) with an immediate reward of v_k and transition to state $(\mathbf{k + 1}, w - w_k)$ in stage $k + 1$. This creates an artificial stage structure with n stages, with the decision in stage k relating only to the inclusion or exclusion of object k, for each $k = 1$ to n. Thus each object's fate is considered in a unique stage, and there can be no confusion in any state what the available decisions in that state are. The available decisions at state $(\mathbf{k}, w)$ and the resulting state transitions are displayed in Figure 12.11.

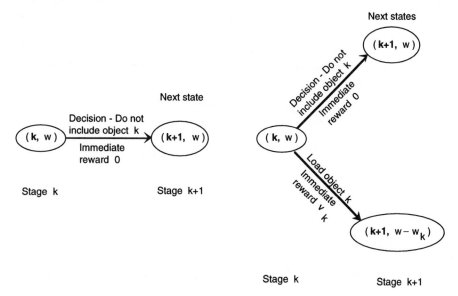

Figure 12.11 On the left is displayed the unique choice in state $(\mathbf{k}, w)$ if $w < w_k$. On the right are displayed the two available choices at state $(\mathbf{k}, w)$ if $w \geqq w_k$.

We now define the OVF in state $(\mathbf{k}, w)$ to be

$f(\mathbf{k}, w)$ = maximum possible value that can be loaded into the knap-   (12.5)
sack if its weight capacity is w, and choice of objects re-
stricted to only those in the set $\{k, k + 1, \ldots, n\}$.

Since only one copy of each object is available, we clearly have

$$f(\mathbf{n}, w) = \begin{cases} 0 & \text{if } w < w_n \\ v_n & \text{if } w > w_n \end{cases} \qquad (12.6)$$

(12.6) are the boundary conditions that the OVF $f(\mathbf{k}, w)$ satisfies in this problem. From the decisions available at state $(\mathbf{k}, w)$ displayed in Figure 12.11, we get
the functional equations satisfied by the OVF to be

$$f(\mathbf{k}, w) = \begin{cases} f(\mathbf{k} + \mathbf{1}, w) & \text{if } w < w_k \\ \max\{f(\mathbf{k} + \mathbf{1}, w), v_k + f(\mathbf{k} + \mathbf{1}, w - w_k)\} & \text{if } w \geqq w_k \end{cases} \qquad (12.7)$$

Using the boundary conditions in (12.6) and the functional equations in (12.7),
the OVF at all states can be evaluated by moving forward one stage at a time
beginning with stage $n - 1$. We compute the OVF at all states in a stage when
we deal with that stage and then move forward to the adjacent stage. At state
$(\mathbf{k}, w)$, the decision is to exclude object k from the knapsack if it happens that
$f(\mathbf{k}, w) = f(\mathbf{k} + \mathbf{1}, w)$ in (12.7); or to load object k into the knapsack if $f(\mathbf{k}, w) = v_k + f(\mathbf{k} + \mathbf{1}, w - w_k)$.

As an example consider the 0−1 knapsack problem with $n = 5$ and the following
data.

Data for a 0−1 knapsack problem

Object i	Weight w_i	Value v_i
1	3	12
2	4	12
3	3	15
4	7	42
5	9	18

Knapsack's weight capacity, $w_0 = 12$

Stage 5: Boundary conditions
$f(\mathbf{5},0)$ to $f(\mathbf{5},8)$ = 0. Opt. decision - terminate.
$f(\mathbf{5},9)$ to $f(\mathbf{5},12)$ = 18. Opt. decision - load object 5 and terminate.
Stage 4
$\quad f(\mathbf{4},w)$ = $f(\mathbf{5},w)$ for $w = 0$ to 6. Opt. decision - exclude object 4
and continue as in state $(\mathbf{5},w)$.
$\quad f(\mathbf{4},7)$ = Max$\{0+f(\mathbf{5},7), 42+f(\mathbf{5},0)\}$ = max$\{0+0, 42+0\}$ = 42.
Opt. decision - load object 4 and continue as in $(\mathbf{5},0)$.

$f(4,8)$ = Max$\{0+f(5,8),\ 42+f(5,1)\}$ = max$\{0+0,\ 42+0\}$ = 42. Opt. decision - load object 4 and continue as in state **(5,1)**.

$f(4,9)$ = Max$\{0+f(5,9),\ 42+f(5,2)\}$ = max$\{0+18,\ 42+0\}$ = 42. Opt. decision - load object 4 and continue as in state **(5,2)**.

$f(4,10)$ = Max$\{0+f(5,10),\ 42+f(5,3)\}$ = max$\{0+18,\ 42+0\}$ = 42. Opt. decision - load object 4 and continue as in state **(5,3)**.

$f(4,11)$ = Max$\{0+f(5,11),\ 42+f(5,4)\}$ = max$\{0+18,\ 42+0\}$ = 42. Opt. decision - load object 4 and continue as in state **(5,4)**.

$f(4,12)$ = Max$\{0+f(5,12),\ 42+f(5,5)\}$ = max$\{0+18,\ 42+0\}$ = 42. Opt. decision - load object 4 and continue as in state **(5,5)**.

Continuing the same way, we get the following OVF values and optimum decisions at states in stages 3, 2.

	Stage 3		Stage 2	
	OVF	Opt.	OVF	Opt.
w	$f(3,w)$	decision	$f(2,w)$	decision
0, 1, 2	$f(4,w)$	Exclude obj. 3 Cont. as in $(4,w)$	$f(3,w)$	Exclude obj. 2 Cont. as in $(3,w)$
3	15	Load obj. 3 Cont. as in **(4,0)**	15	Exclude obj. 2 Cont. as in **(3,3)**
4	15	Load obj. 3 Cont. as in **(4,1)**	15	Exclude obj. 2 Cont. as in **(3,4)**
5	15	Load obj. 3 Cont. as in **(4,2)**	15	Exclude obj. 2 Cont. as in **(3,5)**
6	15	Load obj. 3 Cont. as in **(4,3)**	15	Exclude obj. 2 Cont. as in **(3,6)**
7	42	Exclude obj. 3 Cont. as in **(4,7)**	42	Exclude obj. 2 Cont. as in **(3,7)**
8	42	Exclude obj. 3 Cont. as in **(4,8)**	42	Exclude obj. 2 Cont. as in **(3,8)**
9	42	Exclude obj. 3 Cont. as in **(4,9)**	42	Exclude obj. 2 Cont. as in **(3,9)**
10	57	Load obj. 3 Cont. as in **(4,7)**	57	Exclude obj. 2 Cont. as in **(3,10)**
11	57	Load obj. 3 Cont. as in **(4,8)**	57	Exclude obj. 2 Cont. as in **(3,11)**
12	57	Load obj. 3 Cont. as in **(4,9)**	57	Exclude obj. 2 Cont. as in **(3,12)**

And finally, we have $f(1,12)$ = max$\{0+f(2,12),\ 12+f(2,9)\}$ = max$\{0+57,\ 12+42\}$ = 57, with the optimum decision in state **(1,12)** to be to exclude object 1, and continue as in state (**2**,12). Following the decisions in the various stages, we see that an optimum strategy in the original problem to maximize the value loaded in the knapsack is to load objects 3 and 4 into it. This leads to the maximum value loaded of 57.

Since the value of $f(\mathbf{k}, w)$ has to be evaluated for all $n \geqq k \geqq 1$ and $0 \leqq w \leqq w_0$ in this algorithm, it is not efficient to solve the $0-1$ knapsack problem, in comparison to B&B methods discussed in Chapter 9 when w_0 is large. Our main interest in discussing this algorithm here is to illustrate another application of DP.

12.7   A Resource Allocation Problem

There are K units of a single resource available, which can be distributed among n different activities. K is a positive integer. The problem is to allocate the resource units most profitably among the activities. Assume that resource units can only be allocated to activities in nonnegative integer quantities. Define for $i = 1$ to n

$$x_i \quad = \quad \text{number of units of resource allotted to activity } i \qquad (12.8)$$

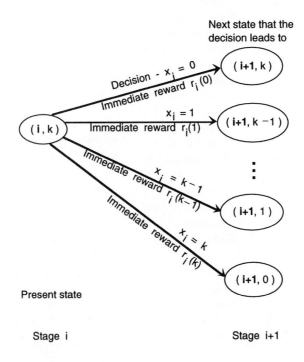

Figure 12.12 State transitions from a state in stage i to stage $i+1$.

Let $r_i(x_i)$ denote the profit or reward realized from an allocation of x_i units of resource to activity i. A table giving the values of $r_i(x_i)$ for $x_i = 0$ to K, $i = 1$ to n, is the data for this problem. We assume that $r_i(x_i) \geq 0$ for all $x_i \geq 0$, $i = 1$ to n. The problem is to choose a nonnegative integer vector $x = (x_1, \ldots, x_n)^T$ so as to maximize the total reward subject to the constraint $x_1 + \ldots + x_n \leq K =$ the units of resource available.

This problem is not really dynamic, but can be posed as a staged sequential decision problem involving n stages, based on the technique used for the $0-1$ knapsack problem. It views this problem as a sequential decision process in which at the ith stage only the value of the variable x_i (the number of units of resource to be allotted to activity i) is determined, $i = 1$ to n. The states of the system in stage i are $(\mathbf{i}, k)$, $0 \leq k \leq K$, where k denotes the number of unallotted units of resource available at this stage. The possible decisions available in state $(\mathbf{i}, k)$ are to select an integral value for the variable x_i between 0 and k, leading to an immediate reward of $r_i(x_i)$ and a transition to state $(\mathbf{i} + 1, k - x_i)$ in stage $i + 1$. These state transitions are illustrated in Figure 12.12.

We now define the OVF in this process to be

$$f(\mathbf{i}, k) \quad = \quad \text{maximum total reward that can be obtained from activities} \quad (12.9)$$
$$i \text{ to } n, \text{ with } k \text{ units of resource that can be allotted among}$$
$$\text{them}$$

This OVF clearly satisfies the following boundary condition

$$f(\mathbf{n}, k) \quad = \quad \max\{r_n(t) : \ 0 \leq t \leq k\} \qquad (12.10)$$

and if t_1 attains the maximum in (12.10), the optimum decision in state $(\mathbf{n}, k)$ is to allot t_1 units of resource to activity n and leave the other $k - t_1$ units of resource unallotted.

Normally the reward function $r_i(k)$ will be monotonic increasing in k for all i (in most real world applications this will be the case since the return is usually an increasing function of the resources committed). In this case, (12.10) becomes $f(\mathbf{n}, k) = r_n(k)$ for all $0 \leq k \leq K$ and the optimum decision in state $(\mathbf{n}, k)$ is to allot all k units of resource to activity n.

From the state transitions illustrated in Figure 12.12, and the principle of optimality, it is clear that

$$f(\mathbf{i}, k) \quad = \quad \max\{r_i(x_i) + f(\mathbf{i} + 1, k - x_i) : 0 \leq x_i \leq k\} \qquad (12.11)$$

for $1 \leq i \leq n$ and $0 \leq k \leq K$. (12.11) are the functional equations satisfied by the OVF in this problem. The optimum decision in state $(\mathbf{i}, k)$ is to make the variable x_i equal to the argument attaining the maximum in (12.11).

Beginning with the known values of $f(\mathbf{n}, k)$, $0 \leqq k \leqq K$ given by the boundary conditions (12.10), the values of the OVF can be evaluated and the optimum decisions at all states in other stages determined, in the order: stage $n - 1, n - 2, \ldots, 1$.

As an example, consider the problem faced by a politician running for reelection for his position in city administration. He has $K = 5$ volunteers who have agreed to help his campaign by distributing posters door to door and talking to residents in his district. We give below estimates of additional votes that would result from assigning these volunteers to 4 different precincts.

No. of volunteers assigned k	$r_i(k)$ = expected additional votes (in 100s) gained by assigning k volunteers to precinct			
	$i = 1$	2	3	4
0	0	0	0	0
1	35	79	130	86
2	42	110	160	120
3	56	130	170	130
4	50	140	180	130
5	50	125	175	125

If too many volunteers knock on the doors people may get irritated and react negatively; that's why in this problem $r_i(k)$ increases as k increases up to a value, and then begins to decrease.

Here the volunteers are the resource and we have 5 of them. The problem is to determine the optimum number of volunteers to allot to the various precincts in order to maximize the total expected additional votes gained by their efforts.

So, we define 4 stages, with stage i dealing with the decision variable $x_i = $ number of volunteers allotted to precinct i, $i = 1$ to 4. For $i = 1$ to 4, the symbol $(\mathbf{i}, k)$ defines the state in stage i of having k volunteers to assign in precincts i to 4. The OVF here is

$f(\mathbf{i}, k)$ $=$ maximum expected additional votes gained by allotting k volunteers in precincts i to 4 optimally.

The boundary conditions for stage 4 are given below.

$f(\mathbf{4}, 0)$ $=$ Max$\{0\} = 0$. Opt. decision - allot 0 volunteers to precinct 4 and terminate.

$f(\mathbf{4}, 1)$ $=$ Max$\{0, 86\} = 86$. Opt. decision - allot 1 volunteer to precinct 4 and terminate.

$f(\mathbf{4}, 2)$ $=$ Max$\{0, 86, 120\} = 120$. Opt. decision - allot 2 volunteers to precinct 4 and terminate.

$f(\mathbf{4}, 3)$ $=$ Max$\{0, 86, 120, 130\} = 130$. Opt. decision - allot 3 volunteers to precinct 4 and terminate.

$f(4,4)$ = Max$\{0, 86, 120, 130, 130\}$ = 130. Opt. decision - allot 4 volunteers to precinct 4 and terminate.

$f(4,5)$ = Max$\{0, 86, 120, 130, 130, 125\}$ = 130. Opt. decision - allot 4 volunteers to precinct 4 and terminate.

We now compute the OVF and the optimum decision in each state in other stages, in the order stage 3, 2, 1, recursively.

Stage 3

$f(3,0)$ = Max$\{0 + f(4,0)\}$ = max$\{0+0\}$ = 0. Opt. decision - allot 0 volunteers to precinct 3 and continue as in state $(4,0)$.

$f(3,1)$ = Max$\{0 + f(4,1), 130+f(4,0)\}$ = max$\{0+86, 130+0\}$ = 130. Opt. decision - allot 1 volunteer to precinct 3 and continue as in state $(4,0)$.

$f(3,2)$ = Max$\{0 + f(4,2), 130+f(4,1), 160+f(4,0)\}$ = max$\{0+120, 130+86, 160+0\}$ = 216. Opt. decision - allot 1 volunteer to precinct 3 and continue as in state $(4,1)$.

$f(3,3)$ = Max$\{0 + f(4,3), 130+f(4,2), 160+f(4,1), 170+f(4,0)\}$ = max$\{0+130, 130+120, 160+86, 170+0\}$ = 250. Opt. decision - allot 1 volunteer to precinct 3 and continue as in state $(4,2)$.

$f(3,4)$ = Max$\{0 + f(4,4), 130+f(4,3), 160+f(4,2), 170+f(4,1), 180+f(4,0)\}$ = max$\{0+130, 130+130, 160+120, 170+86, 180+0\}$ = 280. Opt. decision - allot 2 volunteers to precinct 3 and continue as in state $(4,2)$.

$f(3,5)$ = Max$\{0 + f(4,5), 130+f(4,4), 160+f(4,3), 170+f(4,2), 180+f(4,1), 175+f(4,0)\}$ = max$\{0+130, 130+130, 160+130, 170+120, 180+86, 175+0\}$ = 290. Opt. decision - allot 3 volunteers to precinct 3 and continue as in state $(4,2)$.

Stage 2

$f(2,0)$ = Max$\{0 + f(3,0)\}$ = max$\{0+0\}$ = 0. Opt. decision - allot 0 volunteers to precinct 2 and continue as in state $(3,0)$.

$f(2,1)$ = Max$\{0 + f(3,1), 79+f(3,0)\}$ = max$\{0+130, 79+0\}$ = 130. Opt. decision - allot 0 volunteers to precinct 2 and continue as in state $(3,1)$.

$f(2,2)$ = Max$\{0 + f(3,2), 79+f(3,1), 110+f(3,0)\}$ = max$\{0+216, 79+130, 110+0\}$ = 216. Opt. decision - allot 0 volunteers to precinct 2 and continue as in state $(3,2)$.

Stage 2 contd.

$f(\mathbf{2}, \mathbf{3})$ = Max$\{0 + f(\mathbf{3}, \mathbf{3}), 79+f(\mathbf{3}, \mathbf{2}), 110+f(\mathbf{3}, \mathbf{1}), 130+f(\mathbf{4}, \mathbf{0})\}$
= max$\{0+250, 79+216, 110+130, 130+0\} = 295$. Opt. decision - allot 1 volunteer to precinct 2 and continue as in state $(\mathbf{3}, \mathbf{2})$.

$f(\mathbf{2}, \mathbf{4})$ = Max$\{0 + f(\mathbf{3}, \mathbf{4}), 79+f(\mathbf{3}, \mathbf{3}), 110+f(\mathbf{3}, \mathbf{2}), 130+f(\mathbf{4}, \mathbf{1}), 140+f(\mathbf{3}, \mathbf{0})\}$ = max$\{0+280, 79+250, 110+216, 130+130, 140+0\} = 329$. Opt. decision - allot 1 volunteer to precinct 2 and continue as in state $(\mathbf{3}, \mathbf{3})$.

$f(\mathbf{2}, \mathbf{5})$ = Max$\{0 + f(\mathbf{3}, \mathbf{5}), 79+f(\mathbf{3}, \mathbf{4}), 110+f(\mathbf{3}, \mathbf{3}), 130+f(\mathbf{3}, \mathbf{2}), 140+f(\mathbf{3}, \mathbf{1}), 125+f(\mathbf{3}, \mathbf{0})\}$ = max$\{0+290, 79+280, 110+250, 130+216, 140+130, 125+0\} = 360$. Opt. decision - allot 2 volunteers to precinct 2 and continue as in state $(\mathbf{3}, \mathbf{3})$.

Stage 1

$f(\mathbf{1}, \mathbf{5})$ = Max$\{0 + f(\mathbf{2}, \mathbf{5}), 35+f(\mathbf{2}, \mathbf{4}), 42+f(\mathbf{2}, \mathbf{3}), 56+f(\mathbf{3}, \mathbf{2}), 50+f(\mathbf{2}, \mathbf{1}), 50+f(\mathbf{2}, \mathbf{0})\}$ = max$\{0+360, 35+329, 42+295, 56+216, 50+130, 50+0\} = 364$. Opt. decision - allot 1 volunteer to precinct 1 and continue as in state $(\mathbf{2}, \mathbf{4})$.

By following the optimum decisions beginning with state $(\mathbf{1}, \mathbf{5})$, we see that an optimum strategy is to allot 1 volunteer each to precincts 1, 2, 3, and the remaining 2 volunteers to precinct 4. This yields the maximum expected additional votes of 364 (in units of hundreds).

In this section we discussed a family of simple allocation models involving the distribution of a single resource among various activities. These models can be generalized to encompass situations in which activities require two or more resources, but the number of states needed to represent multiple resource allocation problems for solution by DP grows very rapidly with the number of resources. This unfortunate aspect of DP is called the **curse of dimensionality**.

Summary

In this chapter we introduced the recursive technique of dynamic programming, and illustrated its application to several discrete deterministic optimization problems that can be posed in a sequential decision format. The basic principles behind DP have been in use for many years, but it was R. Bellman who in the 1950s developed it into a systematic tool and pointed out its broad scope. Now, dynamic programming is a powerful technique with many applications in production planning and control, optimization and control of chemical and pharmaceutical batch and continuous processes, cargo loading, inventory control, equipment replacement and maintenance, and in finding optimal trajectories for rockets and satellites. Our treatment of the subject has been very elementary since our aim is mainly to introduce the concepts of systems and their states, optimum value functions, functional equations and the recursive technique for solving them, which are fundamental to DP. The books referenced at the end of this chapter should be consulted for advanced treatments of the subject.

12.8 Exercises

12.1 The US government is worried about increasing unemployment in states on the west coast due to rapid decline of timber-lands in those states by excessive lumbering activity. So, they recently authorized spending an additional $8 mil. over the next 3 calendar years to create new jobs in alternate industries in those states. Because of programs going on already, the effectiveness of additional funds depends on when they are spent. Funds can only be released in integral multiples of $1 mil. for any year. Following table provides important data estimated by a panel of economists, with new jobs measured in units of 100.

Year	\multicolumn{9}{c}{New jobs created if $r mil. are spent in year}
	$r = 0$	1	2	3	4	5	6	7	8
1	0	5	15	40	80	90	95	98	100
2	0	5	15	40	60	70	73	74	75
3	0	4	26	40	45	50	51	52	53

Find an optimum policy for spending the funds over the planning horizon, which maximizes the total number of additional jobs created, using DP.

12.2 There are 4 types of investments. Each accepts investment only in integer multiples of certificates. We have 30 units of money to invest (1 unit = $1000). Following table provides data on rewards obtained from investments in the different types.

Investment type	Cost (units/certificate)	\multicolumn{5}{c}{Reward for buying r certificates}
		$r = 1$	2	3	4	5
1	3	2	3	8	16	23
2	2	1	2	4	7	12
3	4	4	8	15	24	30
4	6	4	9	23	36	42

At least one certificate of each type must be purchased. Use DP to determine the optimum number of certificates of each type to buy to maximize total reward.

12.3 A production process is available for 3 periods. In each period it can produce an integer number of units of a commodity between 0 to 4. A total of 6 units of the commodity must be produced by the end of period 3.

Units produced in period 1 (period 2) have to be stored at a cost of \$2/unit (\$1/unit) till the end of period 3. Those produced in period 3 incur no storage cost. Other data is given below. Determine an optimum production plan to minimize the total cost of production and storage for meeting the requirement.

Period	Production cost (in \$100s) if r units produced				
	$r = 0$	1	2	3	4
1	0	4	8	9	12
2	0	7	10	11	15
3	0	8	11	15	16

12.4 A batch of chemical consisting of 6 tons of it, contains the chemical in particle sizes 1, 2, 3 in equal proportion (size 1 is smaller than size 2 which in turn is smaller that size 3).

The company has 2 sieves. Sieve 1 transfers particles of size 1 to the bottom and leaves everything else on top. Sieve 2 leaves particles of size 3 at the top, but transfers everything else to the bottom. To use either sieve, a minimum of 2 tons of material must be fed. Each use of either sieve costs \$10. Data on selling price of chemical is given below. Determine the maximum amount of money that can be made with the existing batch of chemical.

No.	Chemical containing particle sizes	Price/ton
I	1, 2, 3	\$40
II	1,2 only	\$55
III	1 only	\$60
IV	2, 3 only	\$50
V	3 only	\$70
VI	2 only	\$45

12.5 The major highways in Michigan are US-23, I-94, I-96 and I-75. The state highway department is concerned about the ever increasing number of speed limit violators on these highways. To control the problem they have decided to put 7 new patrol cars on these highways. Following data represents the best estimates of the number of violators ticketed per day.

Highway	Expected no. ticketed/day if r new patrol cars assigned			
	$r = 0$	1	2	3
I-94	30	70	100	140
US-23	20	45	80	115
I-75	10	20	40	65
I-96	20	40	90	110

Determine an optimum allocation of new patrol cars to the various highways (no more than 3 for any highway) using DP.

12.6 A hi-tech company has perfected a process of growing crystalline silicon rods in 10 inch lengths. Profit obtained by selling a silicon rod depends on its length as given below.

Length (in.)	1	2	3	4	5	6	7	8	9	10
Profit ($)	60	125	185	235	260	340	360	400	440	475

The cutting tool only accepts rods whose length in in. is an integer ≥ 2, and it cuts the rod into two pieces whose lengths in in. are integers. Each use of the cutting tool costs $10. The pieces obtained from a cut can be cut again if they satisfy the conditions mentioned above. Determine an optimum cutting policy for each 10 in. rod, to maximize the net profit from it.

12.7 A company has 5 identical machines which it uses to make four products A, B, C, D. Each machine can make any product, and when it is set up to make a product, a production run of one week is scheduled. The following table gives a forecast for the coming week's profit depending on how many machines are scheduled to produce each product.

No. of mcs.	Week's forecasted profit ($10,000 units) from product			
	A	B	C	D
0	0	0	0	0
1	12	17	5	8
2	17	30	12	14
3	25	49	22	25
4	35	64	34	35
5	45	76	48	43

Determine the optimum number of machines to allot to each product for the coming week to maximize the total profit.

12.8 The EPA got into a lot of bad publicity recently about lax monitoring of dioxin contamination of Michigan rivers. EPA divides the state into 3 regions. The following table gives data on the number of tests that can be conducted in each region by allotting some inspectors. EPA is willing to appoint 5 inspectors. Determine how many of these to allot to each region (this should be a nonnegative integer for each region) so as to maximize the total number of tests conducted over the whole state per month.

Region	No. of tests/month if r inspectors allotted				
	$r = 1$	2	3	4	5
1	25	50	80	117	125
2	20	70	130	150	160
3	10	20	35	40	45

12.9 There are 4 objects which can be packed in a vessel. Objects 1, 2, 3 are available in unlimited number of copies; but only four copies of object 4 are available. The weight of each object and the capacity of the vessel are expressed in weight units, and values in money units, in the following table.

object	Wt. per copy	Value/copy if no. included is				
		1	2	3	4	5
1	3	2	3	8	16	23
2	2	1	2	4	7	12
3	4	4	8	15	24	30
4	6	4	9	23	36	

The vessel's weight capacity is 30 weight units. The objective function is total value, and it is additive over objective types. Find the maximum objective value, subject to the constraint that at least one copy of each object must be included.

12.10 A company has four salesmen to allocate to three marketing regions. Their objective is to maximize the total sales volume generated. The sales growth in each region is expected to go up as more salesmen are allocated there, but not linearly. Company's estimates of the sales volume as a function of the number of salesmen allocated to each region are given below.

Region	Sales volume if r salesmen allotted				
	$r = 1$	2	3	4	5
1	25	50	60	80	100
2	20	70	90	100	100
3	10	20	30	50	60

Each salesman has to be allotted to one region exclusively, or his employment can be terminated. Formulate the problem of determining how many salesmen to allot to each region so as to maximize the total sales volume as a DP and solve it.

12.11 A resource may be used on either or both of two processes. Each unit of resource generates $4, $3 when used for a day on the first process, second process, respectively. The resource can be recycled, but in recycling a fraction is lost owing to usage and wastage. Thus, of the units used on the first process (second process) only half (two-thirds) remain for use the following day. 100 units of the resource are available at the start of a 10-day period, at the end of which any units remaining will have no value. Determine how many units should be used on each process (fractions of units are allowed) on each of the ten days in order to maximize the total return. ([P. Dixon and J.M. Norman, 1984])

12.12 A student has final examinations in 3 courses, X, Y, and Z, each worth the same number of credits. There are only 3 days available for study. Assume that the student has to devote a nonnegative integer number of the available days for studying for each course, i.e., a day cannot be split between two courses. Estimates of expected grades based upon various numbers of days devoted for studying for each course are given below.

Course	Expected grade if days of study is				
	0	1	2	3	4
X	0	1	1	3	4
Y	1	1	3	4	4
Z	0	1	3	3	4

(a) Determine the number of available days that the student should devote to each course in order to maximize the sum of all the grades, using DP.

(b) How does the strategy in (a) change if the student has 4 days available to study before the examinations? What is the increase in the optimum objective value?

(c) How do the strategies in (a), (b) change if a new course W is added, with expected grade of 0, 0, 2, 3, and 4 when the number of days devoted to studying it is 0, 1, 2, 3, and 4 respectively? (S. M. Pollock)

12.13 A spaceship is on its way to landing on the moon. At some point during its descent near the moon, it has ϕ units of fuel, a downwards velocity of v towards the surface of the moon, and an altitude z above the surface of the moon. Time is measured in discrete units, and actions are only taken at integer values of time until the spaceship touches down.

At each integer value of time t, you can select an amount y of fuel to use, which will result in new variable values at time $t + 1$ of

$$\begin{aligned} \phi' &= \phi - y \quad \text{(depleted by } y \text{ units of fuel)} \\ v' &= v + 5 - y \quad \text{(the force of gravity is ``5'')} \\ z' &= z - v' \quad \text{(altitude decreases by } v') \end{aligned}$$

If z ever becomes negative, or if $v > 0$ when z becomes 0, the spaceship is fully destroyed.

Given initial (i.e., at time point 0 in this portion of the spaceship's trajectory) fuel, velocity, and altitude values of Φ, V, and Z, solve using DP the problem of reaching the point $(v = 0, z = 0)$ safely, using

(a) the OVF

$$f(\phi, v, z) \quad = \quad \text{maximum amount of fuel remaining when the spaceship safely}$$
$$\text{lands at } (v = 0, z = 0), \text{ given it is at } (\phi, v, z) \text{ at time point 0.}$$

(b) the OVF

$$g(v, z) \quad = \quad \text{minimum amount of fuel required to safely reach } (v = 0, z =$$
$$0), \text{ given it is at } (v, z) \text{ at time point 0.}$$

(c) Assuming that all variables are integer valued, what is the computational effort involved in solving (a)?

(d) Solve the problem numerically when $\Phi = 100$, $V = 20$, $Z = 300$. (S. M. Pollock)

12.14 There are 4 objects available for loading into a knapsack of unlimited weight capacity. Data on the objects is given below.

Object i	1	2	3	4
Value v_i	7	16	19	15
Weight w_i	3	6	7	5

An unlimited number of copies of each object are available for loading into the knapsack. Define

$$g(t) \quad = \quad \text{the minimum total weight of items needed in order to achieve}$$
$$\text{a total value of at least } t \text{ in the knapsack.}$$

Find $g(t)$ and the associated (complete) optimal policy for nonnegative integers $t = 0$ to 100. (S. M. Pollock)

12.15 A person wants to cross an uninhabited desert that is 100 miles wide in a jeep. The jeep is heavy and the sand soft, so he gets only 3 miles/gallon. Gasoline can be purchased in unlimited quantities at the beginning of the desert, but once the jeep enters the desert no gasoline can be purchased until the desert is completely crossed. The jeep has a carrying capacity of 20 gallons of gasoline which includes gasoline consumed while travelling.

The driver plans to cross the desert by using the following procedure. Fill up the jeep at a depot at the beginning of the desert, and drive into the desert to a spot (call it the first "temporary gas dump") where some gasoline is unloaded and stored, and then drive back to the depot to load up again. Continue this process until there is enough gasoline stored up at the first temporary gas dump so that the driver can use this as a new "depot" to continue past it into the desert.

Formulate the problem of minimizing the total quantity of gasoline needed to cross the desert by this procedure, as a DP. Solve your formulation and find the minimum amount of gasoline needed, and the policy that attains it. ([D. Gale, 1970])

12.9   References

E. V. DENARDO, 1982, *"Dynamic Programming Models and Applications"*, Prentice Hall, Englewood Cliffs, NJ 07632.

S. E. DREYFUS and A. M. LAW, 1977, *"The Art and Theory of Dynamic Programming"*, Academic Press, NY.

P. DIXON and J. M. NORMAN, 1984, "An Instructive Exercise in Dynamic Programming", *IIE Transactions*, 16, no. 3, 292-294.

D. GALE, 1970, "The Jeep Once More or Jeeper By the Dozen", *American Math Monthly*, 77, 493-501. Correction published in *American Math Monthly*, 78 (1971) 644-645.

Chapter 13

Critical Path Methods
in Project Management

A project usually refers to an effort that is a one-time effort, one that is not undertaken on a routine production basis. For example, the construction of a skyscraper, a building, a highway, or a manufacturing facility, would be typical (civil engineering) projects. Manufacturing of large items like ships, generators, etc. would be (manufacturing) projects. In addition, the development, planning, and launching of new products; research and development programs; periodic maintenance operations; the development and installation of new management information systems; etc., are all non-routine tasks that can be considered as projects. In this chapter we discuss techniques that help in planning, scheduling, and controlling of projects. The most popular of these methods is the **critical path method (CPM)** which decomposes the project into a number of activities, represents the precedence relationships among them through a project network, and makes a schedule for these activities over time that minimizes the project duration, by applying on the project network the dynamic programming algorithm discussed in Chapter 12 for finding a longest route. We discuss project scheduling using CPM in this chapter.

13.1 The Project Network

A project is usually a collection of many individual **jobs** or **activities**. The words **job, activity** will be used synonymously in this chapter. The first step in CPM is to decompose the project into its constituent activities and determine the precedence relationships among them. These arise from technological constraints that require certain jobs to be completed before others can be started (for example, the job "painting the walls" can only be started after the job "erecting the walls" is completed).

We assume that each job in the project can be started and completed independently of the others within the technological sequence defined by the precedence relationships among the jobs.

If job 2 cannot be started until after job 1 has been completed, then job 1 is known as a **predecessor** or **ancestor** of job 2; and job 2 is known as a **successor** or **descendent** of job 1. If job 1 is a predecessor of job 2, and there is no other job which is a successor of job 1 and predecessor of job 2, then job 1 is known as an **immediate predecessor** of job 2, and job 2 is known as an **immediate successor** of job 1. A job may have several immediate predecessors; it can be started as soon as all its immediate predecessors have been completed. If a job has two or more immediate predecessors, by definition every pair of them must be unrelated in the sense that neither of them is a predecessor of the other.

If 1 is a predecessor of 2, and 2 is a predecessor of 3, then obviously 1 is a predecessor of 3. This property of precedence relationships is called **transitivity**. Given the set of immediate predecessors of each job, it is possible to determine the set of predecessors, or the set of successors of any job, by recursive procedures using transitivity. The predecessor relationships are inconsistent if they require that a job has to be completed before it can be started; so, no job can be a predecessor of itself.

Because of these properties, the precedence relationships define an ordering among the jobs in a project called a **partial ordering** in mathematics.

The planning phase of the project involves the breaking up of the project into various jobs using practical considerations, identifying the immediate predecessors of each job based on engineering and technological considerations, and estimating the time required to complete each job.

Inconsistencies may appear in the predecessor lists due to human error. The predecessor data is said to be **inconsistent** if it leads to the conclusion that a job precedes itself, by the transitivity property. Inconsistency implies the existence of a circuit in the predecessor data, i.e., a subset of jobs $1, \ldots, r$, such that j is listed as a predecessor of $j + 1$ for $j = 1$ to $r - 1$, and r is listed as a predecessor of 1. Such a circuit represents a logical error and at least one link in this circuit must be wrong. As it represents a logical error, inconsistency is a serious problem.

Also, in the process of generating the immediate predecessors for an activity, an engineer may put down more than necessary and show as immediate predecessors some jobs that are in reality more distant predecessors. When this happens, the predecessor data is said to contain **redundancy**. Redundancy poses no theoretical or logical problems, but it unnecessarily increases the complexity of the network used to represent the predecessor relationships. Given the list of immediate predecessors of each job, one must always check it for any inconsistency, and redundancy, and make appropriate corrections.

As an example, we give below the precedence relationships among jobs in the project: **building a hydroelectric power station**. In this example, we have not gone into very fine detail in breaking up the project into jobs. In practice, a job like 11 (dam building) will itself be divided into many individual jobs involved in dam building.

The job duration is the estimated number of months needed to complete the job.

Hydroelectric Power Station Building Project

No.	Job Description	Immediate Predecessors	Job Duration
1.	Ecological survey of dam site		6.2
2.	File environmental impact report and get EPA approval	1	9.1
3.	Economic feasibility study	1	7.3
4.	Preliminary design and cost estimation	3	4.2
5.	Project approval and commitment of funds	2, 4	10.2
6.	Call quotations for electrical equipment (turbines, generators, ...)	5	4.3
7.	Select suppliers for electrical equipment	6	3.1
8.	Final design of project	5	6.5
9.	Select construction contractors	5	2.7
10.	Arrange construction materials supply	8, 9	5.2
11.	Dam building	10	24.8
12.	Power station building	10	18.4
13.	Power lines erection	7, 8	20.3
14.	Electrical equipment installation	7, 12	6.8
15.	Build up reservoir water level	11	2.1
16.	Commission the generators	14, 15	1.2
17.	Start supplying power	13, 16	1.1

We will represent the precedence relationships among activities through a directed network. A **directed network** is a pair of sets $(\mathcal{N}, \mathcal{A})$, where $\mathcal{N}$ is a set of **nodes** (also called **vertices** or **points** in the literature), and $\mathcal{A}$ is a set of directed lines called **arcs**, each arc joining a pair of nodes.

The arc joining node i to node j is denoted by the ordered pair (i, j); it is **incident into** j and **incident out** of i; node i is its **tail**, and j its **head**. For example, (1, 2) is an arc in Figure 13.2 with tail 1 and head 2.

A **chain** $\mathcal{C}$ in the directed network $G = (\mathcal{N}, \mathcal{A})$ from x_1 (**origin** or **initial node**) to x_k (**destination** or **terminal node**) is a sequence of points and arcs alternately

$$\mathcal{C} = x_1, e_1, x_2, e_2, \ldots, e_{k-1}, x_k \tag{13.1}$$

such that for each $r = 1$ to $k-1$, e_k is the arc (x_r, x_{r+1}); i.e., it is a sequence of arcs connecting the points x_1 and x_k, with all the arcs directed towards the destination

x_k. For example, in the directed network in Figure 13.2, $\mathcal{C}_1 = 1$, (1, 2), 2, (2, 5), 5, (5, 8), 8, (8, 13), 13, (13, 17), 17 is a chain from node 1 to node 17 that consists of 5 arcs.

A chain is said to be a **simple chain** if no node or arc is repeated in it. The chain $\mathcal{C}_1$ given above from node 1 to node 17 is a simple chain. A simple chain can be stored using **node labels** called **predecessor indices** or **predecessor labels**. For example, suppose the chain $\mathcal{C}$ in (13.1) from x_1 to $x_k \neq x_1$ is a simple chain. The origin x_1 of $\mathcal{C}$ has no predecessors on $\mathcal{C}$, so its predecessor index is $\emptyset$. For $2 \leq r \leq k$, x_{r-1} is the immediate predecessor of x_r on $\mathcal{C}$, hence x_{r-1} is defined to be the predecessor index of x_r. With predecessor indices defined this way, the simple chain itself can be traced by a backwards trace of these predecessor indices beginning at the terminal node. From the label on the terminal node x_k, we know that the last arc on $\mathcal{C}$, the one incident into x_k, is (x_{k-1}, x_k). Now go back to the predecessor node x_{k-1} of x_k on $\mathcal{C}$, look up the label on it, and continue in the same manner. The trace stops when the node with the $\emptyset$ label, the origin, is reached. As an example, in Figure 13.1 we show a simple chain from node 1 to node 14 and the predecessor labels for storing it. Each node number is entered inside the circle representing it, and its predecessor index on the simple chain is entered by its side. Nodes and arcs in the network not on this chain are omitted in this figure.

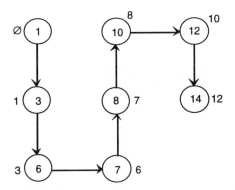

Figure 13.1 A simple chain from node 1 to node 14, and the predecessor labels on the nodes for storing it.

In the application discussed in this chapter, each arc in the network will have its length given, and the problem needs the longest simple chains from the origin node to every other node in the network. In the algorithm for solving this problem, node labels will indicate not only the predecessor indices, but also the actual lengths of the chains.

We now discuss two different ways of representing the precedence relationships among the jobs in a project as a directed network. One leads to the **activity on node (AON) diagram**, and the other the **activity on arc (AOA) diagram** or **arrow diagram** of the project.

Activity on Node (AON) Diagram of the Project

As the name implies, each job is represented by a node in this network. Let node i represent job i, $i = 1$ to n = number of jobs. Include arc (i, j) in the network iff job i is an immediate predecessor of job j. The resulting directed network called the **Activity on Node (AON) diagram**, is very simple to draw, but not too convenient for project scheduling, so we will not use it in the sequel. The AON diagram of the hydroelectric power station building project is given in Figure 13.2.

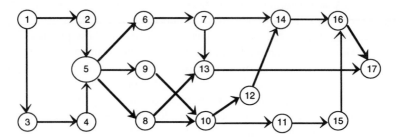

Figure 13.2 AON diagram for the hydroelectric power station building project.

Arrow Diagram of the Project

The **Arrow diagram** or the **Activity on Arc (AOA) diagram** represents jobs by arcs in the network. We refer to the job corresponding to arc (i, j) in this network, as job (i, j) itself. Nodes in the arrow diagram represent **events** over time. Node i represents the event that all jobs corresponding to arcs incident into node i have been completed, and after this event any job corresponding to an arc incident out of node i can be started. The arrow diagram is drawn so as to satisfy the following properties.

Property 1 If (i, j), (p, q) are two jobs, job (i, j) is a predecessor of job (p, q) iff there is a chain from node j to node p in the arrow diagram.

In order to represent the predecessor relationships through Property 1, it may be necessary to introduce **dummy arcs** which correspond to **dummy jobs**. The need for dummy jobs is explained with illustrative examples later on. In drawing the arrow diagram, the following Property 2 must also be satisfied.

Property 2 If (i, j), (p, q) are two jobs, job (i, j) is an immediate predecessor of job (p, q) iff either $j = p$, or there exists a chain from node j to node p in the arrow diagram consisting of dummy arcs only.

In drawing the arrow diagram, we start with an initial node called the **start node** representing the event of starting the project, and represent each job that has no predecessor, by an arc incident out of it. In the same way, at the end we represent jobs that have no successors by arcs incident into a single final node called the **finish node** representing the event of the completion of the project.

A dummy job is needed whenever the project contains a subset $\mathcal{A}_1$ of two or more jobs which have some, but not all, of their immediate predecessors in common. In this case we let the arcs corresponding to common immediate predecessors of jobs in $\mathcal{A}_1$ to have the same head node and then add dummy arcs from that node to the tail node of each of the arcs corresponding to jobs in $\mathcal{A}_1$. As an example consider the following project, the arrow diagram corresponding to which is given in Figure 13.3.

Job	Immediate predecessors
e_1	
e_2	
e_3	
e_4	e_1, e_2
e_5	e_3, e_2

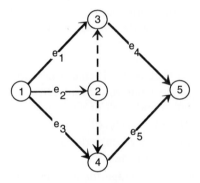

Figure 13.3 Arrow diagram. Dashed arcs represent dummy jobs.

Suppose there are $r \; (\stackrel{\geq}{=} 2)$ jobs, say $1, \ldots, r$, all of which have the same set $\mathcal{A}_1$ of immediate predecessors and the same set $\mathcal{A}_2$ of immediate successors; and there are no other immediate successors for any of the jobs in $\mathcal{A}_1$, or immediate predecessors for any of the jobs in $\mathcal{A}_2$. Then, all jobs in the set $\mathcal{A}_1$ can be represented by arcs incident into a common node, i, say, and all jobs in the set $\mathcal{A}_2$ can be represented by arcs incident out of a common node j, say. Then the jobs $1, \ldots, r$, can be represented by r parallel arcs joining nodes i, j (parallel arcs in a directed network are arcs with the same tail and head nodes). However project engineers do not usually like to deal with parallel arcs, so we introduce additional nodes $i_1, \ldots, i_r$ and represent job h by the arc $(i_h, j), h = 1$ to r; and include dummy arcs (i, i_h) for each $h = 1$ to r. See Figure 13.4.

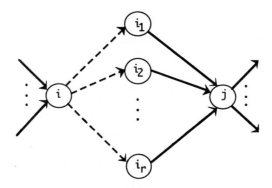

Figure 13.4 Representing jobs with identical sets of immediate predecessors and immediate successors. Arc (i_h, j) represents job h, for $h = 1$ to r. The dashed arcs represent dummy jobs.

If a job b has a single immediate predecessor a, then b can be represented by an arc incident out of the head node of the arc representing a.

If job b has more than one immediate predecessor, let $p_1, \ldots, p_r$ be the head nodes of all the arcs representing its immediate predecessors. If no other job has the same set of immediate predecessors, see if it is possible to represent b by an arc incident out of one of the nodes $p_1, \ldots, p_r$ with dummy arcs emanating from the other nodes in this set into that node. If this is not possible, or if there are other jobs which have identically the same set of immediate predecessors as b, introduce a new node q and represent b and each of these jobs by an arc incident out of q, and include dummy arcs $(p_1, q), \ldots, (p_r, q)$.

If some jobs have identical sets of immediate successors, make the head node of the arcs representing these jobs the same.

We continue this way, at each stage identifying the common immediate predecessors of two or more jobs, and representing these immediate predecessors by arcs with the same head node, and letting dummy arcs issue out of this node if necessary. In introducing dummy arcs, one should always watch out to see that precedence relationships not implied by the original data are not introduced, and those in the original specification are not omitted.

After the arrow diagram is completed this way, one can review and see whether any of the dummy arcs can be deleted by merging the two nodes on it into a single node, while still representing the predecessor relationships correctly. For example, if there is a node with a single arc incident out of it, or a single arc incident into it, and this arc is a dummy arc, then the two nodes on that dummy arc can be merged and that dummy arc eliminated. Other simple rules like these can be developed and used to remove unnecessary dummy arcs.

In this way it is possible to draw an arrow diagram for a project using simple heuristic rules. There are usually many different ways of selecting the nodes and

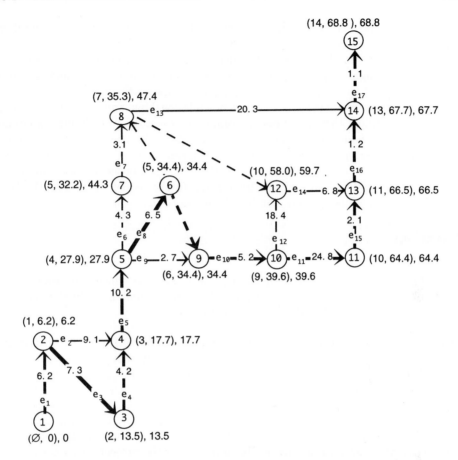

Figure 13.5 Arrow diagram for the hydroelectric power plant building problem. Arc e_j represents job j in the project, with the job duration entered on the arc. Dashed arcs represent dummy jobs. The critical path defined in Section 13.2 is marked with thick arcs. The forward pass label (within parenthesis), and backward pass label; of each node are entered by its side.

dummy arcs for drawing the arrow diagram to portray the specified precedence relationships through Properties 1,2. Any of these that leads to an arrow diagram satisfying Properties 1,2 correctly and completely is suitable for project planning and scheduling computations. One would prefer an arrow diagram with as few nodes and dummy arcs as possible. But the problem of constructing an arrow diagram with the minimum number of dummy arcs is in general a hard problem. In practice, it is not very critical whether the number of dummy arcs is the smallest that it can be or not. Any arrow diagram obtained using the simple rules discussed above is quite reasonable and satisfactory.

As an example, the arrow diagram for the hydroelectric power plant building project discussed above is given in Figure 13.5.

Since dummy arcs have been introduced just to represent the predecessor relationships through Properties 1,2, they correspond to dummy jobs, and the time and cost required to complete any dummy job are always taken to be 0.

The transitive character of the precedence relationships, and the fact no job can precede itself, imply that an arrow diagram cannot contain any circuits (a circuit is a chain from a node back to itself); i.e., it is acyclic. As discussed in Chapter 12, an acyclic numbering of nodes in the arrow diagram is possible, i.e., a numbering such that if (i, j) is an arc in the network, then $i < j$. A procedure for numbering the nodes this way is discussed in Chapter 12. In the sequel we assume that the nodes in the arrow diagram are numbered this way.

Exercises

13.1 Given the predecessor data for a project, develop efficient procedures for checking the data for consistency and for removing redundancies in the specified immediate predecessor lists if the data is consistent.

13.2 Write a practically efficient computer program to derive an arrow diagram for a project, given the list of immediate predecessors of each job. Include in your program simple rules to try to keep the number of nodes and the number of dummy arcs as small as possible.

The first step in CPM is the construction of the arrow diagram that represents the precedence relationships among the activities in the project. This is the most difficult step in CPM. It requires much thought and a very detailed analysis of the work in the project. Once this step is completed, we will have a clear understanding of what must be accomplished to complete the project successfully. This might very well be the greatest benefit of CPM.

13.2   Project Scheduling

Let $G = (\mathcal{N}, \mathcal{A})$ with n nodes, be the arrow diagram for a project with an acyclic numbering for its nodes, and nodes $1, n$ as the start, finish nodes, respectively. For each job $(i, j) \in \mathcal{A}$, let

$$t_{ij} \quad = \quad \text{the time duration required for completing job } (i, j) \ (t_{ij} = 0 \\ \text{if } (i, j) \text{ is a dummy arc)}$$

We assume that $t_{ij} \geqq 0$ for all jobs (i, j). Given these job durations, project scheduling deals with the problem of laying out the jobs along the time axis with the aim of minimizing the project duration. It is concerned with temporal considerations such as

(1) how early would the event corresponding to each node materialize,

(2) how far can an activity be delayed without causing a delay in project completion time,

etc. Make t_{ij} the length of arc (i, j) in the project network G. The minimum time needed to complete the project, known as the **minimum project duration**, is obviously the length of the longest chain from 1 to n in G; a longest chain like that is known as a **critical path** in the arrow diagram. There may be alternate critical paths in G. Any arc which lies on a critical path is called a **critical arc**, it represents a **critical job** or **critical activity**. Jobs which are not on any critical path are known as **slack jobs** in the arrow diagram. For each node $i \in \mathcal{N}$ let

$$t_i \quad = \quad \text{the length of a longest chain from start node 1 to node } i \text{ in G.}$$

t_n, the length of a critical path in G, is the minimum time duration required to complete the project. The quantity t_i is the earliest occurrence time of the event associated with node i assuming that the project has commenced at time 0. For each arc (i, j) incident out of node i, t_i is the earliest point of time at which job (i, j) can be started after the project has commenced; hence it is known as the **early start time of job** (i, j) and denoted by ES(i, j). For all arcs (i, j) incident out of node i, ES(i, j) is the same, and $t_i + t_{ij}$ is the earliest point of time that job (i, j) can be completed. This time is known as the **early finish time of job** (i, j), and denoted by EF(i, j). So, for all jobs $(i, j) \in \mathcal{A}$

$$\text{Early start time for job } (i, j) = \text{ES}(i, j) \quad = \quad t_i$$
$$\text{Early finish time of job } (i, j) = \text{EF}(i, j) \quad = \quad t_i + t_{ij}$$

Since G is acyclic, the t_is can be computed by applying the dynamic programming algorithm discussed in Chapter 12, with appropriate modifications to find the longest (instead of the shortest) chains from the origin node 1 to all the other nodes in G (instead of from every node to a fixed destination node, which was the problem discussed in Chapter 12). The process of computing the longest chains from 1 to all the other nodes in G using the recursive technique of dynamic programming is called **the forward pass** through the arrow diagram. Once the forward pass has been completed, one schedule that gets the project completed in minimum time is to start each job at its early start time. However the forward pass identifies only one critical path, it does not identify all the critical arcs. It will be extremely helpful to the project manager if all the critical jobs can be identified, because if a job is not critical (i.e., it is a slack job) then it can be delayed to a limited extent after its early start time without causing any delay in the whole project. And it is interesting to know how late the starting and completion of a job (i, j) can be delayed without affecting the project completion time. This informs the project management how much leeway they have in scheduling each job and still complete the project in minimum time. For job $(i, j) \in \mathcal{A}$, define

$$
\begin{aligned}
\text{LS}(i,j) \quad = \quad & \text{Late start time of job } (i,j) = \text{latest point of time that this} \\
& \text{job can be started without affecting the project completion} \\
& \text{in minimum time} \\
= \quad & t_n - \text{ length of the longest chain from node } i \text{ to node } n \\[8pt]
\text{LF}(i,j) \quad = \quad & \text{the late finish time of job } (i,j) = \text{LS}(i,j) + t_{ij}.
\end{aligned}
$$

To compute the late finish times, we begin at the finish node at time point t_n and work with backwards recursion; this process is known as the **backward pass** through the arrow diagram.

An arc (i, j) is a critical arc iff $\text{ES}(i,j) = \text{LS}(i,j)$. Hence when both forward and backward passes have been completed, all the critical and slack arcs in the arrow diagram can be identified easily. The combined algorithm comprising the forward and backward passes is described below. In these passes t_{ij} are given data. In the forward pass, node i acquires the **forward label** (L_i, t_i) where t_i is the quantity defined above; it is the earliest event time associated with node i, and L_i is the predecessor index of node i on a longest chain from 1 to i. In the forward pass nodes are labeled in serial order from 1 to n. In the backward pass node i acquires the **backward label** denoted by μ_i; it is the latest event time associated with node i so that the project completion will still occur in minimum time. In the backward pass, nodes are labeled in decreasing serial order beginning with node n.

FORWARD PASS

Step 1   Label the start node, node 1, with the forward label $(\emptyset, 0)$.

General step r , $r = 2$ to n   At this stage, all the nodes 1, $\ldots, r - 1$ would have been forward labeled, let these forward labels be (L_i, t_i) on node $i = 1$ to $r - 1$. Find

$$
t_r = \max\{t_i + t_{ir} : i \text{ is the tail node on an arc incident into } r\} \qquad (13.2)
$$

Let L_r be any of the i that attains the maximum in (13.2). Label node r with the forward label (L_r, t_r). If $r = n$ go to the backward pass, otherwise go to the next step in the forward pass.

BACKWARD PASS

Step 1   Label the finish node, node n, with $\mu_n = t_n$.

General Step r , $r = 2$ to n At this stage all the nodes $n, n-1, \ldots, n-r+2$ would have received backward labels, let these be $\mu_n, \ldots, \mu_{n-r+2}$, respectively. Find

$$\mu_{n-r+1} = \min\{\mu_j - t_{n-r+1,j} : \quad j \text{ is the head node on an arc}$$
$$\text{incident out of } n-r+1\}$$

If $r = n$ terminate; otherwise go to the next step in the backward pass.

Discussion

As mentioned above, the forward pass is an adaptation of the dynamic programming algorithm discussed in Section 12.4, to find longest chains from node 1 to all the other nodes i in the acyclic network G. The backward pass is an adaptation of the same dynamic programming algorithm, to find longest chains from every node i in G to the finish node n, but using the fact that that the longest chain from node 1 to node n has the known length $\mu_n = t_n$.

For any job $(i, j) \in \mathcal{A}$, we have

$$
\begin{aligned}
\text{LF}(i,j) &= \mu_j \\
\text{LS}(i,j) &= \mu_j - t_{ij} \\
\text{ES}(i,j) &= t_i \\
\text{EF}(i,j) &= t_i + t_{ij}
\end{aligned}
$$

The difference $\text{LS}(i,j)$ - $\text{ES}(i,j) = \mu_j - t_{ij} - t_i$ is known as **the total slack** or the **total float** of job (i,j) and denoted by $\text{TS}(i,j)$. Here is a list of some of the activity floats that are commonly used.

$\mu_j - (t_i + t_{ij}) = $ **total slack** or **total float** of job (i,j)
$\mu_j - t_{ij} - \mu_i = $ **safety float** of job (i,j)
$t_j - (t_i + t_{ij}) = $ **free float** or **free slack** of job (i,j)

Job (i,j) is a **critical job** iff $\text{TS}(i,j) = 0$. Hence, after the forward and backward passes, all the critical jobs are easily identified. Any chain from node 1 to n on which all the arcs are critical arcs is a **critical path**. In particular, the chain from node 1 to n traced by the forward pass labels is a critical path. Critical jobs have to start exactly at their early start times if the project has to be completed in minimum time. However, slack jobs can be started any time within the interval between their early and late start times, allowing the scheduler some freedom in choosing their starting times. One should remember that if the start time of a slack job is delayed beyond its early start time, the start times of all its successor jobs are delayed too, and this may affect their remaining total slacks.

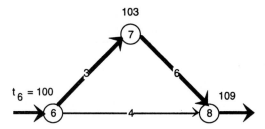

Figure 13.6 An illustration of a job (6, 8) with free slack. Thick arcs are on the critical path.

Free slack can be used effectively in project scheduling. For example, if a job has positive free slack, and its start is delayed by any amount $\leqq$ its free slack, this delay will not affect the start times or slack of succeeding jobs.

A node i is on a critical path iff $t_i = \mu_i$. Two nodes i, j may both be on a critical path, and yet the arc joining them (i, j) may not be a critical arc. An example is given in Figure 13.6. Here, the numbers on the arcs are the job durations, the numbers by the side of the nodes are the t_is, and critical arcs are thick. Even though both nodes 6, 8 are on the critical path, job (6, 8) is not a critical job, and its free slack is $109 - 100 - 4 = 5$. Job (6, 8) has positive float even though both the nodes on it have zero slack. The start time of job (6, 8) can be anywhere between 100 to 105 time units after project start, this delay in job (6, 8) has absolutely no effect on the start times or slack of any of its successors.

Consider the arrow diagram for the hydroelectric dam building project in Figure 13.5. The critical path identified by the forward labels is marked with thick lines. For each node i, the forward pass label, and the backward pass label $(L_i, t_i), \mu_i$ are entered by its side. Minimum project duration is $t_{15} = 68.8$ months. The critical path in this example is unique, as all the nodes not on it satisfy $t_i < \mu_i$. The ES, EF, LF, LS, TS of all the jobs listed under the project (i.e., not the dummy jobs) are given below.

We will now explain how to interpret these results. For example, consider job 3 (corresponding to arc $e_3 = (2, 3)$ in the arrow diagram in Figure 13.5) in this project. Its early start time is $t_2 = 6.2$ months. That means that the earliest time at which this job can be started (this is the time by which all its predecessor jobs would have been completed) is 6.2 months after project commencement. This is the kind of input that project managers need, since it provides information on when to order any special equipment or trained personnel needed to carry out this job, to arrive at project site. And the late start time of this job is $\mu_3 - t_{23} = 6.2$ months, same as its early start time. This means that if job 3 is not started at 6.2 months time after project commencement, the completion of the whole project will be delayed beyond its minimum duration of 68.8 months. Since the early and late start times of job 3 are equal, it is a critical job. In the same way, jobs 1, 4, 5, 8, 10, 11, 15, 16, 17 are also critical jobs, and a similar interpretation can be given to their early start times.

The ES, EF, LF, LS, and TS
for Jobs in the Hydroelectric
Dam Building Project

Job	ES	EF	LF	LS	TS
1	0.0	6.2	6.2	0.0	0.0
2	6.2	15.3	17.7	8.6	2.4
3	6.2	13.5	13.5	6.2	0.0
4	13.5	17.7	17.7	13.5	0.0
5	17.7	27.9	27.9	17.7	0.0
6	27.9	32.2	44.3	40.0	12.1
7	32.2	35.3	47.4	44.3	12.1
8	27.9	34.4	34.4	27.9	0.0
9	27.9	30.6	34.4	31.7	3.8
10	34.4	39.6	39.6	34.4	0.0
11	39.6	64.4	64.4	39.6	0.0
12	39.6	58.0	59.7	41.3	1.6
13	35.3	55.6	67.7	47.4	12.1
14	58.0	64.8	66.5	59.7	1.7
15	64.4	66.5	66.5	64.4	0.0
16	66.5	67.7	67.7	66.5	0.0
17	67.7	68.8	68.8	67.7	0.0

Consider job 2 (corresponding to arc $e_2 = (2, 4)$ in the arrow diagram in Figure 13.5) in this project. Its early and late start times are 6.2 and 8.6 months. Since its late start time is > than its early start time, this job is a slack job. It can be started anytime after 6.2 months (after project commencement); but unless it is started before 8.6 months, project completion will be delayed. The free slack of this job (2, 4) is $t_4 - t_2 - t_{24} = 17.7 - 6.2 - 9.1 = 2.4$. It implies that starting this job any time between 6.2 to 8.6 months after project commencement, has no effect on the early or late start of any succeeding job.

Consider job 12 corresponding to arc (10, 12) in the arrow diagram in Figure 13.5. Its early and late start times are 39.6 and 41.3 months respectively, but its free slack is $t_{12} - t_{10} - t_{10,12} = 58.0 - 39.6 - 18.4 = 0$. It implies that this job can be started any time after 39.6 months (after project commencement), but unless it is started before 41.3 months, the whole project will be delayed. And since its free slack is 0, if it is started some time after 39.6 but before 41.3 months, the early start times of succeeding jobs will be affected (it can be verified that the early start time of job 14 will change depending on the start time of job 12 between 39.6 to 41.3 months).

In the same way, the output from the forward and backward passes of the above algorithm provides extremely useful planning information to the project manager for scheduling the various jobs over time and in evaluating the effects of any unavoidable changes in the schedule on the project completion date.

Summary

In this chapter, we discussed how to represent the precedence relationships among the jobs in a project using a directed project network. Given this project network, and the time durations of the various jobs, we discussed a method for scheduling the jobs over time to complete the project in minimum time. The method is based on the dynamic programming algorithm of Section 12.4 for finding optimal routes in acyclic networks. This is the most basic critical path method, and serves our purpose of exposing the reader to elementary but very important optimization tools for project management.

Sometimes, it may be necessary to complete a project earlier than the minimum duration for it as determined by normal job durations. In this case it will be necessary to complete some jobs in time less than their normal duration, by allowing workers to work overtime, etc. Given the unit cost of expediting each job, there is an algorithm called *project shortening cost minimization algorithm*, which determines the subset of jobs to be expedited, and each by how much, in order to complete the project within the desired duration at minimum shortening cost. Since project managers are often under pressure to complete projects early, this is a very useful algorithm for them. For a discussion of this algorithm, see for example [K. G. Murty, 1992 of Chapter 5].

So far we assumed that the only constraints in scheduling jobs over time are those imposed by the predecessor relationships among them. To carry out jobs in practical project scheduling problems, we require resources such as a crane, or other piece of equipment, or trained personnel, etc. Two or more jobs may require the same resources, and it may not be possible to carry them out simultaneously because of limited supply of resources, even though the precedence constraints do not prevent them from being scheduled simultaneously. The limited availability of resources imposes a new set of constraints. Before starting a job, the project scheduler now has to make sure that all its predecessors have been completed, and also that the resources required to carry it out are available. Problems of this type are known as *resource constrained project scheduling problems*. Practical resource constrained project scheduling normally leads to very large combinatorial optimization problems, for which efficient exact algorithms are not known at the moment. Hence, a variety of heuristic algorithms have been developed for resource constrained project scheduling, see A. Battersby [1967], P. J. Burman [1972], S. Elmaghraby [1977], J. D. Weist and F. K. Levy [1977], and R. J. Willis and N. A. J. Hastings [1976].

Since the 1960s, the network-based CPM has become a part of the language of project management, and has been used extensively in planning, scheduling and controlling large projects. The glamorous successes claimed for their initial applications, and the adoption of these models as standard requirement in contracts by many governments, have added to their importance. Computer packages for these network based techniques specialized to the needs of a variety of industries continue to be the best sellers of all optimization software.

13.3 Exercises

13.3 A project consists of jobs A to L with immediate predecessor (IP) and job duration data as given below.

Activity	A	B	C	D	E	F	G	H	I	J	K	L
IPs				A,B	B	B	C,F	B	E,H	C,D,F,J	K	
Duration	13	8	9	10	6	5	4	7	3	4	8	5

Draw the arrow diagram, and prepare a schedule for the jobs that minimizes the project duration.

13.4 A new product development project The following is the list of activities involved in developing a new product at a company. The expected duration of each activity, in weeks, is also given.

Read the list of activities very carefully, and using your engineering knowledge and judgment, write down what you think are the immediate predecessors of each activity. Justify your choice carefully. Using this information, draw the arrow diagram for the project and prepare a time schedule for the activities to complete the project in minimum time.

No.	Activity	Duration (weeks)	Immediate predecessors
1	Generate marketing plans	3	
2	Assign responsibilities	1	
3	Consolidate plans	1	
4	Review product lines	3	
5	Hire prototype artist	3	
6	Design prototypes	7	
7	Hire layout artist	2	
8	Hire new production crew	4	
9	Train new production crew	7	
10	Review prototypes	1	
11	Final selection	4	
12	Prepare national ads	5	
13	Approve advertising	1	
14	Produce advertising	7	
15	Draft press releases	2	
16	Press ready	1	

([H. J. Thamhain, 1992])

13.5 Coke Depot Project A depot is to be built to store coke and to load and dispatch trucks. There will be three storage hoppers (SH. in abbreviation), a block of bunkers (B. in abbreviation), interconnecting conveyers (abbreviated as C.), and weigh bridges (called WB.). Around the bunkers there will be an area of hard-standing and an access road will have to be laid to the site. Data on this project, and the duration (in weeks) of each job are given below. Draw an arrow diagram for this project, and determine the earliest and latest start and finish times, and the total float of each job. Schedule the jobs so that the project is finished as quickly as possible. (R. J. Willis and N. A. J. Hastings [1976])

No.	Job	IPs	Duration
1.	B. piling		5
2.	Clear site for SH.		8
3.	B. excavation for cols.	1	4
4.	SH. excavations for C.	2, 3	4
5.	Concrete tops of piles for B.	3	3
6.	Place cols. for B.	5	4
7.	Excavate access road	5	4
8.	Put in B.	6	3
9.	Stairways inside B.	6	1
10.	Excavate pit for WB.	4	6
11.	Concrete for SH.	4	12
12.	Main C. foundation	4	4
13.	Brick walls for B.	8, 9	3
14.	Clad in steel for B.	8, 9	1
15.	Install internal equip. in B.	8, 9	6
16.	Erect gantry for main C.	12, 6	1
17.	Install C. under hoppers	11	1
18.	Concrete pit for WB.	11, 10	2
19.	Excavate for hard-standing	7	9
20.	Lay access roadway	7	9
21.	Install outloading equip. for B.	15	2
22.	Line B.	13, 14	1
23.	Install main C.	16	1
24.	Build weighhouse	18	4
25.	Erect perimeter fence	19	4
26.	Install C. to SH.	17, 23	1
27.	Install WB.	24	1
28.	Lay hard-standing	19, 18	6
29.	Commission hoppers	26	1

IP = Immediate predecessors

13.6 Draw the arrow diagram and determine the early and late start and finish times of the various jobs in the following project (A. Kanda and N. Singh [1988]).

No.	Job	IPs	Job duration (months)
	Project: Setting Up a Fossil Fuel Power Plant		
1.	Land acquisition		6
2.	Identi. trained personnel	1	3
3.	Land dev. & infrastructure	1	2
4.	Control room eng.	1	12
5.	Lag in turbine civil works	1	8

contd.

Setting Up a Fossil Fuel Power Plant contd.

No.	Job	IPs	Job duration (months)
6.	Delivery of TG	1	12
7.	Delivery of boiler	1	10
8.	Joining time for personnel	2	3
9.	Boiler prel. civil works	3	2
10.	Control room civil works	4	5
11.	TG civil works	5	9
12.	Training	8	6
13.	Boiler final civil works	9	9
14.	Erection of control room	10	8
15.	Erection of TG	6, 11	10
16.	Boiler erection	7, 13	12
17.	Hydraulic test	16	2
18.	Boiler light up	14, 17	1.5
19.	Box up of turbine	15	3
20.	Steam blowing, safety valve floating	18, 19	2.5
21.	Turbine rolling	20	1.5
22.	Trial run	21	1
23.	Synchronization	22	1

IP = Immediate predecessors

13.7 Draw an arrow diagram for each of the following projects. Prepare a schedule for the various jobs in each project to complete it in minimum time. (R. Visweswara Rao).

(a) Sewer and Waste System Design for a Power Plant

No.	Activity	IPs	Duration (days)
1.	Collection system outline		40
2..	Final design & approval	1	30
3.	Issue construction drawings	2	30
4.	Get sewer pipe & manholes	1	145
5.	Fabricate & ship	3,4	45
6.	Treat. system drawings & approval		70
7.	Issue treat. system construction drawings	6	30
8.	Award contract	7	60
9.	Final construction	8, 5	300

IP = Immediate predecessors

(b) Data Process and Collection System Design for a Power Plant

No.	Activity	IPs	Duration (days)
1.	Prel. Syst. description		40
2.	Develop specs.	1	100
3.	Client approval & place order	2	50

contd.

Data Process and Collection System Design contd.

No.	Activity	IPs	Duration (days)
4.	Develop I/O summary	2	60
5.	Develop alarm list	4	40
6.	Develop log formats	3, 5	40
7.	Software def.	3	35
8.	Hardware requirements	3	35
9.	Finalize I/O summary	5, 6	60
10.	Anal. performance calculation	9	70
11.	Auto. turbine startup anal.	9	60
12.	Boiler guides anal.	9	30
13.	Fabricate & ship	10, 11, 12	400
14.	Software preparation	7, 10, 11	80
15.	Install & check	13, 14	130
16.	Termination & wiring lists	9	30
17.	Schematic wiring lists	16	60
18.	Pulling & term. of cables	15, 17	60
19.	Operational test	18	125
20.	First firing	19	1

IP = Immediate predecessors

(c) Electrical Auxiliary System Design for a Nuclear Plant

No.	Activity	IPs	Duration (days)
1.	Aux. load list		120
2.	13.8 switchgear load ident.	1	190
3.	4.16kv & 480 v. switchgear load ident.	1	45
4.	Vital AC load determination	1	300
5.	DC load determ.	1	165
6.	Voltage drop study	2	84
7.	Diesel gen. sizing	3	77
8.	Inventer sizing	4	20
9.	Battery sizing	5, 8	40
10.	DC fault study	9	80
11.	Prel. AC fault current study	6, 7	20
12.	Power transformer sizing	2, 11	80
13.	Composite oneline diagram	2,3	72

contd.

Electrical Auxiliary System Design contd.

No.	Activity	IPs	Duration (days)
14.	Safety (class 1E) system design	13	200
15.	Non-class 1E system design	13	190
16.	Relaying oneline & metering dia.	13	80
17.	3-line diagram	14, 15, 16	150
18.	Synchronizing & phasing diagrams	17	100
19.	Client review	10, 18	25
20.	Equipment purchase & installation	19	800

13.4   References

A. BATTERSBY, 1967, *Network Analysis for Planning and Scheduling*, Macmillan & Co., London.

P. J. BURMAN, 1972, *Precedence Networks for Project Planning and Control*, McGraw-Hill, London.

D. DIMSDALE, March 1963, "Computer Construction of Minimal Project Network," *IBM Systems Journal*, 2(24-36).

S. E. ELMAGHRABY, 1977, *Activity Networks*, Wiley, NY.

A. C. FISHER, D. S. LIEBMAN, and G. L. NEMHAUSER, July 1968, "Computer Construction of Project Networks," *Communications of the Association for Computing Machinery*, 11(493-497).

A. KANDA and V. R. K. RAO, May 1984, "A Network Flow Procedure for Project Crashing with Penalty Nodes," *European Journal of Operational Research*, 16, no. 2(123 -136).

A. KANDA and N. SINGH, July 1988, "Project Crashing with Variations in Reward and Penalty Functions: Some Mathematical Programming Formulations," *Engineering Optimization*, 13, no. 4(307-315).

J. E. KELLY, Jr., 1961, "Critical Path Planning and Scheduling: Mathematical Basis," *Operations Research*, 9(296-320).

J. E. KELLY, Jr. and M. R. WALKER, Dec. 1959, "Critical Path Planning and Scheduling," *Proceedings of the Eastern Joint Computer Conference*, Boston, MA.

H. J. THAMHAIN, 1992, *Engineering Management Managing Effectively in Technology-Based Organizations*, Wiley-Interscience, NY.

J. D. WEIST and F. K. LEVY, 1977, *A Management Guide to PERT/CPM*, Prentice-Hall, Englewood Cliffs, NJ, 2nd Ed.

R. J. WILLIS and N. A. J. HASTINGS, 1976, "Project Scheduling With Resource Constraints Using Branch and Bound Methods," *Operations Research Quarterly*, 27, no. 2, i(341-349).

Chapter 14

Nonlinear Programming

The title of this subject, **nonlinear programming (NLP)**, may convey the impression to the reader that it includes all optimization problems other than linear programming problems. In this respect the title is misleading. Optimization problems involving $0-1$, integer or discrete valued variables (called variously as integer programs, mixed integer programs, or discrete or mixed-discrete optimization problems) are usually not considered under NLP, but studied separately. There are good reasons for this. Discrete optimization problems require very special techniques (typically of the enumerative type) different from those needed to tackle continuous variable optimization problems[1]. So the term **nonlinear program** usually refers to an optimization problem in which the decision variables are all continuous variables, and the problem is of the following general form:

$$
\begin{aligned}
\text{Minimize} \quad & \theta(x) \\
\text{subject to} \quad & h_i(x) \;=\; 0, \quad i = 1 \text{ to } m \\
& g_p(x) \;\geqq\; 0, \quad p = 1 \text{ to } t
\end{aligned}
\tag{14.1}
$$

where $\theta(x), h_i(x), g_p(x)$ are all real valued continuous functions of the decision variables $x = (x_1, \ldots, x_n) \in \mathbb{R}^n$. If all the functions are smooth (which we will take to mean once or twice continuously differentiable, depending on whether we are planning to use first or second order methods to attack the problem) (14.1) is called a **smooth NLP**. In this chapter we will only discuss smooth NLPs.

As defined earlier, a **linear function** of $x = (x_1, \ldots, x_n)^T$ is a function of the form $c_1 x_1 + \ldots c_n x_n$, where $c_1, \ldots, c_n$ are constants called the **coefficients of the**

---

[1]This needs a clarification. Of course there are techniques for transforming an integer program or a discrete optimization problem into a mathematically equivalent continuous variable optimization problem. We will see one of these transformations later on in this chapter. But these transformations are not practically useful, as they lead to a continuous optimization problem which is very hard to solve.

variables in this function. For example when $n = 4$, $f_1(x) = -2x_2 + 5x_4$ is a linear function of $(x_1, x_2, x_3, x_4)^T$ with the coefficient vector $(0, -2, 0, 5)$.

An **affine function** of $x = (x_1, \ldots, x_n)^T$ is a constant plus a linear function, i.e., a function of the form $f_2(x) = c_0 + c_1 x_1 + \ldots + c_n x_n$. So, if $f_2(x)$ is an affine function, $f_2(x) - f_2(0)$ is a linear function. As an example, $-10 - 2x_2 + 5x_4$ is an affine function of $(x_1, x_2, x_3, x_4)^T$.

A **quadratic form** in the variables $x = (x_1, \ldots, x_n)^T$ is a function of the form $f(x) = \sum_{i=1}^n q_{ii} x_i^2 + \sum_{i=1}^n \sum_{j=i+1}^n q_{ij} x_i x_j$, where the q_{ij} are the coefficients of the terms in this quadratic form. Define

$$d_{ii} = q_{ii} \quad \text{for } i = 1 \text{ to } n$$
$$d_{ij} = d_{ji} = \frac{1}{2} q_{ij} \quad \text{for } i \neq j \text{ and } j > i$$

and let $D = (d_{rs})$ be the $n \times n$ square matrix. As defined above, the matrix D is symmetric. Then the quadratic form $f(x)$ given above is $x^T D x$, and D is known as the symmetric coefficient matrix defining this quadratic form. As an example consider $n = 3$, and $h(x) = 81x_1^2 - 7x_2^2 + 5x_1 x_2 - 6x_1 x_3 + 18x_2 x_3$. Then $x = (x_1, x_2, x_3)^T$, the symmetric matrix defining this quadratic form is D given below and $h(x) = x^T D x$.

$$D = \begin{pmatrix} 81 & 5/2 & -3 \\ 5/2 & -7 & 9 \\ -3 & 9 & 0 \end{pmatrix}$$

For any square matrix $M = (m_{ij})$ of order $n \times n$, whether symmetric or not, $x^T M x = \sum_{i=1}^n m_{ii} x_i^2 + \sum_{i=1}^n \sum_{j=i+1}^n (m_{ij} + m_{ji}) x_i x_j = \frac{1}{2} x^T (M + M^T) x$, and hence the symmetric matrix defining this quadratic form is $(M + M^T)/2$.

A square matrix M of order $n \times n$, whether symmetric or not, is said to be

positive semidefinite (PSD)   iff $x^T M x \geq 0$ for all $x \in \mathbb{R}^n$

positive definite (PD)   iff $x^T M x > 0$ for all $x \neq 0 \in \mathbb{R}^n$

It is possible to check whether a given square matrix of order n is PD, PSD, or not, using at most n pivot steps along its main diagonal. This algorithm is discussed later on.

A **quadratic function** in variables $x = (x_1, \ldots, x_n)^T$ is a function which is the sum of an affine function and a quadratic form in x, i.e.; it is of the form $x^T D x + cx + c_0$ for some symmetric matrix D, and row vector c and constant c_0.

A **convex function** in variables $x = (x_1, \ldots, x_n)^T$, defined over $\mathbb{R}^n$ or over some convex subset of $\mathbb{R}^n$ is a real valued function $f(x)$ satisfying the property that for every x^1, x^2, and $0 \leq \alpha \leq 1$

$$f(\alpha x^1 + (1-\alpha)x^2) \lessgtr \alpha f(x^1) + (1-\alpha)f(x^2) \qquad (14.2)$$

(14.2) defining the convexity of the function $f(x)$ is called **Jensen's inequality** after the Danish mathematician who first defined it. (14.2) is easy to visualize when $n = 1$; it says that if you join two points on the graph of the function by a chord, the function itself lies underneath the chord on the interval joining these points. See Figure 14.1

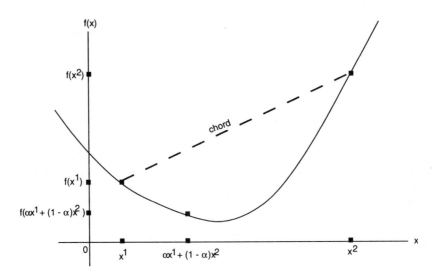

Figure 14.1 A convex function defined on the real line.

A real valued function $g(x)$ defined over $\mathbb{R}^n$ or a convex subset of $\mathbb{R}^n$ is said to be a **concave function** if $-g(x)$ is a convex function as defined above.

The properties of convexity and concavity of functions are of great importance in optimization theory, as we will see later.

From the definition it can be verified that a function $f(x_1)$ of a single variable x_1 is convex (concave) iff its slope is nondecreasing (nonincreasing) in x_1; i.e., iff its second derivative is nonnegative (nonpositive) if the function is twice continuously differentiable. As examples, $x_1^2, x_1^4, e^{-x_1}, e^{x_1}, -\log(x_1)$(over $x_1 > 0$), are all convex functions.

Among functions of several variables, linear and affine functions are both convex and concave. A quadratic function $f(x) = x^T D x + cx + c_0$ defined on $\mathbb{R}^n$ is convex iff the matrix D is PSD, see (K. G. Murty [1988]) for a proof of this. These are special functions for which checking whether they are convex can be carried out efficiently.

Given a general function $f(x)$ of many variables, the definition of convexity given in (14.2) is not very useful to check whether it is convex, because (14.2) has to be verified to hold for all $0 \leqq \alpha \leqq 1$, and every pair of points x^1, x^2 in the space; a very difficult task. That's why checking the convexity of a general function of many variables is usually a very difficult task. The following result is known, its proof can be found in (K. G. Murty [1988]) for example.

THEOREM 14.1 *Let $f(x)$ be a real valued function defined over $\mathbb{R}^n$ which is twice continuously differentiable. It is convex iff its hessian matrix, the $n \times n$ matrix of its second partial derivatives, $H(f(x)) = (\frac{\partial^2 f(x)}{\partial x_i \partial x_j})$, is PSD at all points x.*

Even with the use of Theorem 14.1, checking convexity of a general function is hard, because we need to check whether the hessian matrix of the function is PSD at every point x. Checking convexity of quadratic functions is easy because their hessian matrix is the same at every point.

However, for general functions, Theorem 14.1 is useful in the following sense. If $f(x)$ is a function defined over $\mathbb{R}^n$, and $\bar{x}$ is a point such that the hessian matrix $H(f(\bar{x})$ is PD, then in a small convex neighborhood of the point $\bar{x}$, the function $f(x)$ is convex. This property is called **local convexity**, and it can be put to good use.

The constraints in (14.1) are said to be **linear** if all the functions $h_i(x)$ and $g_p(x)$ are affine functions for all i and p. The optimization problem (14.1) is said to be an

unconstrained minimization problem if there are no constraints on the variables in the statement of the problem;

linear programming problem if all the functions $\theta(x), h_i(x), g_p(x)$ are affine functions;

quadratic programming problem if $\theta(x)$ is a quadratic function, and all $h_i(x), g_p(x)$ are affine functions;

equality constrained problem if there are no inequality constraints in the statement of the problem;

linearly constrained NLP if all the constraint functions $h_i(x), g_p(x)$ are affine functions,

convex programming problem if $\theta(x)$ is a convex function, all $h_i(x)$ are affine functions, and all $g_p(x)$ are concave functions; or

nonconvex programming problem if it is not a convex programming problem as defined above.

In this chapter we present an elementary discussion of smooth nonlinear programming, and some of the issues one encounters in trying to use it in practice.

14.1   Main Differences Between LP and NLP Models

In an LP model, the objective function and the constraint functions are all affine functions. For an NLP model on the other hand, there is an unlimited variety of functional forms for the objective and each of the constraint functions.

In an LP model the data consists of the coefficients of each variable in the objective function and each constraint function; hence if it involves m constraints in n variables, it has $(m+1)(n+1)-1$ pieces of data. Now-a-days the phrase **large scale LP model** usually refers to one involving several thousands of constraints in several tens of thousands of decision variables. Such large scale LP models are being formulated and solved routinely by many companies to optimize their operations.

In an NLP model, the objective and constraint functions should all be given either as mathematical expressions, or in the form of computer subroutines. These functions may not be known explicitly; in this case we may have to develop suitable functional forms from data in order to construct the model. The task of finding the best functional form that fits the data as closely as possible is called a **curve fitting problem**. There are two major components in curve fitting. The first determines a suitable function (usually called the **model function**) perhaps involving one or more unknown parameters. Once the model function is determined, the best values for the parameters in it are selected so as to get the closest fit to observed data; this task is called a **parameter estimation problem**. Parameter estimation problems are usually solved by the least squares method using some unconstrained optimization algorithm. Thus, even the construction of a nonlinear model needs NLP algorithms for unconstrained optimization, for parameter estimation. Because of this, constructing good NLP models is usually much harder than constructing LP models. That's why even a nonlinear model involving 100 variables may be considered a large scale model depending on the degree of nonlinearity in the problem.

Given an LP model, there are many algorithms that one can use to solve it to find a true optimum solution, if one exists for the model, subject only to minor round-off errors introduced during the computation. For NLP models, particularly the nonconvex models that arise often in applications, unfortunately there are no algorithms known that are guaranteed to yield a true optimum solution. When faced with such models, one has to compromise on the quality of the solution expected from existing algorithms. This aspect is discussed in greater detail later on.

14.2 Superdiagonalization Algorithm for Checking PD, PSD

Let M be a square matrix of order n which may or may not be symmetric. By definition, M is PD or PSD iff $D = M + M^T$ is PD or PSD, respectively. To check whether the symmetric matrix $D = (d_{ij})$ is PD or PSD, we can use the algorithms given below. They are based on the following results.

RESULT 14.1 *If D is PD, all its diagonal entries d_{ii}, $i = 1$ to n must be > 0. If D is not PD but PSD, all its diagonal entries d_{ii}, $i = 1$ to n must be $\overset{\geq}{=} 0$.*

RESULT 14.2 *If the symmetric matrix $D = (d_{ij})$ is PSD, and a diagonal entry in it $d_{ii} = 0$, then all the entries in its row and column, i.e., row i and column i, must be zero.*

THEOREM 14.2 *Let $D = (d_{ij})$ be a symmetric matrix with its first diagonal entry $d_{11} \neq 0$. Subtract suitable multiples of row 1 from each of the other rows to convert all entries in column 1 in rows 2 to n to 0, i.e., transform*

$$D = \begin{pmatrix} d_{11} & \cdots & d_{1n} \\ d_{21} & \cdots & d_{2n} \\ \vdots & & \vdots \\ d_{n1} & \cdots & d_{nn} \end{pmatrix} \quad into \quad D_1 = \begin{pmatrix} d_{11} & d_{12} & \cdots & d_{1n} \\ 0 & \tilde{d}_{22} & \cdots & \tilde{d}_{2n} \\ \vdots & \vdots & & \vdots \\ 0 & \tilde{d}_{n2} & \cdots & \tilde{d}_{nn} \end{pmatrix}$$

let E_1 be the matrix of order $(n-1) \times (n-1)$ obtained by deleting column 1 and row 1 from D_1. Then E_1 is also symmetric; and D is PD (PSD) iff $d_{11} > 0$ and E_1 is PD (PSD).

For proofs of Results 14.1, 14.2, and Theorem 14.2, see (K. G. Murty [1988]), for example. The following algorithms for checking PD and PSD are based on repeated use of these results.

SUPERDIAGONALIZATION ALGORITHM
FOR CHECKING WHETHER M IS PD

Let $M = (m_{ij})$ be the matrix of order $n \times n$ being tested for positive definiteness. Let $D = M$ if M is symmetric, otherwise $D = M + M^T$.

Step 1 If any of the principal diagonal elements in D are $\overset{\leq}{=} 0$, D and hence M is not PD, terminate. Otherwise go to Step 2 with D as the current matrix.

Step 2 Subtract suitable multiples of row 1 of D from all the other rows, so that all the entries in column 1 and rows 2 to n of D are made into 0; i.e., transform D into D_1 as in Theorem 14.2. If any diagonal element of D_1 is $\overset{\leq}{=} 0$, D and hence M is not PD, terminate. Otherwise go to Step 3 with the matrix D_1 as the current matrix.

General Step $r + 2$ At this stage the current matrix will be D_r of the following form.

$$
D_r = \begin{pmatrix}
d_{11} & d_{12} & & & & \cdots & d_{1n} \\
0 & \tilde{d}_{22} & & & & \cdots & \tilde{d}_{2n} \\
 & 0 & 0 & \ddots & & \cdots & \vdots \\
 & & & \bar{d}_{rr} & & \cdots & \bar{d}_{rn} \\
 & & & 0 & \hat{d}_{r+1,r+1} & \cdots & \hat{d}_{r+1,n} \\
\vdots & \vdots & & \vdots & \vdots & & \vdots \\
0 & 0 & & 0 & \hat{d}_{n,r+1} & \cdots & \hat{d}_{nn}
\end{pmatrix}
$$

Subtract suitable multiples of row $r + 1$ in D_r from rows i for $i > r + 1$, so that all the entries in column $r + 1$ and rows $i > r + 1$ are transformed into 0. This transforms D_r into D_{r+1}. If any element in the principal diagonal of D_{r+1} is $\overset{\leq}{=} 0$, D and hence M is not PD, terminate. Otherwise, if $r + 2 < n$ go to the next step with D_{r+1} as the current matrix. If $r + 2 = n$, the current matrix is D_{n-1} and it is of the form

$$
D_{n-1} = \begin{pmatrix}
d_{11} & d_{12} & \cdots & d_{1n} \\
0 & \tilde{d}_{22} & \cdots & \tilde{d}_{2n} \\
0 & 0 & & \\
\vdots & \vdots & & \vdots \\
0 & 0 & \cdots & \bar{d}_{nn}
\end{pmatrix}
$$

If no termination has occurred earlier and all the diagonal entries in D_{n-1} are positive, D and hence M is PD, terminate.

If the method goes through Step $n - 1$, the final matrix D_{n-1} is upper triangular. All the work in this method has transformed the original matrix D into the upper triangular matrix D_{n-1}. That's why this method is called the **superdiagonalization algorithm**.

EXAMPLE 14.1
___

Test whether the following matrix M is PD.

$$
M = \begin{pmatrix}
3 & 1 & 2 & 2 \\
-1 & 2 & 0 & 2 \\
0 & 4 & 4 & 5/3 \\
0 & -2 & -13/3 & 6
\end{pmatrix}, \quad
D = M + M^T = \begin{pmatrix}
6 & 0 & 2 & 2 \\
0 & 4 & 4 & 0 \\
2 & 4 & 8 & -8/3 \\
2 & 0 & -8/3 & 12
\end{pmatrix}
$$

All the entries in the principal diagonal of D are > 0. So, apply Step 1 in superdiagonalization getting D_1. Since all elements in the principal diagonal of D_1 are > 0, continue. The matrices obtained in the order are

$$D_1 = \begin{pmatrix} 6 & 0 & 2 & 2 \\ 0 & 4 & 4 & 0 \\ 0 & 4 & 22/3 & -10/3 \\ 0 & 0 & -10/3 & 34/3 \end{pmatrix}, D_2 = \begin{pmatrix} 6 & 0 & 2 & 2 \\ 0 & 4 & 4 & 0 \\ 0 & 0 & 10/3 & -10/3 \\ 0 & 0 & -10/3 & 34/3 \end{pmatrix}$$

$$D_3 = \begin{pmatrix} 6 & 0 & 2 & 2 \\ 0 & 4 & 4 & 0 \\ 0 & 0 & 10/3 & -10/3 \\ 0 & 0 & 0 & 8 \end{pmatrix}.$$

The algorithm terminates now. Since all the diagonal entries in D_3 are > 0, D and hence M is PD.

EXAMPLE 14.2

Check whether $\bar{M} = \begin{pmatrix} 1 & 0 & 2 & 0 \\ 0 & 2 & 4 & 0 \\ 2 & 4 & 4 & 5 \\ 0 & 0 & 5 & 3 \end{pmatrix}$ is PD. M is already symmetric and its

diagonal entries are > 0. Carrying out Step 1 on $D = \bar{M}$ leads to

$$D_1 = \begin{pmatrix} 1 & 0 & 2 & 0 \\ 0 & 2 & 4 & 0 \\ 0 & 4 & 0 & 5 \\ 0 & 0 & 5 & 3 \end{pmatrix}$$

Since the third diagonal entry in D_1 is 0, the matrix $\bar{M}$ here is not PD.

ALGORITHM FOR CHECKING WHETHER M IS PSD

Let $M = (m_{ij})$ be the matrix of order $n \times n$ being tested for positive semidefiniteness. Let $D = M$ if M is symmetric, otherwise $D = M + M^T$.

Step 1   If any of the principal diagonal elements in D are < 0, D and hence M is not PSD, terminate. Otherwise continue.

If any diagonal entries in D are 0, all the entries in the row and column of each 0 diagonal entry must be 0. Otherwise D and hence M is not PSD, terminate. If termination has not occurred, reduce the matrix D by striking off the 0-rows and columns of 0 diagonal entries. We will call the remaining matrix by the same name D. Go to Step 2 with D as the current matrix.

Step 2   Start off by performing row operations as in Step 1 of the above algorithm, i.e., transform D into D_1. If any diagonal entry in D_1 is < 0, D and hence M is not PSD, terminate. Otherwise, let E_1 be the submatrix of D_1 without its row 1 and column 1. If a diagonal entry in E_1 is 0, all entries in its row and column in E_1 must be 0 too; otherwise D and hence M is not PSD, terminate. Continue if termination did not occur. With D_1 as the current matrix go to Step 3.

General Step $r + 2$ At this stage the current matrix will be D_r of the same form as in the above algorithm. Let E_r be the submatrix of D_r with its rows 1 to r and columns 1 to r struck off. If any diagonal entry in E_r is < 0, D and hence M is not PSD, terminate. If any diagonal element of E_r is 0, all the entries in its row and column in E_r must be 0 too; otherwise D and hence M is not PSD, terminate. If termination did not occur, continue.

Let D_{ss} be the first nonzero (and hence positive) diagonal element in E_r. Subtract suitable multiples of row s in D_r from row i for $i > s$, so that all the entries in column s and rows $i > s$ in D_r are transformed into 0. This transforms D_r into D_s, which is the new current matrix. Go to Step $s + 2$ with D_s as the current matrix.

If termination does not occur until D_{n-1} is obtained, and if all the diagonal entries in D_{n-1} are $\geqq 0$, D and hence M is PSD, terminate.

In the process of obtaining D_{n-1}, if all the diagonal elements in D and in all the matrices D_r obtained during the algorithm are > 0, D and hence M is not only PSD but actually PD.

EXAMPLE 14.3

Check whether the following matrix M is PSD.

$$M = \begin{pmatrix} 0 & -2 & -3 & -4 & 5 \\ 2 & 3 & 3 & 0 & 0 \\ 3 & 3 & 3 & 0 & 0 \\ 4 & 0 & 0 & 8 & 4 \\ -5 & 0 & 0 & 4 & 2 \end{pmatrix}, \quad D = M + M^T = \begin{pmatrix} 0 & 0 & 0 & 0 & 0 \\ 0 & 6 & 6 & 0 & 0 \\ 0 & 6 & 6 & 0 & 0 \\ 0 & 0 & 0 & 16 & 8 \\ 0 & 0 & 0 & 8 & 4 \end{pmatrix}$$

$D_{.1}$ and $D_{1.}$ are both zero vectors. So we eliminate them, but will call the remaining matrix by the same name. All diagonal entries in D are $\geqq 0$. So we apply Step 1 of superdiagonalization. This leads to

$$D_1 = \begin{pmatrix} 6 & 6 & 0 & 0 \\ 0 & 0 & 0 & 0 \\ 0 & 0 & 16 & 8 \\ 0 & 0 & 8 & 4 \end{pmatrix}, \quad E_1 = \begin{pmatrix} 0 & 0 & 0 \\ 0 & 16 & 8 \\ 0 & 8 & 4 \end{pmatrix}$$

The first diagonal entry in E_1 is 0, but the column and row of this entry in E_1 are both zero vectors. And all the diagonal entries in D_1 are $\overset{>}{=} 0$. So continue with superdiagonalization. Since the 2nd diagonal element in D_1 is 0, move to the third diagonal element of D_1. This step leads to

$$D_3 = \begin{pmatrix} 6 & 6 & 0 & 0 \\ 0 & 0 & 0 & 0 \\ 0 & 0 & 16 & 8 \\ 0 & 0 & 0 & 0 \end{pmatrix}$$

All diagonal entries in D_3 are $\overset{>}{=} 0$, and it is upper triangular. So, D and hence M are PSD. M is not PD.

EXAMPLE 14.4

Is the matrix $\bar{M}$ in Example 14.2 PSD? We have seen there that it is not PD. Referring to Example 14.2, after Step 1 in superdiagonalization we have

$$E_1 = \begin{pmatrix} 2 & 4 & 0 \\ 4 & 0 & 5 \\ 0 & 5 & 3 \end{pmatrix}$$

The 2nd diagonal entry in E_1 is 0, but its row and column in E_1 are not zero vectors. So, $\bar{M}$ is not PSD.

These algorithms for checking whether a given square matrix is PD or PSD have many uses in NLP. For example, given a quadratic program (QP) in which the objective function $f(x) = x^T D x + c x$ is to be minimized subject to linear constraints, to check whether this problem is a convex QP, we need to check whether the matrix D defining the quadratic form in $f(x)$, is PSD. Given a nonlinear twice continuously differentiable function $g(x)$, a sufficient condition for it to be locally convex at a point $\bar{x}$, is that its hessian matrix at $\bar{x}$ is PD. Also, these algorithms are needed to check whether a given point satisfies the optimality conditions for being a solution to an NLP, as will be seen later.

14.2.1 How to Check if a Quadratic Form is Nonnegative on a Subspace

Let $f(y) = y^T D y$ be a quadratic form in variables $y = (y_1, \ldots, y_n)^T$, where D is the symmetric $n \times n$ matrix of coefficients defining it. Suppose we are given a subspace **T** of $\mathbb{R}^n$ defined by the homogeneous system of equations

$$Ay = 0$$

and wish to check whether $f(y) \geqq 0$ for all $y \in \mathbf{T}$. Of course if D is PSD, then $f(y) \geqq 0$ for all $y \in \mathbb{R}^n$ and hence for all $y \in \mathbf{T}$. However, $f(y)$ may be $\geqq 0$ for all $y \in \mathbf{T}$ even if D is not PSD.

Let the rank of the matrix A in the above system be r. Then the constraints $Ay = 0$ can be used to express r of the variables in the vector y (call them dependent variables in y) in terms of the remaining $n - r$ variables which can be called independent variables. Let z denote the vector of these independent variables in the vector y. By substituting the expressions for the dependent variables in y in terms of the independent variables in the quadratic form $y^T D y$, it can be reduced to a quadratic form in z on $\mathbf{T}$; suppose it is $z^T B z$. Then $y^T D y \geqq 0$ for all $y \in \mathbf{T}$ holds iff the matrix B is PSD, which can be checked efficiently by the algorithm discussed above. Also, if it is desired to check whether $f(y) > 0$ for all $y \neq 0, y \in \mathbf{T}$; this is equivalent to checking whether the matrix B is PD.

As an example, consider the quadratic form $f(y) = -y_1^2 + 2y_2^2 + 2y_3^2 + 3y_4^2 + 2y_1y_2 + 2y_2y_4 - 2y_3y_4$. The matrix of coefficients D of $f(y)$ is

$$D = \begin{pmatrix} -1 & 1 & 0 & 0 \\ 1 & 2 & 0 & 1 \\ 0 & 0 & 2 & -1 \\ 0 & 1 & -1 & 3 \end{pmatrix}$$

It can be verified that D is not PSD, so it is not true that $f(y) \geqq 0$ for all $y \in \mathbb{R}^4$. Suppose we wish to check whether $f(y) \geqq 0$ for all $y \in \mathbf{T}$, where $\mathbf{T}$ is the set of solutions of the following system of homogeneous system of linear equations.

$$\begin{array}{ccccc} y_1 & +y_2 & -y_3 & +y_4 & = & 0 \\ & +y_2 & -y_3 & +y_4 & = & 0 \end{array}$$

By the pivotal methods discussed in Chapter 3, we see that y_1, y_2 can be treated as dependent variables in this system, defined by the expressions

$$\begin{array}{rcl} y_1 & = & 0 \\ y_2 & = & y_3 - y_4 \end{array}$$

in terms of the independent variables y_3, y_4. Substituting these expressions for the dependent variables y_1, y_2 in $f(y)$ we see that on $\mathbf{T}$, $f(y) = 4y_3^2 + 3y_4^2 - 4y_3y_4$ in terms of the independent variables y_3, y_4. So, on $\mathbf{T}$, $f(y) = (y_3, y_4)B(y_3, y_4)^T$ where

$$B = \begin{pmatrix} 4 & -2 \\ -2 & 3 \end{pmatrix}$$

It can be verified that B is not only PSD but PD. So, $f(y) \geqq 0$ on $\mathbf{T}$, and in fact $f(y) > 0$ for all $y \neq 0, y \in \mathbf{T}$.

14.3   Types of Solutions for an NLP

Consider an NLP in which a function $\theta(x)$ is required to be optimized subject to some constraints on the variables $x = (x_1, \ldots, x_n)^T$. Let $\mathbf{K}$ denote the set of feasible solutions for this problem. A feasible solution $\bar{x}$ of this problem is said to be a

local minimum   if there exists an $\epsilon > 0$ such that $\theta(x) \overset{\geq}{=} \theta(\bar{x})$ for all $x \in \mathbf{K} \cap \{x : ||x - \bar{x}|| < \epsilon\}$

strong local minimum   if there exists an $\epsilon > 0$ such that $\theta(x) > \theta(\bar{x})$ for all $x \in \mathbf{K} \cap \{x : ||x - \bar{x}|| < \epsilon\}, x \neq \bar{x}$

weak local minimum   if it is a local minimum, but not a strong one

global minimum   if $\theta(x) \overset{\geq}{=} \theta(\bar{x})$ for all $x \in \mathbf{K}$

local maximum   if there exists an $\epsilon > 0$ such that $\theta(x) \overset{\leq}{=} \theta(\bar{x})$ for all $x \in \mathbf{K} \cap \{x : ||x - \bar{x}|| < \epsilon\}$

strong local maximum   if there exists an $\epsilon > 0$ such that $\theta(x) < \theta(\bar{x})$ for all $x \in \mathbf{K} \cap \{x : ||x - \bar{x}|| < \epsilon\}, x \neq \bar{x}$

weak local maximum   if it is a local maximum, but not a strong one

global maximum   if $\theta(x) \overset{\leq}{=} \theta(\bar{x})$ for all $x \in \mathbf{K}$

stationary point   if some necessary optimality condition for the problem is satisfied at the point $\bar{x}$.

These concepts are illustrated in Figure 14.2 for the one dimensional problem; optimize $\theta(x)$ subject to $a \overset{\leq}{=} x \overset{\leq}{=} b$. $\theta(x)$ is plotted in Figure 14.2. In this problem the points $a, x^5, x^7, x^{10}, x^{12}$ are strong local minima; x^0, x^4, x^6, x^{11}, b are strong local maxima; x^{12} is the global minimum, and x^6 is the global maximum. x^1, x^2 are weak local minima; and x^8, x^9 are weak local maxima. At the point x^3 the derivative of $\theta(x)$ is zero, and so it is a stationary point (satisfies the necessary optimality condition $\frac{d\theta(x)}{dx} = 0$) even though it is neither a local maximum nor a local minimum. In each of the intervals $x^1 \overset{\leq}{=} x \overset{\leq}{=} x^2$, and $x^8 \overset{\leq}{=} x \overset{\leq}{=} x^9$, $\theta(x)$ is a constant; and every point in the interior of these intervals (i.e., points x satisfying $x^1 < x < x^2$ or $x^8 < x < x^9$) is both a weak local minimum and a weak local maximum.

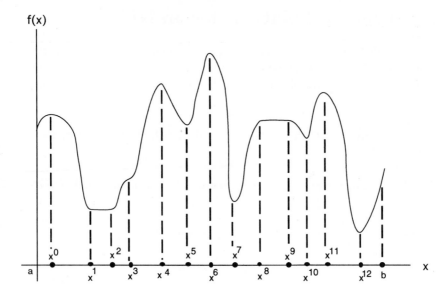

Figure 14.2

14.4   Formulation of NLP Models

We will begin with a few simple nonlinear formulation examples.

EXAMPLE 14.5 Rail car body design

In this problem we need to determine

$$d = \text{length}, \ w = \text{width}, \ h = \text{height}$$

of a rectangular rail car. The top and bottom sheets for the car are of a special gauge that costs \$325 per m^2 (square meter). The siding of the car is made from a thinner gauge sheet that costs \$175 per m^2. The car must hold at least 1000 m^3 of material, and its height must be no more than 3 m; and the area of siding sheet used must be $\leqq$ 75 m^2. The problem of finding a minimum cost design subject to these constraints is the following NLP

$$
\begin{aligned}
\text{Minimize} \quad & 650dw + 350(dh + wh) \\
\text{subject to} \quad 2(dh + wh) \ & \leqq \ 75 \\
h \ & \leqq \ 3 \\
wdh \ & \geqq \ 1000 \\
w, d, h \ & \geqq \ 0
\end{aligned}
$$

EXAMPLE 14.6 A Portfolio Problem

A person has \$10,000 to invest. This person has identified n possible securities in which this money can be invested. The yield from each security is a random variable, but analysis of past data indicates that for $j = 1$ to n

$$\mu_j \quad = \quad \text{expected annual yield per \$ invested in } j\text{th security}$$
$$D \quad = \quad \text{the } n \times n \text{ variance-covariance matrix of annual yields per}$$
$$\text{\$ invested in the various securities}$$

Since the actual annual yield from investments is a random variable, it cannot be predicted exactly. But this person wants to spread his investments among the securities in such a way that the expected total yield is maximized while keeping the variance of this total yield (which measures by how much the actual yield varies around the expected; it is a measure of the risk) as small as possible. Typically higher yields are associated with higher risks. So, there are two objective functions in this problem; one (the expected yield) to be maximized, and the other (the variance of the yield) to be minimized. A convenient goal programming approach to handle this bicriterion optimization problem is to select a target level for the expected yield, and to minimize the variance subject to the constraint that the expected yield is $\geqq$ selected target. For $j = 1$ to n, define the decision variable

$$x_j = \text{\$s invested in the } j\text{th security}$$

Then the total expected annual yield is $\sum_{j=1}^{n} \mu_j x_j$ and the variance of this yield is the quadratic form $x^T D x$ where $x = (x_1, \ldots, x_n)^T$. So, if L = target yield, the problem is the following NLP

$$
\begin{aligned}
\text{Minimize} \quad & x^T D x \\
\text{subject to} \quad & \sum_{j=1}^{n} x_j \quad \leqq \quad 10,000 \\
& \sum_{j=1}^{n} \mu_j x_j \quad \geqq \quad L \\
& x_j \quad \geqq \quad 0, \quad j = 1 \text{ to } n
\end{aligned}
\tag{14.3}
$$

This is a QP (quadratic objective function, linear constraints). This is the simplest type of portfolio model. There are a variety of other portfolio models of varying degrees of complexity that investors use.

EXAMPLE 14.7 A Projectile Problem

Figure 14.3 shows the position of a canon with its barrel at an angle θ from the horizontal line. θ is the firing angle, it can vary between 0 to 90^0. It is required to determine the value of θ that maximizes the range of the projectile (i.e., the distance d of the point where the projectile hits the ground, from the canon). The mass of the projectile is 50 kg., and its initial velocity will be $v_0 = 800$ m/s (meters per second). If the velocity is v, the drag at that time can be assumed to be $0.015v^2$.

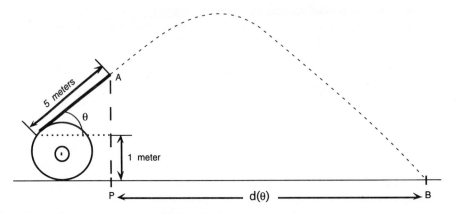

Figure 14.3 The dashed curve is the path of the projectile before it hits the ground at point B.

In this problem, there is only one decision variable θ. The range, the distance between the points P and B in Figure 14.3, is $d(\theta)$ as a function of θ. The problem is to find θ between 0 and 90 that maximizes $d(\theta)$. To find the functional form of $d(\theta)$ we need to use Newton's laws of physics.

Select PB, PA as the axes of coordinates. Let t denote time in seconds, with $t = 0$ corresponding to the instant at which the projectile leaves the tip A of the barrel of the canon. Let t_1 denote the time at which the projectile reaches its maximum height, and let t_2 denote the time at which the projectile hits the ground at point B. Let $(x(t), y(t))$ denote the coordinates of the point denoting the position of the projectile at time t for $0 \stackrel{<}{=} t \stackrel{<}{=} t_2$. Then $(x(0), y(0)) = (0, 1 + 5 \sin \theta)$. The initial velocity of the projectile in the horizontal direction is $v_0 \cos \theta = 800 \cos \theta$ m/s. We have

$$\dot{x}(t) = \frac{dx(t)}{dt} = \quad \text{velocity of the projectile at time } t \text{ in the horizontal direction}$$

$$\ddot{x}(t) = \frac{d^2x(t)}{dt^2} = \quad \text{acceleration of the projectile at time } t \text{ in the horizontal direction}$$

By Newton's laws, $\ddot{x}(t) = (\text{force})/(\text{mass}) = (\text{drag})/(\text{mass}) = -0.15(\dot{x}(t))^2/50$, negative because drag acts opposite to the direction of motion. Using the facts that

$x(0) = 0$, $v(0) = 800 \cos\theta$, and integrating, we get

$$
\begin{aligned}
x(t) &= (50/0.015)\log_e\left(1 + \frac{800 \times 0.015t\cos\theta}{50}\right) \\
&= 3333.33\log_e(1 + 0.24t\cos\theta)
\end{aligned}
$$

Assume that there is no drag in the vertical direction. Then we have $y(0) = 1 + 5\sin\theta$, $\dot{y}(0) = [\frac{dy(t)}{dt}]_{t=0} = v_0\sin\theta = 800\sin\theta$ m/s, and $\ddot{y}(t) = [\frac{d^2y(t)}{dt^2}] = -g$ where g is the gravitational constant. By integrating, we find that $y(t) = (-g/2)t^2 + 800t\sin\theta + (1 + 5\sin\theta)$. The time t_1 at which the projectile is at its highest vertical point is defined by $\dot{y}(t_1) = 0$. From this we have $t_1 = (800/g)\sin\theta$. So the height of the projectile above ground at time point t_1 is $y(t_1) = ((800\sin\theta)^2/2g) + 1 + 5\sin\theta = H$ say.

So, $t_2 - t_1$ is the time it takes the projectile to drop a height of H under the gravitational pull. From Newton's laws this yields $t_2 - t_1 = \sqrt{(2H/g)}$. So, the total time lapse before the projectile hits the ground is

$$
t_2 = (800/g)\sin\theta + \sqrt{\frac{2}{g}\left[\frac{(800\sin\theta)^2}{2g} + 1 + 5\sin\theta\right]}
$$

and the range of the projectile is

$$
d(\theta) = 3333.33\log_e(1 + 0.24t_2\cos\theta)
$$

And the problem is to maximize $d(\theta)$ subject to $0 \leqq \theta \leqq 90$.

14.4.1   Curve Fitting

The problems formulated in Examples 14.5 to 14.7 were all simple. The functions (objective and constraint functions) in them were all obtained through direct arguments, or through the use of well established theory that applied to the system that the problem dealt with. In many real world applications of NLP, the functions used in constructing the mathematical model for the problem have to be developed empirically by examining data. This is the **curve fitting problem**. Each curve fitting problem involves two phases.

Phase I of the curve fitting problem: Selecting an appropriate model function   This phase determines a suitable functional form called the **model function**, that may contain one or more unknown parameters. This is usually a difficult but very important problem. There may be some theory which can help in this effort. Otherwise a clear understanding of the process involved may suggest the form of a model function that may be appropriate. Plots of available data are often very useful in making this choice.

Phase II of curve fitting: Parameter estimation problem   In almost all applications, Phase I leads to a model functions that involve parameters whose values have to be determined using data.

Let $x = (x_1, \ldots, x_n)^T$ be the vector of variables and let the characteristic which we are trying to determine as a function of x be y. Suppose we have observations on the value of y at m points in the x-space, namely that

$$y = y_r \quad \text{when } x = x^r, \, r = 1 \text{ to } m$$

Let $f(a, x)$ be the model function selected, with a as the vector of unknown parameters, for the characteristic y as a function of x. Then the **deviation** between the observed value of the characteristic at the point x^r and the function value is $y_r - f(a, x^r)$ for $r = 1$ to m. The model function can be considered to be a good fit if all these deviations are zero, or close to zero.

To find the best values for the parameters in a that give the closest fit to the data, we need to construct a measure of the closeness of fit. The most commonly used measure is the **least squares measure** $L_2(a)$ defined by

$$L_2(a) \quad = \quad \text{sum of squared deviations} \quad = \quad \sum_{r=1}^{m} (y_r - f(a, x^r))^2$$

The least squares method for parameter estimation determines the parameter vector a to minimize $L_2(a)$. This problem of minimizing $L_2(a)$ is called a **least squares problem**. It is an unconstrained least squares problem if there are no constraints on the parameters a, otherwise it will be a constrained least squares problem. Let $\hat{a}$ denote the optimum solution of the problem of minimizing $L_2(a)$. $L_2(\hat{a})$, the minimum value of $L_2(a)$ is called the **residue**. It measures the error in the fit obtained over the points in the x-space where observations are made.

If the residue is small, $\hat{a}$ is accepted as the best estimate for the parameter vector a, and $f(\hat{a}, x)$ as the function representing the characteristic y as a function of x.

If the residue is unacceptably high, the following reasons can be explored.

i) The model function $f(a, x)$ may not be suitable. Do Phase I of curve fitting again carefully.

ii) The algorithm used to solve the least squares problem may not have obtained even a local minimum. Do local search for a better point, or run the algorithm with a different initial point.

iii) The implementation of the algorithm used may be numerically unstable and may have encountered numerical difficulties.

EXAMPLE 14.8 Parameter estimation for projectile penetration

This problem arises in determining the thickness of the plate used for the body of a tank. An important consideration here is that the tank must have a good chance of surviving enemy attack. So, armor plate should be thick enough that enemy's projectiles will not penetrate it all the way through. Define

$t(v) =$ thickness (inches) of target armor plate that a striking projectile of velocity v (feet/second) will penetrate

To make a valid scientific judgment of what the armor plate thickness of the tank should be, we need the function $t(v)$. We have the following data.

Observation no.	v (ft./sec.)	t (in.) thickness penetrated
1	2300	4.087
2	2800	4.596
3	2850	4.646
4	2900	4.697

From ballistics theory, we know that $t(v)$ is a function of the form $a_1 v + a_2 v^{a_3}$ where $a = (a_1, a_2, a_3)$ are parameters whose values depend on the alloy used to make the target plate, etc. We can determine the values of these parameters to give the best fit to the data given above. The least squares measure for this is

$$L_2(a) = \quad (4.087 - 2300a_1 - a_2(2300)^{a_3})^2 + (4.596 - 2800a_1 - a_2(2800)^{a_3})^2$$
$$+ (4.646 - 2850a_1 - a_2(2850)^{a_3})^2 + (4.697 - 2900a_1 - a_2(2900)^{a_3})^2$$

The best estimate for the parameter vector a is the one which minimizes $L_2(a)$. This can be found by minimizing $L_2(a)$ using an unconstrained minimization algorithm discussed later on. Having obtained it, we can use this function in tank design studies.

Exercise

14.1 Hydrological parameter estimation Engineers who build dams and other water systems need estimates of expected run-off in rivers. Around each river there is a region of land called its catchment area, precipitation on which is the source of water that ends up in the river. We have the following data for a river.

Period	1	2	3	4	5	6	7
Precipitation (in.) in catchment area	3.8	4.4	5.7	5.2	7.7	6.0	5.4
Run-off (acre-feet)	0.5	0.5	1.0	2.1	3.7	4.2	4.3

Tracer studies indicate that it takes up to two periods for some precipitation in the catchment area to end up in the river. Let

$$R_i = \text{run-off (acre-feet) in river in period } i$$
$$P_i = \text{precipitation (inches) in catchment area}$$

Measuring run-off is expensive, so it is desired to estimate it from statistics of precipitation which are collected and published by the government. The appropriate model is

$$R_i = a_0 a P_{i-2} + a_1 a P_{i-1} + a_2 a P_i$$

where the parameters a_0, a_1, a_2, a are defined to be

$a_0 =$ proportion of the precipitation that ends up in the river two periods late

$a_1 =$ proportion of the precipitation that ends up in the river one period late

$a_2 =$ proportion of the precipitation that ends up in the river in the same period

$a =$ a constant depending on the size of the catchment area

Under a 20% evaporation assumption, the parameters satisfy: $a_0 + a_1 + a_2 = 0.8$, $0 \leqq a_0 \leqq a_1 \leqq a_2$, and $a \geqq 0$. Give the constrained least squares formulation of the problem of estimating the best values of the parameters a_0, a_1, a_2, a that give the best fit for the data provided above for this river.

14.4.2   Formulation of a Boiler Shop Optimization Problem

This problem due to C. H. White, arose in the boiler shop of a company which has 5 boilers operating in parallel for generating steam. Data on the boilers is given below.

Boiler i	Boiler load range	
	Lower ℓ_i	upper k_i
1	10 units	60
2	10	60
3	15	120
4	12	112
5	15	135

The unit measures the rate at which steam is produced per unit time. If the ith boiler is kept on, it must be operated within its load range limits ℓ_i, k_i.

At a point of time, the companies steam requirements are 350 units. The problem is to determine how this total load of 350 units should be shared across the 5 boilers so as to minimize the total fuel cost.

It may be possible to get a lower overall cost for generating 350 units of steam by operating only a subset of the 5 boilers. However, the company's steam requirements vary with time. When the demand for steam goes up, if a boiler is kept operating, it is relatively easy to increase its steam output by turning up a few control valves. But turning on a shutdown boiler is an expensive and time consuming operation. In order to be able to meet the varying requirements over time, it was determined that all the 5 boilers must be kept operating.

Let x_i denote the level at which the ith boiler is operated.

A boiler's energy efficiency is defined as a percentage to be

$$\frac{100(\text{energy content of output steam})}{\text{energy content in input fuel}}$$

It tends to increase as the boiler's load moves up from its minimum allowable operating load, and then peaks and drops as the load approaches the upper limit. For $i = 1$ to 5, let

$$f_i(x_i) = \quad \text{efficiency of } i\text{th boiler when it is operated at level } x_i$$

We need the functions $f_i(x_i)$ to model our problem. For this, data was collected on the boiler efficiencies at different operating load levels. The plots indicated that each boiler efficiency function can be approximated by a cubic polynomial in the operating load; i.e., we can approximate $f_i(x_i)$ by $a_{0i} + a_{1i}x_i + a_{2i}x_i^2 + a_{3i}x_i^3$, where $a_{0i}, a_{1i}, a_{2i}, a_{3i}$ are parameters. The best values for these parameters determined using a least squares method are given below for each boiler.

Boiler i	Best fit parameter values			
	a_{0i}	a_{1i}	a_{2i}	a_{3i}
1	56.49	1.67	$-.041$	.00030
2	7137	0.61	$-.016$	.00011
3	23.88	2.05	$-.024$	.00009
4	17.14	2.73	$-.035$	.00014
5	72.38	0.34	$-.003$	.00001

So, the least squares approximation for $f_1(x_1)$ is $56.49 + 1.67x_1 - 0.041x_1^2 + 0.0003x_1^3$, etc. Once the efficiency functions $f_i(x_i)$ are determined, it can be verified that $x_i/f_i(x_i)$ provides a measure of the energy cost of operating the ith boiler at level x_i. So our problem of minimizing the fuel cost of meeting the steam load of 350 units while keeping all the boilers on, is the following linearly constrained NLP.

$$\text{Minimize} \quad \sum_{i=1}^{5}(x_i/f_i(x_i))$$

$$\text{subject to} \quad \sum_{i=1}^{5}x_i = 350$$

$$\ell_i \overset{\leq}{=} x_i \overset{\geq}{=} k_i \qquad i = 1 \text{ to } 5$$

14.4.3   Optimizing Nitric Acid Production Costs

Here we present an NLP model for the problem of minimizing the production cost of nitric acid from (O. Flanigan, W. W. Wilson, and D. R. Sule [1972]). The process in the nitric acid plant where this work was carried out is explained in Figure 14.4.

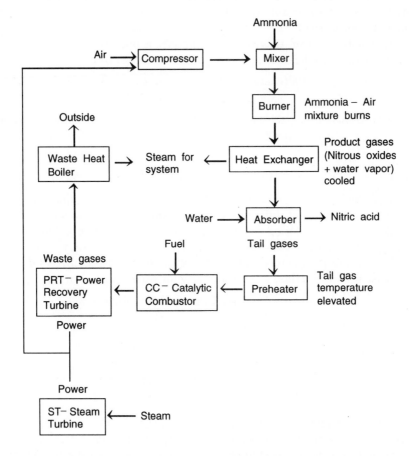

Figure 14.4 The process at the nitric acid plant.

The raw materials for nitric acid manufacture are air and ammonia. Air is drawn from the atmosphere and compressed into a high pressure stream in the compressor. The air is mixed with ammonia (in gaseous form) in the mixer, and the mixture is burnt in the burner. The result is product gases (consisting of nitrous oxides and water vapor) which are lead into the heat exchanger. It traps some of the heat in the hot product gases to produce steam that is used in large quantities all over the plant. The slightly cooled product gases now enter the absorber which mixes the gases with water. Nitric acid is produced when the nitrous oxides in the product gases dissolve in water; this is drawn off, cleaned, packed and shipped out. With the nitrous oxides from them depleted, the gases are now called tail gases.

The nitric acid production process can end with the absorber unit, and if it does the tail gases will be vented out. Even after most of the nitrous oxides from them are dissolved out, the tail gases are still hot, and the plant where this work was done has additional units that help recover some of the waste heat in the tail gases to reduce the production cost of nitric acid.

If the process is to terminate with the absorber unit, the product gases can be allowed to remain in the absorber unit long enough for an almost full recovery of all the usable nitrous oxides in them. This will increase the nitric acid production rate, but make the tail gas temperature too low for any power recovery. To recover waste heat, the tail gases must be drawn off from the absorber at a slightly higher temperature; this reduces the nitric acid production below the maximum possible otherwise. But the slight loss in nitric acid production may be balanced by the gain due to power recovery. Thus there are operating choices, and the optimum policy is not obvious. The aim of the study is to find the ideal operating plan that minimizes the cost per unit nitric acid production.

When the power recovery option is used, the tail gases are drawn out of the absorber into a preheater where their temperature is elevated slightly. Then they are lead into a catalytic combustor (CC) unit which uses fuel to heat them up. The hot tail gases now enter a power recovery turbine (PRT) which uses them to generate power for the system. There is also an independent steam turbine (ST) which also produces any residual power needed for operating the whole system. From the PRT the gases enter a waste heat boiler which generates steam for the system by trapping the heat from them. After the waste heat boiler, the gases are finally vented outside.

The proportion of ammonia in the gases entering the burner unit is a very important characteristic in this process. This has been the subject of many earlier investigations, and has already been optimized. So, it was not considered in this study. The other important process variables are

$P_1 =$ air compressor discharge pressure in psi
$Q_1 =$ flow rate through compressor in cfm at $70^0 F$ and 14.7 psi
$P_2 =$ CC outlet pressure in psi
$Q_2 =$ portion of Q_1 that goes through the PRT in cfm
$T_2 =$ inlet temperature at PRT in $^0 F$

For constructing an NLP model to determine the optimum values of process variables, the following points were considered.

1. Q_1 and P_1 are related by an equation. From an analysis of past data, the linear relationship $P_1 + 0.002032Q_1 = 186.916$ was found to be the best fit for the equation satisfied by them.

2. After PRT and the waste heat boiler, the gases are vented into the atmosphere, so downstream pressure there is a constant 15.2 psi. So, P_2, T_2, Q_2 are related

by an equation. Therefore P_2 can be expressed as a function of T_2 and Q_2 only. Plots of data indicated that a linear relationship is quite suitable, and the best fit was found to be $P_2 - 0.100027T_2 + 0.0234267Q_2 = 683.81$.

3. $P_1 - P_2$, the pressure drop across the process train, can be expressed as a function of Q_1 and T_2. Plots indicated that the effect of T_2 can be neglected, and that $P_1 - P_2$ can be approximated quite closely by an affine function of Q_1. The best fit was found to be $P_1 - P_2 = 0.001104Q_1 + 5.4469$.

4. Let $C_{hp}, PRT_{hp}, ST_{hp}$ denote the horsepowers of the compressor, PRT and ST, respectively. Most of the power in the system is consumed at the compressor. So, $C_{hp} = ST_{hp} + PRT_{hp}$ to a reasonable degree of approximation. The following functional forms for C_{hp} and PRT_{hp} were established by curve fitting.

$$C_{hp} = 0.07706Q_1 P_1^{0.3548} - 0.2Q_1$$
$$PRT_{hp} = 0.0003452Q_2 T_2[1 - (1.7835/P_2^{0.2217})]$$

and so $ST_{hp} = C_{hp} - PRT_{hp}$. There is an upper bound of 2000 on ST_{hp}, this is one of the constraints in the model.

5. One ton of nitric acid is produced from 140,869 cf of air at 70^0F. To maintain consistency of engineering units, the objective function was expressed in terms of \$/cf of air processed. The various costs are

ST steam cost   In \$/cf of air processed, this was estimated to be $1.29066 \times 10^{-4}(ST_{hp}/Q_1)$.

The preheater steam and ammonia costs   These did not vary much with the values of the decision variables in the operating range. So these costs were assumed to be constant.

The CC fuel cost   In \$/cf air processed, this depends on T_2. It has been estimated to be $4.31402 \times 10^{-9}T_2 - 5.86707 \times 10^{-6}$.

The labor cost   In \$/cf of air processed, this is expected to vary inversely with Q_1. A linear fit for labor cost was made using past daily total labor expenditures versus total daily air intake. The best linear fit was $0.1025 \times 10^{-4} - 0.1970 \times 10^{-9}Q_1$.

The other costs   These were also determined similarly. The best linear fit for them was $0.4333 \times 10^{-4} - 0.8055 \times 10^{-9}Q_1$.

The total cost is the sum of all the costs mentioned above.

6. There were bounds (lower and upper) on all the process variables. The overall model is given below.

$$\text{Minimize} \quad 1.290066 \times 10^{-4}(C_{hp} - PRT_{hp})/Q_1$$
$$+4.31402 \times 10^{-9}T_2 - 1.0025 \times 10^{-9}Q_1$$
$$\text{subject to} \quad 100 \leqq P_1 \leqq 140$$
$$85 \leqq P_2 \leqq 100$$
$$24,000 \leqq Q_1 \leqq 28,000$$
$$900 \leqq T_2 \leqq 1250$$
$$Q_1 - Q_2 \geqq 0$$
$$Q_2 \geqq 0$$
$$C_{hp} = 0.07706Q_1P_1^{0.3548} - 0.2Q_1$$
$$PRT_{hp} = 0.0003452Q_2T_2[1 - (1.7835/P_2^{0.2217})]$$
$$0 \leqq C_{hp} - PRT_{hp} \leqq 2000$$
$$P_1 + 0.002032Q_1 = 186.916$$
$$P_2 - 0.100027T_2 + 0.0234267Q_2 = 683.81$$
$$P_1 - P_2 = 0.001104Q_1 + 5.4469$$

When the model was solved it provided an optimum solution which reduced the production cost in \$/ton of acid to 14.6 from 15.3 by the operating conditions existing at the time of the study.

14.4.4   Summary

Consider a real world continuous variable optimization problem. If a linear programming model is appropriate for it, we can consider ourselves lucky since it is much easier to construct an LP model than an NLP model. If an LP model is not appropriate for the problem, then of course one has to build an NLP model for it. This usually involves solving curve fitting problems to determine the functional forms for the objective and constraint functions. It is a difficult task that has to be carried out carefully.

14.5   The Elusive Goal of Finding a Global Optimum

In a convex programming problem (problem of the form (14.1) with $\theta(x)$ convex, $h(x)$ affine, and $g(x)$ concave) every local minimum is a global minimum (see K. G. Murty [1988] for example, for a proof of this statement); that is why these are considered to be the nicest among optimization problems. This class of problems includes all linear programs, convex quadratic programs, and nonlinear convex programs.

It is hard to find a global minimum for a nonconvex program, or even to check whether a given feasible solution is a global minimum for it. Efforts have been made to find global minima by enumerating all local minima, but these methods tend to be very inefficient. The enormity of this task can be appreciated when we realize that some of the most difficult problems in mathematics that have taken many centuries of work by generations of very bright and dedicated researchers to solve, can be posed as nonconvex programming problems. As an example, consider **Fermat's last theorem** which was the object of very intense study worldwide since AD 1630s. It states that the following equation

$$x^n + y^n = z^n$$

has no solution in integers in the region $x \geqq 1, y \geqq 1, z \geqq 1, n \geqq 3$. This deceptively simple conjecture due to the famous French mathematician Pierre de Fermat has challenged, irritated, and daunted researchers for three and a half centuries. After P. Fermat's death in 1665, this statement was found by his son Samuel Fermat while he was going through his father's own papers and books. In a book on Arithmetic, by the side of an exercise which asks "given a number (i.e., a positive integer) which is square, write it as a sum of two other squares," P. Fermat wrote the following note in the margin "on the other hand, it is impossible for a cube to be written as a sum of two cubes, or a fourth power to be written as a sum of two fourth powers, or in general, for any number which is a power greater than the second to be written as a sum of two like powers. I have actually a marvelous demonstration of this proposition which this margin is too narrow to contain." If P. Fermat did indeed have a demonstration of it, it must have been truly marvelous, as it is now more than 350 years since his time, and we still don't have a conclusive proof of its truth.

It is called "Fermat's last theorem" because it is the last of the many unproved theorems that P. Fermat stated, that is still to be settled. See the book (Edwards [1977]) for the technical and colorful history of this theorem up to that time.

P. Fermat did find room (elsewhere) to write down a detailed proof for the case $n = 4$. The Swiss mathematician Leonhard Euler found a proof for $n = 3$ nearly 100 years later. In the 1840s E. E. Kummer initiated the study of algebraic number theory which enabled him to prove Fermat's last theorem for a large number of exponents. By 1992, researchers extended Kummer's approach to include all exponents up to 4 million, using an arsenal of computers. Finally in 1993, building on the work by so many other researchers, Andrew Wiles has written up a proof in a 200-page paper that most definitely would not fit in the margin of Fermat's book. Then a gap was found in this proof, and efforts are continuing to complete it.

Consider the following smooth NLP, where α is some positive data element (say 1), π is the irrational real number which is the length of the circumference of the circle with unit diameter in $\mathbb{R}^2$, and $\cos\theta$ denotes the cosine function of the angle θ measured in radians.

$$\text{Minimize} \quad (x^n + +y^n - z^n)^2 + \alpha((-1 + \cos(2\pi x))^2$$
$$+(-1 + \cos(2\pi y))^2 + (-1 + \cos(2\pi z))^2$$
$$+(-1 + \cos(2\pi n))^2 \qquad (14.4)$$
$$\text{subject to} \quad x, y, z \geq 1, n \geq 3$$

(14.4) is a linearly constrained smooth NLP. It can be verified that Fermat's last theorem is false iff the optimum objective value in (14.4) is zero and attained, since any feasible solution (x, y, z, n) which makes the objective value zero provides a counter example to Fermat's last theorem. (14.4) is a nonconvex programming problem for which every integer feasible solution is a local minimum. It is a penalty formulation for Fermat's last theorem as a smooth optimization problem. Fermat's last theorem is true iff the optimum objective value in this problem is strictly positive.

Since in general it is very difficult to guarantee that a global minimum can be obtained for a nonconvex programming problem, optimization theory has focussed on developing methods that can at least reach a local minimum. In the next section we discuss how difficult it is to find a local minimum.

14.6 Can We at Least Compute a Local Minimum Efficiently?

In the next section we discuss the well known necessary optimality conditions for a local minimum; these are conditions that must be satisfied by a feasible solution $\bar{x}$ if it is a local minimum for the problem. As defined earlier, a feasible solution for a minimization problem, satisfying these optimality conditions, is called a **stationary point**.

For convex programming problems, every stationary point is a local minimum and therefore a global minimum for the problem. To solve a convex programming problem, any algorithm that is guaranteed to find a stationary point, if one exists, is thus adequate. Most of the standard NLP algorithms do converge to a stationary point; and so these algorithms compute local, and thus global minima when applied on convex programming problems.

In a nonconvex program, given a stationary point, it may not even be a local minimum. If it does not satisfy the sufficient optimality conditions for being a local minimum given in the next section, it may be an intractable problem to verify whether it is even a local minimum. In (K. G. Murty and S. N. Kabadi [1987]), the following simple QP is considered.

$$\text{Minimize} \quad x^T D x$$
$$\text{subject to} \quad x \geq 0$$

where D is a given square matrix of order n. When D is not PSD, this problem is the **simplest nonconvex NLP**.

A sufficient condition for 0 to be a local minimum for this problem is that D be PSD. If D is not PSD, there are no efficient methods known to check whether 0 is a local minimum for this problem, and it is a hard problem to check it.

On nonconvex programs, existing algorithms can at best guarantee convergence to a stationary point in general. If the stationary point obtained does not satisfy some known sufficient condition for being a local minimum, it is then hard to check whether it is actually a local minimum.

14.6.1   What are Suitable Goals for Algorithms in Nonconvex NLP?

Much of NLP literature stresses that the goal for algorithms should be to obtain a local minimum. The results mentioned above show that in general this is hard to guarantee.

Many NLP algorithms are iterative in nature; i.e., beginning with an initial point x^0, they obtain a sequence of points $\{x^r: r = 0, 1, \ldots\}$. Algorithms called **descent methods** have the property that the sequence of points obtained is a **descent sequence**, i.e., either the objective function, or a measure of the infeasibility of the current solution to the problem, or some merit function or criterion function which is a combination of both, strictly decreases along the sequence. Given x^r, these algorithms generate a $y^r \neq 0$ such that the direction $x^r + \lambda y^r, \lambda \geqq 0$, is a descent direction for the functions mentioned above. The next point in the sequence x^{r+1} is usually taken to be the point which minimizes the objective or criterion function on the half-line $\{x^r + \lambda y^r : \lambda \geqq 0\}$, obtained by using a line minimization algorithm. On general nonconvex problems, these methods suffer from the same difficulties; they cannot theoretically guarantee that the point obtained at termination is even an approximation to a local minimum. However, it seems reasonable to expect that a solution obtained through a descent process is more likely to be a local minimum, than a solution obtained purely based on necessary optimality conditions. Thus a suitable goal for NLP algorithms is a descent sequence converging to a stationary point. Some algorithms attaining this goal are discussed later on.

A technique often used by practitioners to get a good acceptable solution for a problem is to run such an algorithm with a variety of initial points, and take the best among the terminal points obtained as a reasonably good solution.

14.7   Optimality Conditions for Smooth NLPs

In this section we summarize the most important known optimality conditions for smooth NLPs. **Necessary optimality conditions** are conditions that a feasible solution $\bar{x}$ must satisfy if it is a local minimum for the problem. **Sufficient optimality conditions** are conditions with the property that every point satisfying them is guaranteed to be a local minimum for the problem. For the nice class of convex programming problems, we have conditions which are both necessary and sufficient for a global minimum.

We assume that all the functions are continuously differentiable. If the hessian matrix appears in the statement of conditions, then those conditions apply to problems in which all the functions are twice continuously differentiable. For any function $f(x)$, $\nabla f(x) = (\frac{\partial f(x)}{\partial x_1}, \ldots, \frac{\partial f(x)}{\partial x_n})$ denotes the row vector of its partial derivatives. And $H(f(x)) = (\frac{\partial^2 f(x)}{\partial x_i \partial x_j})$ denotes the $n \times n$ square matrix of second partial derivatives (the **hessian matrix** of $f(x)$ at x).

We only state the optimality conditions here. The reader interested in proofs of these conditions can refer to (K. G. Murty [1988]) or some of the other books referenced at the end of this chapter. We consider various types of problems separately.

14.7.1   Unconstrained Minimization Problems

Consider the problem

$$\text{Minimize} \quad \theta(x) \qquad (14.5)$$

with no constraints on $x \in \mathbb{R}^n$. Here $\theta(x)$ is a differentiable real valued function defined on $\mathbb{R}^n$. Let $\bar{x}$ be a point in $\mathbb{R}^n$.

First order necessary conditions for $\bar{x}$ to be a local minimum for (14.5)

These are

$$\nabla\theta(\bar{x}) = 0 \qquad (14.6)$$

i.e., that the vector of partial derivatives (called the **gradient vector**) at $\bar{x}$ must be zero.

(14.6) is a system of n equations (nonlinear equations in general) in n unknowns. Since every local minimum for (14.5) must satisfy (14.6), some methods for solving (14.5) try to find solutions of (14.6), to consider as candidate points for solution of (14.5). We can try to solve (14.6) using methods for solving nonlinear equations (one of which, called Newton-Raphson method, is discussed later on).

Second order necessary conditions for $\bar{x}$ to be a local minimum for (14.5)

Assuming that $\theta(x)$ is twice continuously differentiable at $\bar{x}$, these are

$$\begin{aligned} \nabla\theta(\bar{x}) &= 0 \\ H(\theta(\bar{x})) &\quad \text{is PSD} \end{aligned} \qquad (14.7)$$

Sufficient conditions for $\bar{x}$ to be a local minimum for (14.5)

Assuming that $\theta(x)$ is twice continuously differentiable, these are

$$
\begin{aligned}
\nabla\theta(\bar{x}) &= 0 \\
H(\theta(\bar{x})) & \quad \text{is PD}
\end{aligned}
\tag{14.8}
$$

If $\theta(x)$ is a differentiable function, and also a convex function, then a point $\bar{x}$ is a global minimum for (14.5) iff (14.6) holds. So, for convex functions $\theta(x)$, (14.6) are necessary and sufficient for a point $\bar{x}$ to be a global minimum.

If $\theta(x)$ is nonconvex, and $\bar{x}$ satisfies (14.7) but not (14.8), then $\bar{x}$ satisfies the necessary conditions for being a local minimum, but not the sufficient condition. In this case we cannot guarantee that $\bar{x}$ is a local minimum for (14.5) from these conditions.

EXAMPLE 14.9

Consider the problem

$$
\text{Minimize} \quad \theta(x) = 2x_1^2 + x_2^2 + x_3^2 + x_1x_2 + x_2x_3 + x_3x_1 - 9x_1 - 9x_2 - 8x_3
$$

over $x \in \mathbb{R}^3$. From the first order necessary optimality conditions, we know that every local minimum for this problem must satisfy

$$
\begin{aligned}
\frac{\partial\theta(x)}{\partial x_1} &= 4x_1 + x_2 + x_3 - 9 = 0 \\
\frac{\partial\theta(x)}{\partial x_2} &= x_1 + 2x_2 + x_3 - 9 = 0 \\
\frac{\partial\theta(x)}{\partial x_3} &= x_1 + x_2 + 2x_3 - 8 = 0
\end{aligned}
$$

This system of equations has the unique solution $\bar{x} = (1,3,2)^T$. The hessian matrix is

$$
H(\theta(\bar{x})) = \begin{pmatrix} 4 & 1 & 1 \\ 1 & 2 & 1 \\ 1 & 1 & 2 \end{pmatrix}
$$

This matrix is PD. So, $\bar{x}$ satisfies the sufficient condition for being a local minimum. Clearly, $\theta(x)$ here is convex, and hence $\bar{x}$ is a global minimum for $\theta(x)$.

EXAMPLE 14.10

Consider the problem

$$\text{Minimize} \quad \theta(x) = 2x_1^2 + x_3^2 + 2x_1x_2 + 2x_1x_3 + 4x_2x_3 + 4x_1 - 8x_2 + 2x_3$$

over $x \in \mathbb{R}^3$. From the first order necessary optimality conditions, we know that every local minimum for this problem must satisfy

$$\frac{\partial\theta(x)}{\partial x_1} = 4x_1 \quad + 2x_2 \quad + 2x_3 \quad +4 \quad = \quad 0$$

$$\frac{\partial\theta(x)}{\partial x_2} = 2x_1 \quad\quad\quad +4x_3 \quad -8 \quad = \quad 0$$

$$\frac{\partial\theta(x)}{\partial x_3} = 2x_1 \quad +4x_2 \quad + 2x_3 \quad +2 \quad = \quad 0$$

This system has the unique solution $\tilde{x} = (-2, -1, 3)^T$. So, $\tilde{x}$ satisfies the first order necessary conditions for being a local minimum for $\theta(x)$. The hessian matrix is

$$H(\theta(\tilde{x})) = \begin{pmatrix} 4 & 2 & 2 \\ 1 & 0 & 2 \\ 1 & 2 & 2 \end{pmatrix}$$

which is not even PSD. Hence $\tilde{x}$ violates the second order necessary conditions for a local minimum. So, the function $\theta(x)$ here does not have a local minimum. It can be verified that $\theta(x)$ is actually unbounded below on $\mathbb{R}^3$.

EXAMPLE 14.11

Let $\theta(x) = -2x_1^2 - x_2^2 + x_1x_2 - 10x_1 + 6x_2$ and consider the problem of minimizing $\theta(x)$ over $x \in \mathbb{R}^2$. The first order necessary conditions for a local minimum are

$$\frac{\partial\theta(x)}{\partial x_1} = -4x_1 \quad +x_2 \quad -10 \quad = \quad 0$$

$$\frac{\partial\theta(x)}{\partial x_2} = \quad x_1 \quad -2x_2 \quad + 6 \quad = \quad 0$$

which has the unique solution $\hat{x} = (-2, 2)^T$. So, $\hat{x}$ is the only point satisfying the first order necessary conditions for being a local minimum for $\theta(x)$. The hessian matrix is

$$H(\theta(\hat{x})) = \begin{pmatrix} -4 & 1 \\ 1 & -2 \end{pmatrix}$$

Since $H(\theta(\hat{x}))$ is not even PSD, $\hat{x}$ violates the second order necessary conditions for being a local minimum for $\theta(x)$. So, $\theta(x)$ does not have a local minimum. In fact it can be verified that the negative of the hessian matrix is PD, so $\hat{x}$ satisfies the sufficient condition for being a local minimum for $-\theta(x)$, or a local maximum for $\theta(x)$. Actually, $\theta(x)$ here is concave, and $\hat{x}$ is a global maximum for $\theta(x)$. It can be verified that $\theta(x)$ is unbounded below on $\mathbb{R}^2$.

An Important Caution to NLP Users

These examples point out one important aspect of NLP applications. One should not blindly accept a solution of the first order necessary optimality conditions as a solution to the problem, if it is a nonconvex programming problem (this caution can be ignored if the problem being solved is a linear or other convex programming problem). An effort should be made to check whether the solution is at least a local minimum by using second order necessary conditions, or the sufficient optimality conditions, or at least through a local search in the neighborhood of the point.

14.7.2   Linearly Constrained NLPs

Consider the NLP

$$\begin{array}{rll} \text{Minimize} & \theta(x) \\ \text{subject to} & Ax & = & b \\ & Dx & \gtreqless & d \end{array} \tag{14.9}$$

The inequality constraints in (14.9) include any sign restrictions, and other lower or upper bound constraints on individual variables. Let A, D be matrices of orders $m \times n$, $t \times n$ respectively. Let $\bar{x}$ be a feasible solution to this problem. For $p = 1$ to t, the pth inequality constraint $D_{p.}x \gtreqless d_p$ is said to be

$$\begin{array}{lll} \textbf{active at the feasible solution } \bar{x} \text{ if} & D_{p.}\bar{x} & = & d_p \\ \textbf{inactive at the feasible solution } \bar{x} \text{ if} & D_{p.}\bar{x} & > & d_p \end{array}$$

Let $\mathbf{P}(\bar{x}) = \{p : 1 \leqq p \leqq t$, and the pth inequality constraint in (14.9) is active at $\bar{x}\}$. So, $\mathbf{P}(\bar{x})$ is the index set of active inequality constraints in (14.9) at $\bar{x}$.

KKT Conditions, or First Order Necessary Optimality Conditions for $\bar{x}$ to be a Local Minimum for (14.9)

These are: if $\bar{x}$ is a local minimum for (14.9), there must exist Lagrange multipliers μ_i for $i = 1$ to m associated with the equality constraints; and π_p for $p = 1$ to t associated with the inequality constraints in (14.9); such that $\mu = (\mu_1, \ldots, \mu_m), \pi = (\pi_1, \ldots, \pi_t)$ together with $\bar{x}$ satisfy

$$
\begin{aligned}
\nabla\theta(\bar{x}) - \mu A - \pi D &= 0 \\
A\bar{x} &= b \\
D\bar{x} &\geq d \\
\pi &\geq 0 \\
\pi_p(D_{p.}\bar{x} - d_p) &= 0 \quad p = 1 \text{ to } t
\end{aligned}
\tag{14.10}
$$

The conditions on the last line in (14.10) are called **complementary slackness optimality conditions**. They state that in the KKT conditions (14.10), the Lagrange multiplier π_p associated with the pth inequality constraint in (14.9) is zero, if that constraint is inactive at $\bar{x}$.

Given the feasible solution $\bar{x}$, checking whether (14.10) holds for it, can be easily carried out by substituting $\bar{x}$ in it, and solving the resulting system of constraints (which are linear in μ, π) by linear programming techniques (Phase I of the simplex method).

Second Order Necessary Conditions for $\bar{x}$ to be a Local Minimum

Define

$$
\mathbf{T} = \{y : Ay = 0, D_{p.}y = 0 \quad \text{for all} \quad p \in \mathbf{P}(\bar{x})\}
\tag{14.11}
$$

Under the assumption that $\theta(x)$ is twice continuously differentiable at $\bar{x}$, these are

$$
(14.10), \quad \text{and} \quad y^T H(\theta(\bar{x}))y \geq 0 \quad \text{for all} \quad y \in \mathbf{T}
\tag{14.12}
$$

The condition $y^T H(\theta(\bar{x}))y \geq 0$ for all $y \in \mathbf{T}$ in (14.12), can be checked using the PSD-testing algorithm as discussed in Section 14.2.1.

Sufficient Conditions for $\bar{x}$ to be a Local Minimum for (14.9)

Suppose $\theta(x)$ is twice continuously differentiable, and $\bar{x}$ is a feasible solution of (14.9) such that there exist Lagrange multiplier vectors $\mu = (\mu_1, \ldots, \mu_m), \pi = (\pi_1, \ldots, \pi_t)$ which together with $\bar{x}$ satisfy the KKT conditions (14.10), and the additional condition

$$y^T H(\theta(\bar{x}))y > 0 \quad \text{for all } y \in \mathbf{T}_1, y \neq 0 \tag{14.13}$$

where $\mathbf{T}_1 = \{y : \nabla h_i(\bar{x})y = 0, i = 1 \text{ to } m, \text{ and} \nabla g_p(\bar{x})y = 0 \quad \text{for all } p \in \mathbf{P}(\bar{x}) \cap \{p : \pi_p > 0\}\}$. $\mathbf{T}_1$ is again a subspace, and if (14.10) holds, we can check whether (14.13) holds, as discussed above, quite efficiently, using the PD-testing algorithm discussed in Section 14.2.1.

If $\theta(x)$ is a differentiable convex function, then a feasible point $\bar{x}$ is a global minimum for (14.9) iff the KKT conditions (14.10) hold; i.e., when $\theta(x)$ is convex in (14.9), the KKT conditions (14.10) are both necessary and sufficient for a feasible point $\bar{x}$ to be a global minimum.

If $\theta(x)$ is nonconvex, sufficient conditions slightly weaker than those in (14.10) and (14.13) are known; but checking whether they hold is a hard problem. That is why we presented (14.10), (14.13) instead.

14.7.3   Equality Constrained NLPs

Consider the NLP

$$\begin{aligned}
\text{Minimize} \quad & \theta(x) \\
\text{subject to} \quad & h_i(x) = 0, \quad i = 1 \text{ to } m
\end{aligned} \tag{14.14}$$

Since the case of linear constraints is discussed above, here we consider problems in which at least one of the constraints is nonlinear.

Let $\bar{x}$ be a feasible solution to the problem. Conditions to check whether $\bar{x}$ is a local minimum for (14.14) have only been derived under some assumptions on $\bar{x}$ called **constraint qualifications**, because these assumptions require the constraints in the statement of the problem to satisfy certain properties at the point $\bar{x}$. Many different constraint qualifications have been developed, but most of them are hard to verify. We will present only one constraint qualification called the **regularity condition** which can be checked efficiently. The feasible solution $\bar{x}$ is said to satisfy the regularity condition (or to be a **regular feasible solution**) for this problem if

$$\{\nabla h_i(\bar{x}) : i = 1 \text{ to } m\} \quad \text{is a linearly independent set of vectors} \tag{14.15}$$

**First Order Necessary Optimality Conditions for a Regular
Feasible Solution of (14.14) to be a Local Minimum**

These are: if $\bar{x}$ is a local minimum for (14.14), and is a regular feasible solution (i.e., (14.15) holds), there must exist Lagrange multipliers μ_i, $i = 1$ to m associated with the constraints in (14.14), such that $\mu = (\mu_1, \ldots, \mu_m)$ together with $\bar{x}$ satisfies

$$\nabla\theta(\bar{x}) - \sum_{i=1}^{m} \mu_i \nabla h_i(\bar{x}) \;=\; 0 \tag{14.16}$$

$$h_i(\bar{x}) = 0, \quad i = 1 \text{ to } m$$

The system of equations (14.16) are the **classical Lagrangian necessary conditions** for the equality constrained NLP (14.14), and the method that tries to identify candidate points for the solution of (14.14) through operations on the system of equations (14.16) is the **classical Lagrange multiplier technique**. Examples providing illustration of the Lagrange multiplier technique are given later on. If we treat the Lagrange multipliers μ also as variables, (14.16) is a system of $(n + m)$ equations in $(n + m)$ unknowns, and we can try to solve it by methods for solving systems of nonlinear equations.

Also, given a feasible solution $\bar{x}$ for (14.14), we can first check whether it is a regular point by checking whether (14.15) holds. If $\bar{x}$ is regular, then we can substitute $\bar{x}$ in the system of equations in the first line of (14.16), and try to solve the resulting system of linear equations in μ. If $\bar{x}$ is a regular feasible solution, and this system has no solution in μ, $\bar{x}$ cannot be a local minimum for (14.14). On the other hand, if there is a solution μ for this system, then $\bar{x}$ satisfies the first order necessary conditions together with that μ.

If $\bar{x}$ is actually a local minimum for (14.14) at which a constraint qualification such as the regularity condition (14.15) does not hold, then a Lagrange multiplier vector μ which together with $\bar{x}$ satisfies the first order necessary conditions (14.16) may or may not exist. Optimization theory has not been able to develop simple conditions to check whether a feasible solution at which the constraint qualification does not hold, is a local minimum for the problem.

Second Order Necessary Optimality Conditions for a Regular Feasible Solution of (14.14) to be a Local Minimum

If $\theta(x)$, and $h_i(x)$ for $i = 1$ to m are all twice continuously diffenrentiable, and $\bar{x}$ is a regular feasible solution (14.14) which is a local minimum for it, then by the above, there must exist a lagrange multiplier vector $\mu = (\mu_1, \ldots, \mu_m)$ which together with $\bar{x}$ must satisfy (14.16). Having found that μ, define $L(x) = \theta(x) - \sum_{i=1}^{m} \mu_i h_i(x)$. In this case, $\bar{x}$ and μ together must satisfy the following.

$$(14.16), \quad \text{and} \quad y^T H(L(\bar{x}))y \geqq 0 \quad \text{for all } y \in \mathbf{T} \tag{14.17}$$

where $\mathbf{T} = \{y : \nabla h_i(\bar{x})y = 0, \quad i = 1 \text{ to } m\}$ and $H(L(\bar{x}))$ is the hessian matrix of $L(x)$ at $\bar{x}$.

The additional condition in (14.17) over those in (14.16), can be checked using the algorithm for PSD-testing as described in Section 14.2.1.

If (14.16) holds for the regular feasible solution $\bar{x}$, but not (14.17), then it cannot be a local minimum for (14.14).

**Sufficient Conditions for the Feasible Solution $\bar{x}$
to be a Local Minimum for (14.14)**

Suppose $\theta(x)$, and all the $h_i(x)$ are twice continuously differentiable, and $\bar{x}$ is a feasible solution for (14.14). If there exists a Lagrange multiplier vector $\mu = (\mu_1, \ldots, \mu_m)$ which together with $\bar{x}$ satisfies the following conditions (14.18), where $L(x) = \theta(x) - \sum_{i=1}^{m} \mu_i h_i(x)$, $H(L(\bar{x}))$ is the hessian matrix of $L(x)$ at $\bar{x}$, and $\mathbf{T}$ is the subspace defined above, then $\bar{x}$ is a local minimum for (14.14).

$$(14.16), \quad \text{and} \quad y^T H(L(\bar{x}))y > 0 \quad \text{for all} \quad y \in \mathbf{T}, y \neq 0 \tag{14.18}$$

The last condition in (14.18) can be checked using the PD-testing algorithm as explained in Section 14.2.1.

EXAMPLE 14.12 Illustration of the Lagrange Multiplier Technique

---

Consider the following NLP in two variables $x = (x_1, x_2)^T$.

$$\begin{aligned}
\text{Minimize} \quad \theta(x) &= 3x_1 - x_1^3 \tag{14.19} \\
\text{subject to} \quad h_1(x) &= x_1^4 - 13x_1^2 + x_2^2 + 36 = 0
\end{aligned}$$

Denoting the Lagrange multiplier associated with the single constraint by μ_1, the first order necessary optimality conditions for a local minimum for this problem are: $h_1(x) = 0$ and

$$\begin{aligned}
3 - 3x_1^2 - \mu_1(4x_1^3 - 26x_1) &= 0 \tag{14.20} \\
-2\mu_1 x_2 &= 0
\end{aligned}$$

From the second equation in (14.20) we see that either $\mu_1 = 0$ or $x_2 = 0$ in any solution of this system.

If $\mu_1 = 0$, from the first equation in (14.20) we have $3(1 - x_1^2) = 0$, i.e., $x_1 = \pm 1$. Substituting $x_1 = \pm 1$ in the constraint $h_1(x) = 0$ we get $1 - 13 + x_2^2 + 36 = x_2^2 + 24 = 0$, which has no feasible real solution. So, $\mu_1 = 0$ does not lead to any solution to the system of first order necessary optimality conditions for this problem.

So, we try $x_2 = 0$. Substituting this in the constraint $h_1(x) = 0$, we get $x_1^4 - 13x_1^2 + 36 = 0$, which has solutions $x_1 = \pm 2$ or ± 3. So, the first order necessary optimality conditions for this problem have four different solutions, listed below.

$$\begin{aligned}
x^1 &= (-2, 0)^T \quad , \quad \mu_1 = -9/20 \\
x^2 &= (2, 0)^T \quad , \quad \mu_1 = 9/20 \\
x^3 &= (-3, 0)^T \quad , \quad \mu_1 = 4/5 \\
x^4 &= (3, 0)^T \quad , \quad \mu_1 = -4/5
\end{aligned}$$

$\nabla h_1(x) = (4x_1^3 - 26x_1, 2x_2)$, and we verify that $\nabla h_1(x) \neq 0$ at all the four points x^1 to x^4, so all these solutions are regular feasible solutions. Letting $L(x) = 3x_1 - x_1^3 - \mu_1(x_1^4 - 13x_1^2 + x_2^2 + 36)$, we see that the hessian of $L(x)$ is

$$H = \begin{pmatrix} -6x_1 - 12\mu_1 x_1^2 + 26\mu_1 & 0 \\ 0 & -2\mu_1 \end{pmatrix}$$

The subspace $\mathbf{T} = \{y = (y_1, y_2)^T : \nabla h_1(x)y = 0\}$ is $\{y = (0, y_2)^T : y \text{ real}\}$ at all the four points x^1 to x^4. We verify that $y^T H y \geq 0$ for all $y \in \mathbf{T}$ requires $\mu_1 < 0$. From this we verify that the solutions x^1 and x^4 satisfy both the second order necessary conditions for being a local minimum, and also the sufficient conditions (14.18); while x^2, x^3 do not even satisfy the second order necessary conditions.

So, x^1, x^4 are both local minima for (14.19). The objective values at x^1, x^4 are 2, -18 respectively. So, the best local minimum for (14.19) is $x^4 = (3,0)^T$ with an objective value of -18.

We now provide two more examples. To keep them simple, the constraints in these examples are taken to be linear equations.

EXAMPLE 14.13

Consider the problem

$$\text{Minimize} \quad \theta(x) = x_1 x_2$$
$$\text{subject to} \quad h_1(x) = x_1 + x_2 - 2 = 0$$

Denoting the Lagrange multiplier by μ_1, the first order necessary conditions are

$$x_2 - \mu_1 = 0$$
$$x_1 - \mu_1 = 0$$
$$x_1 + x_2 - 2 = 0$$

This system has the unique solution $x^1 = (1,1)^T, \mu_1 = 1$. Defining $L(x) = \theta(x) - \mu_1 h_1(x) = x_1 x_2 - (x_1 + x_2 - 2)$, its hessian is $H = \begin{pmatrix} 0 & 1 \\ 1 & 0 \end{pmatrix}$.

The subspace $\mathbf{T} = \{y : \nabla h_1(x)y = 0\} = \{y : y_1 + y_2 = 0\}$. In this subspace $y^T H y = y_1 y_2 = -y_2^2 < 0$ whenever $y \neq 0$. So, the second order necessary optimality conditions for being a local minimum are violated by x^1. Since x^1 is the only solution for the first order necessary conditions, this implies that $\theta(x)$ has no local minimum in the feasible region. In fact it can be verified that x^1 satisfies the sufficient conditions for being a local maximum for $\theta(x)$ in the feasible region. And $\theta(x)$ is unbounded below in the feasible region.

EXAMPLE 14.14

Consider the problem

$$\text{Minimize} \quad \theta(x) \;=\; 2x_1^3 + (1/2)x_2^2 + x_1x_2 + (1/24)x_1$$
$$\text{subject to} \quad h_1(x) \;=\; x_1 + x_2 - 2 = 0$$

Denoting the Lagrange multiplier by μ_1, the first order necessary conditions for being a local minimum are

$$6x_1^2 + x_2 + (1/24) - \mu_1 \;=\; 0$$
$$x_2 + x_1 - \mu_1 \;=\; 0$$
$$x_1 + x_2 \;=\; 2$$

This system has the unique solution $x^1 = (1/12, 23/12)^T, \mu_1 = 2$. The subspace $\mathbf{T} = \{y : \nabla h_1(x)y = 0\} = \{y : y_1 + y_2 = 0\}$. Defining $L(x) = \theta(x) - \mu_1 h_1(x) = 2x_1^3 + (1/2)x_2^2 + x_1x_2 + (1/24)x_1 - 2(x_1 + x_2 - 2)$, its hessian is

$$H = \begin{pmatrix} 12x_1 & 1 \\ 1 & 1 \end{pmatrix} = \begin{pmatrix} 1 & 1 \\ 1 & 1 \end{pmatrix} \quad \text{at } x = x^1$$

So, $y^THy = (y_1 + y_2)^2 \geqq 0$ for all $y \in \mathbf{T}$, but $y^THy \not> 0$ for $y \in \mathbf{T}, y \neq 0$. Thus x^1 satisfies the second order necessary conditions for being a local minimum for this problem, but not the sufficient conditions.

14.7.4   NLPs Involving Inequality Constraints

Consider the NLP

$$\text{Minimize} \quad \theta(x)$$
$$\text{subject to} \quad h_i(x) \;=\; 0, \quad i = 1 \text{ to } m \qquad (14.21)$$
$$g_p(x) \;\geqq\; 0, \quad p = 1 \text{ to } t$$

The inequality constraints in (14.21) include all sign restrictions, and any lower or upper bounds on individual variables, and any linear or nonlinear inequality constraints.

Let $\bar{x}$ be a feasible solution to the problem. For $p = 1$ to t, the inequality constraint $g_p(x) \geqq 0$ is said to be

$$\textbf{active at } \bar{x} \quad \text{if} \quad g_p(\bar{x}) = 0$$

$$\textbf{inactive at } \bar{x} \quad \text{if} \quad g_p(\bar{x}) > 0 \tag{14.22}$$

Let $\mathbf{P}(\bar{x}) = \{p : 1 \leqq p \leqq t, g_p(\bar{x}) = 0\}$; i.e., it is the index set of active inequality constraints in (14.21) at the feasible solution $\bar{x}$.

The feasible solution $\bar{x}$ for (14.21) is said to satisfy the **regularity condition for (14.21)** (and said to be a **regular feasible solution** for it) if the set of vectors

$$\{\nabla h_i(\bar{x}) : i = 1 \text{ to } m\} \cup \{\nabla g_p(\bar{x}) : p \in \mathbf{P}(\bar{x})\} \quad \text{is linearly independent} \tag{14.23}$$

Conditions to check whether a feasible solution is a local minimum have only been derived under some constraint qualification such as the regularity condition.

KKT Conditions or First Order Necessary Optimality Conditions for a Regular Feasible Solution of (14.21) to be a Local Minimum

These are: if $\bar{x}$ is a local minimum for (14.21) and is a regular feasible solution (i.e., (14.23) holds), there must exist Lagrange multipliers μ_i associated with the ith equality constraint in (14.21), $i = 1$ to m; and π_p associated with the pth inequality constraint for $p = 1$ to t; such that $\mu = (\mu_1, \ldots, \mu_m), \pi = (\pi_1, \ldots, \pi_t)$ together with $\bar{x}$ satisfy

$$
\begin{aligned}
\nabla\theta(\bar{x}) - \sum_{i=1}^{m} \mu_i \nabla h_i(\bar{x}) - \sum_{p=1}^{t} \pi_p \nabla g_p(\bar{x}) &= 0 \\
h_i(\bar{x}) = 0, \quad i &= 1 \text{ to } m \\
g_p(\bar{x}) \geqq 0, \quad p &= 1 \text{ to } t \\
\pi_p \geqq 0, \quad p &= 1 \text{ to } t \\
\pi_p g_p(\bar{x}) = 0, \quad p &= 1 \text{ to } t
\end{aligned}
\tag{14.24}
$$

The conditions in the last line of (14.24) which require that for each $p = 1$ to t, either $\pi_p = 0$ or $g_p(\bar{x}) = 0$; are known as **complementary slackness optimality conditions**. They require that $\pi_p = 0$ for all $p \in \{p : g_p(\bar{x}) > 0\} = \{1, \ldots, t\} \backslash \mathbf{P}(\bar{x})$.

Given a regular feasible solution $\bar{x}$, it satisfies the conditions in lines 2, 3 of (14.24) automatically. The remaining system is

$$\nabla\theta(\bar{x}) - \sum_{i=1}^{m} \mu_i \nabla h_i(\bar{x}) - \sum_{p=1}^{t} \pi_p \nabla g_p(\bar{x}) = 0$$

$$\pi_p \overset{\geq}{=} 0, \quad p \in \mathbf{P}(\bar{x}) \tag{14.25}$$

$$\pi_p = 0, \quad \text{for all} \quad p \notin \mathbf{P}(\bar{x})$$

and it is a system of linear constraints in μ, π. Hence we can try to solve it using linear programming techniques (Phase I of the simplex method). If μ, π satisfying (14.25) exist, then $\bar{x}$ satisfies the first order necessary conditions together with that μ, π; in this case $\bar{x}$ is said to be a **KKT point** for the problem (14.21).

Second Order Necessary Optimality Conditions for a Regular Feasible Solution of (14.21) to be a Local Minimum

If $\theta(x)$ and all the $h_i(x), g_p(x)$ are twice continuously differentiable, and $\bar{x}$ is a regular feasible solution for (14.21) which is a local minimum, then by the above, there must exist Lagrange multiplier vectors $\mu = (\mu_1, \ldots, \mu_m), \pi = (\pi_1, \ldots, \pi_t)$ which together with $\bar{x}$ satisfy (14.24). Having found μ, π, define

$$L(x) = \theta(x) - \sum_{i=1}^{m} \mu_i h_i(x) - \sum_{p=1}^{t} \pi_p g_p(x)$$

In this case, $\bar{x}, \mu, \pi$ together must satisfy the following

$$(14.24) \quad \text{and} \quad y^T H(L(\bar{x}))y \overset{\geq}{=} 0 \quad \text{for all } y \in \mathbf{T} \tag{14.26}$$

where $\mathbf{T} = \{y : \nabla h_i(\bar{x})y = 0, \quad i = 1 \text{ to } m; \nabla g_p(\bar{x})y = 0, p \in \mathbf{P}(\bar{x})\}$ and $H(L(\bar{x}))$ is the hessian matrix of $L(x)$ at $\bar{x}$.

The additional condition in (14.26) over those in (14.24), can be checked using the algorithm for PSD-testing as described in Section 14.2.1.

Sufficient Conditions for the Feasible Solution $\bar{x}$ to be a Local Minimum for (14.21)

Suppose $\theta(x)$, and all the $h_i(x), g_p(x)$ are twice continuously differentiable, and $\bar{x}$ is a feasible solution for (14.21). If there exists Lagrange multiplier vectors $\mu = (\mu_1, \ldots, \mu_m), \pi = (\pi_1, \ldots, \pi_t)$ which together with $\bar{x}$ satisfy the following conditions (14.27), with $L(x)$ as defined above, $H(L(\bar{x}))$ as the hessian matrix of $L(x)$ at $\bar{x}$, and $\mathbf{T}_1 = \{y : \nabla h_i(\bar{x})y = 0, i = 1 \text{ to } m; \nabla g_p(\bar{x})y = 0, p \in \mathbf{P}(\bar{x}) \cap \{p : \pi_p > 0\}\}$, then $\bar{x}$ is a local minimum for (14.21).

$$(14.24), \quad \text{and} \quad y^T H(L(\bar{x}))y > 0 \quad \text{for all} \quad y \in \mathbf{T}_1, y \neq 0 \tag{14.27}$$

The last condition in (14.27) can be checked using the PD-testing algorithm as explained in Section 14.2.1.

If (14.21) is a convex programming problem (i.e., $\theta(x)$ is convex, all $h_i(x)$ are affine, and all $g_p(x)$ are concave), and $\bar{x}$ is a regular feasible solution for it, it is a global minimum iff there exist Lagrange multiplier vectors $\mu = (\mu_1, \ldots, \mu_m), \pi = (\pi_1, \ldots, \pi_t)$ which together with $\bar{x}$ satisfy the KKT conditions (14.24).

If $\theta(x)$ is nonconvex, sufficient conditions slightly weaker than those in (14.27) are known; but checking whether they hold is a hard problem, that is why we presented (14.27) instead.

EXAMPLE 14.15

Consider the problem

$$\text{Minimize} \quad \theta(x) = x_1^2 + (1/2)x_2^2 + x_3^2 + (1/2)x_4^2$$
$$\text{subject to} \qquad x_1 + x_2 = 5$$
$$-x_2 + x_3 + x_4 = 4$$
$$x_1, x_2, x_3, x_4 \geqq 0$$

Clearly this is a convex quadratic program. Introducing the Lagrange multipliers $\mu_1, \mu_2; \pi_1$ to π_4; the KKT conditions for this problem are

$$2x_1 - \mu_1 - \pi_1 = 0$$
$$x_2 - \mu_1 + \mu_2 - \pi_2 = 0$$
$$2x_3 - \mu_2 - \pi_3 = 0$$
$$x_4 - \mu_2 - \pi_4 = 0$$
$$\pi_1 \text{ to } \pi_4 \geqq 0$$
$$\pi_1 x_1 = \pi_2 x_2 = \pi_3 x_3 = \pi_4 x_4 = 0$$

It can be verified that $\bar{x} = (3, 2, 2, 4)^T$, $\mu = (6, 4)$, $\pi = 0$, together satisfy these KKT conditions. Hence $\bar{x}$ is a global minimum for this problem.

14.7.5 An Important Caution

We have so far discussed several necessary optimality conditions for a given point to be a local minimum to a variety of NLP models. If the NLP is a convex program, any point satisfying these necessary optimality conditions is not only a local minimum, but actually a global minimum. Unfortunately, many NLP models that arise in real world applications tend to be nonconvex, and for such a problem, a point satisfying the necessary optimality conditions may not even be a local minimum. Algorithms for NLP are usually designed to converge to a point satisfying the necessary optimality conditions, and as mentioned earlier, one should not blindly accept such a point as an optimum solution to the problem without checking (by the second order necessary optimality conditions, or by some local search in the vicinity of the point) that it is at least better than other nearby points. Also, the system of necessary optimality conditions may have many solutions. Finding alternate solutions of this system, and selecting the best among them, usually leads to a good point to investigate further.

We will illustrate the importance of this with the story of US Air Force's controversial B-2 **Stealth** bomber program. There were many design variables such as the various dimensions, distribution of volume between the wing and the fuselage, flying speed, thrust, fuel consumption, drag, lift, air density, etc., that could be manipulated for obtaining the best range (i.e., the distance it can fly starting with full tanks, without refueling). The problem of maximizing the range subject to all the constraints was modeled as an NLP in a secret Air Force study going back to the 1940s. A solution to the necessary optimality conditions of this problem was found; it specified values for the design variables that put almost all of the total volume in the wing, leading to the **flying wing** design for the B-2 bomber. After spending billions of dollars, building test planes, etc., it was found that the design solution implemented works, but that its range was too low in comparison with other bomber designs being experimented subsequently in the US and abroad.

A careful review of the model was then carried out. The review indicated that all the formulas used, and the model itself, are perfectly valid. However, the model was a nonconvex NLP, and the review revealed a second solution to the system of necessary optimality conditions for it, besides the one found and implemented as a result of earlier studies. The second solution makes the wing volume much less than the total volume, and seems to maximize the range; while the first solution that is implemented for the B-2 bomber seems to actually minimize the range. In other words, the design implemented was the aerodynamically worst possible choice of configuration, leading to a very costly error.

For an account, see the research news item "Skeleton Alleged in the Stealth Bomber's Closet," *Science*, Vol. 244, 12 May 1989 issue, pages 650-651.

14.8   The Most Useful NLP Algorithms

Research on algorithms for nonlinear programming has been going on ever since the days of Newton and Lebnitz, and the literature on it is very very vast, deep, and fascinating. In this section we will present an elementary description of one or two most popular methods for each problem type, with no proofs of convergence etc., just to provide a flavor of the methods and to whet the appetite of the reader.

Many algorithms for NLP use iterations of the following form: given a point x^r, and a direction $y^r \neq 0$, find a real number α called the **step length** that

$$\text{Minimizes} \quad p(\alpha) = f(x^r + \alpha y^r) \quad \text{over} \quad \alpha \geqq 0 \qquad (14.28)$$

In this iteration, $f(x)$ is typically either the objective function in the original NLP, or a composite function known as a **merit function** which is usually the sum of the objective function plus a penalty function representing an **infeasibility**

measure (a measure of the infeasibility of the point x to the system of constraints in the problem). Since x^r and y^r are given vectors, $f(x^r + \alpha y^r)$ is a function of the single variable α, which we denote by $p(\alpha)$. In (14.28), the vector y^r will be a **descent direction** for $f(x)$ at the point x^r; i.e., $f(x^r + \alpha y^r) < f(x^r)$ whenever α is positive and sufficiently small; this guarantees that as we move from the point x^r in the direction y^r, $f(x)$ decreases initially; that's why α is restricted to be $\overset{\geq}{=} 0$ in (14.28). y^r will be a descent direction at x^r for $f(x)$ if $\nabla f(x^r)y^r < 0$. In the point $x^r + \alpha y^r$, α is known as the step length. If $\bar{\alpha}$ is an optimum solution of (14.28), it is known as the **optimum step length** for (14.28). A problem of the form (14.28) is known as a **line search problem**, it is an NLP involving only one variable, the only constraint on which is the nonnegativity restriction. Since line searches are the backbone of most NLP algorithms, we begin with a discussion of two most popular line search methods.

14.8.1 Methods for Line Search

The most popular algorithms for NLP can be characterized as **line search descent methods**, i.e., in each step they solve a line search problem of the form

$$\text{Minimizes} \quad p(\alpha) = f(x^r + \alpha y^r) \quad \text{over} \quad \alpha \overset{\geq}{=} 0 \qquad (14.29)$$

where $f(x)$ is a differentiable real valued function defined over $\mathbb{R}^n$; $x^r \in \mathbb{R}^n, y^r \neq 0$ in $\mathbb{R}^n$ are given point, and descent direction for $f(x)$ at x^r respectively (i.e., $\nabla f(x^r)y^r < 0$).

As an example, consider the real valued function $f(x) = (x_1 - 2)^4 + (x_1 - 2x_2)^2$ defined over $\mathbb{R}^2$. Let $x^1 = (0, 3)^T, y^1 = (1, 0)^T$. $\nabla f(x) = (4(x_1 - 2)^3 + 2(x_1 - 2x_2), -4(x_1 - 2x_2))$, and so $\nabla f(x^1) = (-44, 24)$, $\nabla f(x^1)y^1 = -44 < 0$, hence y^1 is a descent direction for $f(x)$ at x^1. $p(\alpha) = f(x^1 + \alpha y^1) = (\alpha - 2)^4 + (\alpha - 6)^2$ in this example problem. The optimum step length in this problem is the α which minimizes $p(\alpha)$ over $\alpha \overset{\geq}{=} 0$.

A method which tries to find the optimum step length for a line search problem is called an **exact line search method**. In the past it was considered to be very important to choose the step length in each step in an NLP method close to an optimum step length. However, exact line searches usually require too many function evaluations, making them computationally expensive. Research has shown that in order to guarantee the convergence of the overall algorithm, it is enough to adopt an **inexact line search technique** that guarantees a sufficient degree of accuracy or descent in the function value. The following Armijo's procedure is a popular inexact line search procedure for selecting the step length.

ARMIJO'S INEXACT LINE SEARCH PROCEDURE FOR (14.29)

The procedure uses two parameters ϵ_1 and ϵ_2, $0 < \epsilon_1 < 1$, $\epsilon_2 > 1$, which respectively help keep the step length from being too large or too small (commonly used values are $\epsilon_1 = 0.2, \epsilon_2 = 2$). The derivative of $p(\alpha)$ at $\alpha = 0$ is $\frac{dp(0)}{d\alpha} = \nabla f(x^r)y^r$. The first order Taylor approximation of $p(\alpha)$ is $p(0) + \alpha\frac{dp(0)}{d\alpha}$. Define the affine function in the single variable α, $\ell(\alpha) = p(0) + \alpha\epsilon_1\frac{dp(0)}{d\alpha}$. A step length $\bar{\alpha}$ in (14.29) is considered to be acceptable (i.e., having a sufficient rate of descent) if $p(\bar{\alpha}) \stackrel{<}{=} \ell(\bar{\alpha})$. However, to prevent $\bar{\alpha}$ from being too small, this procedure also requires $p(\epsilon_2\bar{\alpha}) > \ell(\epsilon_2\bar{\alpha})$. A step length satisfying both these conditions can be found by the method listed below.

Step 1   Select a fixed step length parameter $\hat{\alpha}$ (say 1). If $p(\hat{\alpha}) \stackrel{<}{=} \ell(\hat{\alpha})$ go to Step 2. If $p(\hat{\alpha}) > \ell(\hat{\alpha})$ go to Step 3.

Step 2   In this case, the step length $\bar{\alpha}$ can be taken to be either $\hat{\alpha}$, or $\epsilon_2^t\hat{\alpha}$ where t is the largest integer $t \stackrel{>}{=} 0$ for which $p(\epsilon_2^t\hat{\alpha}) \stackrel{<}{=} \ell(\epsilon_2^t\hat{\alpha})$. Terminate.

Step 3   In this case, the step length $\bar{\alpha}$ is taken to be $\hat{\alpha}/\epsilon_2^t$ where t is the smallest integer > 1 for which $p(\hat{\alpha}/\epsilon_2^t) \stackrel{<}{=} \ell(\hat{\alpha}/\epsilon_2^t)$. Terminate.

THE QUADRATIC FIT LINE SEARCH METHOD FOR (14.29)

This is another popular line search method that we will present. To initiate this method, we need three values of α, $0 \stackrel{<}{=} \alpha_1 < \alpha_2 < \alpha_3$ such that $p(\alpha_2) \stackrel{<}{=} \min\{p(\alpha_1), p(\alpha_3)\}$, as shown in Figure 14.5. The interval $[\alpha_1, \alpha_3]$ with α_2 in-between is then called a **three point bracket (TPB)** for the minimum of $p(\alpha)$, which we will denote by the symbol $(\alpha_1, \alpha_2, \alpha_3)$.

An initial TPB can be selected by the following procedure. Begin with $\alpha_0 = 0$, and choose a positive step length Δ satisfying $p(\alpha_0 + \Delta) < p(\alpha_0)$ (this is possible because $\alpha_0 = 0$, and $p(\alpha)$ strictly decreases as α increases from 0). Compute $p(\alpha_0)$ and $p(\alpha_1)$, where $\alpha_1 = \alpha_0 + \Delta$. So, $p(\alpha_1) < p(\alpha_0)$. Define $\alpha_r = \alpha_{r-1} + 2^{r-1}\Delta$ for $r = 2, 3, \ldots$ as long as the value of $p(\alpha)$ at these points keeps on decreasing, until a value k for r is found such that $p(\alpha_{k+1}) > p(\alpha_k)$. In this case we have $\alpha_{k-1}, \alpha_k, \alpha_{k+1}$ satisfying $p(\alpha_k) < p(\alpha_{k-1})$, $p(\alpha_{k+1}) > p(\alpha_k)$. Among the four points $\alpha_{k-1}, \alpha_k, (\alpha_k + \alpha_{k+1})/2$, and α_{k+1}, drop either α_{k-1} or α_{k+1}, whichever is farther from the point in the pair $\{\alpha_k, (\alpha_k + \alpha_{k+1})/2\}$ that yields the smallest value for $p(\alpha)$. Let the remaining points be called $\alpha_a, \alpha_b, \alpha_c$, where $\alpha_a < \alpha_b < \alpha_c$. These points are equi-distant, and $p(\alpha_b) \stackrel{<}{=} p(\alpha_a)$, $p(\alpha_b) \stackrel{<}{=} p(\alpha_c)$. So $(\alpha_a, \alpha_b, \alpha_c)$ is a TPB.

Having found an initial TPB, this method approximates the original function $p(\alpha)$ by a simpler convex quadratic function $q(\alpha)$ by curve fitting, and then uses the minimum of $q(\alpha)$ as an approximation to the minimum of $p(\alpha)$. The steps in the method are described below.

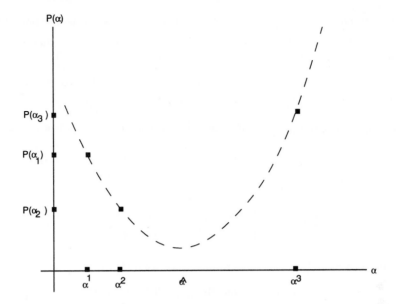

Figure 14.5 A quadratic curve through three points $(\alpha_1, P(\alpha_1)), (\alpha_2, P(\alpha_2)), (\alpha_3, P(\alpha_3))$, where $(\alpha_1, \alpha_2, \alpha_3)$ form a TPB for $P(\alpha)$. $\hat{\alpha}$ is the point where the quadratic fit has its minimum.

Step 1 Let $(\alpha_1, \alpha_2, \alpha_3)$ be the current TPB satisfying $p(\alpha_2) \stackrel{\leq}{=} \min\{p(\alpha_1), p(\alpha_3)\}$. α_2 is the current best point.

The method now constructs the unique convex quadratic function $q(\alpha) = a\alpha^2 + b\alpha + c$ whose values agree with those of $p(\alpha)$ at $\alpha = \alpha_1, \alpha_2, \alpha_3$. These conditions provide three equations, namely

$$a\alpha_s^2 + b\alpha_s + c = p(\alpha_s), \quad s = 1, 2, 3$$

for determining the values of the coefficients a, b, c in $q(\alpha)$. The property of $\alpha_1, \alpha_2, \alpha_3$ that they form a TPB guarantees that $q(\alpha)$ is convex and has a unique minimum which will be attained at the point α satisfying $\frac{dq(\alpha)}{d\alpha} = 2a\alpha + b = 0$; i.e., it is $\hat{\alpha} = -b/2a$. It can be verified that $\hat{\alpha}$ is given by the following formula directly.

$$\hat{\alpha} = \frac{(\alpha_2^2 - \alpha_3^2)q(\alpha_1) + (\alpha_3^2 - \alpha_1^2)q(\alpha_2) + (\alpha_1^2 - \alpha_2^2)q(\alpha_3)}{2[(\alpha_2 - \alpha_3)q(\alpha_1) + (\alpha_3 - \alpha_1)q(\alpha_2) + (\alpha_1 - \alpha_2)q(\alpha_3)]}$$

Go to Step 2 if $\hat{\alpha} > \alpha_2$, to Step 3 if $\hat{\alpha} < \alpha_2$, and to Step 4 if $\hat{\alpha} = \alpha_2$.

Step 2   Here $\hat{\alpha} > \alpha_2$. If $p(\hat{\alpha}) \overset{\geq}{=} p(\alpha_2)$, choose $(\alpha_1, \alpha_2, \hat{\alpha})$ as the new TPB. If $p(\hat{\alpha}) \overset{\leq}{=} p(\alpha_2)$, choose $(\alpha_2, \hat{\alpha}, \alpha_3)$ as the new TPB. If $p(\hat{\alpha}) = p(\alpha_2)$, either of these choices for the new TPB are permissible. Go to Step 5.

Step 3   Here $\hat{\alpha} < \alpha_2$. As in Step 2, choose the new TPB to be $(\hat{\alpha}, \alpha_2, \alpha_3)$ if $p(\hat{\alpha}) \overset{\geq}{=} p(\alpha_2)$, or $(\alpha_1, \hat{\alpha}, \alpha_2)$ if $p(\hat{\alpha}) \overset{\leq}{=} p(\alpha_2)$. Go to Step 5.

Step 4   Here $\hat{\alpha} = \alpha_2$. In this case, the minimization of the quadratic fit $q(\alpha)$ has not produced a new point. If $\alpha_3 - \alpha_1 \overset{\leq}{=} \epsilon$ for some positive tolerance ϵ, stop with the current best point α_2 as the step length for (14.29) and terminate. Otherwise, define

$$\text{New } \hat{\alpha} = \begin{cases} \alpha_2 + \epsilon/2 & \text{if } \alpha_3 - \alpha_2 \overset{\geq}{=} \alpha_2 - \alpha_1 \\ \alpha_2 - \epsilon/2 & \text{if } \alpha_3 - \alpha_2 \overset{\leq}{=} \alpha_2 - \alpha_1 \end{cases}$$

and go to Step 2 or Step 3 as appropriate with this new value of $\hat{\alpha}$.

Step 5   Now let $(\alpha_1, \alpha_2, \alpha_3)$ denote the new TPB obtained under Step 2 or Step 3. If any one of the following termination conditions is satisfied

$$\max\{p(\alpha_1), p(\alpha_3)\} - p(\alpha_2) \quad \overset{\leq}{=} \quad \epsilon_1$$
$$\text{or} \quad \alpha_3 - \alpha_1 \quad \overset{\leq}{=} \quad \epsilon$$
$$\text{or} \quad |\frac{dp(\alpha_2)}{d\alpha}| \quad \overset{\leq}{=} \quad \epsilon_2$$

where $\epsilon_1, \epsilon, \epsilon_2$ are some positive tolerances, terminate with α_2 as the step length for (14.29). Otherwise, with $(\alpha_1, \alpha_2, \alpha_3)$ as the current TPB, go back to Step 1.

Discussion

The quadratic fit method is also known as the quadratic interpolation method in the literature. It usually produces very satisfactory results after only a few repetitions of Step 1.

We discussed only two popular line search methods. There are many others, see the NLP books referenced at the end of this chapter.

14.8.2   Newton-Raphson Method for Solving a System of n Nonlinear Equations in n Unknowns

For $i = 1$ to n, let $f_i(x)$ be a differentiable real valued function defined on $\mathbb{R}^n$. Consider the following system of equations

$$f_i(x) = 0, \quad i = 1 \text{ to } n \tag{14.30}$$

Since (14.30) contains as many equations as there are unknowns, it is called a **square system of equations.** Here we discuss the classical Newton-Raphson method for solving a square system of equations.

A system in which the number of equations is less (greater) than the number of variables, is called an under-determined (over-determined) system of equations. For methods to solve under- or over-determined systems of equations, see (J. E. Dennis and R. B. Schnabel [1983]).

Starting with an initial point x^0, the Newton-Raphson method generates a sequence of points $\{x^0, x^1, x^2, \ldots\}$ which under certain conditions can be shown to converge to a solution of (14.30). The method is also called **Newton's method,** or **Newton's method for solving equations** in the literature.

Let x^r be the current point for some r. If x^r satisfies (14.30) within some acceptable tolerance, terminate with x^r as an approximate solution for (14.30). Otherwise, denote the solution of (14.30) by $x^r + y$, where y represents the unknown correction vector to add to x^r to make it into a solution for (14.30). For each $i = 1$ to n, the first order Taylor approximation for $f_i(x^r + y)$ around x^r is $f_i(x^r) + \nabla f_i(x^r)y$. So, a first order approximation of (14.30) is the following system of linear equations in y (because x^r is a given point in (14.31), the only variables there are the y).

$$f_i(x^r) + \nabla f_i(x^r)y = 0, \quad i = 1 \text{ to } n \tag{14.31}$$

Denote the column vector of functions $(f_1(x), \ldots, f_n(x))^T$ by $f(x)$.

$\nabla f_i(x^r)$ is the row vector of partial derivatives of $f_i(x^r)$ at x^r. The $n \times n$ matrix with these as row vectors, denoted by $J(f(x^r))$ is called the **Jacobian matrix** of $f(x)$ at x^r.

$$J(f(x^r)) = \begin{pmatrix} \nabla f_1(x^r) \\ \vdots \\ \nabla f_n(x^r) \end{pmatrix} \tag{14.32}$$

If $J(f(x^r))$ is nonsingular, the unique solution of (14.31) is $-(J(f(x^r)))^{-1}f(x^r) = y$. If the Jacobian matrix $J(f(x^r))$ is singular, its inverse does not exist, and the method is unable to proceed further. Otherwise, it takes the next point in the sequence to be

$$x^{r+1} = x^r - (J(f(x^r)))^{-1}f(x^r) \tag{14.33}$$

and continues in the same way with x^{r+1} as the current point.

The method is called an iterative method with the iterative formula given by (14.33).

As an example, consider the following system of two equations in two unknowns $x = (x_1, x_2)^T$.

$$f_1(x) = x_1^2 + x_2^2 - 1 = 0$$
$$f_2(x) = x_1^2 - x_2 = 0$$

The Jacobian matrix $J(f(x))$ is

$$J(f(x)) = \begin{pmatrix} 2x_1 & 2x_2 \\ 2x_1 & -1 \end{pmatrix}$$

Let $x^0 = (1,0)^T$ be the initial point. Then $f(x^0) = (0,1)^T$, and $J(f(x^0)) = \begin{pmatrix} 2 & 0 \\ 2 & -1 \end{pmatrix}$. This leads to the next point

$$x^1 = x^0 - (J(f(x^0)))^{-1} f(x^0) = \begin{pmatrix} 1 \\ 0 \end{pmatrix} + \begin{pmatrix} 2 & 0 \\ 2 & -1 \end{pmatrix}^{-1} \begin{pmatrix} 0 \\ 1 \end{pmatrix} = \begin{pmatrix} 1 \\ 1 \end{pmatrix}$$

It can be verified that $x^2 = (5/6, 2/3)^T$, and so on. The actual solution in this example can be seen from Figure 14.6.

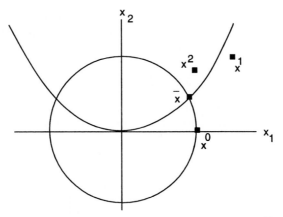

Figure 14.6 The circle is the set of all points $(x_1, x_2)^T$ satisfying $x_1^2 + x_2^2 - 1 = 0$. The parabola is the set of all points satisfying $x_1^2 - x_2 = 0$. The two intersect in two points (solutions of the system) one of which is $\bar{x}$. Beginning with x^0, the Newton-Raphson method obtains the sequence $x^1, x^2, \ldots$ converging to $\bar{x}$.

Even when the method is able to generate the entire sequence without encountering a singular Jacobian at any stage, the generated sequence can be shown to converge to a solution only under some conditions (this usually requires that the initial point x^0 is very close to an actual solution of (14.30)). There are several modifications proposed to this method to guarantee that it converges to a solution of (14.30) when one exists, under weaker conditions, see (M. S. Bazaraa, H. D. Sherali, C. M. Shetty [1993], and J. E. Dennis Jr. and R. B. Schnabel [1983]).

14.8.3   A Method for Unconstrained Minimization Over $\mathbb{R}^n$

Let $\theta(x)$ be a differentiable real valued function defined over $\mathbb{R}^n$. Here we consider the problem

$$\text{Minimize} \quad \theta(x) \quad \text{over} \quad x \in \mathbb{R}^n \qquad (14.34)$$

We describe a method for this problem known as the **BFGS method** named after the four people (Broyden, Fletcher, Goldfarb, and Shanno) who developed it, belonging to the family of variable metric methods. At present this method is considered to be the most effective method for unconstrained optimization, and is very widely used. It is based on line searches. Beginning with an initial point x^0 it generates a sequence of points $\{x^0, x^1, \ldots\}$ along which the objective value decreases. The method maintains a square matrix of order n which is updated in each step. The sequence of matrices obtained is denoted by $\{D_0, D_1, \ldots\}$ where the first one D_0 is usually an initial PD symmetric matrix, typically the unit matrix I.

Step 1 of the method begins with the initial point x^0 and the matrix D_0. We will now describe the general step in the method.

General step $r+1$ for $r \geqq 0$ This step begins with the current point x^r, and matrix D_r. Define

$$y^r = -D_r(\nabla\theta(x^r))^T$$

and solve the line search problem

$$\text{Minimize} \quad \theta(x^r + \alpha y^r) \quad \text{over} \quad \alpha \geqq 0$$

It is not necessary to find an optimum step length for this line search problem; solving it using a fast inexact line search method is fine. Let α_r be the step length obtained as a solution for this line search problem. Define

$$x^{r+1} = x^r + \alpha_r y^r$$

If one or more of the following termination conditions are satisfied

$$
\begin{aligned}
|\theta(x^{r+1}) - \theta(x^r)| &< \epsilon_1 \\
\|x^{r+1} - x^r\| &< \epsilon_2 \\
\|\nabla\theta(x^{r+1})\| &< \epsilon_3
\end{aligned}
$$

where $\epsilon_1, \epsilon_2, \epsilon_3$ are suitably chosen tolerances (small positive numbers), terminate with x^{r+1} as the computed solution of (14.34). Otherwise, define

$$
\begin{aligned}
\xi^r &= x^{r+1} - x^r = \alpha_r y^r \\
\eta^r &= (\nabla\theta(x^{r+1}) - \nabla\theta(x^r))^T \\
D_{r+1} &= D_r + \left(1 + \frac{(\eta^r)^T D_r \eta^r}{(\xi^r)^T \eta^r}\right)\left(\frac{\xi^r(\xi^r)^T}{(\xi^r)^T \eta^r}\right) - \frac{\xi^r(\eta^r)^T D_r + D_r\eta^r(\xi^r)^T}{(\xi^r)^T \eta^r}
\end{aligned}
$$

With x^{r+1}, D_{r+1} as the current point and matrix, go to the next step.

Discussion

Here we described one of the most commonly used methods for unconstrained optimization. There are several other methods, and the subject of unconstrained optimization is highly developed and very beautiful. This subject is now a mature field of optimization, serving the needs of many practitioners; see the references at the end of this chapter.

14.8.4   Penalty Function Methods for Constrained Optimization

These methods transform a constrained optimization problem into an equivalent unconstrained optimization problem, which is then solved using the methods for unconstrained optimization discussed above. The basic technique is to eliminate the constraints, but add a penalty term corresponding to each of them in the objective function to be minimized. The penalty term corresponding to a constraint is always zero at every point satisfying that constraint, and positive at every point violating it. And the value of the penalty term increases rapidly as the point moves farther away from the feasible region for that constraint.

Consider an equality constraint: $h_1(x) = 0$. Penalty terms corresponding to it are: $(h_1(x))^2$, or $(h_1(x))^4$, or $(|h_1(x)|)^r$ for $r = 1, 2, 3$, etc. Clearly these penalty terms satisfy the properties mentioned above. $(h_1(x))^2$ is called the **quadratic penalty term**.

Consider an inequality constraint: $g_1(x) \geqq 0$. Penalty terms corresponding to it are $(\max\{0, -g_1(x)\})^r$ for $r = 1, 2, 3, \ldots$

Consider the general NLP

$$
\begin{aligned}
\text{Minimize} \quad & \theta(x) \\
\text{subject to} \quad & h_i(x) = 0, \quad i = 1 \text{ to } m \\
& g_p(x) \geqq 0, \quad p = 1 \text{ to } t
\end{aligned}
\tag{14.35}
$$

A penalty function corresponding to (14.35) is the sum of penalty terms corresponding to each of the constraints in it. A typical penalty function for this problem is of the form

$$P(x) = \sum_{i=1}^{m} (|h_i(x)|)^r + \sum_{p=1}^{t} (\max\{0, -g_p(x)\})^s \qquad (14.36)$$

where r, s are positive integers. Usually values of 1 or 2 are used for r and s. When $r = 2$, $(|h_i(x)|)^2$ is the same as $(h_i(x))^2$. The function

$$f(\lambda, x) = \theta(x) + \lambda P(x) \qquad (14.37)$$

where λ is a positive penalty parameter, is known as the **auxiliary function** or **composite function**. In the feasible region for (14.35), $f(\lambda, x) = \theta(x)$. And for every point x which is infeasible to (14.35), $\lambda P(x) > 0$ is the penalty in $f(\lambda, x)$ corresponding to its infeasibility. As the penalty parameter λ increases, the penalty for an infeasible point increases rapidly.

Since the penalty function is positive at every infeasible point, and 0 at every feasible point for (14.35); an optimum solution that minimizes $f(\lambda, x)$ over $x \in \mathbb{R}^n$ tends to be feasible to (14.35) if (14.35) has feasible solutions; particularly as the penalty parameter λ becomes large.

For $\lambda > 0$, let $x(\lambda)$ denote an unconstrained minimum of $f(\lambda, x)$ over $x \in \mathbb{R}^n$. Assuming that (14.35) has an optimum solution, it can be proved that $x(\lambda)$ converges to an optimum solution of (14.35) as $\lambda \to \infty$.

As an example, consider the following NLP.

$$\begin{aligned} \text{Minimize} \quad & x_1^2 + 2x_2^2 \\ \text{subject to} \quad & x_1 + x_2 - 2 = 0 \end{aligned} \qquad (14.38)$$

Taking the quadratic penalty term corresponding to the equality constraint, the auxiliary function is $f(\lambda, x) = x_1^2 + 2x_2^2 + \lambda(x_1 + x_2 - 2)^2$. The unconstrained minimum of $f(\lambda, x)$ over $x \in \mathbb{R}^n$ is

$$x(\lambda) = \left(\frac{2\lambda^2}{2 + \lambda + \lambda^2} , \frac{4}{2 + \lambda + \lambda^2} \right)^T$$

This point $x(\lambda)$ is infeasible to (14.38) for all $\lambda > 0$ finite, but $\lim_{\lambda \to \infty} x(\lambda) = (2, 0)^T$, is the optimum solution of (14.38).

The exterior penalty function method for solving (14.35), selects an increasing sequence of values of the penalty parameter λ diverging to ∞, $\{\lambda_1, \lambda_2, \ldots\}$. In Step 1, the method finds an unconstrained minimum of $f(\lambda_1, x)$ over $x \in \mathbb{R}^n$ beginning with an appropriate initial point x^0. Let it be $x(\lambda_1)$. The general step is described below.

General step $r + 1$ for $r \geqq 1$ Let $x(\lambda_r)$ be the unconstrained minimum of $f(\lambda_r, x)$ over $x \in \mathbb{R}^n$, obtained in the previous step. Beginning with $x(\lambda_r)$ as the initial point, find an unconstrained minimum of $f(\lambda_{r+1}, x)$ over $x \in \mathbb{R}^n$. Let it be $x(\lambda_{r+1})$. If the penalty function at $x(\lambda_{r+1})$

$$P(x(\lambda_{r+1})) \overset{\leq}{=} \epsilon \qquad (14.39)$$

for some positive tolerance ϵ, terminate with $x(\lambda_{r+1})$ as an approximate optimum for (14.35). If (14.39) does not hold, go to the next step.

Because of its sequential nature, this method is also called a sequential unconstrained minimization technique.

14.9   Exercises

14.2 A company has made commitments to provide fuel to n customers in a region on a monthly basis. For $j = 1$ to n the jth customer is located at the given point (a_i, b_j) on the two dimensional Cartesian plane and needs d_j units of fuel delivered monthly.

The problem is to determine the location, say (x_i, y_j) on the Cartesian plane, $i = 1$ to m, of m depots to serve these customers, and the quantities w_{ij} of fuel shipped from depot i to customer j. Assuming that the cost c for shipping fuel per unit per unit distance is given, formulate this problem as an NLP. Write the first order necessary optimality conditions for this problem, and check whether they are necessary everywhere.

14.3 Production Smoothing   Many companies seek a smooth production pattern with minimum fluctuations in production levels from period to period, to avoid the costs associated with setups and hiring and firing, and the negative public reaction and union opposition. However a smooth production usually means higher inventories and hence higher inventory carrying charges. So, managements are faced with two conflicting objectives of minimizing the fluctuations in the level of output from period to period, and also minimizing the production and inventory holding costs.

Consider the case of a product for which d the forecasted demand (in units) over the next 13 periods is given below. The production cost c per unit varies from period to period because of seasonal fluctuations in the cost of raw materials; this data is also given in the following table. Inventory holding cost for the finished product is estimated to be 4 units of money per unit from one period to the next. No backlogging is allowed, i.e., demand in each period has to be met exactly.

Formulate as an LP the problem of determining a production and inventory holding schedule that meets the demands at minimum total cost (sum of production costs and inventory holding costs). Solve this LP using an available LP software package and determine the optimum objective value α.

Formulate the problem of minimizing the sum of squared deviations in the production levels of consecutive periods, subject to all the constraints in the above model and the additional constraint that the total production and inventory holding costs should be no more than 1.1α.

Period i	1	2	3	4	5	6	7
d_i	100	180	220	150	100	200	250
c_i	45	45	46	47	48	50	55
Period i	8	9	10	11	12	13	
d_i	300	260	250	240	210	140	
c_i	56	57	55	50	48	46	

$$d_i = \text{Expected demand (in units) in period } i$$
$$c_i = \text{Expected production cost per unit}$$

([N. Demokan and A. H. Land, June 1981]).

14.4 Farm Planning   A farmer has 12 units of farm land (1 unit = 1000 acres) on which he can cultivate wheat, oats, or corn. Due to water availability constraints, the amount of land devoted to corn and oats put together can be at most 8 units. Each unit of land devoted to wheat, oats, corn needs an investment of 30, 20, 40 units of money (for seed, fertilizer, and pesticides) respectively; and 5, 5, 8 units of labor respectively. The farmer only has 400 units of money for investment, and 80 units of labor.

Crop yields are highly dependent on weather conditions, and therefore cannot be predetermined with certainty. Depending on crop yields and political conditions, crop prices are also subject to random fluctuations, perhaps with an upward trend. Analysis of past data indicates that the expected net revenue from a unit of land devoted to wheat, oats, corn is 72, 54, 89 units of money respectively. And the variance-covariance matrix of these returns has been estimated to be

$$\sigma = \begin{pmatrix} 3600 & 1655 & 3907 \\ 1655 & 1980 & 2470 \\ 3907 & 2470 & 5476 \end{pmatrix}$$

The farmer would like to make sure that the expected total net return is $\geq$ 890 money units. Formulate the problem of determining the allocation of land to the crops that minimizes the variance of the total net return, subject to all the constraints. ([N. Demokan, and A. H. Land, June 1981]).

14.5 Portfolio Analysis   There are 8 investment opportunities with data as shown below.

Investment i	1	2	3	4	5	6	7	8
μ_i	41	40	67	17	12	16	14	13

$$\mu_i = \text{expected return per 250 invested}$$

The variance-covariance matrix (σ_{ij}) of returns from these investment opportunities is given below.

$j =$	1	2	3	4	5	6	7	8
$i = 1$	688	881	2744	−104	3	−159	79	100
2		2425	6790	−299	26	−319	78	201
3			26381	393	156	−739	−401	63
4				366	−1	94	−196	−134
5					4	−2	−2	−1
6						74	−70	−51
7							161	96
8								86

It is required to find how to divide 250 units of available funds among these investment opportunities, so as to ensure an expected return of 35 while minimizing the variance of the return. Formulate as a quadratic program. ([N. Demokan and A. H. Land, June 1981]).

14.6 The cost of bunker fuel is the major component in the operating expense of a ship. Fuel savings gained by operating a ship at reduced speed may be substantial, but the additional sea days represent loss of alternative profits. Optimal speed is the one that maximizes total net profit.

(i) Consider the problem of determining the optimal speed during a voyage leg which generates income, when the income is independent of the cruising speed. Let x denote the cruising speed during the leg in nautical miles/hour. We are given the following data.

$$
\begin{aligned}
x_0 &= \text{nominal (i.e., maximum) cruising speed} \\
F_0 &= \text{daily fuel consumption by main engines at nominal speed, in tons} \\
c &= \text{cost of bunker fuel (i.e., fuel consumed by main engines) in \$/ton} \\
d &= \text{fixed daily cost for the vessel that is independent of cruising speed} \\
&\quad \text{(this consists of the costs for crew, supplies, insurance, mainte-} \\
&\quad \text{nance, and fuel for the small auxiliary engines)} \\
\ell &= \text{distance of the voyage leg in nautical miles} \\
r &= \text{gross revenue from the leg} \\
T_p &= \text{port-days for the leg (i.e., the number of days vessel spends in the} \\
&\quad \text{port, which is independent of the cruising speed)}
\end{aligned}
$$

The total duration of the voyage leg is T_p+ the travel time for the voyage in days.

From past experience it is known that the actual bunker fuel consumption (i.e., fuel used by the main engines) is $F_0(x/x_0)^3$, and that the feasible range of speeds is $(1/2)x_0$ to x_0 (i.e., from about half the nominal cruising speed, to the nominal cruising speed).

Using this information, find an expression for net profit from this voyage leg (= gross revenue − the fixed costs for the duration of the voyage − cost of bunker fuel for the main engines, ignoring port charges which are constant for a specified voyage) as a function of the cruising speed. Then describe a procedure for determining the optimal cruising speed in its feasible range, which maximizes the net daily profit = (net profit from the voyage leg)/(duration of the voyage leg in days).

(ii) Now consider a positioning (i.e., empty) leg in which the ship is empty (in ballast). Such a leg is usually a return leg. It does not generate any income but costs money, and therefore the cost of this leg should be minimized. Every day by which the duration of this leg is extended owing to slow steaming the company incurs a loss equal to the alternative daily value of the ship (= the opportunity value of the ship, i.e., the net daily operating profit that can be gained by chartering out the ship).

Let v = alternative daily value of the ship, and L = distance of the positioning leg in nautical miles. Let x_0, F_0, c, x have the same definitions as in (i). The total cost of this empty leg can be taken to be the cost of the ship's time plus the cost of the bunker fuel. Express this total cost as a function of the cruising speed x. Using it, determine the optimum speed that minimizes the total cost of this leg.

([D. Rones, Nov. 1982]).

14.7 A system consists of a number of working components. The probability density function of the time to failure of a component is a Weibull distribution with parameters $\beta = 5, \theta = 25$. The maintenance policy for this system is as follows. The entire system is replaced with a new system after every T units of time (this is called preventive replacement). In between preventive replacements, if a component fails, just that failed component is quickly replaced. From the properties of the Weibull distribution, it is known that the expected number of component failures to be repaired in between preventive replacements is $(T/\theta)^\beta$. The cost of each preventive replacement is $c_p = 200$ units of money. The cost of repairing a single failed component is $c = 10$ units of money. Formulate the problem of finding an optimum value of T that minimizes the expected total cost per unit time, and find an optimum solution for this problem.

14.8 Finding A Lost Kid A couple are searching for their kid who is lost. It is known that the kid is likely to remain stationary, and that his location is equally likely to be at any point on the circumference of a circle (see Figure 14.7). Their search begins at point A, and the husband agrees to search the left of the boundary of the circle, while the wife agrees to search the right half. The walking speed of the two searchers is the same, and they each expect to take 2 hours to search their half of the circumference of the circle. They agree to meet at the center of the circle after their search is over. From any point on the circumference of the circle, it takes each of them 0.2 hours to reach the center. One possible search strategy is for each searcher to look in their half, and then return to the agreed meeting place, the center. This guarantees that the kid and the parents will be reunited in 2.2 hours. Can they do better than this by checking back more frequently if the other person has already found the child?

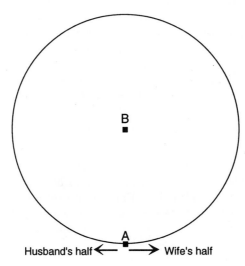

Figure 14.7 A is the starting point of the search; and B, the center, is the agreed meeting point after the search.

Consider the strategy indexed by the positive integer k, the number of times the searchers plan to return to the meeting point to check if the target has been found. Let x_i denote the proportion of the circumference searched by the two searchers between the $(i-1)$th and the ith planned return to the center (clearly $x_1 + \ldots + x_k$ should be 1 and all $x_i > 0$). Let $f_k(x_1, \ldots, x_k)$ denote the expected time until both the parents and the kid are reunited under this strategy.

Find the function $f_k(x_1, \ldots, x_k)$ in terms of $x_1, \ldots, x_k$. Find the optimum solution $(\bar{x}_1, \ldots, \bar{x}_k)$ that minimizes $f_k(x_1, \ldots, x_k)$. Then discuss a method for finding the optimum value of k that minimizes the time before the searchers and their kid are reunited at the center under this strategy. ([L. C. Thomas, June 1992]).

14.9 Ambulance Allocation   A county is divided into 10 sectors. Each sector is uniquely characterized as urban, rural, or mixed, primarily according to geographic characteristics and distinguishable population areas.

In general urban sectors are characterized by high average demand for ambulances, shorter travel distances and total service times, higher costs per call, and slower travel speeds. Alternatively, rural sectors are characterized by lower demand, larger geographical area, lower costs per call, and higher travel speeds.

It is assumed that the EMS ambulance units will be relatively centrally located within each sector. The following data on the various sectors is given, where

$$\lambda_i = \text{average number of calls for an ambulance per hour in sector } i$$
$$A_i = \text{area of sector } i \text{ in square miles}$$
$$v_i = \text{average velocity (mph) at which an ambulance travels in sector } i$$
$$c_i = \text{a correction factor for metric used based on shape of sector } i, \text{ to determine travel distance}$$
$$s_i = \text{average service time in minutes (not including travel time), per call, in sector } i$$

Sector i	Type	λ_i	A_i	v_i	c_i	s_i
1	Urban	4.0	30	20	0.5	4.0
2	Urban	2.1	25	20	0.5	4.3
3	Mixed	1.3	75	30	0.4	5.3
4	Mixed	1.3	75	30	0.4	5.6
5	Mixed	1.3	60	25	0.5	5.6
6	Rural	0.6	100	30	0.4	8.2
7	Rural	0.6	100	30	0.4	8.2
8	Rural	0.6	110	30	0.4	8.2
9	Rural	0.6	110	30	0.4	8.2
10	Rural	0.1	125	30	0.4	8.2

It can be assumed that calls for ambulance in sector i occur randomly according to the poisson distribution with average rate of calls per hour equal to λ_i. Within each sector the system of calls for ambulances can be assumed to be a queueing system with poisson arrivals, and negative exponential service times with number of servers equal to the number of ambulances allocated to the sector.

T_i, the average travel time per call in sector i, can be estimated from the formula $T_i = (c_i/v_i)\sqrt{A_i}$. The average service time per call in sector i is $s_i + T_i$, and its inverse, $1/(s_i + t_i)$ is μ_i, the service rate in sector i.

Let x_i denote the number of ambulances assigned to sector i, $i = 1$ to 10.

The total number of ambulances deployed in all the counties is required to be ≤ 26. From queueing theory we know that the average ambulance utilization factor in the ith sector is $\rho_i = \lambda_i/(x_i\mu_i)$. This is required to lie between 0.25 and 0.9 for each i.

Let w_i be the dispatch time, i.e., the average time a call must wait in queue before an ambulance is dispatched for it, in sector i. From queueing theory we know that the probability that $w_i \geq t$ is $\rho_i e^{-\mu_i(1-\rho_i)t}$. This probability must be at most 0.05 for $t = 15$ in urban and mixed sectors, and for $t = 25$ in rural sectors.

There are fixed and variable costs in ambulance operations. The fixed costs include the salaries of paramedics and emergency medical technicians which must be paid regardless of the demand for ambulance work. This is expected to be $30,000, $27,000, $20,000 per month per ambulance in the urban, mixed, and rural sectors. The variable costs of ambulance operation depend on the number of calls, and are expected to be $145, $135, $110 per call in the urban, mixed, and rural sectors. The total cost of ambulance operations in the county is the sum of the fixed and variable costs.

It is required to determine the optimal allocation of ambulances to the various sectors in the county so as to minimize the total cost of ambulance operations in all the sectors, subject to all the constraints mentioned above. Formulate this problem treating x_is as continuous variables (i.e., ignore the integer requirements on these decision variables). ([J. R. Baker, E. R. Clayton, and B. W. Taylor III, May 1989]).

14.10 The recently discovered comet 1968 Tentax is moving within the solar system. We have the following data on its observed positions in a certain polar coordinate system (r, ϕ) in which ϕ = angle in degrees.

Observation no.	1	2	3	4	5
r	2.7	2.0	1.61	1.20	1.02
ϕ	48	67	83	108	126

By Kepler's first law the comet should move in a plane orbit of elliptic form if perturbations from the planets are neglected. Then the coordinates must satisfy

$$r = \frac{x}{1 - y \cos(\phi)}$$

where y is the eccentricity parameter, and x is another parameter. It is required to find the best values for the parameters x, y that give the closest fit to the observed data, using the least squares method. Formulate the least squares problem.

14.11 A company manufactures a product for which they can achieve superior process capability and quality by having higher unit production costs. Let x = expected fraction acceptable, which is a decision variable. Then the unit production cost is $p(x) = a/(1 - x)^b$ where a, b are given positive constants.

To keep the problem simple, assume that there is constant demand for this product at the rate of d units per unit time that the company has to meet. The company's policy is to have a production run of y units of product, where y is also a decision variable. Again to keep the problem simple, assume that once the production run is set up, production and inspection take place instantaneously. The acceptable product is stored in inventory from which demand is met. The moment the stock level in inventory becomes 0, the next production run occurs and this same process repeats again and again.

So, the expected stock level in inventory at the start of a production run will be xy, and since the demand rate is d units/unit time, this production run is expected to cover the demand over a period of length xy/d. Since demand is constant and uniform over time, the stock level in inventory in this period varies as in Figure 14.8. Hence, the average inventory at any point of time will be the area of the triangle in Figure 14.8.

Assume that the inventory carrying cost is $\$h$ per unit per unit time, and that the setup cost to start a production run is $\$s$. Find the total cost per unit time as a function of the decision variables x and y. Find the optimum values for x and y which minimize the total cost per unit time.

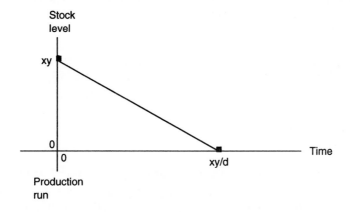

Figure 14.8 Stock level during the period covered by a production run.

14.12 A person has the habit of taking daily D vitamin tablets of a particular kind. This person can purchase these vitamin tablets through mail order at a purchase price of $\$Kx^\alpha$ for x tablets, and for the sake of simplicity assume that x is a continuous variable, and that delivery always takes place the day after an order is placed. These tablets deteriorate rapidly if left at room temperature, so any unused tablets must be kept frozen at a cost of $\$h$ per tablet per day. The parameters $D > 0, K > 0, 0 \overset{\leq}{=} \alpha \overset{\leq}{=} 1, h \overset{\geq}{=} 0$ are the data elements in this problem. Given values of all these data elements, determine the optimum order quantity (the number of to be purchased at a time) which minimizes the total daily cost of taking these vitamin tablets.

Let $D = 10, \alpha = 0.8$, and consider several possible values for (K, h), say, (20, 0.6), (30, 0.9), (40, 1.2), (50, 1.5), (60, 1.8). In each case compute the optimum order quantity.

Determine the optimum order quantity as a function of D, K, h when $\alpha = 1$ (i.e., purchase cost is \$$K$ per tablet), and $\alpha = 0$ (i.e., take all you can take for a fixed price of \$$K$).

Assuming $D = 10$, $K = 20$, $h = 0.6$ sketch a plot of the optimum order quantity as a function of α, and explain it. (M. Singh)

14.13 We are given a perfect cylindrical glass tube with a sealed bottom but open top. The tube is resting on level ground and has water in it to a height of h cm. We want to puncture this cylinder at a height, x say, from the bottom. When the hole is created, water will flow out through it along a parabolic arc and will hit the ground at a distance, $d(x)$ say, from the cylinder. The objective is to find x which maximizes $d(x)$ in the interval $0 \leqq x \leqq h$.

A drop of water emerging from the hole just as it is created, is subject to two forces that influence it independently. One is the gravitational force pulling it down. Starting at a height of x cm from the ground at time point 0, the height of the drop at time point t seconds will be $(x - \frac{980}{2}t^2)$ cm, because of this gravitational pull, until it hits the ground.

The second force acting on the drop is the horizontal velocity that the drop gets as it emerges from the hole. This velocity depends on the height of the water column above the drop, $h - x$, at time point 0; and is $\sqrt{980(h - x)}$ cm/sec. You can assume that this horizontal velocity of the drop continues to be the same until the drop hits the ground.

Formulate the problem of finding the height x that maximizes $d(x)$, and solve the model to find an optimum x.

14.14 Let S_1, S_2 be the balls with centers at $a = (a_1, a_2, a_3)^T$, $b = (b_1, b_2, b_3)^T$ and radii r_1, r_2 respectively in $\mathbb{R}^3$. $H_1 = \{x : p_1x_1 + p_2x_2 + p_3x_3 = p_0 = p_1a_1 + p_2a_2 + p_3a_3\}$, $H_2 = \{x : q_1x_1 + q_2x_2 + q_3x_3 = q_0 = q_1b_1 + q_2b_2 + q_3b_3\}$, are hyperplanes in $\mathbb{R}^3$ through a, b respectively. Let $D_1 = S_1 \cap H_1$, $D_2 = S_2 \cap H_2$ be the circular discs. It is required to find points $x \in D_1$, $y \in D_2$ which are the nearest pair of points (nearest by Euclidean distance) from these discs. Formulate this problem and write the first order necessary optimality conditions for it. Discuss an efficient algorithm for solving this problem (H. Al-Mohammad).

14.15 Let C_1, C_2 be two spherical cylinders in $\mathbb{R}^3$. The axis of C_1 is the line segment $\{(x_1, x_2, x_3)^T = (a_1, a_2, a_3)^T + \lambda(b_1, b_2, b_3)^T : 0 \leqq \lambda \leqq 1\}$, and that for C_2 is $\{(y_1, y_2, y_3)^T = (p_1, p_2, p_3)^T + \lambda(q_1, q_2, q_3)^T : 0 \leqq \lambda \leqq 1\}$. It is required to find points $x \in C_1$, $y \in C_2$ which are the nearest pair of points (nearest by Euclidean distance) from these cylinders. Formulate this problem and write the first order necessary optimality conditions for it. Discuss an efficient algorithm for solving this problem (H. Al-Mohammad).

14.16 Consider the following nonlinear program

$$\text{Minimize} \quad x_1 \log(x_1/30) + x_2 \log(x_2/40) + x_3 \log(x_3/40)$$
$$+ x_4 \log(x_4/25) + x_5 \log(x_5/35)$$

subject to

x_1	$+x_2$	$+x_3$			$\leqq$	23
	x_2	$+x_3$	$+x_4$		$\leqq$	20
	x_2		$+x_4$	$+x_5$	$\leqq$	22
x_1			$+x_4$	$+x_5$	$\leqq$	26

$$x_j \geqq 0 \text{ for all } j$$

Write the first order necessary optimality conditions for this problem.

Assuming that all the variables are strictly positive at an optimum solution to the problem, discuss how to find it using the first order necessary optimality conditions.

14.10 References

J. R. BAKER, E. R. CLAYTON, and B. W. TAYLOR III, May 1989, "A Nonlinear Multi-criteria Programming Approach for Determining County Emergency Medical Service Ambulance Allocations", *Journal of the Operational Research Society*, 40, no. 5 (423-432).

M. S. BAZARAA, H. D. SHERALI, and C. M. SHETTY, 1993, *Nonlinear Programming Theory and Algorithms*, 2nd edition, Wiley-Interscience, NY.

N. DEMOKAN, and A. H. LAND, June 1981, "A Parametric Quadratic Program to Solve a Class of Bicriterion Decision Problems", *Journal of the Operational Research Society*, 32, no. 6 (477-488).

J. E. DENNIS Jr., and R. B. SCHNABEL, 1983, *Numerical Methods for Unconstrained Optimization and Nonlinear Equations*, Prentice Hall, Englewood Cliffs, NJ.

H. M. EDWARDS, 1977, *Fermat's Last Theorem: A Genetic Introduction to Algebraic Number Theory*, Springer-Verlag, NY.

O. FLANIGAN, W. W. WILSON, and D. R. SULE, 1972, "Process-Cost Reduction Through Linear Programming", *Chemical Engineering*, (68-73).

R. FLETCHER, 1987, *Practical Methods of Optimization*, 2nd edition, Wiley-Interscience, NY.

D. G. LUENBERGER, 1984, *Linear and Nonlinear Programming*, 2nd edition, Addison-Wesley, Reading, MA.

G. P. MCCORMICK, 1983, *Nonlinear Programming Theory, Algorithms, and Applications*, Wiley-Interscience, NY.

K. G. MURTY, 1988, *Linear Complementarity, Linear, and Nonlinear Programming*, Heldermann Verlag Berlin.

K. G. MURTY, and S. N. KABADI, 1987, "Some NP-Complete Problems in Quadratic and Nonlinear Programming", *Mathematical Programming*, 39(117-129).

J. NOCEDAL, 1991, "Theory of Algorithms for Unconstrained Optimization", *Acta Numerica*, (199-242).

D. RONES, Nov. 1982, "The Effect of Oil Price on the Optimal Speed of Ships", *Journal of the Operational Research Society*, 33, no. 11, (1035-1040).

L. C. THOMAS, June 1992, "Finding Your Kids When They Are Lost", *Journal of the Operational Research Society*, 43, no. 6 (637-639).

Index